C000154508

1 MONTH OF
FREE
READING
at
www.ForgottenBooks.com

By purchasing this book you are eligible for one month membership to ForgottenBooks.com, giving you unlimited access to our entire collection of over 1,000,000 titles via our web site and mobile apps.

To claim your free month visit:

www.forgottenbooks.com/free96987

* Offer is valid for 45 days from date of purchase. Terms and conditions apply.

ISBN 978-1-5284-7319-4
PIBN 10096987

This book is a reproduction of an important historical work. Forgotten Books uses state-of-the-art technology to digitally reconstruct the work, preserving the original format whilst repairing imperfections present in the aged copy. In rare cases, an imperfection in the original, such as a blemish or missing page, may be replicated in our edition. We do, however, repair the vast majority of imperfections successfully; any imperfections that remain are intentionally left to preserve the state of such historical works.

Forgotten Books is a registered trademark of FB &c Ltd.
Copyright © 2018 FB &c Ltd.
FB &c Ltd, Dalton House, 60 Windsor Avenue, London, SW19 2RR.
Company number 08720141. Registered in England and Wales.

For support please visit www.forgottenbooks.com

WEST VIRGINIA

GEOLOGICAL SURVEY

Boone County

By

C. E. KREBS, Assistant Geologist,

and

D. D. TEETS, JR., Field Assistant.

Part IV—Paleontology, by W. ARMSTRONG PRICE, Paleontologist.

I. C. WHITE, State Geologist.

WHEELING NEWS LITHO. CO.
WHEELING, W. VA.
1915

GEOLOGICAL SURVEY COMMISSION.

HENRY D. HATFIELD............................*President*
GOVERNOR OF WEST VIRGINIA.

E. L. LONG..................................*Vice President*
TREASURER OF WEST VIRGINIA.

FRANK B. TROTTER..............................*Member*
ACTING PRESIDENT, WEST VIRGINIA UNIVERSITY.

E. DWIGHT SANDERSON................*Executive Officer*
DIRECTOR, STATE AGRICULTURAL EXPERIMENT STATION.

STATE BOARD OF CONTROL.

JAMES S. LAKIN...................................*President*

WM. M. O. DAWSON............................*Treasurer*

J. M. WILLIAMSON..............................*Auditor*

STATE OF WEST VIRGINIA
JUNE 20 1863
MONTANI SEMPER LIBERI

SCIENTIFIC STAFF.

I. C. WHITE.................................*State Geologist*

SUPERINTENDENT OF THE SURVEY.

G. P. GRIMSLEY........................*Assistant Geologist*

RAY V. HENNEN.........................*Assistant Geologist*

CHARLES E. KREBS...................*Assistant Geologist*

DAVID B. REGER.......................*Assistant Geologist*

W. ARMSTRONG PRICE......................*Paleontologist*

ROBERT M. GAWTHROP.................*Field Assistant*

D. D. TEETS, JR..........................*Field Assistant*

BERT H. HITE............................*Chief Chemist*

JAN B. KRAK............................*Assistant Chemist*

J. LEWIS WILLIAMS..........................*Chief Clerk*

RIETZ C. TUCKER.........................*Stenographer*

LETTER OF TRANSMITTAL.

To His Excellency, Hon. Henry D. Hatfield, Governor of West Virginia, and President of the West Virginia Geological Survey Commission:

SIR:

I have the honor to transmit herewith the Detailed Report, and Soil, Topographic and Geologic Maps covering the County of Boone. The Report and Geologic Map were prepared by Assistant C. E. Krebs and D. D. Teets, Jr., his Field Assistant, while the Soil Report and Map were prepared and published by the Bureau of Soils of the U. S. Department of Agriculture in cooperation with the West Virginia Geological Survey, all the field work having been done by W. J. Latimer, U. S. soil expert, who has been engaged in the study of the soils of our State for several years. This Boone County Soil Report marks a change in the plan of publication over that heretofore adopted; viz, that instead of republishing each Soil Report in the volume devoted to the economic and geologic features of each county, an edition of 2,500 copies of each Soil Report to be hereafter issued in cooperation with the U. S. Department of Agriculture will be purchased direct from the Public Printer of the United States and distributed along with the corresponding Detailed County Reports of our Survey. This method, which was first adopted for Boone and the Logan-Mingo Reports, will effect a saving in expenditure of $100 to $200 for each Soil Report and in addition will give the reader a copy of the original Soil Reports and accompanying maps of the several counties just as they were published by the U. S. Department of Agriculture.

The development of the coal resources of Boone County, especially its cannel coal deposits, was undertaken as early as 1840 through the building of a series of locks and dams on Coal River between its mouth at Saint Albans on the Kanawha River and Peytona in Boone by the Coal River Navigation Company under the superintendency of Captain Rosecrans, later to become a distinguished General in the Union army. During the period from 1840 to 1860, a considerable amount of excellent cannel coal was mined in the vicinity of Peytona and shipped to the markets reached by the Kanawha and Ohio Rivers through these Coal River locks and dams, but when the Civil War supervened these improvements were neglected, with the result that many of them were soon washed out in the annual floods of the river, and hence when the Civil War ended, so many of the locks and dams of the Coal River Navigation Company had been destroyed that the navigation of Coal River was never resumed, and therefore the further development of Boone County's coal resources remained quiescent until the Coal River Railroad, a branch of the Chesapeake and Ohio Railway System, was extended well into the interior of that county on Little Coal River, about 5 years ago, since which time the coal output of the county has so rapidly increased that with further extensions of the railway lines up Little Coal River, and the Clear, Marsh, and other Forks of the main river, it gives promise of soon rivaling in its coal production the older coal yielding counties of Kanawha, Fayette, Logan, Mingo, and others. The Kanawha Series attains a very rich development in Boone, as may be perceived in the immense amount of detailed information given by Messrs. Krebs and Teets in this Report, and skilfully brought into orderly sequence for the convenience of the reader. The accompanying structural map will also reveal the direction and rate of dip of the several coal beds, and can not fail to prove of vast benefit to the coal operators in the future development of the Boone County mines, many of which will be shafting propositions, especially to the No. 2 Gas and the Eagle Coals, and those associated closely with these important seams. To the farmers of the county the accompanying excellent Soil Report and Map must also prove invaluable in the scientific development of the agricultural resources of Boone. A large area of virgin timber—poplar, oak, walnut and other hard wood varieties—still remains uncut in this county, and these, together with coal—splint, gas, and cannel—building stone, clays and shales, natural gas, and possibly petroleum, together with her soils, constitute the natural wealth of Boone. Neither the New River nor Pocahontas

Series of coals exists under Boone, since they thin out northwestward and disappear as minable seams before reaching the latitude of this county, so that her coals belong entirely to the Kanawha and Allegheny Measures, one, apparently No. 5 Block, in the latter series attaining locally a thickness exceeding 25 feet with its included slates, in the celebrated "Cook seam" of the Bald Knob region. This exceptional development, however, covers only a very restricted area.

I. C. WHITE, *State Geologist.*

Morgantown, W. Va., May 1st, 1915.

CONTENTS.

Page.
Members of Geological Survey Commission and State Board of Control III
Members of Scientific Staff V
Letter of Transmissal .... VI-VIII
Table of Contents .... IX-XII
Illustrations .... XIII-XIV
Author's Preface .... XV-XVIII

PART I. HISTORY AND PHYSIOGRAPHY.

Chapter I.—Historical and Industrial Development 1-10
Location and History 1
General Description 1-6
Miscellaneous Items 1-4
Towns and Industries 4-6
Indian Mounds 7
History of Transportation 7-10
Water Ways 7
Steam Railroads 7-8
Projected Railroads 8-9
Highways 9-10

Chapter II.—Physiography 11-36
Table of Stream Data 12-13
Drainage Area 14-15
Description of Drainage Basins 15-34
Coal River and Tributaries 15-29
Little Coal River and Tributaries 29-34
Mud River and Trace Fork of Big Creek 34
Topography of the Land Area 34-36

PART II. GEOLOGY.

Chapter III.—Structure 37-42
Introduction 37-38
Methods of Representing Structure 38-39
Detailed Geologic Structure 40-42

Chapter IV.—Stratigraphy—General Sections 43-203
Generalized Section 44-47
Sections 47-202
Peytona District 47-66
Scott District 66-84
Washington District 85-149
Sherman District 149-169
Crook District 170-202
Summary 203

	Page.
Chapter V.—Stratigraphy—The Conemaugh Series	204-206
Description of the Conemaugh Formations	204-206
The Pittsburgh Red Shale	205
The Saltsburg Sandstone	205-206
The Bakerstown (Barton) Coal	206
The Buffalo Sandstone	206
The Brush Creek Coal	206
The Mahoning Sandstone	206
Chapter VI.—Stratigraphy—The Allegheny Series	207-224
Description of the Allegheny Formations	207-224
The Upper Freeport Coal	207-208
The Upper Freeport Sandstone	208
The Lower Freeport Coal	208
The East Lynn Sandstone	208-211
The Upper Kittanning (North Coalburg) Coal	211
Middle and Lower Kittanning (No. 5 Block) Coal	211-224
Flora of the Allegheny Series	224
Chapter VII.—Stratigraphy—The Pottsville Series	225-515
Description of the Kanawha Series or Upper Pottsville Formations	226-514
The Homewood Sandstone	226-231
The Stockton-Lewiston-Belmont Coal	231-250
The Coalburg Sandstone	250-254
The Coalburg Coal	254-277
The Little Coalburg Coal	277
The Lower Coalburg Sandstone	277-278
The Buffalo Creek Coal	278
The Buffalo Creek Limestone	278
The Upper Winifrede Sandstone	278-284
The Winifrede Coal	284-313
The Lower Winifrede Sandstone	313
The Chilton "A" Coal	313-314
The Upper Chilton Sandstone	314-319
The Chilton Coal	320-336
The Lower Chilton Sandstone	336
The Little Chilton Coal	336-338
The Hernshaw Sandstone	339
The Hernshaw Coal	339-359
The Naugatuck Sandstone	359-360
The Dingess Coal	360
The Williamson Sandstone	360
The Dingess Limestone	360-370
The Williamson Coal	370-375
The Upper Cedar Grove Sandstone	375
The Cedar Grove Coal	376-403
The Middle Cedar Grove Sandstone	403
The Lower Cedar Grove Coal	403
The Lower Cedar Grove Sandstone	403-404
The Alma Coal	404-443
The Logan (Monitor) Sandstone	443-444
The Little Alma Coal	444
The Malden Sandstone	444
The Campbell Creek Limestone	444-446
The Campbell Creek (No. 2 Gas) Coal	446-492
The Brownstown Sandstone	492
The Powellton Coal	492-497

	Page.
The Stockton Limestone	497
The Bald Knob Shale	497
The Matewan Coal	497-501
The Matewan Sandstone	501
The Eagle "A" Coal	501
The Eagle Sandstone	501-502
The Eagle Coal	502-511
The Bens Creek Sandstone	511-512
The Bens Creek Coal	512
The Decota Sandstone	512
The Little Eagle Coal	512-513
The Eagle Limestone	513-514
Diamond Drill Borings	514
Summarized Record of Diamond Core Test Holes	514A
Coals in the Lower Pottsville Series	515

PART III. MINERAL RESOURCES.

Chapter VIII.—Petroleum and Natural Gas	516-560
The Oil and Gas Horizons of West Virginia	517
Description of Sands	518-519
Oil and Gas Development	519-560
Early History	519
Well Records	519-560
Summarized Record of Wells	520A
Peytona District	521-528
Prospective Oil and Gas Territory	528
Scott District	528-535
Prospective Oil and Gas Area	535
Washington District	536-542
Prospective Oil and Gas Areas	541-542
Sherman District	542-555
Prospective Oil and Gas Areas	555
Crook District	555-560
Prospective Oil and Gas Area	560
Chapter IX.—The Coal Resources of Boone County	561-582
Statistics of Coal Production	561-563
Coals in Boone County	564-573
Coals in the Allegheny Series	565-566
Coals in the Kanawha Series	566-573
Summary of Available Coal in Boone County	570-573
Table Showing Total Available Coal in Boone County	571
Table Showing Available Coal by Districts	571-572
Comparative Calorific Value of the Coals in Boone County	573-582
Table of Coal Analyses	574-578
Page Reference to Detailed Description and Sections of Coal Openings and Mines	579-582
Chapter X.—Clays, Road Materials, Building Stones, Sand, Iron Ore, Forests, and Carbon Black	583-590
The Clays in Boone County	583-585
Clays in the Conemaugh Series	583
Clays in the Allegheny Series	584
Clays in the Kanawha Series	584-585
Recent Clays	585
Road Materials	585
Gravel Pits	585

Page.
Building Stones 585-586
Sands 586
Iron Ore 587
Forests 587-590
Operating Saw Mills.................................. 589-590
Carbon Black Industry.................................. 590

PART IV. PALEONTOLOGY.

Chapter XI.—Notes on the Paleontology of Boone County..... 591-619
Scope of the Investigation.............................. 591-593
Faunal Horizons 593-597
Table of Fossiliferous Members of the Kanawha Series.. 594
List of Fossils from Boone, Logan and Mingo Counties..... 595
Register of Localities.................................. 595-597
Description of Species.................................. 597-617
Description of Plate XLII................................ 618
Index to Species... 619

Appendix.—Levels Above Mean Tide in Boone County........ 620-627

Index 628-648

ILLUSTRATIONS.

Maps I and II in Atlas (Under Separate Cover).

Map I.—Topographic Map of Boone County.

Map II.—Map Showing Economic Geology and Structure Contours of the Campbell Creek (No. 2 Gas) Coal.

Map III.—Soil Map of the Boone County Area (attached to the Soil Report of this area).

Plates.

No.		Facing Page.
	View of Madison on Little Coal River looking South, showing the Topography of the Kanawha Series	Frontispiece
I.	—East Lynn Sandstone on head of Whites Branch, Crook District	16
II.	—Junction of Pond and Spruce Forks, illustrating topography of Kanawha Series	24
III.	—Old Stone House at Peytona	40
IV.	—Mouth of West Fork of Pond Fork, and topography of Kanawha Series	48
V.	—Malden Sandstone at Peytona	64
VI.	—Bald Knob, elevation 2334' above tide, showing topography of Allegheny Series	72
VII.	—View of Big Coal River at mouth of Tony Creek, illustrating topography of Kanawha Series	88
VIII.	—Looking up West Fork from mouth of Little Ugly Branch; Kanawha Series	96
IX.	—View from head of Whites Branch looking west, illustrating topography of Allegheny Series	112
X.	—View of Seth, also showing topography of Kanawha Series, capped with the Allegheny	120
XI.	—View at mouth of Laurel Creek, looking west, at Seth, Kanawha Series	136
XII.	—Old-fashioned water-power mill on head of Pond Fork	144
XIII.	—View at head of Whites Branch, looking east over Kanawha Series	160
XIV.	—View at mouth of Robinson Creek, showing topography of Kanawha Series	168
XV.	—View where Dingess Limestone Fossil Horizon goes under Pond Fork, 1 mile north of Lantie	184
XVI.	—View looking up West Fork from point ¾ mile south of mouth of Browns Branch, and topography of Kanawha Series	192
XVII(a).	—View on Coal River at Dartmont	200
XVII(b).	—The twin trees, birch and maple, on Haggle Branch of Coal River	200
XVIII(a).	—Irwin Green coal opening, Campbell Creek (No. 2 Gas) Coal, on Drawdy Creek	216
XVIII(b).	—"Turtleback" Limestone concretions on Turtle Creek	216
XIX.	—Workman coal opening, No. 5 Block Coal, 1 mile north of Bald Knob, Crook District	224

XX.—Band Saw Mill of Lackawanna Coal & Lumber Co., showing mill pond, logs and conveyor line, at Seth, W. Va........ 240
XXI.—Winifrede Coal opening on hill south of Bald Knob...... 248
XXII.—Upper Winifrede Sandstone on hill to south of Bald Knob 264
XXIII(a).—Kanawha Series 0.5 mile north of Orange P. O...... 272
XXIII(b).—View of River Terrace south of Madison............ 272
XXIII(c).—Winifrede Coal opening, one mile east of Mud, at mouth of Ballard Fork of Mud River...................... 272
XXIV.—Winifrede Coal in mine of Anchor Coal Co., at High Coal 288
XXV.—Hernshaw Coal opening on Little Ugly Branch of West Fork 296
XXVI.—Tipples of Peytona Block Coal Company, and topography of Kanawha Series.. 312
XXVII.—Crop of Alma Coal in grade of Coal River Railroad at Ottawa 320
XXVIII.—Campbell Creek (No. 2 Gas) Coal opening at Uneeda.. 336
XXIX.—Dingess Fossiliferous Limestone and Shale below Lantie on Pond Fork.. 344
XXX.—View where Dingess Limestone Fossil Horizon goes under Pond Fork, 1.5 miles south of mouth of Robinson Creek.... 360
XXXI.—View where the Dingess Fossiliferous Limestone Horizon comes out of Pond Fork above Lantie, 1.5 miles south of mouth of West Fork.. 368
XXXII.—Alma Coal where it comes out of Pond Fork at mouth of Beaver Pond Branch.................................. 384
XXXIII.—Sandstone over Upper Campbell Creek (No. 2 Gas) Coal on Pond Fork, below Grapevine Branch.................. 400
XXXIV.—View of Malden Sandstone at mouth of Grapevine Branch where Campbell Creek (No. 2 Gas) Coal comes above water level.. 416
XXXV.—Outcrop of Eagle Fossiliferous Limestone and Shale Horizon, just north of Madison.......................... 432
XXXVI.—View of sandstone about 30 feet above the Eagle Fossils in roadside just below Madison....................... 448
XXXVII.—Malden Sandstone at Peytona....................... 464
XXXVIII.—Campbell Creek (No. 2 Gas) Coal opening below mouth of Joes Creek.................................... 480
XXXIX.—View where Campbell Creek (No. 2 Gas) Coal goes under on Short Creek, and topography of Kanawha Series.... 496
XL.—Tobacco barn on head of Big Creek...................... 512
XLI.—Coalburg Sandstone, northeast of Gordon P. O., on head of Whites Branch.. 528
XLII.—Fossils from the Kanawha Series....................... 618

Figures.

1. Map Showing Progress of Topographic and Detailed County Surveys XVI
2. Map Showing Location of the Boone County Area............ XVI
3. Map Showing Progress of Studies of Invertebrate Fossils.... 592

AUTHOR'S PREFACE.

The ultimate purpose of this report is to assemble the present knowledge, including a large amount of unpublished data collected by the writer, his assistant and others in the field, not only of the general geology of the County, but also a brief history of its settlement and growth, along with a description of the physiography and economic resources, and to present the facts in a form convenient to those who are interested in their study either for scientific purposes or for development.

The Report gives (1) A brief history of the County and its development; (2) A study of its drainage system and other surface features; (3) The geologic structure with a contour map of the base of the Campbell Creek (No. 2 Gas) Coal; (4) Four Chapters on the general geology and detailed stratigraphy, with a map showing the outcrop of the different divisions of the rock column, according to the generally accepted classification of geologists; (5) A description of the oil and gas fields therein, with suggestions for their future development, along with a map showing the accurate location of the oil and gas wells and dry holes; (6) Minable coals, with a table showing the chemical composition, calorific value and fuel ratio, with a summary exhibiting the approximate available tonnage of the County; (7) Clays, road materials, building stones, sands, forests and carbon black industry of the County; (8) A Chapter on the Paleontology of the area; and (9) An Appendix showing railroad and U. S. Geological Survey levels above tide at numerous localities in every portion of the County.

Special attention is called to the structure map accompanying this Report, whereon are shown by means of contour lines the tidal elevations of the Campbell Creek (No. 2 Gas) Coal horizon at all points in the County. These contour lines, separated by 25 feet in elevation, exhibit at a glance the approximate position of the horizon of this great coal bed in all parts of the County, also the direction and location of the

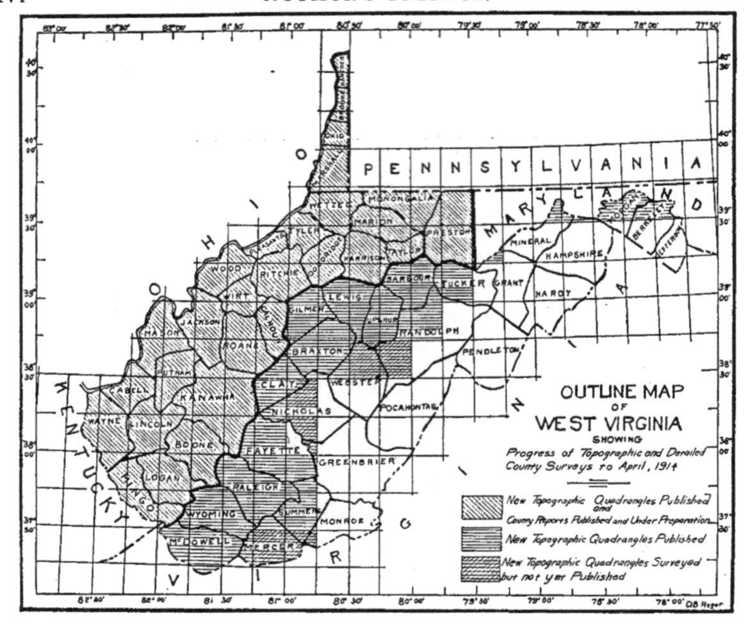

Figure 1. See explanation on figure.

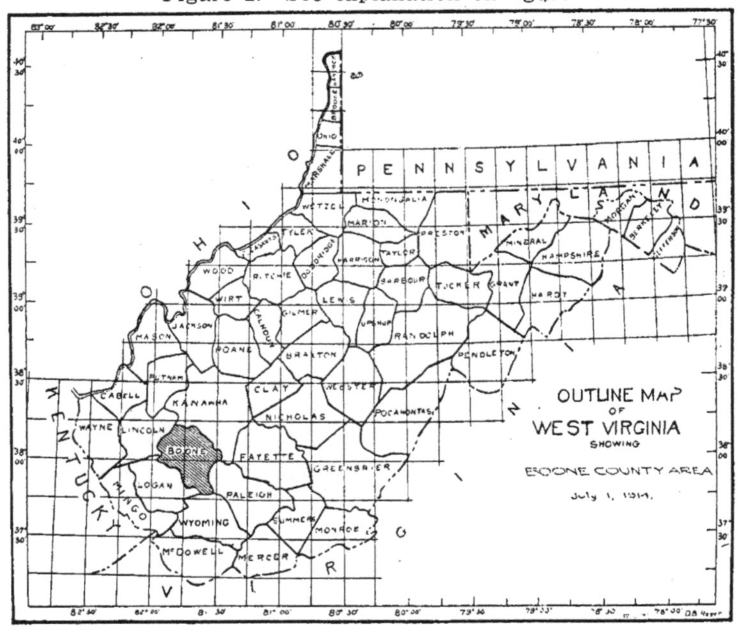

Figure 2. See explanation on figure.

anticlines and synclines, as well as the dip and strike of the rock strata at any point, a knowledge of which is of great value not only for the future development of oil and gas pools therein, but also for the future mining of the several coal seams where the latter are of minable thickness, purity and regularity. Special attention is also called to the several areas outlined by the writer as "Prospective Oil and Gas Territory," and the page references in the Index under this heading by magisterial districts.

The Chapters on the general geology and detailed stratigraphy, though quite technical, give a large fund of data concerning the Conemaugh, Allegheny and Kanawha Series of the Pennsylvanian group of rocks. Therein some errors of correlation in former State Reports are corrected, the writer ever keeping in mind the generally accepted classification of the strata which permits comparison with the formations in other parts of West Virginia.

The Chapter on coals gives the general distribution of the several minable beds along with approximate estimates of the available area and tonnage of each vein, with a final summary of the total available coal for the county. The commercial mines and openings are referred to in the table of analyses therein by serial numbers, the same corresponding to numbers assigned to the symbol designating the accurate location of the mine on the map referred to above.

Chapter X gives a description of the clays, road materials, building stones, sand, forests and carbon black industry.

The accompanying Report on the soil and its products, prepared by W. J. Latimer, of the Bureau of Soils of the U. S. Department of Agriculture, well trained and versed in his profession, can not fail to interest the progressive farmers of the County.

Three maps of the entire area accompany this Report (Maps I and II in Atlas under separate cover and Map III attached to the Soil Report), one of which shows by appropriate symbols the character of the surface, the roads, streams, railroads, etc.; the second, by the same means, the general and economic geology, with several items of special economic interest; and the third, in a similar way, the character, classification and distribution of the soils.

The writer and his Assistant, D. D. Teets, Jr., spent a portion of the field season of 1913, and a short time in 1914, gathering data for this volume, and opportunity is here taken to mention that the accurate, painstaking and faithful discharge of all duties assigned to Mr. Teets, both in the office and in the field, has been of great assistance in the completion of this Report.

Much valuable aid and assistance were given by residents of the area, as well as by officials of the several companies engaged in the development of the oil, gas and coal fields therein. Due credit and acknowledgment have been given in the text for all such data obtained.

The chemical analyses and heat determinations were made in the Survey Laboratory by Jan B. Krak, Assistant Chemist, under the supervision of B. H. Hite, Chief Chemist.

The writer also takes the opportunity to express his obligations to Dr. I. C. White, State Geologist, whose writings and suggestions have added greatly to the value of this Report.

C. E. KREBS.

Charleston, W. Va., June 15, 1914.

PART I.

The History and Physiography of Boone County.

CHAPTER I.

THE HISTORICAL AND INDUSTRIAL DEVELOPMENT.

LOCATION AND HISTORY.

The portion of the State of West Virginia discussed in detail in this Report, Boone County, lies between the parallels of 37° 45′ and 38° 15′ North latitude, and the meridians 81° 30′ and 82° 00′ West longitude from Greenwich.

GENERAL DESCRIPTION.

Miscellaneous Items.

Boone County lies southwest of the central part of West Virginia, about 25 miles south of the capital of the State, and is drained by Coal River and its tributaries, running in a northwest direction through the entire county. It is bounded on the north by Kanawha County, on the east by Kanawha and Raleigh Counties, on the south by Wyoming and Logan Counties, and on the west by Logan and Lincoln Counties.

Area—Its area given by districts as computed from accurate topographic sheets of the U. S. Geological Survey is as follows:

Districts.	Sq. Miles.
Crook	137.80
Peytona	60.05
Scott	70.83
Sherman	136.15
Washington	101.17
Total	506.00

Relief.—The general surface of the county varies in elevation from 595 feet above tide at the mouth of Fork Creek, 1 mile south of Brounland, where the county line comes to Coal River, to 3375 feet above tide in the Boone-Wyoming County Line on Guyandot Mountain, three-fourths mile southwest of Pilot Knob, a range in elevation of 2780 feet.

Population.—The population in 1900 was 8,194, of which 8,059 were white, 135 colored and 7 foreign born. The census of 1910 gives the total population as 10,331, an increase of 25 per cent.

The following table shows the changes in population of the county by districts during the last twenty years as given by the census of 1910:

Population of Boone County.

Districts	1910	1900	1890
Crook	1,466	1,503	1,189
Peytona	1,896	1.295	1,131
Scott	2,673	1,751	1,403
Sherman	1,619	1,596	1,522
Washington	2,677	2,049	1,640
Totals	10,331	8,194	6,885

The above figures show that Crook and Sherman Districts have made very little gain in the last twenty years, owing to the fact that very little farming has been done in these districts, and very little industrial work has been done until recently.

Formation.—Boone County was formed from Kanawha, Cabell and Logan Counties by an "Act of the General Assembly" of Virginia, passed in 1847, and was named in honor of the celebrated hunter and explorer, Daniel Boone, who was a resident of the Kanawha Valley for several years.

The county as first formed had the following boundary lines:

"Beginning at the mouth of Cobb's Creek, a branch of Little Coal River on the point between said creek and river, and thence, running with the top of the ridge, on the upper side of said creek, to the dividing ridge between said creek and the waters of Mud River; thence, with and along the top of said dividing ridge to the head of Horse Creek and the Laurel Fork of Mud River; thence, taking the dividing ridge between said Laurel Fork and the waters of the Sugar Tree Fork of said Mud River to the mouth of said Laurel Fork; thence, crossing the main fork of Mud River. a straight line to the top of the dividing ridge between the headwaters of Ugly Creek and Mud River; thence, with and along the top of said ridge to the head of the Grassy Fork of Ugly Creek; thence, down said Grassy Fork to Main Ugly Creek; thence, a straight line, crossing Main Ugly Creek and Big Creek at William Martin's, so as to include said Martin on Big Creek; thence, up a small branch to the top of the dividing ridge between Godby Fork and the Middle Fork; thence, with and along the top of the said dividing ridge, passing the head of Vannater Branch and taking or containing the said ridge above said branch to Hewett Creek, crossing said creek above the house of Obadiah Bias, Senior, so as to include said Bias and James Bias within the new county; thence, a straight line to the point ridge above the mouth of Big Laurel Creek on the Spruce Fork of Coal River, with and along the top of said ridge to the head of Big Laurel Creek; thence, with and along the top of the said dividing ridge between the waters of Guyandot River and Little Coal River to the dividing ridge between the Marsh Fork of Big Coal River and Little Coal River, with and along the top of said dividing ridge to the mouth of the said Marsh Fork; thence, crossing Big Coal River at the mouth of Marsh Fork of said Big Coal River, up the mountain to the top of the dividing ridge between Big Coal River and Kanawha River; thence, with and along the top of said dividing ridge down to the head of Bull Creek, with and along the top of the ridge below Bull Creek to Big Coal River, thence down said river as it meanders to the mouth of Fork Creek; thence, taking the point below said Fork Creek to the head of Dick Creek; and thence, with and along said creek as it meanders to the beginning."

Products.—Very little farming is done in Boone County, as the lands contain considerable virgin forests, and are rich in coal and gas, and possibly oil. Its principal products are corn, hay, beef cattle, hogs, poultry, fruit, natural gas, coal and lumber.

The quality and character of the soil and its products, as well as the mineral wealth, will be discussed in detail in subsequent Chapters of this Report.

Property Valuation.—The State Auditor gives the following property valuations for Boone County for the year 1913:

	Assessed Valuation.	State Tax.
Real Estate	$7,604,380.00	$4,562.63
Personal Property	1,515,706.00	872.09
Totals	$9,120,086.00	$5,434.72

The personal property is only about 20 per cent. of that of the real estate, owing to the fact that very few improvements have been made in the county.

The School tax of Boone County for the year 1913, as given in the records of the Sheriff's office, was as follows:

Teachers' Fund	$21,336.19
Building Fund	9,162.81
Total	$30,499.00

Towns and Industries.

Boone County has no large towns or cities. The principal towns are as follows: Madison, Danville, Peytona, Highcoal, Webb, Racine, Sterling, Whitesville, Seth, Clothier, Greenview, Jeffery, Ramage, Ottawa, Altman, Mistletoe, Woodville, Bald Knob, Pond, Gordon, Chap, Van, Lantie, Low Gap, Turtle Creek and Uneeda.

Madison.

Madison is the county seat of Boone County, and is located northwest of the center of the county at the junction of Pond and Spruce Forks of Big Coal River. It is incorporated and has two banks, 4 stores, 3 hotels, 1 drug store, a school, and 3 churches. Its population in 1910 was 295, and this has increased to 360 in 1914.

Danville.

Danville is situated on Little Coal River opposite the mouth of Turtle Creek. It is a thriving village, and has a high school, one church, 6 stores, one bank, and two hotels. It was incorporated in 1911, and has a population of 300 people.

Peytona.

Peytona is located on Big Coal River, at the mouth of Drawdy Creek. It was named after William Peyton, one of the pioneers in the coal business on Coal River. It is a mining town, and has two stores, two hotels, one church and one school. It is not incorporated and its population fluctuates. January 1, 1914, it had 300 population.

Highcoal.

Highcoal is a mining village on Seng Creek of Big Coal River. It is the mining town for the Anchor Coal Company, has one store and one school, its population fluctuating from 100 to 250.

Webb.

Webb is situated on Seng Creek of Big Coal River, and is a mining town with a population of 150 to 250. The town has one store and one school house.

Racine.

Racine is a small village at the mouth of Short Creek, on Big Coal River. The town has one store, one school, and 100 population.

Sterling.

Sterling is located on Indian Creek of Big Coal River, and is the mining town of the Hickory Ash Coal Company. It has one store and school. Its population varies from 100 to 300.

Whitesville.

This town is located on Big Coal River, one mile south of the mouth of Seng Creek. It has a population of about 300 and is a mining town.

Seth.

Seth is a lumber town, located at the mouth of Laurel Creek, on Big Coal River. It has one hotel, two stores, a school, and a large band saw mill. Its population is about 200.

Clothier.

Clothier is located on Spruce Fork of Little Coal River on the Boone-Logan County Line. It is a mining and lumber town, and has two stores, one school, and one hotel. Its population varies from 200 to 300.

Greenview.

Greenview is a small town located on Spruce Fork of Little Coal River, four miles north of Clothier. It has one hotel, one store, one church, and one school. Its population is about 100.

Jeffery, Ramage and Ottawa.

Jeffery, Ramage and Ottawa are small towns on Spruce Fork of Little Coal River between Greenview and Clothier.

Mistletoe.

Mistletoe is a small town on Horse Creek. It has one store, one school, and a large lumber plant.

Altman.

Altman is located on Little Coal River at the mouth of Horse Creek. It is a lumber town, the Leatherwood Lumber Company being located here. Its population is about 200, and it has one store and one school.

The remaining villages and small towns have a post-office, and usually one store and a few houses.

INDIAN MOUNDS.

Numerous small artificial mounds are located in the Coal River Valley. The mounds are usually small, from 25 to 40 feet in diameter at base, and from ten to fifteen feet in height. Two of these mounds are located on Whites Branch of West Fork at mouth of Indian Hollow, and several are located on Pond Fork and Big Coal River.

HISTORY OF TRANSPORTATION.

Water Ways.

Coal River.—Coal River was formerly used to float logs to Saint Albans, where large piers were built and dams placed across the stream to stop the logs from floating into the Kanawha River. These logs were transported into Coal River from its tributaries by building large splash dams and collecting a large body of water, and then lowering the dams quickly by means of large gates, thus flooding the small stream and floating the logs to the river.

Coal River was also used in the early part of the nineteenth century for transporting cannel coal from the mines at Peytona. Since the construction of the Coal River Railroad this method of transportation has been abandoned.

Steam Railroads.

Coal River Branch of the Chesapeake and Ohio Railway.—This railroad leaves the main line of the Chesapeake and Ohio at Saint Albans, and extends to Sproul, where it branches, one branch extending up Big Coal River to Seth at the mouth of Laurel Creek, where the Lackawanna Coal & Lumber Company has recently extended a branch up Laurel Creek and Hopkins Fork of same, for about 9.2 miles to mouth of Lotts Branch. The other branch extends up Little Coal River to Clothier, where it branches again, one branch extending up Laurel Fork of Spruce and the other up Main Spruce Fork into Logan County.

The road was first constructed by local capital; but in 1905 was absorbed by the Chesapeake and Ohio Railway Company. The road has an extent of 18 miles on Big Coal River in Boone County, and 24 miles on Little Coal River in the same county. A branch route also leaves the Coal River Branch at Altman, and extends up Coal River to Mistletoe, a distance of 5 miles.

Cabin Creek Branch of **the Chesapeake and** Ohio **Railway.** This railroad leaves the main line of the Chesapeake and Ohio Railway Company at Cabin Creek Junction, and extends up Cabin Creek and its tributaries, entering Boone County through a tunnel 4000 feet long, on to Seng Creek, a branch of Coal River, and extending down Seng Creek to Coal River, and up Coal River and Clear Fork to Colcord. It was first constructed by private capital in 1894 as far as Acme, and later, in 1902, was absorbed by the Chesapeake and Ohio Railway, and extended later to the heads of the different streams and finally to Coal River. It extends 7 miles in Boone County.

It is a freight carrying road and is the outlet for the Cabin Creek coal field and a portion of the Coal River field.

Projected Railroads.

Big **Coal River.**—Surveys have been made up Coal River, from Seth to Jarrolds Valley, and the right of way has been bought by the Chesapeake and Ohio Railway to connect the Cabin Creek Branch with the Coal River Branch at Whitesville, a distance of 12.5 miles. This road will doubtless be constructed before 1915.

Pond Fork **Railroad.**—Surveys have been made up Pond Fork of Coal River, from Madison to Bald Knob, a distance of 24 miles, by the Chesapeake and Ohio Railway Company for the extension of its line. This branch will doubtless be built very shortly, in order to market the vast timber area and rich coal deposits along Pond Fork and West Fork of Coal River.

Spruce-Laurel Creek Railroad.—The Spruce-Laurel Branch of the Chesapeake and Ohio Railway Company has been surveyed for about twelve miles up Spruce-Laurel Creek, from the end of its present track, and will be built very shortly in order to transport to market the vast timber and rich deposits of coal occurring along said waters.

Highways.

Interest in the value of good roads is being awakened by the advent of the automobile and the new road laws recently passed by the State Legislature, and hence steel bridges are being constructed across the streams, and roads are being projected by the State Road Engineer. In the early history of Boone, Logan and Mingo Counties nearly all the freight imported into these counties was hauled in wagons from the Kanawha River at Marmet (formerly Brownstown). This route extended from the Kanawha to the Guyandot River, near Chapmanville, along the creeks and across the low divides, starting at Marmet, thence going up Lens Creek and crossing the dividing ridge and down Short Creek to Racine, on Coal River; thence, down the north side of said river to the mouth of Drawdy Creek; thence, crossing the river and up Drawdy Creek and across the divide and down Rock Creek for about four miles; thence, crossing the divide to the south, to Price Branch, and down same to Little Coal River at Danville; thence by two courses, one by way of Turtle Creek to Big Creek of Guyandot River, and the other by way of Little Coal River and Spruce Fork to mouth of Low Gap Creek, and up same to Big Creek, and thence down Big Creek for about six miles, and thence crossing the divide to Chapmanville.

Another route for southern Boone and the northern part of Wyoming Counties extends from Racine up Coal River to mouth of Laurel Creek at Seth, and up Seth Creek to mouth of Coal Fork, and up that branch to cross the divide to Whites Branch of West Fork of Coal River, and down same to Pond Fork and thence up Pond Fork.

Very little work has been done on these roads since the completion of the railroads, and they are now used very little.

Until the construction of the railroad into Boone, no steel bridges were built within the limits of the county. Very little work was ever done in constructing roads and keeping them in repair. But within the past year, 1913, two steel highway bridges have been constructed across Big Coal River, one at Peytona, and the other at Seth, and two wire suspension foot bridges have been constructed across the same river, one at Racine, and the other at Whitesville.

A steel highway bridge has just been completed across Little Coal River at Danville. A wire suspension foot bridge was built at Danville across Little Coal River two years ago.

CHAPTER II.

THE PHYSIOGRAPHY OF THE BOONE COUNTY AREA.

A general description of the physiography of any region is of interest to those who are engaged in a scientific study of Nature's work. The general principles of Physiography are ably set forth by Prof. G. P. Grimsley in the Detailed Report on Ohio, Brooke and Hancock Counties, pages 18 to 23, to which the reader is referred.

After an examination of the drainage system of Boone County is made, it will be found that Big and Little Coal Rivers and their tributaries drain about 95 per cent. of the entire area of the county, while Big Creek of Guyandot River drains 3 per cent., and Mud River drains the remaining area of the county. The general trend of Coal River and its tributaries is to the north, toward the Kanawha River at Saint Albans.

The following table shows in a graphic manner not only the rate of fall per mile of the principal streams of Boone County, but also their departure from a straight line course, and the ratio of the total distance between the points of the same, measured by the meanders of the stream, to the air line distance between the same points:

Table of Stream Data.

Streams.	Tota Fal. Feet.	Tota Distance. Miles.	Rate of Fal Per Mile. Feet.	Air Line Distance. Miles.	Ratio T. D. to A. L. D.
Big Coal River, from Jarrolds Valley to Brounland..................	227.0	35.3	6.4	23.0	1.5
Little Coal River, from Madison to MacCorkle	54.5	17.5	3.1	11.2	1.5
Pond Fork, from mouth of Skin Creek to mouth of Pond Fork..	73.5	29.7	24.8	21.6	1.4
West Fork of Pond Fork, from mouth of Mats Creek to mouth of West Fork	421.0	10.5	40.0	8.0	1.3
Spruce Fork, from Clothier to mouth	94.0	12.6	7.5	7.5	1.7
Spruce Laurel Fork, from mouth of Dennison Fork to mouth.......	625.0	14.7	42.5	9.1	1.6
Mud River in Boone Co., from head to Boone-Lincoln Co. Line......	445.0	.0	74.	4.1	1.5
Fork Creek, from head to mouth...	615.0	.0	102.	4.4	1.4
Brush Creek, from head to mouth...	960.0	.0	160.	4.5	1.3
Whiteoak Creek, from head to mouth	630.0	6.0	210.8	2.4	1.2
Drawdy Creek, from Andrew P. O. to mouth......................	205.0	4.1	50.0	3.1	1.3
Bull Creek from head to mouth.....	800.0	4.6	173.9	4.4	1.0
Lick Creek, from head to mouth....	635.0	4.5	141.1	4.0	1.1
Short Creek, from head to mouth..	620.0	3.0	206.7	2.6	1.2
Tony Creek, from head to mouth....	760.0	3.0	253.3	2.6	1.2
Joes Creek, from head to mouth....	900.0	7.5	120.0	5.8	1.3
Upper Whiteoak Creek, from mouth of Right Fork to mouth.........	300.0	.6	65.2	3.	1.3
Seng Creek, from head to mouth...	1480.0	.5	269.1	4.	1.2
Elk Run, from head to mouth......	1325.0	4.9	270.4	4.5	1.2
Little Elk Creek, from head to mouth	1075.0	3.0	358.3	2.4	1.2
Laurel Creek of Big Coal River, from mouth of Little Laurel Creek to mouth	300.0	8.2	36.6	6.1	1.3
Hopkins Fork, from mouth of Jarrolds Fork to mouth of Logan Fork	200.0	6.1	32.	4.2	1.5
Logan Fork, from head to mouth...	1045.0	5.6	186.	4.7	1.2
Jarrolds Fork, from head to mouth.	995.0	6.5	153.	4.9	1.4
Sand Lick Creek, from head to mouth	970.0	4.9	198.	4.4	1.1
Indian Creek from head to mouth..	965.0	4.0	191.	3.4	1.5
James Creek, from head to mouth...	1150.0	3.5	328.	3.2	1.1
James Branch, from head to mouth.	1810.0	4.8	377.8	4.1	1.2
Mats Creek, from head to mouth...	1145.0	3.9	293.6	3.3	1.2

Table of Stream Data.—Continued.

Streams.	Total Fall. Feet.	Total Distance. Miles.	Rate of Fall Per Mile. Feet.	Air Line Distance. Miles.	Ratio T. D. to A. L. D.
Skin Creek, from head to mouth....	1850.0	3.0	416.	2.7	1.1
Rocklick Branch, from head to mouth	1690.0	.0	563.	2.7	1.1
Jasper Workman Branch, from head to mouth..........................	1225.0	3.1	395.8	2.6	1.2
Grapevine Branch, from head to mouth	1410.0	4.1	343.9	3.8	1.1
Cow Creek, from head to mouth.....	1090.0	4.0	272.5	3.3	1.2
Casey Creek, from head to mouth...	1270.0	6.0	211.7	4.8	1.3
Bull Creek, from head to mouth....	1005.0	4.8	209.4	3.8	1.3
Jacks Branch, from head to mouth..	815.0	3.9	209.0	2.3	1.7
Robinson Creek, from head to mouth	935.0	6.0	158.3	3.1	1.9
Whites Branch, from head to mouth	790.0	5.8	136.2	3.7	1.6
Brown Branch, from head to mouth..	1090.0	4.8	227.1	3.6	1.3
Skin Poplar Branch, from head to mouth	1075.0	4.5	238.9	3.5	1.3
Hewett Creek, from Boone-Logan Co. Line to mouth..................	90.0	3.1	29.	2.5	1.2
Sixmile Creek, from head to mouth	310.0	4.8	64.	4.4	1.1
Turtle Creek, from head to mouth..	460.0	6.9	66.0	6.4	1.1
Trace Fork of Big Creek of Guyandot River, from head to Logan-Boone County Line..................	400.0	5.0	80.0	4.3	1.2
Lick Creek of Little Coal River, from head to mouth.................	800.0	3.6	222.2	3.3	1.1
Rock Creek, from source to mouth..	215.0	5.6	38.4	4.2	1.3
Horse Creek, from source to Boone-Lincoln County Line............	140.0	3.7	37.8	2.8	1.2
Little Horse Creek, from head to mouth	625.0	3.4	183.8	2.8	1.2
Camp Creek, from head to mouth...	555.0	5.0	111.0	4.5	1.1
Hewett Creek of Little Coal River, from head to mouth...........	565.0	3.8	148.7	3.2	1.2

In the last column of the foregoing table is given the ratio of the total distance (T. D.) measured by meanders of the stream to the air line distance (A. L. D.). In each instance it is very evident that the nearer this ratio approaches unity, the greater the ratio of fall.

DRAINAGE AREA.

The following table shows the area of Boone County drained by the different streams:

STREAMS.	AREA DRAINED Square Miles.
Big Coal River, with tributaries	192.20
Little Coal River, with tributaries	285.80
Mud River, with tributaries	11.00
Big Creek, with tributaries	17.00
Total	506.00
Big Coal River tributaries in Boone County:	
Fork Creek	13.8
Brush Creek	9.0
Whiteoak Creek	1.7
Drawdy Creek	8.3
Bull Creek	8.4
Lick Creek	5.3
Short Creek	3.3
Tony Creek	2.6
Joes Creek	14.3
Whiteoak Creek	18.9
Seng Creek	4.3
Elk Run	5.3
Little Elk Creek	2.1
Laurel Creek	49.0
Hopkins Fork	24.6
Logan Fork	5.1
Jarrolds Fork	7.0
Sand Lick Creek	7.6
Indian Creek	4.7
Little Coal River tributaries in Boone County:	
Pond Fork and tributaries (except West Fork)	138.4
West Fork	43.5
James Creek	5.0
James Branch	4.7
Mats Creek	3.8
Browns Branch	3.4
Roach Branch	3.0
Whites Branch	6.7
Skin Fork	3.1
Rocklick Branch	2.6
James Branch	3.2
Grapevine Branch	2.5
Cow Creek	4.4
Beaver Pond Branch	2.5
Casey Creek	5.0
Bull Creek	4.3
Jacks Branch	7.6
Spruce Fork and tributaries in Boone Co.	96.6
Skin Poplar Branch	4.9
Spruce Laurel Creek	32.4
Hewett Creek	8.7
Sixmile Creek	4.6

STREAMS.	AREA DRAINED Square Miles.
Big Branch	2.0
Low Gap Branch	2.0
Turtle Creek	11.8
Lick Creek	5.5
Rock Creek	13.6
Horse Creek	15.6
Little Horse Creek	3.8
Camp Creek	6.4
Hewett Creek	4.2
Trace Fork of Big Creek of Guyandot River	6.3

DESCRIPTION OF DRAINAGE BASINS.

Coal River.

Coal River with its tributaries, as already stated, drains 95 per cent. of the area of Boone County. It is formed by the junction of Clear and Marsh Forks at Jarrolds Valley on the Boone-Raleigh County Line, and flows in a northwestern direction for 59 miles, emptying into the Kanawha River at Saint Albans. It passes 31.5 miles through Boone County, 3.2 miles as dividing line between Boone and Kanawha Counties, then 6 miles through Kanawha County, then 4 miles as dividing line between Lincoln and Kanawha Counties, and 14.3 miles through Kanawha County, before emptying into the Kanawha River.

Its fall from Jarrolds Valley to Saint Albans is 265 feet, or 4.5 feet per mile. It falls 227 feet from Jarrolds Valley to Brounland, a distance of 35.3 miles, or 6.4 feet to the mile. Its drainage area in Boone County is 174.5 square miles. Its entire length, measured by the meanders, is 59 miles, while the air line distance between its source and its mouth is but 34.3 miles. Passing up the river, the first 12 miles has a very crooked course and has almost reached base level. It is very probable that it once left its present course about 0.5 mile above Lower Falls P. O. and flowed across a low divide southwest to Saint Albans, reaching Saint Albans at the same place, in 1.4 miles instead of 5.5 miles, measured by its present course.

Coal River was locked and dammed in 1840-1860 by the Coal River Navigation Company under Captain (later General) Rosecrans, for the purpose of transporting cannel coal from Peytona to market. During the Civil War these improvements were neglected and the river washed them out.

In order to determine the surface water supply of the Ohio River Basin, the United States Geological Survey established several gaging stations on the many large tributaries of the latter in West Virginia. One of these stations is located on Coal River at Brushton, Boone County, 12.6 miles above the junction of Big and Little Coal Rivers at the Forks of Coal.

The following interesting data obtained at Brushton are taken from Water Supply Paper No. 263 of the United States Geological Survey, page 98:

Coal River at Brushton, W. Va.

"This station, which is located at the Chesapeake & Ohio Railway bridge at Brushton Station, near Cobbs, W. Va., was established June 23, 1908, to obtain data for water power, water supply, flood control, and storage purposes.

"The drainage area above the station is about 379 square miles. Brush Creek enters a short distance below the station.

"The datum of the chain gage attached to the railroad bridge has remained unchanged. The records are reliable and accurate. Sufficient data have not yet been collected to enable estimates of the flow to be made."

Discharge Measurements of Coal River at Brushton, W. Va., in 1909.

Date	Hydrographer.	Width	Area of Section.	Gage Height.	Discharge.
		Feet	Sq. Ft.	Feet	Sec. Ft.
March 19....	A. J. Jackson	136	236	2.68	423
November 7*.	A. H. Horton	...	...	1.20	23

*Made by wading.

PLATE I.—East Lynn Sandstone on head of Whites Branch, Crook District.

Daily Gage Height, in Feet, of Coal River at Brushton, W. Va., for 1909. (G. W. Fitzpatrick, Observer.)

Day.	Jan.	Feb.	Mar.	Apr.	May.	June.	July.	Aug.	Sept.	Oct.	Nov.	Dec.
1	2.75	2.15	2.95	3.35	6.8	1.7	3.25	2.1	1.0	1.2	1.3	1.3
2	2.35	2.15	3.25	3.15	6.2	1.6	2.9	2.2	1.0	1.2	1.3	1.3
3	2.15	2.1	2.75	3.15	4.7	1.65	2.65	2.05	1.0	1.2	1.3	1.3
4	2.0	2.1	3.25	3.1	3.95	1.75	2.25	1.85	1.0	1.1	1.25	1.3
5	2.0	2.05	3.65	3.0	3.65	1.8	2.05	1.7	1.1	1.1	1.2	1.3
6	2.15	2.2	3.8	2.9	3.4	1.75	2.65	1.6	1.15	1.1	1.2	1.3
[illegible]	[illegible]	[illegible]	[illegible]	[illegible]	[illegible]	[illegible]	.5	[illegible]	[illegible]	[illegible]	[illegible]	1.35
[illegible]	[illegible]	. 5	[illegible]	. 5	[illegible]	[illegible]	5.4	. 5	[illegible]	[illegible]	[illegible]	1.4
[illegible]	[illegible]	[illegible]	[illegible]	[illegible]	[illegible]	. 5	3.95	. 5	[illegible]	[illegible]	. 5	1.35
1	[illegible]	. 5	. 5	[illegible]	. 5	. 5	3.05	[illegible]	-. 5	[illegible]	[illegible]	1.4
11	2.0	4.3	5.6	2.4	2.85	3.75	2.6	1.4	3.15	1.05	1.3	1.4
12	2.0	3.45	4.25	2.35	2.8	3.45	2.35	1.3	2.6	1.1	1.3	1.5
13	1.9	3.1	3.65	2.35	2.7	3.1	2.45	1.3	2.2	1.1	1.3	1.5
14	2.0	2.75	3.6	2.65	2.6	2.95	2.7	1.35	1.95	1.2	1.3	1.5
15	3.85	2.55	3.5	2.85	2.45	3.2	2.6	2.0	1.75	1.2	1.3	2.05
16	4.55	4.65	3.3	2.8	2.4	2.9	2.5	1.95	1.7	1.3	1.3	2.05
17	4.4	5.2	3.15	2.7	2.25	2.6	2.25	1.9	1.6	1.2	1.3	1.85
18	3.95	4.15	2.85	2.6	2.1	2.8	2.1	1.65	1.6	1.3	1.25	1.75
19	3.1	3.5	3.0	2.5	2.0	2.7	2.0	1.6	1.5	1.3	1.2	1.65
20	2.7	3.45	2.55	3.35	2.0	2.35	1.9	1.5	1.45	1.3	1.2	1.6
21	2.45	3.6	2.6	3.7	2.0	2.2	1.75	1.45	1.4	1.2	1.2	1.5
22	2.25	[illegible]	2.75	6.05	1.95	2.15	1.65	1.35	1.4	1.2	1.2	1.5
23	2.2	[illegible]	3.1	5.55	1.85	2.1	2.0	1.3	1.4	1.25	1.35	1.4
24	2.05	[illegible]	2.85	5.3	1.75	2.2	2.0	1.3	1.4	1.3	1.4	1.4
25	2.0	[illegible]	3.15	4.4	1.8	2.2	1.9	1.2	1.3	1.5	1.3	1.4
26	2.15	4.2	4.85	3.85	2.0	2.05	1.8	1.2	1.3	1.6	1.3	1.5
27	2.2	[illegible]	4.75	3.3	2.0	1.9	1.8	1.15	1.3	1.6	1.35	1.5
28	2.2	. 5	5.4	3.2	2.05	1.8	2.0	1.1	1.25	1.5	1.35	1.6
29	2.25	....	5.45	3.05	1.95	2.15	1.95	1.1	1.2	1.45	1.35	1.55
30	2.3	....	4.35	3.75	1.9	2.45	2.25	1.05	1.2	1.4	1.3	1.5
31	2.1	....	3.7	....	1.8	....	2.05	1.0	....	1.35	....	1.5

NOTE.—During the latter part of December, ice formed in pools above and above the gage; no ice at gage. Ice affects the gage heights very slightly, if any.

Discharge Measurements of Coal River at Brushton, W. Va., in 1910.

Date.	Hydrographer.	Width.	Area of Section.	Gage Height.	Discharge.
		Feet	Sq. Ft.	Feet	Sec. Ft.
March 4..........	C. T. Bailey......	136	292	3.12	666
March 4..........	do	136	292	3.12	668
August 9	Bailey and Dort..	45	40.0	1.37	*40.1

*Measurement made by wading and not at regular section.

Daily Gage Height, in Feet, of Coal River at Brushton, W. Va., for 1910. (Geo. W. Fitzpatrick, Observer.)

Day.	Jan.	Feb.	Mar.	Apr.	May.	June.	July.	Aug.	Sept.	Oct.	Nov.	Dec.
1	1.55	2.7	2.8	1.9	2.55	2.7	2.05	1.7	1.65	1.2	1.5	2.3
2	1.75	2.6	3.45	1.8	[illegible].45	[illegible].45	1.9	1.6	[illegible].55	[illegible].[illegible]5	1.[illegible]	2.05
3	3.0	2.65	3.35	1.85	.3	.3	2.25	1.55	.2	.	1.	[illegible].[illegible]
4	3.1	3.35	3.1	2.0	.25	.15	4.95	1.5	.4	.	1.	.
5	2.7	3.7	2.9	1.9	.2	.2	5.8	1.45	.7	.	1.	. 5
6	2.9	3.35	2.75	2.1	.1	.2	4.5	1.4	.45	.	1.	. 5
7	5.15	2.85	2.65	2.05	.1	.5	4.35	1.4	.15	.	1.	.
8	4.15	2.75	2.55	1.95	.25	3.3	4.25	1.4	.0	. 5	1.	. 5
9	3.15	2.7	2.45	1.9	.8	2.95	3.5	1.4	.85	.	1.	. 5
1[illegible]	2.55	2.8	2.35	1.8	.95	[illegible].15	3.0	1.4	.9	.	1.	.
[illegible]3	2.25	3.2	2.65	1.75	.55	.0	2.7	1.35	.05	.	1.	.
	2.3	3.1	3.0	1.95	.7	.0	2.4	1.3	.9	.	1.	.
	2.1	2.9	3.3	2.4	.95	.0	2.4	1.3	.75	.	1.	.
4	2.6	2.7	3.35	3.2	.05	.1	2.4	1.3	.65	.	1.	.
[illegible]	3.0	2.8	3.1	3.05	.5	.4	2.3	1.2	.5	.	1.	. 5
	2.9	3.75	2.85	2.8	.1	.3	2.2	1.2	.5	. 5	1.	.
	2.7	4.65	2.75	2.7	.9	.15	2.35	1.3	.5	.	1.	.
	3.05	5.25	2.55	2.65	.75	.35	2.75	1.25	.4	.	1.	.
	6.1	4.3	2.45	2.6	.65	.3	3.2	1.2	.4	.	1.	. 5
	4.35	3.7	2.4	2.6	.5	.4	2.9	1.2	.4	.	1. 5	.
	5.1	3.35	2.3	3.8	.4	.4	2.55	1.1	.3	.	1.	.
	5.5	3.35	2.25	5.15	.4	.8	2.35	1.1	.3	.	1.	.
	4.05	3.65	2.2	4.25	.35	.7	2.15	1.1	.3	.	1.	* . 5
	3.45	3.6	2.1	3.7	.4	.7	2.2	1.1	.2	.	1.	. 5
	3.15	3.3	2.1	3.5	.45	.75	2.0	1.1	.2	.	1.	.
	3.05	3.0	2.1	3.4	.1	.25	2.15	1.0	.2	.	1.	.
	3.05	2.8	2.0	3.15	.45	.6	2.05	1.0	.2	.	1.	.
	3.25	2.7	2.0	2.[illegible]5	.95	.45	2.0	1.0	.2	.	1.	. 5
	3.25	....	1.9	.	.6	.25	1.9	0.9	.25	.	2.	.
	2.9	....	1.9	.	.5	.15	1.8	1.0	.25	. 5	2.5	.
31	2.75	....	1.9	....	.95	....	1.8	1.3	....	1.5	....	.

*Observer reported ice 0.3 foot thick above and below gage Dec. 23; river open at gage.

Discharge Measurements of Coal River at Brushton, W. Va., in 1911.

Date.	Hydrographer.	Gage Height	Discharge
		Feet	Sec. Feet
October 4........	C. T. Bailey......	2.75	468
October 4........	do	2.54	35

Daily Gage Height, in Feet, of Coal River at Brushton, W. Va., for 1911. (G. W. Fitzpatrick, Observer.)

Day.	Jan.	Feb.	Mar.	Apr.	May.	June.	July.	Aug.	Sept.	Oct.	Nov.	Dec.
1	3.35	5.5	2.35	2.9	3.2	1.7	1.35	0.95	1.25	1.14	1.60	2.52
2	4.95	5.1	2.3	3.0	3.65	1.65	1.25	.95	1.1	[illegible]	1.[illegible]	2.40
3	5.5	4.15	2.25	3.8	3.65	1.5	1.15	.95	1.0	.	.	2.[illegible]
4	4.8	4.2	2.25	6.7	3.3	1.45	1.15	.95	1.0	.	.	.
5	3.55	4.1	2.2	8.1	3.0	1.5	1.15	.95	1.15	.	.	.
6	3.05	3.8	7.05	6.7	2.8	1.5	1.05	1.0	1.8	1.89	1.83	1.99
7	2.75	3.45	7.75	5.5	2.6	1.55	1.05	1.0	1.6	1.84	2.62	1.94
8	2.55	3.25	7.3	4.45	2.55	1.5	1.75	.9	1.4	2.38	2.95	1.90
9	2.35	3.15	5.25	5.55	2.45	1.45	1.8	.85	1.3	1.98	2.60	1.88
10	2.25	3.35	4.3	5.25	2.4	1.35	1.4	1.5	1.35	2.06	2.36	1.84
11	2.2	3.25	3.65	4.25	2.3	1.3	1.45	1.35	1.45	2.90	2.18	1.83
12	2.2	3.15	3.3	3.65	2.2	1.25	1.8	1.25	1.3	2.[illegible]	2.[illegible]	1.[illegible]2
13	2.2	2.9	3.15	3.3	2.15	1.25	2.0	1.1	1.25	.	.	1.82
14	2.15	2.75	3.1	3.05	2.05	1.25	1.75	1.1	1.2	.	.	1.83
15	2.2	2.6	3.0	3.2	1.95	1.15	1.55	1.05	1.35	.	.	1.[illegible]6
16	2.4	2.45	2.85	3.5	1.9	1.15	1.4	1.1	1.5	2.32	2.42	2.24
17	2.65	2.4	2.75	3.45	1.85	1.15	1.5	1.25	1.3	[illegible]	2.[illegible]4	2.[illegible]
18	2.65	2.4	2.65	3.25	1.8	1.3	1.6	1.05	1.25	.	. 1	.
19	2.55	2.35	2.65	3.05	1.75	1.45	1.55	1.0	1.3	.	. 2	.
20	2.55	3.9	2.95	2.7	1.7	1.35	1.45	.9	1.4	.	. 9	.
21	3.25	4.7	3.35	3.5	1.65	1.3	1.35	1.0	1.4	2.45	2.76	2.24
22	6.0	3.9	3.25	3.95	1.65	1.25	1.25	1.0	1.[illegible]5	2.[illegible]	2.[illegible]	[illegible]
23	5.0	3.45	3.1	4.55	1.6	1.15	1.25	.8	1. 5	.	.	.
24	3.85	3.1	2.9	4.0	1.55	1.15	1.35	.8	1.	.	.	.
25	3.35	2.9	2.75	3.6	1.45	1.15	1.3	.8	1.	.	.	.
26	3.25	2.7	2.65	3.2	1.4	1.15	1.15	.9	1.4	1.94	2.85	3.40
27	3.7	2.6	2.65	3.0	1.35	1.2	1.[illegible]5	.[illegible]	1.25	1.[illegible]	2.[illegible]	[illegible]
28	3.65	2.5	2.55	2.85	1.3	1.25	.	1. 5	.	.	.	.
29	3.9	....	2.6	2.65	1.25	1.45	. 5	1.	.	.	.	.
30	11.35	....	2.6	2.7	1.25	1.5	. 5	.	.	.	.	.
31	6.1	....	2.7	....	1.35	....	. 5	.	....	.	....	.

NOTE.—Observer made no report relative to ice. Relation of gage height to discharge probably not affected by ice during 1911.

An examination of the above tables readily shows that the low water period for Coal River, for the year 1909, occurred during the months of September, October, November, and December; for the year 1910, during the months of August, September, October and November; and for the year 1911, from June to November, inclusive.

Another station located at Tornado in Kanawha County, 19.6 miles north of Brushton, and 11 miles north of the Boone-Kanawha County Line, where the waters from both branches of Coal River pass through, gives the following results:

Coal River at Tornado, W. Va.

"This station is located at the highway bridge near Tornado, W. Va. It was established June 24, 1908, to obtain data for water-power, water supply, flood-control, and storage problems.

"The datum of the chain gage attached to the bridge has not been changed. The records are reliable and accurate. The low-water gage heights may be affected by a dam a short distance below the station. Sufficient data have not yet been collected to enable estimates of the discharge to be made."

Discharge Measurements of Coal River at Tornado, W. Va., in 1909.

Date.	Hydrographer.	Width.	Area of Section.	Gage Height.	Discharge.
		Feet	Sq. Ft.	Feet	Sec. Ft.
March 19......	H. J. Jackson.......	168	922	3.08	904
November 6*..	A. H. Horton.......			2.50	20

*Weir measurement.

Daily Gage Height, in Feet, of Coal River at Tornado, W. Va., for 1909. (J. F. Burdette, Observer.)

Day.	Jan.	Feb.	Mar.	Apr.	May.	June.	July.	Aug.	Sept.	Oct.	Nov.	Dec.
1	....	2.90	3.30	3.54	9.00	2.42	3.46	2.72	2.37	2.22	2.61	2.50
[illegible]	....	2.90	3.15	3.38	7.95	2.40	3.30	2.69	2.41	2.35	2.51	2.54
	....	2.95	3.15	3.31	5.20	2.50	3.38	2.69	2.39	2.42	2.47	2.51
	....	3.00	3.35	3.31	4.18	2.68	3.03	2.57	2.42	2.39	2.45	2.52
	....	2.90	3.75	3.26	3.82	3.06	2.83	2.46	2.82	2.45	2.45	2.58
	....	3.00	4.05	3.18	3.64	2.80	3.32	2.39	2.78	2.43	2.52	2.56
	....	3.10	6.15	3.14	3.36	2.56	13.02	2.30	2.71	2.49	2.58	2.59
	3.00	3.35	4.75	3.06	3.30	2.66	7.00	2.48	2.68	2.47	2.50	2.58
9	2.95	3.65	5.85	2.98	3.29	3.20	4.26	2.30	2.92	2.46	2.75	2.58
10	2.85	4.30	8.40	2.99	3.30	3.94	3.58	2.30	3.72	2.52	2.60	2.58
11	2.90	4.65	5.65	2.92	3.28	3.86	3.02	2.29	3.39	2.52	2.45	2.58
12	2.80	3.90	4.30	2.86	3.10	3.88	2.88	2.37	2.61	2.47	2.50	2.60
13	2.85	3.65	3.90	2.85	3.08	3.60	2.83	2.48	2.51	2.41	2.54	2.64
14	3.05	3.35	4.85	3.22	3.02	3.48	2.80	2.83	2.75	2.41	2.58	2.90
15	4.08	4.25	4.75	3.70	3.12	3.56	3.00	2.67	2.59	2.28	2.58	2.86
16	4.98	5.80	4.65	3.40	2.72	3.50	2.87	2.79	2.42	2.35	2.59	2.85
17	4.40	5.95	3.45	3.35	2.68	3.40	2.80	2.76	2.35	2.45	2.58	2.82
18	4.05	4.35	3.15	3.18	2.60	3.43	2.67	2.68	2.30	2.46	2.54	2.74
19	4.55	3.90	3.08	3.12	2.72	3.38	2.46	2.76	2.38	2.39	2.51	2.65
20	3.30	3.70	3.00	3.82	2.70	3.28	2.28	2.77	2.39	2.45	2.51	2.65
21	3.10	3.70	3.00	4.48	2.71	3.18	2.19	2.64	2.28	2.48	2.51	2.65
22	2.95	4.25	3.00	7.02	2.65	3.10	2.42	2.41	2.43	2.45	2.52	2.65
23	2.95	4.05	3.25	7.45	2.60	3.06	3.46	2.39	2.38	2.38	2.52	2.64
24	2.85	4.45	3.08	6.86	2.52	3.13	3.60	2.49	2.36	2.38	2.54	2.62
25	2.75	4.95	3.40	5.08	2.54	3.08	3.73	2.34	2.31	2.41	2.54	2.52
26	2.80	4.25	4.42	3.98	2.68	2.50	3.26	2.55	2.41	2.42	2.56	2.68
27	[illegible]	[illegible]	[illegible]	[illegible]	2.70	2.86	2.56	2.57	2.31	2.38	2.58	2.65
28	[illegible]	[illegible]	[illegible]	[illegible]	2.70	2.76	2.62	2.51	2.30	2.46	2.60	2.65
29	[illegible]	....	[illegible]	[illegible]	2.70	2.70	2.60	2.55	2.29	2.53	2.58	2.62
30	[illegible]	....	4.25	3.80	2.60	2.88	2.60	2.56	2.22	2.59	2.56	2.61
31	[illegible]	....	3.78	....	2.52	....	2.84	2.32	....	2.58	....	2.59

NOTE—No ice conditions January to April. Ice conditions December 10 to 31. Gage read to top of ice except on December 19. December 27, thickness of ice 0.3 foot; gage height to water surface, 2.30 feet. December 31, thickness of ice 0.5 foot.

Discharge Measurements of Coal River at Tornado, W. Va., in 1910.

Date.	Hydrographer.	Width.	Area of Section.	Gage Height.	Discharge.
		Feet	Sq. Ft.	Feet	Sec. Ft.
March 3.......	C. T. Bailey.........	1[illegible]5	9[illegible]6	3.[illegible]	*1,46[illegible]
March 3.......	do	1 5	9 6	3.	,4[illegible]
August 8......	Bailey and Dort.....	0	2	2.	1

*Measurements made by wading and not at regular section.

Daily Gage Height, in Feet, of Coal River at Tornado, W. Va., for 1910. (G. C. Hoy, Observer.)

Day	Jan.	Feb.	Mar.	Apr.	May.	June.	July.	Aug.	Sept.	Oct.	Nov.	Dec.
1	2.64	3.02	3.19	2.42	2.82	2.89	2.62	2.45	2.78	2.42	2.92	3.30
2	2.84	2.97	3.32	2.36	2.74	2.76	2.46	2.33	2.86	2.50	2.93	3.08
3	3.18	2.99	3.39	2.40	2.59	2.63	2.34	2.34	3.10	2.52	2.88	2.96
4	3.54	3.25	3.34	2.46	2.59	2.49	2.88	2.43	3.49	2.66	3.00	2.92
5	3.51	3.58	3.21	2.52	2.57	2.46	8.74	2.30	3.59	2.67	2.90	2.91
6	3.81	3.57	3.12	2.54	2.44	4.09	4.96	2.15	3.18	2.68	2.88	3.30
7	4.98	3.24	3.01	2.53	2.27	4.29	4.28	2.16	2.88	2.64	2.82	3.85
8	4.43	3.16	2.93	2.45	2.60	3.66	4.16	2.15	2.90	2.82	2.82	3.64
9	3.58	3.06	2.83	2.40	3.18	3.53	3.71	2.24	2.85	2.95	2.89	3.00
10	3.00	3.04	2.75	2.36	4.02	3.93	3.25	2.25	2.80	2.70	2.86	3.00
11	3.34	3.15	2.92	2.25	3.68	3.89	3.00	2.51	2.78	2.66	2.90	2.85
12	2.84	3.22	3.17	2.44	6.56	3.93	2.71	2.51	2.64	2.70	2.84	2.72
13	2.84	3.19	3.46	3.05	6.48	3.49	2.76	2.42	2.47	2.70	2.82	2.60
14	3.47	3.23	3.49	3.36	4.44	3.38	2.80	2.34	2.30	2.70	2.81	2.26
15	3.53	3.44	3.35	3.40	3.70	3.53	2.59	2.32	2.24	2.71	2.80	2.23
16	3.41	4.12	3.15	3.30	3.40	3.87	2.40	2.34	2.12	2.66	2.78	2.35
17	3.26	5.22	3.03	3.22	3.20	4.49	3.06	2.35	2.10	2.67	2.68	2.15
18	3.68	5.76	2.92	3.26	3.09	4.27	3.47	2.48	2.00	2.68	2.70	2.30
19	5.41	4.70	2.83	3.32	2.96	3.67	3.94	2.47	2.00	2.65	2.69	2.55
20	4.91	3.89	2.67	3.37	2.86	3.68	3.39	2.30	1.96	2.62	2.62	2.60
21	4.65	3.66	2.65	4.15	2.88	3.61	2.99	1.98	1.95	2.64	2.68	2.50
22	6.21	3.90	2.60	5.85	2.78	3.22	2.69	2.05	1.90	2.67	2.66	2.58
23	4.25	3.84	2.52	4.43	2.64	3.04	2.45	2.30	2.95	2.90	2.75	2.61
24	3.68	3.84	2.51	3.82	2.79	3.88	2.30	2.33	1.88	2.65	2.79	2.67
25	3.47	3.65	2.44	3.58	3.26	3.30	2.25	2.45	2.06	2.64	2.82	3.10
26	3.34	3.36	2.42	3.45	4.08	3.02	2.49	2.38	2.15	2.70	2.80	3.40
27	3.29	3.20	2.36	3.22	3.57	9.88	2.77	2.46	1.95	2.82	2.90	3.30
28	3.34	3.08	2.32	3.14	3.26	3.90	2.75	2.40	1.90	2.77	3.05	3.10
29	3.41	....	2.45	3.02	2.97	3.18	2.75	2.46	1.88	2.82	3.28	3.08
30	3.23	....	2.51	2.92	2.81	2.88	2.55	2.43	2.29	2.90	3.55	3.80
31	3.09	....	2.48	....	2.94	....	2.45	2.43	....	2.90	....	4.42

NOTE.—Relation of gage height to discharge affected by ice, Jan. 1 to 5, and December 18 to 28.

Discharge Measurements of Coal River at Tornado, W. Va., in 1911.

Date.	Hydrographer.	Gage Height.	Discharge.
		Feet	Sec. Feet.
March 9......	C. T. Bailey........	6.09	5,730
October 3.....	do	4.46	3,010

Daily Gage Height, in Feet, of Coal River at Tornado, W. Va., for 1911. (G. C. Hoy, Observer.)

Day	Jan.	Feb.	Mar.	Apr.	May.	June.	July.	Aug.	Sept.	Oct.	Nov.	Dec.
1	3.96	5.56	2.67	2.90	3.35	2.76	1.56	1.40	2.07	1.52	1.78	2.95
2	4.46	5.19	2.61	2.91	3.80	2.20	1.54	1.40	1.88	2.75	1.76	2.83
3	4.76	4.69	2.57	3.32	3.70	1.93	1.50	1.40	1.62	4.19	1.62	2.75
4	4.96	4.25	2.48	7.90	3.45	1.88	1.49	1.41	1.57	3.10	1.70	2.68
5	3.81	4.46	2.44	8.60	3.24	1.76	1.48	1.41	1.50	2.65	1.70	2.58
6	3.33	4.07	5.48	8.41	3.05	1.71	1.45	1.40	1.53	2.25	1.85	2.45
7	3.08	3.72	9.45	5.74	2.92	1.78	1.44	1.40	1.86	2.20	3.22	2.30
8	2.90	3.42	10.12	4.62	2.84	1.77	1.45	1.40	1.81	2.82	3.25	2.26
9	2.86	3.37	6.07	5.82	2.71	1.70	1.98	1.51	1.72	2.54	2.96	2.21
10	2.73	3.43	4.36	5.44	2.61	1.66	1.95	1.45	1.70	3.00	2.78	2.20
11	2.62	3.38	3.74	4.35	2.60	1.60	1.69	1.68	1.76	3.32	2.57	2.09
12	2.68	3.37	3.50	3.80	2.49	1.61	1.61	1.56	1.85	3.42	2.68	2.08
13	2.76	3.15	3.29	3.48	2.35	1.70	2.08	1.49	1.82	2.94	3.08	2.18
14	2.68	2.99	3.22	3.25	2.28	1.60	2.15	1.48	1.75	2.63	3.15	2.16
15	2.75	2.82	3.12	3.44	2.18	1.58	1.81	1.45	1.88	2.15	3.02	2.34
16	2.76	2.55	3.10	3.50	2.00	1.52	1.72	1.45	2.40	2.60	2.88	3.20
17	2.85	2.63	2.90	3.64	2.00	1.50	1.65	1.46	1.90	3.45	2.68	3.42
18	2.96	2.61	2.85	3.38	1.94	1.70	1.61	1.42	1.80	5.70	2.72	3.19
19	2.90	2.59	2.84	3.28	1.88	1.96	1.82	1.42	1.75	4.18	4.02	2.75
20	2.89	3.64	2.95	3.35	1.86	1.82	1.78	1.40	1.78	3.30	3.60	2.88
21	3.35	4.76	3.15	3.58	1.78	1.70	1.68	1.37	1.72	2.85	3.25	2.83
22	6.59	4.12	3.20	3.90	1.72	1.61	1.58	1.33	1.85	2.59	3.02	2.80
23	5.19	3.12	3.16	4.50	1.80	1.60	1.52	1.38	1.88	2.43	2.82	3.12
24	4.06	3.37	3.02	4.08	1.69	1.59	1.50	1.38	1.81	2.25	2.90	3.54
25	3.62	3.21	2.86	3.70	1.65	1.54	1.50	1.38	1.76	2.15	3.12	3.72
26	3.73	3.05	2.82	3.45	1.68	1.50	1.51	1.39	1.72	2.08	3.28	3.85
27	3.52	2.89	2.79	3.22	1.61	1.50	1.50	1.37	1.68	2.00	3.23	4.85
28	3.61	2.75	2.70	3.02	1.58	1.52	1.42	1.37	1.60	1.92	3.18	5.95
29	4.39	....	2.70	2.94	1.53	1.55	1.40	2.25	1.56	1.87	3.14	4.45
30	13.07	....	2.82	2.92	1.64	1.58	1.41	3.02	1.55	1.82	3.04	3.65
31	8.02	....	2.89	....	2.10	....	1.40	2.66	....	1.80	....	3.60

NOTE.—Relation of gage height to discharge probably not affected by ice during 1911.

Another station located one mile below mouth of Fuqua Creek, and 16.6 miles below the Boone-Kanawha County Line, and 3 miles above Tornado, gives the following results for the latter part of the year 1911, published in Water Supply Paper 303, page 61:

Coal River at Fuqua, W. Va.

"**Location**—At W. C. Hoy's passenger ferry half a mile below Fuqua railroad station and 1 mile below the mouth of Fuqua Creek.

Records Available—October 12, to December 31, 1911.

Drainage area—Not measured.

Gage—Staff gage in two sections on right bank.

Channel—Firm sand.

Discharge measurements—Made from boat 300 feet above gage or by wading.

Winter flow—The relation of gage height to discharge may be affected by ice for short periods.

Accuracy—Gage-height record is considered reliable. Sufficient data have not been obtained to permit estimates of discharge to be made."

"The following discharge measurement was made by Bailey and Perwien, October 13, 1911: Gage height, 2.85 feet; discharge, 700 second-feet."

Daily Gage Height, in Feet, of Coal River at Fuqua, W. Va., for 1911. (W. C. Hoy, Observer.)

Day	Oct.	Nov.	Dec.	Day	Oct.	Nov.	Dec.	Day	Oct.	Nov.	Dec.
1	....	1.40	2.95	11	....	2.34	1.80	21	2.95	3.48	2.69
2	....	1.30	2.78	12	3.60	2.39	1.78	22	2.50	3.01	2.60
3	....	1.30	2.62	13	2.85	2.80	1.78	23	2.25	2.70	3.48
4	....	1.20	2.50	14	2.30	3.38	1.79	24	2.10	2.72	3.94
5	....	1.20	2.32	15	2.10	3.18	1.95	25	1.95	3.50	4.38
6	....	2.65	2.12	16	2.30	2.86	3.25	26	1.85	3.49	4.64
7	....	3.50	2.02	17	3.88	2.53	3.67	27	1.70	3.52	7.72
8	....	3.60	1.96	18	9.25	2.80	3.32	28	1.60	3.40	9.80
9	....	3.08	1.89	19	6.10	5.32	3.04	29	1.55	3.30	6.15
10	....	2.64	1.84	20	3.85	4.30	2.85	30	1.50	3.12	4.74
								31	1.40	....	4.55

NOTE.—Relation of gage height to discharge probably not affected by ice during 1911.

A. H. Horton, District Engineer, Water Resources Branch, U. S. Geological Survey, has prepared, at the instance of Dr. I. C. White, a valuable paper on the "Water Power Resources of West Virginia," for the Semi-Centennial Commission of West Virginia, and published same in the History of West Virginia, prepared by James Morton Callahan. On page 409 of this History, he gives the following concerning Coal River:

PLATE II.—Junction of Pond and Spruce Forks, illustrating topography of Kanawha Series.

COAL RIVER.

"The basin of Coal River lies in the south-central part of West Virginia, south of the Kanawha River. The river rises in the central part of Raleigh County, flows northwesterly across Boone County and enters the Kanawha River near St. Albans in Kanawha County. The drainage area is about 900 square miles.

"The elevation of its source is about 2,000 feet; the slope from Clear Fork to the mouth averages about 6 feet to the mile. The basin is roughly a right triangle in shape with the hypotenuse parallel to the Kanawha River. At the headwaters the topography is rough and mountainous. It is not cut up to such an extent as the tributaries on the north side of the Kanawha, whose sources are at a much greater elevation. The headwaters are forested to a considerable extent. The mean annual rainfall at the sources is about 45 inches, decreasing towards the mouth of the river.

"The principal tributaries are Clear Fork and Little Coal River.

"The Chesapeake and Ohio Railway parallels the river from St. Albans to Peytona, and the Little Coal from the mouth to Seng in Logan County. The Chesapeake and Ohio also enters the basin at the sources of the Coal, following along Clear Fork to Lawson in Raleigh County."

The following table shows the power on the Coal River and its tributaries, prepared by Horton:

TABLE SHOWING INDICATED HORSEPOWER DEVELOPED BY COAL RIVER.

Section of River. From	Section of River. To	Length. Miles.	Mean Drainage Area. Sq. Mi.	Minimum Discharge. Sec.-Ft.	Assumed Discharge for Maximum Development. Sec.-Ft.	Total Fall. Feet.	Minimum Horsepower.	Assumed Maximum Development. Horsepower.
Source	Below Clear Fk.	35	a232	8	77	1,600	294	2,840
Below Clear Fk.	Below Little Coal R.	36	339	12	113	270	298	2,800
Below Little Coal R.	Mouth	17	865	31	288	70	200	1,850
THE COAL RIVER.								
Source	Below Spruce Fk.	35	a272	10	91	2,200	506	4,600
Below Spruce Fk.	Coal River	25	330	12	110	100	110	1,010
	Totals						1,408	13,100

a Total Area.

The principal tributaries to **Coal River** in Boone County, ascending, are: Fork, Bull, Lick, Brush, Drawdy, Short, Indian, Tony, Joes, Laurel, Upper Whiteoak and Seng Creeks. Data of Coal River and its tributaries in regard to drainage area, length and fall are given on preceding pages of this Report.

Fork **Creek.**—Fork Creek heads near the Scott-Peytona District Line, about 4 miles southwest from MacCorkle, and flows in a general northern direction, emptying into Coal River at the Kanawha-Boone County Line, about 0.7 mile south of Brounland.

Bull Creek.—Bull Creek is formed by the confluence of Road and Rich Forks, in the northern part of Boone County, near the Kanawha-Boone County Line, and flows almost due west, emptying into Coal River at Dartmont.

Lick **Creek.**—Lick Creek rises in the Kanawha-Boone County Line at the Peytona-Sherman District corner, 2.8 miles northeast from Racine, and flows in an eastern direction, emptying into Coal River at Cobbs. For drainage data, see preceding drainage tables.

Brush Creek.—Brush Creek has its source near the Peytona-Scott District Line in the southern portion of Peytona District, and flows just a little west of north to Cabell P. O., where it deflects to the northeast, emptying into Coal River at Brushton.

Drawdy Creek.—Drawdy Creek rises in the Scott-Peytona District Line in a direct line between Peytona and Madison and almost exactly half-way between these points, and flows in an eastern direction to the mouth of Morgan Branch, where it turns to the northeast, flowing in a general northeastern direction, emptying into Coal River at Peytona.

Short Creek.—Short Creek has its source at the Peytona-Sherman District corner in the Kanawha-Boone County Line, and flows in a southwestern direction, emptying into Coal River at Racine.

Indian Creek.—Indian Creek has its source in the Sherman-Peytona District Line, 3 miles south from Peytona, and flows in a northeastern direction, emptying into Coal River one-half mile below Racine.

Tony Creek.—Tony Creek rises near the Kanawha-Boone County Line, about one mile east of the Peytona-Sherman District corner, and flows in a general southeastern direction for about 1.5 miles; then it deflects to the west and flows in a southwestern direction, emptying into Coal River 1¼ miles east of Racine.

Joes Creek.—Joes Creek rises at the Kanawha-Boone County Line near the head of Slaughter Creek in Kanawha County, 3 miles west of Ohley, Kanawha County, and flows in a general southwestern direction for about 3 miles, then it deflects to the northwest and flows in a northwestern direction, emptying into Coal River 2.5 miles southeast from Racine. For drainage data, see preceding drainage tables.

Laurel Creek.—Laurel Creek is formed by the confluence of Mudlick and Stollings Forks at a point two miles southeast from the common corner of Scott, Crook and Sherman Districts, and flows in a southwestern direction for one mile, then it makes a right angled deflection to the northeast and flows in a general northeastern direction, emptying into Coal River at Seth.

The principal tributaries to **Laurel Creek** are Sandlick Creek and Hopkins Fork.

Sandlick Creek.—Sandlick Creek rises in the western portion of Sherman District near the Scott-Peytona and Sherman District corner, and flows in a northeastern direction, emptying into Laurel Creek, about 1¼ miles southwest from Seth.

Hopkins Fork.—Hopkins Fork has its source near the Sherman-Crook District Line, at a point 2 miles northwest from Sherman-Crook District corner in the Raleigh-Boone County Line, 3½ miles, just a little south of west, from Jarrolds Valley, and flows in a general northern direction, emptying into Laurel Creek at Nelson P. O. For drainage data, see preceding drainage tables. Its principal tributaries are Logan and Jarrolds Fork.

Upper Whiteoak Creek.—Upper Whiteoak Creek rises near the Kanawha-Boone County Line, one mile west of Kayford, Kanawha County, and flows in a general northwestern direction to the mouth of Left Fork, where it makes a right angled turn to the southwest and flows in a southwestern direction for about 1½ miles, where it deflects to the northwest and follows a general northwestern direction, emptying into Coal River at Orange P. O.

Seng Creek.—Seng Creek rises at the common corner of Raleigh, Kanawha and Boone Counties, about 2 miles south of Kayford, and flows in a general western direction, emptying into Coal River at Whitesville. Its length is about seven miles.

Little Coal River.

Little Coal River is formed by the junction of Pond and Spruce Forks just south of Madison, from which place it flows in a general northern direction 11.1 miles through Boone County, then 6.1 miles as a dividing line between Lincoln and Kanawha Counties, to Forks of Coal, where it empties into Coal River. The tables on preceding pages give its rate of fall and drainage area in Boone County. Little Coal River and its tributaries drain all of Washington, Scott and Crook Districts in Boone County, nearly 286 square miles, or more than one-half of Boone County. About one mile above Ivy Branch, this river makes a loop of one and one-half miles in length, while the C. & O. Ry. Co. connected the same points by tunnel—Pinnacle Tunnel—a distance of one-tenth of a mile.

The principal tributaries of Little Coal River in Boone County, ascending, are: Horse, Little Horse, Hewett, Camp, Rock, Lick, Turtle Creeks, Spruce and Pond Forks. The data in regard to drainage area, gage and fall of the waters are given in preceding tables.

Horse Creek.—Horse Creek rises in the northwestern part of Boone County at the junction of Spruce and Jule Webb Forks and flows in a general northern direction 3.7 miles through Boone County, to the Lincoln-Boone County Line; thence, 2.5 miles through Lincoln County to a point just northeast of Woodville, where it makes a right angled turn to the east and flows in a general eastern direction 2.3 miles through Boone County into Little Coal River at Altman. See preceding tables for drainage data.

Little Horse Creek.—Little Horse Creek heads in the northwestern part of Scott District and flows in a general northern direction, emptying into Little Coal River at Julian.

Hewett Creek—(of Little Coal River).—Hewett Creek rises four miles northeast from Harless at a very sharp corner in the Kanawha-Boone County Line, and flows in a general southwestern direction, emptying into Little Coal River at Harless.

Camp Creek.—Camp Creek heads in the dividing line between Scott and Peytona Districts, about 4 miles southwest from Peytona and flows in a northern direction, emptying into Little Coal River about one-fourth mile above Lowrey.

Pond Fork.

The principal tributaries of Pond Fork, ascending, are: Robinson Creek, Jacks Branch, Bull Creek, West Fork, Casey Branch, and Skin Creek. Data in regard to drainage area, length and fall of Pond Fork and its tributaries, are given in the table on preceding pages of this volume.

Robinson Creek.—Robinson Creek heads in the northwestern part of Crook District, just west of the common corner of Scott, Sherman and Crook Districts, flowing in a general southwestern direction to the mouth of Cabin Fork, where it makes a right angled turn to the west, and flows directly west, emptying into Pond Fork about 5 miles southeast from Madison.

Jacks Branch.—Jacks Branch rises near the Crook-Washington District Line, about 3 miles northeast of Clothier, and after flowing for 1.7 miles in a northwestern direction, makes a right angled turn, and flows for 2½ miles in a northeastern direction, emptying into Pond Fork 6¼ miles southeast from Madison.

Bull Creek.—Bull Creek rises near the Crook-Washington District Line, about 5 miles east from Clothier, and flows in a general northern direction, emptying into Pond Fork 7 miles southwest from Madison.

Casey Creek.—Casey Creek rises near the Crook-Washington District Line, about four miles due west of Bald Knob P. O., and flows in a general northern direction, almost paralleling the Crook-Washington District Line and just east of the same, emptying into Pond Fork 2½ miles south of the mouth of West Fork.

Cow Creek.—Cow Creek rises near the Crook-Washington District Line and flows in a northern direction, emptying into Pond Fork about 3 miles below Bald Knob P. O.

Rock Creek.—Rock Creek is formed by the junction of the Right and Left Forks in the southwestern portion of Scott District and flows in a general western direction, emptying into Little Coal River at Rock Creek Station.

Lick Creek.—Lick Creek heads near the Scott-Washington District Line, 3.5 miles west of Hopkins Station, and flows in an eastern direction through Scott District, emptying into Little Coal River at Hopkins.

Turtle Creek.—Turtle Creek heads in the western portion of Washington District, and flows in a northeastern direction emptying into Little Coal River at Danville. It takes its name from the turtle shaped limestone concretions which are found in profusion along its bed.

Pond Fork.—Pond Fork of Little Coal River heads in the extreme southwestern corner of Boone County on Pilot Knob of Guyandot Mountain, the common corner of Raleigh, Wyoming and Boone Counties, at an elevation of about 3300 feet. It flows in a northwesterly direction for 35 miles to Madison, where it empties into Little Coal River at an elevation of about 690 feet, a fall of 2610 feet. Its drainage area, as calculated by planimeter, is 282.0 square miles. Its fall from source to mouth of Skin Creek is 1880 feet in 5.3 miles or 354.7 feet to the mile. From the mouth of Skin Creek to mouth of Jasper Workman Branch is 32.4 feet in 6 miles, or 5.4 feet per mile; from the mouth of Jasper Workman Branch to mouth of West Fork is 290 feet in 11.9 miles, or 24.3 feet per mile; from the mouth of West Fork to mouth of Pond Fork is about 121 feet in 11.8 miles, or 10.3 feet per mile. Pond Fork and its tributaries drain the entire area of Crook District and about two square miles of Scott District.

James Branch.—James Branch takes its course just west of the head of West Fork, in a section known as the "Flats," in the southwestern portion of Boone and southeastern end of Cook Mountain, and flows in a northwestern direction, emptying into Pond Fork about one mile southeast from Echart P. O.

West Fork.—West Fork of Pond Fork rises in Cherry Pond Mountain, 3¼ miles northwest of Pilot Knob of Guyandot Mountain, at an elevation of about 2700 feet, and flows in a general northwest direction for 16.8 miles, emptying into Pond Fork about 9 miles southeast from Madison, at an elevation of 805 feet. The fall in the first 2.4 miles from the head is 1000 feet or about 416 feet per mile; in the next 3.9 miles down to the mouth of Mats Creek, the fall is 423 feet, or 108.5 feet per mile; from the mouth of Mats Creek to the mouth of West Fork, the fall is 421 feet in 10.5 miles, or 40 feet per mile. Its principal tributaries are Whites Branch, Browns Branch, James and Mats Creeks.

Whites Branch.—Whites Branch has its source near the Crook-Sherman District Line, about 9 miles southeast from Madison, and flows in a general southwestern direction, emptying into the West Fork 0.7 mile east from the mouth of West Fork.

Browns Branch.—Browns Branch has its source near the Crook-Sherman District Line, also near the northern end of Cherry Pond Mountain, 4¼ miles almost due west from Jarrolds Valley and flows in a general northwestern direction, emptying into West Fork two miles southeast from mouth of Whites Branch.

James Creek.—James Creek rises in Cherry Pond Mountain, at the Boone-Raleigh County Line, about 5.5 miles southwest from Jarrolds Valley, and flows in a general northwestern direction, emptying into the West Fork, ½ mile southeast from Chap P. O.

Mats Creek.—Mats Creek takes its origin in Cherry Pond Mountain, at the Boone-Raleigh County Line, 8 miles just a little west of south of Jarrolds Valley, and flows in a general northwestern direction, emptying into West Fork, 3 miles northeast from Bald Knob P. O.

Spruce Fork.

Spruce Fork is formed by the junction of Brush and Laurel Forks in the northeastern portion of Logan County and flows in a general northern direction for 12.1 miles through Logan County, crossing the Logan-Boone County Line at Clothier; thence for 12.6 miles through Boone County, emptying into Little Coal River just south of Madison. Its air line distance from source to mouth is 14.5 miles, while its length, measured with the meanders of the stream, is 24.7 miles. Its fall from source to mouth is approximately 440 feet, or 17.8 feet per mile, while from Clothier to its mouth the distance is 12.6 miles and the fall is 94 feet, making a fall of 7.5 feet per mile. Data in regard to the drainage, length and fall of Spruce Fork and its tributaries are given in the tables on preceding pages of this volume.

The principal tributaries of Spruce Fork in Boone County, ascending, are Sixmile and Hewett Creeks and Spruce Laurel Fork.

Sixmile Creek.—Sixmile Creek has its source in the southwestern portion of Boone County, at the head of Trace Fork of Big Creek of Guyandot River, and flows in a general eastern direction, emptying into Spruce at Ramage. .

Hewett Creek.—Hewett Creek heads in Logan County, about 5 miles northeast from Logan Court House, and flows in a general northeastern direction 5.2 miles through Logan County, crossing the Logan-Boone County Line about one mile southwest from Hewett P. O., and thence for 3 miles through Boone County, emptying into Spruce Fork at Jeffery.

Spruce Laurel Fork.—Spruce Laurel Fork has its source in the extreme southern portion of Boone, at the Logan-Boone County Line, near the corner of Crook-Washington District Line, and flows in a general western direction to the mouth of Dennison Fork, where it turns to the north and flows in a general northern direction to the mouth of Whiteoak Branch, at which point it changes its course to a northwestern one to the mouth of Burnt Cabin Creek, where it turns to the west and continues on that course to its mouth at Clothier. Its general course is rudely parallel with the Logan-Boone County Line from its source to mouth of Burnt Cabin Branch. Its fall from source to mouth of Dennison Fork is about 950 feet; from mouth to Dennison Fork to mouth of Skin Poplar Branch, a distance of 4.5 miles, it falls 320 feet, or 71.1 feet per mile; from mouth of Skin Poplar to mouth of Whiteoak Branch, a distance of four miles, the fall is 152 feet, or 38 feet per mile; from mouth of Whiteoak Branch to mouth of Spruce Laurel, the fall is about 150 feet in 6.2 miles, or 24.2 feet per mile.

Mud River.

Mud River has its source in Washington District at Mud Gap, 4½ miles just a little south of west from Madison, and it flows in a general northwestern direction for 6.5 miles to the mouth of Sugartree Branch, where it crosses the Boone-Lincoln County Line; thence it flows in a general northwestern direction for 75 miles, emptying into the Guyandot River at Barboursville, Cabell County. It meanders much and the character of its flow is very sluggish, since it has very little fall—328 feet in 77 miles, or about 4.3 feet per mile. The river probably derives its name from the fact that its waters are always muddy, owing to the reddish alluvial deposits derived from the Conemaugh Series along its banks.

Trace Fork of Big Creek.

Trace Fork of Big Creek has its source in the western portion of Washington District and flows in a general western direction for 5 miles, where it crosses the Logan-Boone County Line. Its fall in the 5 miles is 400 feet, or 80 feet per mile. For its drainage area in Boone County, see preceding drainage tables.

TOPOGRAPHY OF THE LAND AREA.

Like all other counties in the southern part of West Virginia, Boone County is a highly dissected plateau ranging in height from 1000 to 3300 feet above sea level. The agencies of erosion have reduced the plateau practically all to slope. The streams generally flow in narrow, deeply indented "V" shaped valleys. The numerous ridges and knobs capped with harder layers of rock strata, ranging from 1000 to 3300 feet above tide, bear testimony of the existence of this plateau.

Big Coal River has cut a deep gorge from 1/8 to 1/4 mile in width and 500 to 1500 feet in depth, in a northwest direction through the northeast part of the county. Little Coal River has done likewise in a north and south direction, through the western part of the county. The valley walls are quite steep and rough, caused by the outcrop of great rugged sandstone ledges.

The flood plains of Big and Little Coal Rivers are represented by narrow strips of bottom land, fairly fertile, along both shores that widen out first on one side and then on the other.

River Terraces.—Several fairly pronounced terraces occur along the valley walls of both Big and Little Coal Rivers. The most persistent terraces occur at elevations ranging from 700 to 1000 feet above tide. It is more than probable that the highest terrace was formed when the glacial ice came down from the North, changing the course of the Kanawha River from its ancient channel through Teays Valley by way of Huntington, and thence in a northern direction across the state of Ohio to the St. Lawrence drainage, to a northern direction to the newly formed Ohio River, which it joined at Point Pleasant.

At this time a great inland lake was formed along the Coal River Valley and other tributaries of the Kanawha River by backwater from the glacial dam across southern Ohio.

This old terrace is well defined along Big and Little Coal Rivers, the town of Madison being partly built upon it. Near the mouths of large streams emptying into Coal River, these terraces are very pronounced.

Mountains.

Three ranges of mountains occur in the southern part of Boone County; viz, Cook Mountain, Cherry Pond Mountain, and Guyandot Mountain.

Cook Mountain.—This range extends in a northeast and southwest direction, between the waters of Pond and West Forks of Coal River in Crook District, for a distance of about 12 miles. The highest peaks range in elevation from 2200 feet in the northern part to 3200 feet above tide in the southern part.

Cherry Pond Mountain.—These mountains extend in a northeast and southwest direction, forming the watershed between West Fork of Pond Fork, and Marsh Fork of Big Coal River, for a distance of about 11 miles. The highest peaks range in elevation form 2100 to 3200 feet above tide.

Guyandot Mountain.—This range forms the southern boundary of Boone County and is the divide between the waters of Coal River and the waters of Guyandot River, and extends in an eastern and western direction, ranging in elevation from 3000 to 3300 feet above tide.

PART II.

The Geology of Boone County.

CHAPTER III.

STRUCTURE.

INTRODUCTION.

Geological Structure treats of the general pitch or lay of the strata of the earth's surface. The original position of the rock beds, however formed, whether by lava flow or sedimentation, is normally horizontal or tangential to the curvature of the earth, at the point of contact. The original lay or attitude of the rocks, however, is much changed by lateral and tangential pressure, produced by the contraction of the earth's interior, and the rocks composing the crust of the latter are bent and warped by pressure into numerous approximately parallel wrinkles or folds. In Boone County these folds all have a distinct northeast-southwest trend. In the description of these folds the upward bending arch is called the **anticline** and the downward bending a **syncline.** The **axis** of any **fold** is the line joining the highest points of any anticline or the lowest points of any syncline, and from which the strata dip is an **anticline** and to which they dip is a **syncline.** The **strike** is the direction of the horizontal edge of dipping beds and is frequently, although not always, parallel to the axis of a fold, diverging from it when the axis is not horizontal. The sudden fall of the axis of the anticline forms what is called a **nose** of the fold. The original position of rock beds of any sedimentary formation is nearly horizontal. These deposits, however, may take on a slope of considerable pitch. Earth movements cause modification of this original position and produce the above described structural forms generally used in geology.

METHODS OF REPRESENTING STRUCTURE.

Cross Sections.—There are two methods that can be used in representing geologic structure. One of these is by the cross sections at right angles to the strata, which shows the rocks as they would appear if a deep canal or section were dug perpendicularly across the entire area under discussion. This method can be used where the dip of the rock is very heavy and is perceptible to the eye, but in the Boone area it would not be practical nor satisfactory without greatly exaggerating the vertical scale of the cross section in comparison to the horizontal scale; also this method would only give an idea of the structure along certain lines and would not give the slope of the arches or basins.

Structure Contours.—The method of representing structure which has been adopted by the West Virginia Geological Survey, for areas of gentle dip, consists in the representation by contour lines of the position of some particular stratum. This stratum is generally the one that is well known throughout the area by its exposures in the outcrop, its relation to some other bed above or below it, or its wide use as a "key-rock" by the drillers for oil and gas. These contour lines show in a general way the form and the size of the folds into which the key-stratum has been distorted and its altitude above the level of the sea at all points, and are of great value in the commercial development of the area for the future mining of its coal and the exploitation for possible oil and gas pools.

The bottom of the **Campbell Creek or No. 2 Gas Coal** has been selected as the "key-rock" on which to base the structure, since it is very persistent and widely known in the area, both from exploitation by miners and by core drill holes. The elevation of the base of the Campbell Creek (No. 2 Gas) Coal over a large part of Boone County was obtained by levels directly on its crop; also from the records of core drill holes and the records of wells drilled for oil and gas throughout the area.

On Map II the structure contour lines are printed in green, and these exhibit not only the approximate elevation above sea level of the base of the Campbell Creek (No. 2 Gas) Coal, but also the horizontal contours of the **troughs, arches** and **domes,** and the direction of the **dip** of the **strata.** By examining this map it will be readily seen whether or not the coal is above or below drainage anywhere in the county by noting the elevations of the land surface at the point desired, as exhibited by the topographic contours, and the elevations of the coal as shown by the structure contours at the same place. As an illustration, suppose the position of the Campbell Creek (No. 2 Gas) Coal was desired at Seth. This map shows the elevation of mouth of Laurel Creek at about 685′, and the elevation of the coal, as exhibited by the contour lines, about 570′, therefore the bed should lie about 115 feet below drainage.

As has already been mentioned in preceding County Reports, these structure contours are only approximately correct from the assumed fact that over small areas the rocks maintain a uniform thickness. However, this assumption is often erroneous, since it has frequently been established that the interval between two easily determined formations will vary many feet in a very short distance. While a very large number of elevations were determined with spirit level, a much greater number were obtained with the aneroid barometer, checked as frequently as possible on spirit level elevations of the U. S. Geological Survey marked at the forks of public highways and other conspicuous places. These checks kept down errors in barometric elevations to the extent that it is believed over most of the area their sum is less than one contour interval—that is, less than 25 feet.

DETAILED GEOLOGIC STRUCTURE.

Boone County is traversed by the Warfield Anticline, passing through the area in a northeast and southwest direction. About one-third of the area lies north and the remaining two-thirds lie south of this anticline.

The following are the most important folds:

Anticline.	**Syncline.**
The Warfield.	The Coalburg.

The accurate location and shape of the above folds are exhibited on Map II.

The Warfield Anticline.—This great anticline enters West Virginia from Kentucky at Warfield on the Kentucky side and Kermit on the West Virginia side, and extends in a general northeastern direction through Mingo and Logan Counties, entering Logan near the head of Big Hart Creek and crossing the Guyandot River near Chapmanville, and the Boone-Logan County Line at the head of Curry Fork of Garrett Creek and thence crossing the head of Left Fork of Meadow Fork of Hewett Creek and the divide between Hewett and Sixmile Creeks in a northeastern direction to Sixmile Creek, 1.3 miles northwest of Havana P. O.; thence, N. 50° E. 3.8 miles to Spruce Fork, 0.3 mile southeast of Low Gap Station; thence, it changes its course to N. 41½° E. 7.7 miles, crossing Pond Fork 1.2 miles southeast of Madison, the Right and Left Forks of Rock Creek, 0.9 mile southeast of Foster P. O., and the divide between Rock and Drawdy Creeks at the head of Drawdy Creek, 1.3 miles west of Andrew P. O. at the top of divide between Drawdy and Brush Creeks; thence, it changes its course to N. 48½° E. for 3.2 miles to Big Coal River, crossing said river, 1 mile northwest of the mouth of Drawdy Creek; from this point its course is changed to N. 31° E. for 2.7 miles, crossing the head of Curtis Branch of Lick Creek, 0.4 mile southwest of mouth of Falling Rock; thence, N. 41° E. 2 miles to the Boone-Kanawha County Line at the head of Right Fork of Bull Creek, entering Kanawha County at the head of Mary Branch of Spruce

PLATE III.—Old Stone House at Peytona.

Fork of Four Mile Fork of Lens Creek. It passes about one-half mile south of Hernshaw; thence in the same general direction as Lens Creek and just south of said creek, intersecting the Kanawha River at the mouth of Simmons Creek, 3/4 mile southeast of Marmet; thence about parallel with the general course of this creek to its source, where it crosses the Cabin Creek-Malden District Line; and thence to the head of the Left Fork of Witchers Creek, where it again crosses the Cabin Creek-Malden District Line and apparently dies out near Eight P. O. at the mouth of Eight Mile Fork of Campbell Creek.

It is possible that further careful tracing to the northeast may demonstrate that this Warfield Anticline connects with the Chestnut Ridge Anticline from the northern part of the State, but the present facts do not yet definitely prove this assumption.

The **axis** of the Warfield Anticline rises rapidly to the southwest from Campbell Creek in Kanawha County to where it crosses the Kanawha River, where the Campbell Creek (No. 2 Gas) Coal is about 760 feet; thence it gradually rises to where it crosses Big Coal River, where the Campbell Creek (No. 2 Gas) Coal is about 860 feet B., and from thence to Low Gap, where it crosses Spruce Fork of Little Coal River, the crest rises rapidly so that the Campbell Creek (No. 2 Gas) Coal horizon is 1130 feet at Low Gap; and thence the rise of the axis becomes less rapid to the Boone-Logan County Line, rising 50 feet in six miles.

The Coalburg Syncline.—This trough is the largest in the southern part of West Virginia, lying to the southeast of the Warfield Anticline and almost paralleling it. It originates to the south of Boone County in Mingo and Logan Counties, entering Boone County 0.4 mile southwest of Clothier, crossing Spruce Laurel Fork, and extending N. 62° E. 6 miles to Pond Fork, 0.3 mile east of the mouth of Bull Creek: from thence it changes its course to the left, going N. 34° E. for 6 miles, passing one-fourth mile west of Nelson P. O., and crossing Big Coal River 0.3 mile southwest of Seth, to north side of Coal River; from thence, the general course is N. 60°

E. crossing Joes Creek just south of the mouth of Left Fork and the Boone-Kanawha County Line, 2.0 miles south of Winifrede, near the head of North and South Hollows of Fields Creek, and extending in a northwestern direction touching the sharp bend in Slaughter Creek, ¾ mile south of Chelyan, crossing the Kanawha River near the mouth of Cabin Creek between Cabin Creek Junction and Coalburg at the town from which it derives its name; and from thence parallels said river to Monarch, where it leaves the river, deflecting slightly more to the north, and dies out just above the mouth of Little Ugly Branch of Witchers Creek.

The elevation of the Campbell Creek (No. 2 Gas) Coal in the trough of the syncline, where it enters Boone County from the southwest, is about 450 feet B., just south of Clothier, and the axis of the trough rises on Pond Fork, bringing this coal to 570 feet, and it remains about the same level until it reaches Seth on Big Coal River, where it rises again to the Boone-Kanawha County Line, where the elevation is about 640 feet.

CHAPTER IV.

STRATIGRAPHY—GENERAL SECTIONS.

The outcropping stratified rocks of Boone County are included wholly in the Carbonic System, and the exposed beds extend from the middle of the Conemaugh Series down through the Allegheny and Upper Pottsville or Kanawha Series. The following table illustrates the several divisions of the stratified rocks of Boone County, some portions of which crop at the surface within this area, or have been penetrated by the oil and gas well prospecting:

Table of Geological Formations.

Formation	Thickness	
Pennsylvanian.		
Conemaugh Series	....500 to 600 feet,	
Allegheny Series	......200 to 350 feet;	
Pottsville Series	......250 to 3,000 feet;	
Mississippian.		
Mauch Chunk Red Shales	.... 40 to 400 feet;	
Greenbrier Limestone	.150 to 250 feet;	
Pocono Sandstones	....450 to 600 feet;	
Devonian.		
Catskill Sandstones (Venango Oil Group)	300 to 500 feet,	probably not represented in Boone
Chemung and Hamilton Shales.		

Some general sections will now be given to illustrate the order and character of the several formations.

The following is a generalized section of the Conemaugh, Allegheny, and Kanawha Series, found in the study of the different strata in Boone County. In the Kanawha Series several of the coals and sandstones are named from a generalized section found in the study of Logan and Mingo Counties by Messrs. Hennen and Reger, from whose manuscript these names are obtained:

Generalized Section, Conemaugh, Allegheny and Kanawha Series, Boone County.

	Thickness Feet.	Total Feet.
Conemaugh Series (510')		
Pittsburgh Coal, found in the top of Sugar Tree Knob, Lincoln County, 3.5 miles east of the Boone-Lincoln County Line.	2	2
Interval to **Elk Lick Coal**	215	217
Fire clay and shale	20	237
Limestone, dark gray, **Elk Lick**	2	239
Red and sandy shale, **Birmingham**	11	250
Sandstone, grayish buff, **Grafton**	20	270
Shells, gray and sandy	35	305
Limestone, dark gray, impure, **Ames horizon**	2	307
Red and sandy shales, **Pittsburgh Reds**	45	352
Sandstone, massive, buff, medium coarse grained, **Saltsburg**	38	390
Coal, Bakerstown	1	391
Fire clay	2	393
Sandstone, massive, gray and white, **Buffalo**	40- 50	443
Coal, Brush Creek	2	445
Shale, gray and sandy	25	470
Sandstone, massive, **Mahoning**	40	510
Allegheny Series (175')		
Coal, Upper Freeport	2	512
Fire clay	2	514
Sandstone and sandy shale	30	544
Fire clay, **Lower Freeport**	2	546
Sandstone, massive, pebbly, **East Lynn**	50-110	656
Coal, Upper Kittanning (North Coalburg?)	8	664
Shales, gray	2	666
Coal, Middle Kittanning, No. 5 Block	6	672
Shales gray	3	675
Coal, Lower Kittanning, No. 5 Block, Lower Bench	5	680
Sandy shale	5	685
Kanawha Series (1844')		
Sandstone, **Homewood,** massive	75-100	785
Shale, sometimes holding coal (**"Black Flint" horizon**)	0- 10	795
Coal, Stockton-Lewiston, always multiple bedded and in two to four divisions	0- 10	805
Sandy shale or impure fire clay	10- 20	825
Sandstone, **Upper Coalburg,** coarse, massive, gray, often weathering into "Chimney Rock" columns on the summits	50- 80	905
Shale, dark gray	0- 10	915
Coal, **Coalburg,** multiple bedded, splinty layers sometimes containing cannel coal on top	5- 10	925
Fire clay, impure and sandy shale	0- 20	945
Coal, Little Coalburg,	0- 3	948

	Thickness Feet.	Total Feet.
Shale and impure fire clay	0- 22	970
Sandstone, **Lower Coalburg**, often forming great cliffs	20- 40	1010
Sandy shale	3- 9	1019
Coal, Buffalo Creek, multiple bedded, hard, splinty	0- 6	1025
Fire clay and shale	20- 25	1050
Limestone, **Buffalo Creek**, gray and hard, lenticular, sometimes containing marine fossils	0- 2	1052
Sandstone, **Upper Winifrede**, massive, yellowish gray, micaceous	40- 60	1112
Shale	2- 3	1115
Coal, Winifrede, multiple bedded, hard, splinty	1- 10	1125
Fire clay, impure, and sandy shale	2- 5	1130
Sandstone, **Lower Winifrede**, massive	20- 40	1170
Shale	2	1172
Coal, Chilton "A", multiple bedded	0- 3	1175
Fire clay, impure, and shale	10- 18	1193
Sandstone, **Upper Chilton**, medium grained and massive, micaceous	20- 40	1233
Coal, Chilton "Rider," splinty, multiple bedded	0- 4	1237
Fire clay and gray shale	0- 20	1257
Coal, Chilton, multiple bedded, splinty	1- 8	•1265
Fire clay, impure, and shaly	0- 5	1270
Sandstone, **Lower Chilton**, massive	0- 30	1300
Coal, Little Chilton, double bedded	0- 2	1302
Sandstone, **Hernshaw**	20- 49	1351
Coal, Hernshaw, multiple bedded	0- 4	1355
Fire clay and sandy shale	1- 5	1360
Sandstone, **Naugatuck**	15- 21	1381
Coal, Dingess, multiple bedded, gas coal, once mined at Dingess	1- 4	1385
Shale	1- 5	1390
Sandstone, **Williamson**	5- 20	1410
Shale	1- 5	1415
Limestone, **Dingess**, gray and hard, frequently brown and silicious, lenticular, and ferriferous, with **marine fossils**	0- 2	1417
Shale, dark green, sandy, iron ore nodules and plant fossils	0- 30	1447
Coal, Williamson, multiple bedded, splinty	1- 8	1455
Fire clay, impure, shaly	1- 5	1460
Sandstone, **Upper Cedar Grove**	10- 40	1500
Shale, dark gray, with iron ore nodules and plant fossils	20- 45	1545
Shale, dark gray, slaty limestone, Seth, marine fossils, many **producti**	0- 3	1548
Shale, dark	0- 2	1550
Coal, Upper Cedar Grove, splinty at bottom, **Island Creek** bed; also same as **Red Jacket** or **Upper Thacker**	2- 5	1555
Fire clay and shale	0- 10	1565

	Thickness Feet.	Total Feet.
Sandstone, **Middle Cedar Grove**, sometimes holding a thin coal	0- 60	1625
Coal, Lower Cedar Grove, multiple bedded, splinty at top, and is **lower bench** of the **Cedar Grove** proper, and **Island Creek** bed; also same as **Lower Thacker**	2- 5	1630
Fire clay, impure, and shale	1- 10	1640
Sandstone, **Lower Cedar Grove**	20- 30	1670
Coal, Alma "A"	0- 1	1671
Shale, dark gray, with iron ore nodules and plant fossils	5- 9	1680
Coal, Alma, multiple bedded, splinty layers, same as **Draper** bed of **Logan**	2- 5	1685
Fire clay, impure, and shale	0- 5	1690
Sandstone, **Logan**, massive, bluish gray, medium grained	20- 40	1730
Shale, sandy	1- 5	1735
Coal, Little Alma, multiple bedded, slaty	0- 3	1738
Sandstone, massive, **Malden**	15- 29	1767
Shale, sandy, gray and flaggy	1- 5	1772
Limestone, **Campbell Creek**, dark gray, hard, silicious, lenticular, **marine fossils** found at Bald Knob and High Coal	0- 2	1774
Shale, flaggy and sandy, with iron ore nodules and plant fossils	10- 20	1794
Coal, Campbell Creek, Upper Bench, multiple bedded, gas type and **Upper War Eagle** on Turkey Creek; also same as **No. 2 Gas** of Kanawha Valley	2- 6	1800
Shale	1- 3	1803
Sandstone, **Lower Malden**, massive	8- 30	1833
Shales, gray	1- 5	1838
Coal, Campbell Creek (No. 2 Gas), Lower Bench, multiple bedded, gas type	1- 3	1841
Shale, gray	1- 3	1844
Sandstone, **Brownstown**, massive	21- 51	1895
Coal, Powellton "A," double bedded	0- 1	1896
Shale, sandy, flaggy and laminated	10- 20	1916
Coal, Powellton, double bedded, same as **"Hatfield Tunnel"** vein	0- 3	1919
Shale, dark, laminated	10- 20	1939
Limestone, **Stockton**, silicious, lenticular	0- 4	1943
Sandstone, shale, dark, and sandstone	25- 34	1977
Coal, Matewan, double bedded	0- 2	1979
Sandstone, **Matewan**	20- 40	2019
Coal, Eagle "A", gas coal	0- 2	2021
Sandstone, **Eagle**	20- 37	2058
Shale	5- 10	2068
Coal, Eagle, gas, and same as **Middle War Eagle** of Turkey Creek	1- 6	2074
Fire clay, impure and shale	0- 5	2079
Sandstone, **Bens Creek**	0- 27	2106

	Thickness Feet.	Total Feet.
Coal, Bens Creek, gas, soft, columnar, multiple bedded and probably a split off of main **Eagle**	0- 3	2109
Fire clay and shale	5- 10	2119
Sandstone, **Decota,** massive	40- 57	2176
Shale	5- 10	2186
Coal, Little Eagle, double bedded gas coal	1- 3	2189
Sandstone, flaggy and shaly	0- 20	2209
Coal, Cedar, multiple bedded gas coal, same as once mined at Cedar, Mingo County, and probably a split off of **Little Eagle**	0- 4	2213
Sandstone, **Grapevine**	25- 30	2243
Slate, black, laminated, with marine fossils, **Eagle**	15- 20	2263
Limestone, **Eagle,** dark, brittle, hard, lenticular, marine fossils	0- 2	2265
Slate, black, iron ore nodules, marine fossils, **Eagle**	10- 25	2290
Coal, Little Cedar, gas coal	0- 1	2291
Sandstone, **Lower War Eagle**	20- 30	2321
Shale	5- 10	2331
Coal, Lower War Eagle, gas, soft, multiple bedded	0- 3	2334
Shale	1- 5	2339
Sandstone, **Upper Gilbert,** grayish white, massive, medium grained	40- 50	2389
Shale, black with iron ore nodules	10- 15	2404
Sandstone, bluish gray, laminated	10- 15	2419
Coal, Glenalum Tunnel, multiple bedded, much split up with slate partings, 1" to 3' thick, soft gas type	0- 15	2434
Sandstone, **Lower Gilbert,** massive, grayish white	15- 80	2514
Shale	1- 4	2518
Coal, Gilbert, double bedded, gas	0- 1	2519
Shale, sandy	5- 10	2529

Some general sections will now be given to illustrate the order and character of the several formations composing the rock column in Boone County, made from exposures at crop and from records of the numerous borings for coal, petroleum and natural gas throughout the county.

SECTIONS.

Peytona District.

The following section was obtained with spirit level by Krebs, descending from a high knob eastward to Lens Creek,

0.6 mile south of Hernshaw P. O., Loudon District, Kanawha County, and two miles east of the Boone-Kanawha County Line:

Section 0.6 Mile South of Hernshaw, Loudon District, Kanawha County.

		Thickness Feet.	Total Feet.	
Pennsylvanian (830′)				
Allegheny Series (86′)				
Sandstone, massive, making cliff, **East Lynn** (1436′ L.)		66	66	
Concealed in slope		15	81	
Concealed in bench, **No. 5 Block Coal horizon**		5	86	86.0′
Kanawha Series (744′)				
Concealed in slope		35	121	
Concealed in bench		0	121	
Concealed in slope		20	141	
Black slate (thickness concealed) (1354′ L.)		...	141	
Concealed		55	196	
Sandstone, massive		5	201	
Coal, hard, splint....1′ 7½″	(6′ 6″)			
Slate, gray..........0 2				
Coal, slaty, laminated 2 6	**Stockton** ...	6.5	207.5	121.5′
Coal, hard, splint....1 5	(1303.8′ L.)			
Slate, gray..........0 0½				
Coal, softer..........0 9				
Slate, gray, concealed, with sandstone		49	256.5	
Coal, (4″), local		0.3	256.8	
Slate, dark brown, with iron ore nodules		4	260.8	
Coal, (5″), local		0.4	261.2	
Interval		15.1	276.3	
Coal, cannel..........0′ 4″	(3′ 1″)			
Slate 0 5	**Coalburg** ...	3.1	279.4	71.9′
Coal 2 4				
Interval		25	304.4	
Coal, Buffalo Creek		1.9	306.3	
Concealed and sandstone		54.2	360.5	
Coal ....0′ 5″	(3′ 9″)			
Slate 1 10	**Winifrede** ..	3.8	364.3	84.9′
Coal 1 6				
Concealed and sandstone		51	415.3	
Coal, Chilton "A"		1.8	417.1	52.8′
Sandstone and concealed		58.8	475.9	
Coal, Chilton (2′ 3″)		2.2	478.1	61.0′
Concealed		33.1	511.2	
Coal ....1′ 10″	(6′ 3″)			
Slate and fire clay....4 0	**Little Chilton**	5.4	516.6	38.5′
Coal 0 5				
Interval		22.6	539.2	
Coal, Hernshaw (2′ 10″)		2.8	542.0	25.4′
Unrecorded		15.3	557.3	
Coal, (5″), local		0.4	557.7	
Unrecorded, containing **Dingess marine limestone** and shale fossils		58.5	616.2	

PLATE IV.—Mouth of West Fork of Pond Fork, and topography of Kanawha Series.

	Thickness Feet.	Total Feet.	
Coal, Williamson, dirty, (1' 2")	1.2	617.4	75.4'
Unrecorded	40.0	657.4	
Coal ... 1' 2" / Slate and shale ... 0 2 / Coal ... 2 3 } (3' 7") Cedar Grove.	3.6	661.0	
Unrecorded	31.0	692.0	
Coal, Alma (1' 6")	1.5	693.5	
Unrecorded	85.2	778.7	
Coal ... 0' 2¾" / Slate ... 0 3¼ / Coal ... 3 10½ / Fire clay ... 0 7½ / Coal ... 2 3½ } (7' 3½") Campbell Creek or No. 2 Gas	7.3	786.0	
Concealed to creek	44.0	830.0	

The above section has been verified by Messrs. Hennen and Reger from their study of the coals in Mingo and Logan Counties.

The foregoing section illustrates the succession of the coals in the northern part of Boone County.

The following aneroid section was measured by Teets, descending from a high point in the Peytona-Sherman District Line at head of Roundbottom Creek, joined to well No. 1 (13) Columbus Gas & Fuel Company:

Section 1.4 Miles Northeast of Racine, Peytona District.

	Thickness Feet.	Total Feet
Pennsylvanian (1735')		
Allegheny Series (70')		
Sandstone, massive, coarse grained, buff color, **East Lynn**	50	50
Sandy shale	20	70
Kanawha Series (880')		
Sandstone, massive, medium coarse grained, gray buff color, micaceous, **Homewood**	60	130
Sandy shale	30	160
Sandstone, massive, buff color, coarse grained **Coalburg**	60	220
Sandy shale and sandstone	40	260
Sandy shale and concealed	50	310
Sandstone, massive, medium coarse grained	19	329
Coal blossom, Winifrede	1	330
Sandy shale	15	345
Sandstone, massive	42	387
Coal blossom, Chilton "A"	3	390
Sandstone, massive, micaceous, **Chilton**	37	427
Coal blossom (Chilton)	3	430
Sandstone and concealed	56	486
Coal opening, fallen shut, **Hernshaw**	4	490
Sandstone and concealed	80	570

		Thickness Feet.	Total Feet.
Sandy shale		10	580
Sandstone and concealed		30	610
Coal, gas..........0' 7"	Cedar Grove...		
Coal, impure........0 2			
Coal, block.........0 10			
Slate 0 1			
Coal, gray splint....1 3			
Slate, dark..........0 1		4-4"	614-4"
Coal, gray splint....0 3			
Bone 0 2			
Coal, gray splint....0 4			
Slate, gray..........0 1			
Coal, bony..........0 6			
Concealed and sandstone		86-8"	701
Coal, block..........0' 9"	**Campbell Creek (No. 2 Gas) Upper Bench...**		
Slate 0 2			
Coal, gray splint....0 5		3-9"	704-9"
Coal, gas...........1 0			
Bone 0 1			
Coal, gray splint....1 4			
Fire clay and shale, to top of well No. 1 (13), (875' L.)		5-3"	710
Gravel, brown		27	737
Sand, white		53	790
Slate, black		60	850
Sand, white		60	910
Slate, black		40	950
Middle and Lower Pottsville Series (785')			
Sand, white, (Nuttall)		160	1110
Slate, black		45	1155
Sand, white		55	1210
Slate, black		25	1235
Sand, white		75	1310
Slate, black		20	1330
Lime, black		40	1370
Slate, black		60	1430
Sand, Salt, white		265	1695
Slate, black		30	1725
Sand, white		10	1735
Mississippian (856')			
Mauch Chunk (190')			
Red rock		45	1780
Slate, white		25	1805
Sand, Maxton		45	1850
Little Lime, black		70	1920
Pencil cave, black		5	1925
Greenbrier Limestone (180')			
Big Lime, white		180	2105
Pocono Sandstones (486')			
Big Injun, red		50	2155
Slate and shells		414	2569
Shale, brown		7	2576
Berea Grit		15	2591
Devonian (19')			
Slate, gray		19	2610

In the foregoing section the Upper Bench of the Campbell Creek or No. 2 Gas Coal is mined near the top of the well. The Lower Bench of the same seam is about 20 feet under the Upper, so that the interval between the Campbell Creek or No. 2 Gas Coal and the top of the Big Lime is 1200 feet at this point.

The following aneroid section was measured by Teets, descending from a high point on the north side of Coal River, about midway between Peytona and Racine, southward to top of Peytona Well No. 1 (14) located on south side of Coal River, near mouth of Andrew Creek, drilled by the Crude Oil Company:

Section 0.7 Mile East of Peytona, Peytona District.

	Thickness Feet.	Total Feet.
Pennsylvanian (1395')		
Pottsville Series (1395')		
Sandstone, **Winifrede**, massive, coarse grained	45	45
Sandy shale and concealed, **Winifrede Coal horizon**	20	65
Sandstone	20	85
Sandy shale and concealed	15	100
Sandstone, **Lower Winifrede**, massive, micaceous, fine grained	45	145
Sandy shale, bench	10	155
Sandstone, **Chilton**, massive, medium coarse grained	65	220
Sandy shale and concealed	25	245
Sandstone	55	300
Sandy shale and concealed	50	350
Coal blossom, Cedar Grove	1	351
Sandy shale	9	360
Sandstone and sandy shale	35	395
Coal blossom, Alma	2	397
Sandy shale	13	410
Sandstone, massive, medium coarse grained	45	455
Sandy shale	20	475
Coal 0' 8" / Slate 0 3 / **Coal, gas, visible** 1 9 } **Campbell Creek (No. 2 Gas) Coal Upper Bench**	2-8"	477-8"
Sandstone	12-4"	490
Coal blossom, Campbell Creek (No. 2 Gas) Coal, Lower Bench	2	492
Sandstone, massive, medium coarse	38	530
Sandy shale and concealed to top of well, (765' B.)	40	570
Earth	13	583
Sand	80	663
Slate	90	753

	Thickness Feet.	Total Feet.
Sand	181	934
Slate	20	954
Lime	156	1110
Sand	100	1210
Lime	83	1293
Sand	70	1363
Slate	32	1395
Mississippian (602')		
Mauch Chunk (322')		
Sand	15	1410
Lime	20	1430
Sand, Maxton	90	1520
Rock	10	1530
Sand	132	1662
Little Lime	55	1717
Greenbrier Limestone (195')		
Big Lime	195	1912
Pocono Sandstone (85')		
Sand and red rock, Big Injun	60	1972
Slate	25	1997

The foregoing section shows the interval between the base of the Campbell Creek or No. 2 Gas Coal and the top of the Big Lime to be 1239 feet.

The following aneroid section was measured by Teets, descending from a high point just northwest of Racine, southward to Coal River, at the mouth of Indian Creek, ½ mile south of Racine P. O.:

Section ½ Mile South of Racine P. O., Peytona District.

	Thickness Feet.	Total Feet.	
Pennsylvanian (631')			
Pottsville Series (631')			
Sandy shale and concealed	10	10	
Sandstone, **Winifrede**, massive, coarse grained	60	70	
Coal blossom, Winifrede	7	72	72'
Sandstone, **Lower Winifrede**, massive, medium coarse grained	88	160	
Sandy shale	15	175	
Sandstone, massive	65	240	
Coal, Hernshaw, impure, prospect opening	2	242	140'
Sandstone, **Hernshaw**, massive, medium coarse grained	78	320	
Sandy shale and concealed	20	340	
Sandstone, **Lower Cedar Grove**, massive	40	380	

		Thickness Feet.	Total Feet.	
Coal, tough, gnarly....1' 7" Slate, gray............0 1 **Coal**, block...........0 7 Slate, gray............0 1 **Coal**, gnarly..........0 4 Slate, gray............0 2 **Coal**, gray splint.......0 7 Slate, black...........0 3 **Coal**, gas.............0 5 **Coal**, bony............0 8	**Alma**	4-9"	384-9"	143'
Fire clay, visible		1-3"	386	
Sandstone and concealed		59	445	
Sandy shale and concealed		10	455	
Sandstone and sandy shale		23	478	
Coal blossom, Lower Bench, Campbell Creek or No. 2 Gas Coal		2	480	95'
Sandstone, massive, fine grained		65	545	
Sandy shale		20	565	
Sandstone, massive, fine grained		60	625	
Coal, block...........0' 6" Slate, gray............0 3 **Coal**, block...........0 3	**Eagle**	1	626	146'
Fire clay, (elevation 664' L.)		5	631	

The following aneroid section was measured by Teets, descending from a high point southward to Falling Rock Branch of Lick Creek, and combined with Coal River **Mining** Company's "A" Well No. 1 (9), drilled by the Columbus Gas & Fuel Company 3.2 miles north of Racine:

Section 3.2 Miles North of Racine, Peytona District.

	Thickness Feet.	Total Feet.	
Pennsylvanian (1574')			
Allegheny Series (15')			
Sandy shale	15	15	
Kanawha Series (941')			
Sandstone, **Homewood**, massive	60	75	
Sandy shale and concealed	30	105	
Sandstone, **Coalburg**, massive, coarse grained, buff color	65	170	
Sandy shale	15	185	
Sandstone, **Winifrede**, massive, coarse grained	95	280	
Sandy shale and sandstone	105	385	
Sandstone, **Hernshaw**	45	430	
Sandy shale	12	442	
Fire clay, **Hernshaw Coal horizon**	3	445	445'
Sandstone, **Naugatuck**, massive	70	515	
Sandy shale and sandstone	88	603	
Coal blossom, Alma	2	605	160'
Sandy shale	10	615	
Sandstone, massive	57	672	

	Thickness Feet.	Total Feet.	
Coal blossom, Campbell Creek (No. 2 Gas), Upper Bench	3	675	70′
Sandy shale and sandstone	18	693	
Coal blossom, Campbell Creek (No. 2 Gas), Lower Bench	2	695	20′
Sandstone, **Brownstown,** massive, to top of well "A," No. 1 (9), at 819′ L.	21	716	
Gravel	12	728	
Slate	15	743	
Sand	133	876	
Slate	80	956	
Middle and Lower Pottsville Series (618′)			
Sand	75	1031	
Slate	15	1046	
Sand	255	1301	
Slate	40	1341	
Sand	233	1574	
Mississippian (882′)			
Mauch Chunk (200′)			
Slate and red rock	99	1673	
Sand, Maxton	60	1733	
Little Lime	38	1771	
Pencil cave	3	1774	
Greenbrier Limestone (193′)			
Big Lime	193	1967	
Pocono Sandstone (489′)			
Slate	4	1971	
Red sand	60	2031	
Slate and shells	399	2430	
Brown shale	16	2446	
Sand, Berea	10	2456	

The foregoing section gives the interval between the Campbell Creek (No. 2 Gas) Coal and top of the Big Lime as 1079 feet.

At the head of Bull Creek in the northern part of Boone County, the following section was measured by Krebs in the northern corner of Peytona District, descending from the summit of a high knob on the Boone-Kanawha County Line, four miles northeast of Dartmont, southwestward to Road Fork of Bull Creek and joined to Dart well No. 1 (1):

Section Four Miles Northeast of Dartmont, Peytona District.

	Thickness Feet.	Total Feet.
Pennsylvanian (1640′)		
Allegheny Series (100′)		
Sandy shale	20	20
Sandstone, massive, **East Lynn,** full of iron ore?	75	95
Fire clay and **coal blossom, No. 5 Block**	5	100

	Thickness Feet.	Total Feet.	
Kanawha Series (945')			
Sandstone, coarse grained, **Homewood**, to bench	75	175	
Sandstone and concealed	75	250	
Coal blossom, Stockton	5	255	
Sandstone and concealed, Coalburg	100	355	
Bench, **Coalburg Coal horizon**	0	355	
Sandstone and concealed	99	454	
Coal blossom, Winifrede	1	455	355'
Sandstone and concealed	159	614	
Slate, gray	1	615	
Sandstone and concealed	68	683	
Coal, blossom, Williamson	2	685	267'
Sandstone and concealed	59	744	
Coal blossom, Cedar Grove	1	745	60'
Sandstone and concealed	20	765	
Soil (top of well, elevation 815' B.)	15	780	
Sand	82	862	
Coal, Campbell Creek (No. 2 Gas)	2	864	119'
Sand	81	945	
Slate	100	1045	
Middle and Lower Pottsville (595')			
Sand (Nuttall)	120	1165	
Slate	8	1173	
Sand	97	1270	
Slate	10	1280	
Sand	355	1635	
Slate	5	1640	
Mississippian (865')			
Mauch Chunk (153')			
Red rock	5	1645	
Black slate	6	1651	
Lime	25	1676	
Red rock	35	1711	
Slate	7	1718	
Lime	7	1725	
Sand, Maxton	44	1769	
Slate	3	1772	
Lime and pencil cave	21	1793	
Greenbrier Limestone (192')			
Lime 118', Sand 20, Lime 54 } Big Lime	192	1985	
Pocono Sandstone (520')			
Sand, Big Injun	90	2075	
Slate	60	2135	
Sand, Squaw	50	2185	
Slate and shells	293	2478	493'
Sand, Berea Grit	27	2505	
Devonian (40')			
Black slate to bottom (1780')	40	2545	

The foregoing section gives the interval between the No. 5 Block Coal and the Campbell Creek (No. 2 Gas) Coal as 764 feet, and that of the Winifrede and the Campbell Creek (No. 2 Gas) Coal, 409 feet, comparing with 422 feet in the Hernshaw Section; and the interval between the last coal and the top of the Big Lime, 929 feet, compared with 1239 feet in the Peytona Section, an increase in sediments of 310 feet in 5.6 miles to the south, or 55 feet per mile.

At the mouth of Bull Creek near Dartmont, the following section was measured by Teets, descending from a high point, ¾ mile southeast of the mouth of Bull Creek of Coal River, westward down the hill, and joined to the Dart well No. 4 (8), drilled by Leschen Oil & Gas Company, ½ mile southwest of Dartmont:

Section ½ Mile Southeast of Dartmont Station, Peytona District.

	Thickness Feet.	Total Feet.
Pennsylvanian (1435′)		
Kanawha Series (950′)		
Sandstone, massive	30	30
Concealed	20	50
Sandstone, **Coalburg**, massive	40	90
Concealed	15	105
Sandstone and concealed	50	155
Concealed	20	175
Sandstone, **Winifrede**, massive	45	220
Concealed, **Winifrede Coal horizon**	15	235
Sandstone and concealed	60	295
Concealed and sandy shale	130	425
Sandstone and sandy shale	60	485
Concealed	20	505
Sandy shale	40	545
Concealed	10	555
Sandy shale and concealed to top of well (elevation, 715′ B.)	95	650
Soil	23	673
Coal, Campbell Creek (No. 2 Gas) horizon	2	675
Sand	75	750
Slate	200	950
Middle and Lower Pottsville Series (485′)		
Sand	35	985
Slate	10	995
Sand	430	1425
Slate	10	1435
Mississippian (278′)		
Mauch Chunk (255′)		
Black lime	30	1465
Red rock	25	1490
Black lime	20	1510
Sand	10	1520
Slate	80	1600
Black lime	10	1610

	Thickness Feet.	Total Feet.	
Slate	10	1620	
Lime	50	1670	
Sand, grit	20	1690	255′
Greenbrier Limestone (20′)			
Lime, Big Lime	20	1710	
Unrecorded, to bottom	3	1713	

The foregoing section gives the interval between the Campbell Creek (No. 2 Gas) Coal and top of Big Lime as 1015 feet.

Two miles east of Dartmont, on Lick Creek, the following aneroid section was measured by Krebs, descending from a high summit southward to Lick Creek, 1.5 miles east of Cobb P. O.:

Section 1.5 Miles East of Cobb P. O., Peytona District.

	Thickness Feet.	Total Feet.	
Pennsylvanian (705′)			
Kanawha Series (705′)			
Sandy shale	5	5	
Sandstone, massive, **Homewood**, ferruginous	40	45	
Sandy shale, **Stockton-Lewiston Coal horizon**	5	50	
Sandstone, coarse, **Coalburg**	75	125	
Bench, **Coalburg Coal horizon**	0	125	
Sandstone, massive	50	175	
Sandstone, flaggy	52	227	
Coal blossom, Winifrede	3	230	230′
Sandstone and concealed to bench	50	280	
Sandy shale and concealed	50	330	100′
Bench, **Hernshaw Coal horizon**	0	330	
Sandstone and concealed to bench	60	390	
Sandstone and concealed	85	475	
Bench, **Cedar Grove Coal horizon**	0	475	145′
Sandy shale and concealed	107	582	
Coal blossom, Campbell Creek (No. 2 Gas)	3	585	
Sandstone, shelly	32	617	
Coal blossom, Campbell Creek (No. 2 Gas), Lower Bench	3	620	
Sandstone and concealed	43	663	
Slate, **Powellton Coal horizon**	2	665	
Sandstone and concealed to creek, (710′ B.)	40	705	

The section was taken to the rise so that the interval between the Winifrede Coal and the Campbell Creek or No. 2 Gas is greater than 390 feet shown in the section.

On Lick Creek, the following section was measured by Teets from the Peytona-Sherman District corner in the Kanawha-Boone County Line, descending southwestward to the head of Lick Creek, 2.7 miles northeast of Racine:

Section 2.7 Miles Northeast of Racine, Peytona District.

Kanawha Series (475')	Thickness Feet.	Total Feet.
Sandstone and concealed	85	85
Sandy shale	15	100
Sandstone, **Upper Chilton**, massive	38	138
Coal blossom, Chilton	2	140
Sandy shale and concealed	75	215
Bench, **Hernshaw Coal horizon**	0	215
Sandstone and concealed	50	265
Sandy shale and concealed	25	290
Sandstone and concealed	35	325
Sandy shale and concealed	33	358
Coal blossom, Cedar Grove	2	360
Sandstone and concealed	30	390
Sandy shale and concealed	82	472
Coal blossom, Campbell Creek (No. 2 Gas), (875' B.)	3	475

One mile south of Emmons Station, the following section was measured by Teets, descending westward to Coal River, to a point 0.4 mile south of the C. & O. mile post marked "St. A. 19," Washington District, Kanawha County:

Section 1 Mile North of Emmons Station, Washington District, Kanawha County.

Kanawha Series (454')		Thickness Feet.	Total Feet.
Sandy shale and concealed		30	30
Sandstone, **Coalburg**, massive		27	57
Sandy shale		11	68
Sandstone, **Upper Winifrede**, massive		99	167
Sandy shale		27	194
Coal blossom, Winifrede		2	196
Sandstone, **Lower Winifrede**, massive		33	229
Coal ....0' 8"	**Chilton "A"**		
Slate, gray 0 3			
Coal, splint 1 6		4	233
Slate, gray 0 2			
Coal, splint 1 5			
Slate		1	234
Sandy shale		10	244
Sandstone, **Chilton**, massive		66	310
Sandy shale		17	327
Sandstone, **Hernshaw**, massive		37	364
Sandy shale		5	369

	Thickness Feet.	Total Feet.
Sandstone, **Naugatuck**	30	399
Coal blossom, Dingess	1	400
Sandstone	5	405
Gray shale	19	424
Slate and shale	8	432
Limy shale, Dingess, marine fossils	1	433
Slaty shale	1-4"	434-4"
Sandy shale	1-8"	436
Coal 0' 7" / Fire clay 0 2 / **Coal** 0 11½ } **Williamson**	1-8½"	437-8½"
Gray shale	10-3½"	448
Coal blossom	1	449
Sandstone, massive, to 641' L.	5	454

The coal at 233 feet has been mined by Jefferson Gillispie for local fuel use, and correlates with the Chilton "A."

The Dingess fossils appear in the railroad grade in the Dingess Limestone.

The following aneroid section was measured by Krebs, descending from a high point ½ mile south of Olcott Station, northward to Brier Creek at Olcott, Washington District, Kanawha County, and one mile north of the Boone-Kanawha County Line:

Section at Olcott, Washington District, Kanawha County.

	Thickness Feet.	Total Feet.	
Kanawha Series (510')			
Sandy shale and concealed	115	115	
Sandstone, **Homewood**, massive, pebbly	50	165	165'
Fire clay and concealed, **Stockton Coal horizon**	5	170	
Sandstone, massive and concealed	130	300	
Coal, Coalburg	3	303	138'
Sandstone and concealed	30	333	
Slate	1	334	
Coal, splint 0' 6" / **Coal**, splint, hard 0 4 / **Coal**, block, hard 2 9 } **Winifrede**	3-7"	337-7"	
Slate	1-5"	339	
Sandstone and concealed	161	500	
Coal blossom, Hernshaw	3	503	
Concealed, to 790' B.	7	510	272'

The foregoing section shows the base of the Winifrede Coal 172 feet below the base of a pebbly, massive sandstone, which correlates with the Homewood Sandstone

The following aneroid section was measured by Krebs, descending from a high summit northeast of Brounland, southwestward to Coal River at mouth of Brier Creek, Washington District, Kanawha County, ¾ mile north of the Boone-Kanawha County Line:

Section at Brounland, Washington District, Kanawha County.

	Thickness Feet.	Total Feet.	
Pennsylvanian (560')			
Conemaugh Series (180')			
Sandstone, conglomeratic	25	25	
Sandy shale and concealed	125	150	
Sandstone, massive, conglomeratic	30	180	180'
Allegheny Series (146')			
Sandy shale and concealed	70	250	
Coal blossom, cannel? North Coalburg	2	252	
Sandy shale and concealed	20	272	
Sandstone, massive, **East Lynn**	40	312	
Slate	1	313	
Coal, No. 5 Block (2' 3")	3	316	
Sandy shale and concealed	10	326	146'
Kanawha Series (234')			
Sandstone, massive, **Homewood**	100	426	
Sandy shale and concealed	25	451	
Sandstone, shale	4	455	
Sandy shale, horizon of **Kanawha Black Flint**	13	468	
Coal, slaty 0' 9"			
Fire clay 0 7			
Coal, impure 0 9			
Coal, cannel 1 6			
Coal and slate 1 7			
Fire clay 0 11 — **Stockton-Lewiston**	12	480	
Coal, hard, block 1 6			
Slate and shale 1 5			
Coal, hard 1 3			
Coal, bone 0 4			
Coal, hard, visible 1 5			
Sandy shale and concealed	10	490	
Sandstone, massive, grayish 20' / Sandy shale and concealed 10 — **Coalburg Sandstone**	30	520	
Coal, Coalburg	2-6"	522-6"	
Sandstone and concealed to Coal River, (590' B.)	37-6"	560	234'

The following section was measured by Teets from a high point on the north side of River Fork of Fork Creek of Coal River, descending southward to top of core drill hole No. 1 (103) of Holly and Stephenson, 1.6 miles up River Fork, and 3½ miles south from Brounland:

Section 3½ Miles Southeast of Brounland, Peytona District.

Kanawha Series (400')	Thickness Feet.	Total Feet.	
Sandstone and concealed, Homewood	92	92	
Coal blossom, Stockton-Lewiston	3	95	95'
Sandstone and concealed	93	188	
Coal, good, Coalburg	2	190	95'
Sandstone and concealed	78	268	
Coal blossom, Winifrede	2	270	80'
Sandstone and concealed to top of core drill hole, at 860' B	130	400	

The writer was unable to get a correct log of the core test hole.

The following aneroid section was measured, descending from a high summit between Locust Fork and Fork Creek, northeastward to mouth of Locust Fork, where it was joined to a diamond core hole (100) 1.2 miles west of Emmons:

Section 1.2 Miles West of Emmons P. O., Peytona District.

Pennsylvanian (965')		Thickness Feet.	Total Feet.
Allegheny Series (78')			
Sandy shale		20	20
Sandstone, massive, cliff rock, large pebbles near base		50	70
Coal and shale......2' 6"	No. 5 Block	6-1"	76-1"
Coal, block.........2 0			
Slate 0 7			
Coal, block.........1 0			
Slate		1-11"	78
Kanawha Series (887')			
Sandstone and concealed		87	165
Coal blossom, Stockton-Lewiston		3	168
Sandstone and concealed		67	235
Coal ....2' 0"	Coalburg	4-7"	239-7"
Shale, gray.........1 5			
Coal 1 2			
Slate		1-5"	241
Sandstone and concealed		65	306
Coal ....0' 10"	Winifrede Upper Bench	4-3"	310-3"
Slate 0 3			
Coal 0 8			
Slate 1 4			
Coal 1 2			
Sandstone and sandy shale		16-9"	327
Coal, splint........1' 3"	Winifrede Lower Bench	2-7"	329-7"
Coal, softer........0 3			
Coal, splint........1 1			
Sandstone and concealed		130-5"	460
Coal, Little Chilton		1-8"	461-8"

				Thickness Feet.	Total Feet.
Concealed to top core drill hole (100) at 680′ B				18-4″	480
Coaly gravel				23	503
Shale				4	507
Sandstone				2	509
Shale				8	517
Sandstone				28	545
Shale				2	547
Sandstone				10	557
Shale				15	572
Sandstone				0-8″	572-8″
Coal	0′	4″	**Cedar Grove**		
Fire clay	1	0	**Cedar Grove**	1-8″	574-4″
Coal	0	4	**Cedar Grove**		
Fire clay				0-8″	575
Shale				12	587
Sandstone				4	591
Shale				1	592
Sandstone				11	603
Shale				8	611
Sandstone				13	624
Coal, Alma				1-6″	625-6″
Fire clay				1-6″	627
Shale				17-6″	644-6″
Coal	0′	6″	**Little Alma**		
Fire clay	1	6	**Little Alma**	4-10″	649-4″
Shale	1	6	**Little Alma**		
Coal	1	4	**Little Alma**		
Fire clay				2	651-4″
Shale				55-8″	707
Coal	0′	4″	**Campbell Creek (No. 2 Gas)**		
Fire clay	1	8	**Campbell Creek (No. 2 Gas)**	3	710
Coal	1	0	**Campbell Creek (No. 2 Gas)**		
Sandstone				19	729
Shale				12	741
Clay				1	742
Shale				12	754
Sandstone				3	757
Shale				11	768
Sandstone				12	780
Sandy shale				7	787
Sandstone				7	794
Shale				7	801
Sandstone				144	945
Shale				17-6″	962-6″
Coal and shale				0-6″	963
Shale, to bottom of boring				2	965

In the foregoing section the interval between the base of the No. 5 Block Coal and the Campbell Creek (No. 2 Gas) Coal is 630′ 11″, compared with 700 feet at Hernshaw.

In going up Fork Creek southward for two miles to the mouth of Jimmy Fork, the measures rise faster than the bed of Fork Creek, as is shown in the following aneroid section, measured by Teets, descending from a high point east of mouth of Jimmy Fork of Fork Creek of Coal River, westward down the hill and joined to shot-core test hole No. 1 (102), drilled by the Forks Coal & Land Company, at mouth of Jimmy Fork, 2 miles southwest of Emmons:

Section 2 Miles Southwest of Emmons P. O., Peytona District.

	Thickness Feet.	Total Feet.	
Pennsylvanian (807′ 2″)			
Allegheny Series (160′)			
Sandstone and concealed	75	75	
Sandy shale and concealed	20	95	
Sandstone, **East Lynn,** massive, **Upper Bench**	40	135	
Sandy shale and concealed, **No. 5 Block Coal horizon**	25	160	
Kanawha Series (647′ 2″)			
Sandstone, **Homewood,** massive, Lower Bench	30	190	
Sandy shale	20	210	115′
Sandstone and concealed	50	260	
Bench, **Stockton Coal horizon**	0	260	
Sandstone and concealed	108	368	
Coal blossom, prospect opening, **Coalburg**	2	370	
Sandstone and concealed	95	465	
Coal blossom, prospect opening, **Winifrede**	1	466	96′
Sandstone and concealed	89	555	
Coal, impure, prospect opening, **Chilton**	4	559	
Concealed to top of core test hole, (755′ B.)	51	610	
Surface	10	620	
Shale	5	625	
Coal	0-4″	625-4″	66′-4″
Shale	17-8″	643	
Sandstone, **Cedar Grove**	27-6″	670-6″	
Shale	19-7″	690-1″	
Coal, Cedar Grove	0-5″	690-6″	67′-2″
Fire clay	4-6″	695	
Shale with layers of sandstone	7-6″	702-6″	
Shale with mottled sandstone and **coal** streaks	13-9″	716-3″	
Shale	27-9″	744	
Shale, streaky	2	746	
Sandstone	30	776	
Shale	21-1″	797-1″	
Coal ....1′ 10″ / Shale ....0 1 / **Coal** ....3 9 } **Campbell Creek (No. 2 Gas)**	5-8″	802-9″	
Shale	0-4″	803-1″	
Sandstone, to bottom of boring	4-1″	807-2″	

The interval between the No. 5 Block Coal horizon and the base of the Campbell Creek or No. 2 Gas Coal is 642 feet

9 inches, compared with 630 feet 11 inches at the mouth of Locust Fork in the section 1.2 miles west of Emmons.

In passing over the divide from Fork Creek to Brush Creek at Cabell, the Campbell Creek (No. 2 Gas) Coal comes above water level, as shown in the following aneroid section, measured by Krebs, descending from a high summit northward to Brush Creek, two miles south of Cabell P. O.:

Section 2 Miles South of Cabell P. O., Peytona District.

	Thickness Feet.	Total Feet.	
Kanawha Series (620')			
Sandy shale	30	30	
Sandstone, **Coalburg**, massive, to **bench**	70	100	
Sandstone and concealed	105	205	
Bench, **Winifrede Coal horizon**	...	205	
Sandstone and concealed	65	270	
Coal blossom, Chilton	2	272	
Sandstone, massive, to **bench**	68	340	
Sandstone, massive	60	400	
Bench, **Hernshaw Coal horizon**	...	400	
Sandstone and concealed	117	517	
Coal blossom, Cedar Grove	3	520	117'
Sandy shale and concealed	19	539	
Coal, hard, block, **Alma**, visible	1	540	20'
Sandy shale and concealed	65	605	
Coal blossom, Campbell Creek—No. 2 Gas	3	608	112'
Sandy shale and concealed to creek (915' B.)	12	620	

The above section was taken near the crest of the Warfield Anticline, and shows the succession of the different coals at that point.

Passing to the east to the head of Whiteoak Creek, the following section was measured by Teets, descending from a high point at head of Whiteoak Creek, northward to Whiteoak Creek, 3 miles southwest of Peytona P. O.:

Section 3 Miles Southwest from Peytona P. O., Peytona District.

	Thickness Feet.	Total Feet.
Kanawha Series (623')		
Sandstone and concealed	50	50
Sandy shale and concealed	60	110
Sandstone, **Winifrede**, massive, coarse grained, gray buff color	60	170
Sandy shale, **Winifrede Coal horizon**	30	200

PLATE V.—Malden Sandstone at Peytona.

	Thickness Feet.	Total Feet.	
Sandstone, Lower Winifrede, medium coarse grained, micaceous	40	240	
Sandy shale and concealed	20	260	
Sandstone, **Upper Chilton**, medium coarse grained	30	290	
Fire clay, **Chilton Coal horizon**	3	293	
Sandy shale and sandstone	97	390	
Sandy shale and concealed	40	430	
Sandstone and concealed	80	510	
Sandy shale and concealed	79	589	
Coal blossom, Campbell Creek (No. 2 Gas) Upper Bench	4	593	
Sandstone and concealed	27	620	
Coal, opening fallen shut, **Campbell Creek (No. 2 Gas), Lower Bench**, (920′ B.)	3	623	30′

Near the head of Drawdy Creek, the following aneroid section was measured by Krebs and Teets, descending from a point in the Peytona-Scott District Line, southward along the county road, two miles west of Andrew P. O.:

Section 2 Miles West of Andrew P. O., Peytona District.

		Thickness Feet.	Total Feet.	
Kanawha Series (595′)				
Sandstone, massive, **Coalburg**		55	55	
Sandy shale and concealed		15	70	
Sandstone, massive, **Winifrede**		85	155	
Sandy shale and concealed		30	185	
Coal blossom, Winifrede		4	189	189′
Sandstone, massive, blue		21	210	
Coal blossom, Chilton "A"		2	212	23′
Sandstone, **Upper Chilton**, massive, medium coarse grained, buff color		63	275	
Sandy shale		25	300	
Coal blossom, Chilton		3	303	91′
Sandstone, **Hernshaw**, medium coarse grained		82	385	
Coal, Hernshaw		1	386	83′
Sandy shale and sandstone		26	412	
Coal	0′ 4″			
Slate, gray	0 4			
Coal, block	0 5			
Slate	0 0½			
Coal	0 2			
Slate	0 4	**Williamson** 3-4″	415-4″	30′
Coal	0 3			
Slate	0 1			
Coal	0 3			
Slate	0 1			
Coal	1 0½			
Slate, visible		0-8″	416	
Sandstone and sandy shale		68	484	

		Thickness Feet.	Total Feet.	
Coal, gray splint.......1' 0"				
Coal, gas..............0 2				
Coal, block............0 8	Cedar Grove..	3	487	72'
Coal, gas..............0 2				
Coal, block............1 1				
Sandstone, massive........................		90	577	
Slaty shale................................		2	579	
Coal ....0' 9"	Campbell Creek			
Slate 0 4	(No. 2 Gas),			
Coal, hard..........2 2	Upper Bench...	4-2"	583-2"	96'
Bone 0 2				
Coal 0 9				
Fire clay, visible..........................		1-10"	585	
Concealed		8	593	
Coal, Campbell Creek (No. 2 Gas) Lower Bench (940' B.)........................		2	595	

Scott District.

Scott District lies west of Peytona and occupies the northwestern portion of Boone. It is bounded on the north by Lincoln County, where the base of the Conemaugh Series occurs in the highest hills. The Warfield Anticline passes through the southeastern part of the district, so that the lower strata of the Kanawha Series crop along Little Coal River near Madison, and south to Low Gap where the axis crosses Spruce Fork.

The following section was obtained by Teets, descending from a high summit westward to Little Coal River, one mile north of Altman, Scott District:

Section One Mile North of Altman, Scott District.

		Thickness Feet.	Total Feet.	
Pennsylvanian (330')				
Allegheny Series (53' 6")				
Sandstone, massive, East Lynn.............		50	50	
Coal, gray splint.......1' 8"				
Coal, bony.............0 8	No. 5 Block..	3-6"	53-6"	
Coal, gray splint, visible 1 2				
Kanawha Series (276' 6")				
Sandstone, Homewood......................		71-6"	125	
Coal blossom, Stockton-Lewiston...........		2	127	73'-6"
Sandstone, massive, Coalburg...............		63	190	
Sandy shale................................		20	210	
Sandstone		20	230	
Coal blossom, Winifrede...................		1	231	157'-6"
Sandstone, massive, Lower Winifrede.......		49	280	
Concealed to river, (630" B.)...............		50	330	

The Standard Fuel Company of Charleston, W. Va., owns about 12,000 acres of coal land in Lincoln County on the waters of Cobbs Creek, and has recently sunk a diamond core test hole on this property, the record of which has been kindly furnished the Survey by J. C. Blair, President of the Company.

The following hand-leveled section was measured by Teets, descending from a high point on divide between Wolf-pit Fork and Cobbs Creek to core test hole No. 1 (143 L.) drilled by the Standard Fuel Company, located **N.** 75° W. 2.2 miles from MacCorkle:

Section 2.2 Miles Northwest of MacCorkle, Washington District, Lincoln County.

	Thickness Ft.	In.	Total Ft.	In.
Pennsylvanian (553′ 1″)				
Conemaugh Series (270′ 8″)				
Sandstone and sandy shale	27	0	27	0
Red shale	17	0	44	0
Sandy shale and concealed to top of core test hole, 978′ B., and thence with hole	66	0	110	0
Surface	44	4	154	4
Shale	10	5	164	9
Sandstone	13	7	178	4
Shale	15	3	193	7
Fire clay	0	5	194	0
Brush Creek.. — **Coal** 2′ 2″; Interlaminated coal and bone 0′ 3″; **Coal** 0′ 7″; Slate 0′ 2″; Coal 0′ 7″	3	9	197	9
Fire clay	2	0	199	9
Sandy shale	10	4	210	1
Fire clay	1	5	211	6
Slate	1	0	212	6
Shaly slate	4	0	216	6
Shale	1	3	217	9
Sandy shale	6	0	223	9
Sandstone	12	3	236	0
Shale	14	2	250	2
Fire clay	3	0	253	2
Shale	3	6	256	8
Sandstone	10	0	266	8
Shale	4	0	270	8
Allegheny Series (177′ 4″)				
Fire clay, Upper Freeport Coal horizon	2	0	272	8
Shale	4	8	277	4
Sandstone	17	5	294	9
Fire clay	3	0	297	9
Coal, Lower Freeport	0	3	298	0

				Thickness Ft.	In.	Total Ft.	In.
Fire clay				5	0	303	0
Sandy shale				6	0	309	0
Sandstone				101	0	410	0
Interlaminated coal and bone	0′	7″	Middle? Kittanning...				
Fire clay	1	5		2	7	412	7
Interlaminated coal and bone	0	7					
Shale				1	0	413	7
Sandstone				19	1	432	8
Interlaminated coal and bone	1′	2″	No. 5 Block, Lower Kittanning				
Fire clay	2	6					
Shale	5	0		14	3	446	11
Coal	4	3					
Bone	0	2					
Coal	1	2					
Sandy shale				1	1	448	0
Kanawha Series (105′ 1″)							
Sandstone				25	9	473	9
Shale				1	9	475	6
Sandstone				37	0	512	6
Shaly slate				9	6	522	0
Coal				0	6	522	6
Sandy shale				11	2	533	8
Interlaminated coal and bone				1	0	534	8
Shale				8	10	543	6
Bone				0	1	543	7
Coal	0′	2 ″	Stockton-Lewiston..				
Shale	0	5					
Coal	2	2					
Sandy shale	1	4					
Coal	2	8		7	6	551	1
Bone	0	0½					
Coal	0	4½					
Bone	0	2½					
Coal	0	1½					
Fire clay				2	0	553	1

The coal encountered at the base of the core test hole is the Stockton-Lewiston Seam that is mined locally at MacCorkle, on Ivy Branch and on Horse Creek. The No. 5 Block Coal is mined on Cobbs Creek for locomotive fuel, and exhibits nearly the same structure as shown in core test hole.

Another section was measured by Teets, descending from a high point eastward to Coal River, opposite Altman Station, as follows:

Section at Altman Station, Scott District.

	Thickness Feet.	Total Feet.	
Pennsylvanian (515')			
Conemaugh Series (50')			
Sandstone and concealed, **Buffalo**	50	50	
Allegheny Series (156')			
Sandy shale	20	70	
Sandstone and concealed	50	120	
Sandy shale	20	140	
Sandstone, **East Lynn**, massive, coarse grained	60	200	
Coal, gray splint.....0' 6" / Slate, gray.....0 0½ / **Coal**, gray splint.....2 3 / Bony slate.....0 2 / **Coal**, block.....0 7 } **No. 5 Block**	3-6½"	203-6½"	
Slate, visible	2-5½"	206	156'
Kanawha Series (309')			
Sandstone and concealed, **Homewood**	94	300	
Sandy shale and concealed	60	360	
Sandstone and concealed	50	410	
Sandy shale	20	430	
Sandstone and concealed	30	460	
Sandy shale and sandstone, **Lower Winifrede**, (635' B.)	55	515	

The measures rise very rapidly, going up Coal River from Altman to Julian at the mouth of Little Horse Creek, as is shown in the following aneroid section measured by Teets descending from a high point westward to Coal River opposite Julian Station:

Section Opposite Julian Station, Scott District.

	Thickness Feet.	Total Feet.	
Pennsylvanian (480')			
Conemaugh-Allegheny Series (218')			
Sandstone and concealed	110	110	
Sandstone, massive, coarse grained, **East Lynn**	105	215	
Coal, opening fallen shut, No. 5 Block	3	218	218'
Kanawha Series (262')			
Sandstone, **Homewood**, massive, medium coarse grained	67	285	
Sandy shale and concealed	55	340	
Sandstone, **Coalburg**	30	370	
Sandy shale and concealed	20	390	
Sandstone and concealed	40	430	
Sandstone, massive, **Winifrede**	40	470	
Coal blossom, Winifrede	1	471	255'
Sandy shale and concealed to railroad level, (660' L.)	9	480	

On Horse Creek to the west of Altman one mile, the following aneroid section was measured by Krebs, descending from a point on the divide between Ivy Branch and Horse Creek, southward to Horse Creek:

Section One Mile West of Altman, Scott District.

	Thickness Feet.	Total Feet.	
Pennsylvanian (416′)			
Allegheny Series (160′)			
Sandy shale	80	80	
Sandstone, massive, coarse grained, **East Lynn**	80	160	160′
Kanawha Series (256′)			
Sandy shale and concealed	20	180	
Sandstone, **Homewood**, massive	65	245	
Shale	1	246	
Coal, splint 1′ 4″			
Shale, gray 0 5			
Coal, impure 1 0			
Coal, hard splint 1 0			
Coal, splint 0 4			
Coal, block 2 0 } **Stockton-Lewiston**	18-9″	264-9″	104′-9″
Coal, impure 0 8			
Shale, gray 8 0			
Coal, hard splint 3 0			
Shale, gray 0 6			
Coal 0 6			
Shale	1-3″	266	
Sandstone, **Coalburg**, current bedded	49	315	
Coal, hard splint 1′ 0″ } **Coalburg**	4	319	53′
Coal, impure 3 0			
Sandstone and concealed	76	395	
Coal, impure, Winifrede	1	396	
Sandy shale to Horse Creek, (650′ B.)	20	416	77′

The following aneroid section was measured by Krebs, descending from a high summit at the head of Brushy Fork, southward to Peter Cave Fork of Horse Creek, two miles west of Woodville, Duval District, Lincoln County, and one mile north of the Boone-Lincoln County Line:

Section 2 Miles West of Woodville, Duval District, Lincoln County.

	Thickness Feet.	Total Feet.
Pennsylvanian (482′)		
Conemaugh Series (180′)		
Sandstone	65	65
Sandy shale	15	80
Limy shale	5	85
Sandy shale	30	115

	Thickness Feet.	Total Feet.	
Red shale	5	120	
Sandstone, friable, coarse, **Buffalo-Mahoning**	60	180	
Allegheny Series (169')			
Fire clay and **coal blossom, Brush Creek**	3	183	
Sandstone, ferruginous, **Mahoning**	37	220	220'
Fire clay and **coal, Upper Freeport**	3	223	
Sandstone	20	243	
Coal blossom, Lower Freeport	2	245	
Sandstone, **Freeport**	50	295	
Coal, hard	2	297	
Sandstone, **East Lynn**	50	347	
Coal, No. 5 Block	2	349	129'
Kanawha Series (133')			
Sandstone, hard, **Homewood**	50	399	
Coal 4' 4" / Slate 8 4 / **Coal** 3 4 } **Lewiston**	16	415	
Slate	1	416	66'
Sandstone, hard, irregular	50	466	
Cannel coal and slate, Coalburg	2	468	52'
Fire clay	3	471	
Sandstone, (722' L.)	11	482	122'

The following aneroid section was measured in the western part of Scott District, descending from a high point on the Boone-Lincoln County Line, eastward to Wash Hill Fork of Horse Creek, three miles west of Mistletoe P. O.:

Section 3 Miles West of Mistletoe P. O., Scott District.

	Thickness Feet.	Total Feet.	
Pennsylvanian (480')			
Conemaugh Series (163')			
Red limy shale	5	5	
Sandstone, buff	29	34	
Red and sandy shale	20	54	
Sandstone, **Saltsburg**	40	94	
Fire clay and coal blossom	3	97	
Sandstone, **Buffalo**, friable, buff	30	127	
Fire clay	3	130	
Sandstone, **Mahoning**, massive	33	163	
Allegheny Series (167')			
Coal blossom, Upper Freeport	2	165	
Sandy shale and concealed	28	193	
Fire clay, **Lower Freeport**	2	195	
Sandstone and concealed	64	259	
Fire clay	1	260	
Sandstone, **East Lynn**, full of iron ore	60	320	
Sandy shale and **No. 5 Block Coal**	10	330	167'
Pottsville Series (150')			
Sandstone, **Homewood**, massive	115	445	
Coal blossom, Stockton	5	450	120'
Sandstone and concealed, (810' B.)	30	480	

The following aneroid section was measured by Teets from a high point just northeast of Vancamp, descending southwest to the Pryor and Allen gas well (25), drilled by the Crude Oil Company, on Browns Branch of Little Coal River, ½ mile southeast of Vancamp:

Section ½ Mile Southeast of Vancamp, Scott District.

			Thickness Feet.	Total Feet.
Pennsylvanian (1730')				
Allegheny Series (125')				
Sandstone, **East Lynn**, massive, coarse grained, buff color, makes a great cliff			85	85
Sandy shale and sandstone			40	125
Kanawha Series (955')				
Sandstone, **Homewood**, massive, cliff rock			90	215
Coal, impure	0' 7"	**Upper Stockton-Lewiston**		
Coal, splint, block	0 1			
Coal, gray splint	2 10			
Slate, gray	1 0			
Coal, very hard	0 4			
Slate and nigger-head	0 6		9-1"	224-1"
Coal, block	0 8			
Slate, gray	0 11			
Coal, gray splint, very hard, visible	2 2			
Sandstone, massive			30-11"	255
Coal, gray splint	1' 11"	**Lower Stockton-Lewiston**		
Fire clay	1 0			
Coal, block	0 8			
Slate, gray	0 2			
Coal, block, hard	0 9		7-7"	262-7"
Slate and nigger-head	0 10			
Coal, gray splint, visible	1 5			
Sandstone, massive, cliff rock, medium coarse grained			42-5"	305
Coal, cannel	1' 5"	**Coalburg**		
Slate, cannel	0 6			
Coal, cannel	0 4		4-5"	309-5"
Slate, semi-cannel	1 1			
Coal, cannel, impure, visible	1 0			
Sandstone and concealed			100-7"	410
Sandy shale			20	430
Sandstone, massive, and concealed to bench			80	510
Sandstone and concealed			70	580
Coal	0' 1"	**Williamson**		
Slate, dark	0 4			
Coal,	0 1			
Slate, gray	0 0½		1-7"	581-7"
Coal, block	0 10½			
Slate, dark	0 1			
Coal	0 1			
Fire clay, visible			1-5"	583
Concealed and sandstone			87	670

PLATE VI.—Bald Knob, elevation 2334′ above tide, showing topography of Allegheny Series.

	Thickness Feet.	Total Feet.	
Concealed	30	700	
Coal, hard splint......1' 0" / Slate, gray............0 1 / Coal, block, hard.......2 2 } **Cedar Grove.**	3-3"	703-3"	
Concealed to Pryor and Allen gas well (25), (750' B.)	26-9"	730	
Conductor	16	746	
Sand, **Malden**, white	84	830	126'-9"
Slate, black	70	900	
Sand, white	50	950	
Slate, black	130	1080	
Middle and Lower Pottsville Series (650')			
Sand, white, **Nuttall**	250	1330	
Slate, white	150	1480	
Sand, Salt, white	180	1660	
Slate, black	70	1730	
Mississippian (865')			
Mauch Chunk (150')			
Rock, red	20	1750	
Slate, white	20	1770	
Rock, red	10	1780	
Slate, white	20	1800	
Sand, **Maxton**, white	40	1840	
Slate, black	6	1846	
Little Lime	26	1872	
Slate, black	8	1880	
Greenbrier Limestone (225')			
Big Lime	225	2105	
Pocono Sandstones (490')			
Sand, **Big Injun**	60	2165	
Slate and shells, white	390	2555	
Slate	20	2575	
Sand, **Berea**, white	20	2595	
Devonian (9')			
Slate to bottom	9	2604	

The foregoing section exhibits several of the important coals in the Kanawha Series. The interval between the base of the Allegheny Series and the top of the Big Lime is 1755 feet.

The following section was measured by Teets descending from a high point northward to Camp Creek, 1.5 miles northeast of Foster P. O.:

Section 1.5 Miles Northeast of Foster P. O., Scott District.

	Thickness Feet.	Total Feet.
Kanawha Series (757')		
Sandstone and concealed, **Homewood**	120	120
Coal, opening fallen shut, Stockton, reported	7	127

	Thickness Feet.	Total Feet.
Sandstone and concealed	38	165
Sandy shale	20	185
Sandstone and concealed	110	295
Sandy shale	30	325
Sandstone and concealed	100	425
Sandy shale, bench	20	445
Sandstone and concealed	40	485
Sandy shale and concealed, bench	40	525
Sandstone and concealed	80	605
Coal, blossom, Alma	3	608
Sandstone and concealed	82	690
Coal 0' 6" / Slate 0 6 / **Coal** 2 1 } **Campbell Creek (No. 2 Gas)**	3-1"	693-1"
Fire clay, Little Alma, Coal horizon	3	343
Sandy shale and concealed	39	735
Coal, impure, Powellton	1-8"	736-8"
Sandy shale and concealed to creek, (768' L.)	20-4"	757

The above section shows the interval between the base of the Stockton and the Campbell Creek (No. 2 Gas) Coals to be 563 feet.

The following aneroid section was measured by Teets from a high point on a divide between Camp and Rock Creeks, 3¼ miles northeast of Danville, descending southwestward along a private road to a branch of Rock Creek:

Section 3¼ Miles Northeast of Danville, Scott District.

	Thickness Feet.	Total Feet.	
Pottsville Series (410')			
Sandy shale	10	10	
Coal blossom, Winifrede	1	11	11'
Sandy shale and sandstone, mostly sandstone	89	100	
Coal blossom, Chilton	3	103	92'
Sandstone and sandy shale	67	170	
Coal blossom, Hernshaw	2	172	69'
Sandstone, **Naugatuck**, massive, coarse grained	58	230	
Sandy shale	30	260	
Coal blossom, Upper Cedar Grove	4	264	92'
Sandstone, massive	76	340	
Fire clay, **Little Alma Coal horizon**	3	343	
Sandstone, **Malden**, massive, friable	65	408	
Coal blossom, Campbell Creek (No. 2 Gas), (800' B.)	2	410	

Another section was measured by Teets, descending southwestward along the county road to Camp Creek, 2.4 miles north of Foster P. O.:

Section 2.4 Miles North of Foster P. O., Scott District.

	Thickness Feet.	Total Feet.	
Pennsylvanian (620')			
Allegheny Series (45')			
Sandy shale and concealed	45	45	
Kanawha Series (575')			
Sandstone, **Homewood**, medium coarse grained, Upper Bench	40	85	
Sandy shale	20	105	
Sandstone, **Homewood,** Lower Bench	35	140	
Snale, soft	30	170	
Sandstone, **Coalburg**, massive, coarse grained, gray buff color	50	220	
Sandy shale	15	235	
Sandstone, **Winifrede,** massive, coarse grained, buff color	70	305	
Sandy shale, **Winifrede Coal horizon**	10	315	
Sandstone, **Lower Winifrede,** massive, friable	30	345	
Sandy shale and sandstone	50	395	
Sandstone, **Upper Chilton,** massive, medium coarse grained	38	433	
Coal blossom, Chilton	2	435	435'
Sandy shale	40	475	
Sandstone	67	542	
Fire clay	3	545	
Sandstone, **Cedar Grove**	63	608	
Blue slaty shale	4	612	
Coal, hard 0' 9" Slate 0 4 **Coal,** black, glossy, gas 0 11 Slate 0 3 **Coal,** hard 1 6 } **Cedar Grove**	3-9"	615-9"	180'-9"
Fire clay, visible, (895' B.)	4.2	620	

Near the head of Rock Creek, the following aneroid section was measured by Krebs, descending from the divide between Drawdy Creek and Hubbard Fork of Rock Creek, two miles east of Foster P. O., Scott District:

Section 2 Miles East of Foster P. O., Scott District.

	Thickness Feet.	Total Feet.	
Kanawha Series (440')			
Sandstone and concealed	42	42	
Fire clay and **coal blossom, Winifrede**	3	45	45'
Sandy shale and concealed	78	123	
Coal blossom, Chilton	2	125	80'
Sandy shale and concealed	95	220	
Fire clay	2	222	
Sandstone, massive and concealed	80	302	
Coal, Cedar Grove	3	305	180'
Sandstone and concealed	115	420	

			Thickness Feet.	Total Feet.	
Coal ....0'	8 "	Campbell Creek (No. 2 Gas)...	4-9½"	424-9½"	120'
Slate 0	0½				
Coal 0	1				
Shale 0	3				
Coal, splint.......2	4				
Coal, gas..........0	2				
Coal, splint.......1	3				
Shale			1-2½"	426	
Concealed to creek, (930' B.)..............			14	440	

The above section exhibits four of the coals of the Kanawha Series.

Near the head of Left Fork of Rock Creek, the following aneroid section was measured by Krebs, descending northward to Left Fork of Rock Creek, two miles southeast of Foster P. O., Scott District:

Section 2 Miles Southeast of Foster P. O., Scott District.

			Thickness Feet.	Total Feet.	
Kanawha Series (700')					
Sandstone and concealed to bench.........			130	130	
Sandstone and concealed..................			150	280	
Large bench, Winifrede?..................			0	280	280'
Sandstone and concealed to bench.........			165	445	
Sandstone and concealed to bench.........			65	510	
Sandstone and concealed to bench.........			95	605	
Sandstone and concealed..................			70	675	
Coal ....1'	1"	Cedar Grove, Upper and Middle Benches......	12	687	407'
Shale, gray............0	2				
Coal, splint...........1	1				
Shale, gray............0	4				
Coal, splint...........0	4				
Sandstone and concealed 7	0				
Coal, splint...........2	0				
Concealed			2	689	
Sandstone, massive....................			4	693	
Shale, gray...........................			0-10"	693-10"	
Coal, splint.........1'	4"	Cedar Grove, Lower Bench...	4-4"	698-2"	
Shale, gray.........0	6				
Coal, splint.........2	6				
Concealed to creek, (1054' L.).............			1-10"	700	13'

The following aneroid section was measured by Krebs, descending from the divide between the waters of Price Branch and Pedee Branch of Rock Creek, eastward to Pedee Branch of Rock Creek, two miles south of Foster P. O.:

Section 2 Miles South of Foster P. O., Scott District.

Kanawha Series (546')	Thickness Feet.	Total Feet.	
Sandstone and concealed, **Coalburg**	80	80	
Bench, **Coalburg Coal horizon**	0	80	
Sandstone and concealed	10	90	
Sandy shale	15	105	
Sandy shale, yellowish	4	109	
Sandy shale	60	169	
Coal blossom, Winifrede	1	170	170'
Sandstone and concealed to **bench**	60	230	
Sandstone, massive, friable	30	260	
Sandstone and concealed	43	303	
Coal blossom, Hernshaw	2	305	135'
Sandstone and concealed	93	398	
Fire clay	2	400	
Sandstone and concealed	40	440	
Bench, **Cedar Grove Coal horizon**	0	440	135'
Sandstone and concealed	105	545	
Limestone, Campbell Creek, (930' B.)	1	546	

The above section was measured to the rise so that the intervals given between the different coals are less than they really are.

The following section was measured by Teets, descending from a high point at head of Price Branch of Coal River, and 1¾ miles northeast of Danville, along slope and county road northeastward to the Pryor and Allen well No. 2, drilled on branch of Rock Creek by the Columbus Gas & Fuel Company, and connected with said well:

Section 1¾ Miles Northeast of Danville, Scott District.

Pennsylvanian (1765')	Thickness Feet.	Total Feet.
Allegheny Series (25')		
Sandy shale and concealed	25	25
Kanawha Series (1035')		
Sandstone, **Homewood,** massive, coarse grained	70	95
Sandy shale and fire clay	10	105
Sandstone, **Coalburg,** massive, buff color	35	140
Sandy shale	25	165
Sandstone, **Upper Winifrede,** massive, coarse grained	85	250
Sandy shale and concealed	25	275
Sandstone	15	290
Sandy shale and concealed	35	325
Sandstone, **Lower Chilton,** massive, micaceous	40	365
Sandy shale and sandstone	49	414

	Thickness Feet.	Total Feet.	
Coal blossom, Hernshaw	1	415	415′
Sandy shale and sandstone	38	453	
Coal blossom	2	455	
Sandstone, massive	40	495	
Sandy shale and lime, Dingess, fossiliferous	2	497	
Sandy shale	8	505	
Coal 0′ 8″ / Shale 0 2 / **Coal** 0 4 } **Williamson**	1-2″	506-2″	
Slate and fire clay	5-10″	512	
Sandstone with shale and limestone	30	542	
Sandstone, massive	6	548	
Fire clay	16-10″	564-10″	
Coal 1′ 6″ / Shale 0 2 / **Coal** 0 6 } **Cedar Grove**	2-2″	567	55′
Shale and fire clay	5	572	
Concealed	8-6″	580-6″	
Coal, Lower Cedar Grove	1-6″	582	
Shale and concealed	21	603	
Sandstone, massive	4	607	
Shale	9	616	
Coal, Alma	1	617	50′
Concealed and fire clay	10	627	
Shale, brownish	15	642	
Concealed	30	672	
Shale, brownish	10	682	
Sandstone, massive	13-6″	695-6″	
Coal, Campbell Creek (No. 2 Gas)	0-6″	696	79′
Shale and fire clay	10	706	
Sandstone, massive	25	731	
Shale and concealed	18	749	
Sandstone	2	751	
Shale, buff	8-6″	759-6″	
Coal, Powellton	1-6″	761	65′
Sandstone, massive	20	781	
Sandstone, flaggy	24	805	
Limestone	1	806	
Sandy shale and concealed to top of well, (880′ L.)	14	820	
Clay, yellow	16	836	
Slate, white	34	870	
Sand, white	30	900	
Slate, white	40	940	
Slate, black	120	1060	
Middle and Lower Pottsville (705′)			
Sand, white	24	1084	
Lime, white	26	1110	
Slate, black	5	1115	
Sand, white	280	1395	
Coal, Sewell?	3	1398	
Sand, Salt, white	367	1765	
Mississippian (885′)			
Mauch Chunk (200′)			
Slate, white		1772	

	Thickness Feet.	Total Feet.
Lime, white	6	1778
Red rock	7	1785
Slate, white	43	1828
Red rock	10	1838
Slate, white	26	1864
Little Lime	97	1961
Pencil cave, black	4	1965
Greenbrier Limestone (205')		
Big Lime, white	205	2170
Pocono Sandstones (480')		
Rock, red, **Big Injun**	30	2200
Sandy shells, gray	80	2280
Slate, white	40	2320
Shells and slate, white	300	2620
Shale, brown	18	2638
Sand, Berea, gray	12	2650
Devonian (31')		
Slate, white	31	2681

The interval between the base of the Campbell Creek (No. 2 Gas) Coal and top of Big Lime is 1269′ as shown in the foregoing section.

The following aneroid section was measured by Teets from a point east of Danville, descending southwestward and joined on to the Mandaville Hopkins well No. 1 (31), drilled by Shields Oil & Gas Company, on Hopkins Branch, one mile east of Danville:

Section 1 Mile East of Danville, Scott District.

	Thickness Feet.	Total Feet.	
Pennsylvanian (1525')			
Kanawha Series (837')			
Sandstone, **Coalburg**, coarse grained, capping high point	55	55	
Sandy shale	10	65	
Sandstone, **Winifrede**, massive, medium coarse grained	95	160	
Sandy shale, **Winifrede Coal horizon**	20	180	
Sandstone, medium coarse grained, Lower **Winifrede**	60	240	
Sandy shale	5	245	
Sandstone and concealed	50	295	
Sandstone, **Hernshaw and Naugatuck**, massive, cliff rock, medium coarse grained	105	400	
Sandy shale	15	415	
Sandstone and concealed	70	485	
Sandy shale	15	500	
Coal, Alma	3	503	503′
Sandstone and concealed	37	540	
Sandy shale	10	550	

	Thickness Feet.	Total Feet.	
Sandstone, **Malden**, massive, medium grained	50	600	
Sandy shale, **Campbell Creek (No. 2 Gas) Coal horizon**	5	605	102′
Sandstone, **Brownstown**, massive, medium coarse grained	85	690	
Sandy shale	10	700	
Sandstone, massive, concealed	35	735	
Sandy shale, sandstone and concealed, to top of well (elevation, 700′ B.)	90	825	
Clay	12	837	
Middle and Lower Pottsville (688′)			
Sand	100	937	
Lime	38	975	
Sand	40	1015	
Slate	5	1020	
Sand	85	1105	
Slate	10	1115	
Sand	60	1175	
Slate	100	1275	
Sand, Salt	200	1475	
Slate	3	1478	
Sand	47	1525	
Mississippian (974′)			
Mauch Chunk Series (200′)			
Slate	15	1540	
Lime	10	1550	
Slate	1	1551	
Sand	44	1595	
Slate	10	1605	
Red rock	10	1615	
Slate	30	1645	
Sand	30	1675	
Slate	50	1725	
Greenbrier Limestone (260′)			
Big Lime	260	1985	
Pocono Sandstones (514′)			
Sand, Big Injun	60	2045	
Slate	419	2464	
Slate, dark	15	2479	
Sand, Berea	20	2499	
Devonian (2′)			
Slate	2	2501	

The above section shows the thickness of the Pottsville Series to be 1525 feet, and the interval between the Campbell Creek (No. 2 Gas) Coal and top of Big Lime 1120 feet.

The following section was measured by Teets, descending from a high point north of Workman Branch southward to Pond Fork:

Section 1.2 Miles Southeast of Madison, Scott District.

Kanawha Series (1225′)	Thickness Feet.	Total Feet.	
Sandstone, **Homewood**, massive, coarse grained, current bedded, buff color, capping high point	75	75	
Sandy shale and concealed	15	90	
Sandstone, **Coalburg**	70	160	
Sandy shale and concealed	10	170	
Sandstone, massive, medium coarse grained	65	235	
Coal blossom, Coalburg	2	237	237′
Sandstone, **Upper Winifrede**, medium coarse grained, ferriferous, buff color	68	305	
Sandy shale and concealed, **Winifrede Coal horizon**	10	315	
Sandstone, **Lower Winifrede**, massive, medium coarse grained, cliff rock	50	365	
Sandy shale and concealed, **Chilton "A" Coal horizon**	10	375	
Sandstone, **Upper Chilton**, massive, medium coarse grained, cliff rock	90	465	
Sandstone, **Lower Chilton**, sandy shale and concealed	45	510	
Coal blossom, Little Chilton	1	511	274′
Sandstone, **Hernshaw**, massive, medium coarse grained	96	607	
Coal blossom, Upper Bench, Hernshaw	3	610	
Sandstone	10	620	
Coal, cannel, Hernshaw	5	625	114′
.Sandstone, massive, and concealed	100	725	
Coal blossom, Cedar Grove	2	727	102′
Sandstone and concealed	53	780	
Coal, and fire clay, **Alma**	3	783	56′
Sandstone, **Malden**, medium coarse grained and concealed	77	860	
Fire clay, **Campbell Creek (No. 2 Gas) Coal horizon**	2	862	79′
Sandstone, **Brownstown**, massive, medium coarse grained, gray buff color	98	960	
Fire clay, **Powellton Coal horizon?**	2	962	
Sandstone, massive, medium coarse grained	53	1015	
Sandstone	60	1075	
Slaty shale with "turtle" limestone, **Eagle fossil horizon**	8	1083	
Sandstone and concealed (690′ B.)	142	1225	

The above section was measured near the crest of the Warfield Anticline, and about on the strike of the measures, so that the intervals given in same are practically correct.

A diamond core test hole was drilled at Danville, near the mouth of Price Branch, for Croft & Stolling by the Sullivan Machinery Company of Chicago, the record of which has been kindly furnished by Mr. C. M. Croft of Huntington:

Croft & Stolling Diamond Core Test Hole (No. 108A), Scott District.

Located on the north side of Price Branch at Danville, about 350 feet above where the Coal River Railroad crosses said branch; authority, S. M. Croft; hole completed, 1907; elevation, 682' L.

	Thickness Feet.	Total Feet.
Pottsville Series (500')		
Clay, sand and boulders	9	9
Blue clay, **Eagle Limestone horizon**	8	17
Dark shale	17	34
Sand shale	3	37
Coal, Little Cedar (0' 6")	0.5	37.5
Sand shale	10.5	48
Dark shale	1	49
Coal, saved core (0' 6")	0.5	49.5
Dark shale	1.5	51
Sand shale	4	55
Slate, **Lower Eagle**	4	59
Coal, saved core (0' 6")	0.5	59.5
Sandstone, **Upper Gilbert**	35.5	95
Dark shale	1	96
Coal, saved core (0' 6")	0.5	96.5
Dark shale, **Glenalum Tunnel**	1.5	98
Slate	10	108
Sandstone	26	134
Slate	6	140
Sandstone 41'; Conglomerate sandstone 4; Sandstone 43 — **Sewell?**	88	228
Slate	9	237
Sandstone	7	244
Slate	1	245
Sandstone	14	259
Slate	1	260
Sandstone	9	269
Conglomerate sandstone (13' 9")	13.7	282.7
Coal, Sewell? (1' 3")	1.3	284
Sandstone with black seams	28	312
Sandy shale	14	326
Sandstone with black seams	3	329
Black shale	1	330
Gray shale	1	331
Sandstone with black seams	7	338
Sand shale	2	340
Hard sandstone	1	341
Sand shale	2	343
Sandstone, hard	30	373
Slate (0' 6")	0.5	373.5
Sandstone, hard (1' 6")	1.5	375
Slate (0' 8")	0.7	375.7
Sandstone, hard 90' 4"; Conglomerate sandstone 12 0; Hard sandstone 18 0; Conglomerate sandstone 3 0; Hard sandstone, to bottom 1 0	124.3	500

The foregoing core test hole began about 10 to 15 feet above the Eagle Limestone, so that all the coals in the Kanawha Series were above the mouth of the hole, and the New River Coals were not reached in the depth penetrated, unless the one at 282.7′ could be the Sewell bed. However, it is doubtful if any of these coals would be of commercial thickness at this point. The great sandstone mass at the bottom of the boring very probably represents the Raleigh Sandstone horizon of Fayette County.

West of Coal River, the following aneroid section was measured by Teets, descending from a point in low gap at head of Slippery Cut Branch of Little Coal River, eastward along county road to said Branch:

Section 1.5 Miles Northwest of Rock Creek, Scott District.

	Thickness Feet.	Total Feet.	
Kanawha Series (338′)			
Coal blossom, Coalburg, (1112′ B.)	3	3	3′
Sandstone and sandy shale	35	38	
Sandy shale	15	53	
Sandstone and sandy shale	57	110	
Sandy shale	15	125	
Sandstone, **Lower Chilton**	60	185	
Sandy shale	10	195	
Sandstone and sandy shale	15	210	
Sandy shale	15	225	
Sandstone, **Williamson,** massive	60	285	
Sandy shale and sandstone	50	335	
Coal, block ....0′ 10″ / Slate, gray ....0 2 / **Coal,** hard, block ...1 5 / **Coal,** gray splint ....0 2 / **Coal,** gas ....0 4 } **Cedar Grove** (750′ B.)	2-11″	337-11″	335′

This section was taken to the rise. The coal has probably risen 40 feet in the one-half mile which it covers.

The following aneroid section was measured by Krebs, descending from a high summit on divide between Lavender Fork of Horse Creek and Lick Creek, two miles southwest of Hopkins, Scott District:

Section 2 Miles Southwest of Hopkins, Scott District.

Kanawha Series (647′)	Thickness Feet.	Total Feet.	
Sandy shale	10	10	
Sandstone, **Homewood,** dark, coarse grained	40	50	
Sandy shale	50	100	
Bench, **Coalburg Coal horizon**	0	100	100′
Sandstone and concealed	75	175	
Bench, **Winifrede Coal horizon**	0	175	
Sandstone and concealed to bench	115	290	
Sandstone and concealed	100	390	
Bench, **Hernshaw Coal horizon**	0	390	290′
Sandstone and concealed	90	480	
Shale, gray, full of plant fossils	5	485	
Coal, semi-cannel 0′ 1″			
Slate 0 0½			
Coal, block 0 2			
Slate 0 3			
Coal, splint 0 3			
Shale, mixed with coal 0 5			
Coal 0 3			
Shale, gray 0 3			
Coal, block 0 10			
Slate 0 0½			
Coal, splint 0 8			
Cedar Grove, Upper Bench (above coals and partings)	3-3″	488-3″	98′-3″
Slate	1-9″	490	
Sandstone and concealed	70	560	
Coal blossom, Alma	2	562	
Sandstone and concealed to creek	73	635	
Interval to **Campbell Creek Coal** (715′ B.)	12	647	

The following aneroid section was measured by Krebs, descending from a high point between Mud River and Lick Creek, northward to Pigeon Roost Fork of Lick Creek, one mile north of Mud Gap, Scott District:

Section 1 Mile North of Mud Gap, Scott District.

Kanawha Series (420′)	Thickness Feet.	Total Feet.	
Sandstone, **Homewood,** coarse grained	100	100	
Lower bench, **Stockton Coal horizon**	0	100	
Sandstone and concealed, to bench	35	135	
Sandstone and concealed, **Coalburg**	100	235	
Fire clay	5	240	
Sandstone and concealed	115	355	255′
Bench, **Winifrede Coal horizon**	0	355	
Sandy shale and concealed	60	415	
Coal, block, **Chilton**	2	417	
Concealed to creek (955′ B.)	3	420	

The sandstone at the top of the section is very massive and forms rugged cliffs.

Washington District.

Washington District lies south of Scott District. The Warfield Anticline passes in a northeast-southwest direction through the northern part of the district, where the following aneroid section was measured by Teets, descending from a high point north to Mud Gap, and thence southeastward to head of Mud Fork of Turtle Creek, and joined to Little Coal Land Co. well No. 2 (39A):

Section 1¾ Miles West of Turtle Creek P. O., Washington District.

	Thickness Feet.	Total Feet.	
Pennsylvanian (1802')			
Allegheny Series (60')			
Sandstone, **East Lynn**, massive, coarse grained	60	60	60'
Kanawha Series (1167')			
Sandy shale	15	75	
Sandstone, **Homewood**, massive, medium coarse grained	75	150	
Sandy shale, **Stockton-Lewiston Coal horizon**	15	165	105'
Sandstone, **Coalburg**, massive and concealed	90	255	
Sandy shale, **Coalburg Coal horizon**	10	265	100'
Sandstone, **Winifrede**, massive, coarse grained, friable	75	340	
Sandy shale, **Winifrede Coal horizon**	10	350	
Sandstone	71	421	
Coal blossom, Chilton "A"	4	425	160'
Sandstone	20	445	
Coal blossom, Chilton	3	448	23'
Sandstone and sandy shale	74	522	
Coal, Little Chilton "Rider"	3	525	77'
Sandstone and shale	8	533	
Coal blossom, Little Chilton	2	535	10'
Sandstone and sandy shale	85	620	
Shale, dark, limy, **Dingess fossiliferous horizon**	2	622	
Sandstone	28	650	
Coal, Williamson	1	651	116'
Sandstone	29	680	
Coal blossom, Upper Cedar Grove	2	682	31'
Sandstone and sandy shale	23	705	
Coal, Lower Cedar Grove	2	707	25'
Sandstone	18	725	
Coal	1	726	
Sandstone	22	748	
Coal, Alma	2	750	43'
Sandstone and sandy shale	50	800	
"Turtle" limestone and dark shale, **Campbell Creek (No. 2 Gas) Coal horizon**	10	810	60'

	Thickness Feet.	Total Feet.
Sandstone and sandy shale to top of Little Coal Land Co. well No. 2 (39A), drilled by Columbus Gas & Fuel Co. (833' L.)	17	827
Slate	35	862
Lime	5	867
Coal? Powellton	5	872
Slate	40	912
Sand	65	977
Black slate	50	1027
Lime	45	1072
Coal? Little Eagle	5	1077
Slate	20	1097
Lime	130	1227
Middle and Lower Pottsville Series (575')		
Sand	200	1427
Black slate	65	1492
Sand, 1st Salt	85	1577
Slate	10	1587
Sand, 2nd Salt	140	1727
Slate, 3rd Salt	10	1737
Sand	65	1802
Mississippian (900')		
Mauch Chunk (135')		
Red rock	10	1812
White slate	20	1832
Lime	10	1842
Black slate	15	1857
Red rock	10	1867
White slate	40	1907
Black lime	20	1927
White slate	10	1937
Greenbrier Limestone (250')		
Black lime	30	1967
White lime	59	2026
Big Lime	161	2187
Pocono Sandstones (515')		
Unrecorded	490	2677
Sand, Berea	25	2702

The above section was taken to the rise so that the intervals between the coals are too small.

The interval between the Campbell Creek Coal and top of Big Lime is 1127 feet.

The following aneroid section was measured by Teets from a point at the head of Ballard Fork of Wash Hill Fork of Horse Creek of Little Coal River, descending southwestward along lumber railroad, to head of a branch of Sugartree Branch of Mud River, three miles southwest of Mistletoe P. O.:

Section 3 Miles Southwest of Mistletoe P. O., Washington District.

	Thickness Feet.	Total Feet.
Pennsylvanian (300′)		
Conemaugh-Allegheny Series (220′)		
Sandstone and sandy shale	93	93
Coal blossom, Lower Freeport	2	95
Sandstone, **East Lynn**, massive	50	145
Sandy shale and concealed	49	194
Coal blossom, Upper Kittanning	1	195
Sandstone	9	204
Slate	6	210
No. 5 Block: **Coal** 0′ 6″; Slate, soft, dark 1 2; Bone 0 3; **Coal** 0 11; Bone 0 2; **Coal** 0 2; Slate, dark 0 0½; **Coal, gray splint** 1 9; **Coal, block** 1 2½; Slate 0 6; Sandstone 1 0; **Coal, reported** 1 8	9-4″	219-4″
Slate	0-8″	220
Kanawha Series (80′)		
Sandstone, **Homewood**	19	239
Coal blossom	1	240
Sandstone and sandy shale	57	297
Coal blossom, Stockton-Lewiston (920′ B.)	3	300

The following section was measured by Teets, descending from a high point north of Mud Gap, descending southward to Mud Gap, thence along county road to head of Mud River, one mile west of Mud Gap:

Section 1 Mile West of Mud Gap, Washington District.

	Thickness Feet.	Total Feet.	
Pennsylvanian (615′)			
Allegheny Series (75′)			
Sandstone, **East Lynn**, massive, coarse grained, capping high point	60	60	
Sandy shale	15	75	
Kanawha Series (540′)			
Sandstone, **Homewood**, massive, medium coarse grained	75	150	
Sandy shale	15	165	
Sandstone, **Coalburg**, massive, and concealed	90	255	
Sandy shale	10	265	
Sandstone, **Upper Winifrede**, coarse grained	75	340	
Sandy shale	10	350	
Sandstone, **Lower Winifrede**	71	421	
Coal blossom, Chilton "A"	4	425	425′

	Thickness Feet.	Total Feet.	
Sandstone	20	445	
Coal blossom, Chilton	3	448	23′
Sandstone and sandy shale	32	480	
Sandstone, **Hernshaw**, massive, medium coarse grained	78	558	
Coal, Hernshaw	2	560	112′
Sandstone, friable, coarse grained	28	588	
Coal, Dingess	2	590	30′
Sandstone, massive	13	603	
Sandy and slaty shale	7	610	
Limestone, fossiliferous, marine fossils	0½	610½	
Dark slaty shale (1045′ A. T.)	4½	615	

The following aneroid section was measured by Teets, descending from a point at head of Lukey Fork of Mud River, at the Lincoln-Boone County Line, about one mile northwest from the common corner of Logan, Lincoln and Boone Counties, southwestward along slope and county road to Big Ugly Creek of Guyandot River:

Section 1 Mile Northwest of Logan-Lincoln-Boone County Corner, Washington District.

	Thickness Feet.	Total Feet.
Kanawha Series (620′)		
Concealed to bench	50	50
Sandstone and concealed to bench	85	135
Sandstone and concealed	55	190
Sandstone and concealed to bench	100	290
Sandstone, **Lower Winifrede**, massive, medium coarse grained	65	355
Sandy shale	42	397
Fire clay, **Chilton Coal horizon**	3	400
Sandstone and concealed	45	445
Sandy shale	15	460
Sandstone and concealed	50	510
Sandy shale	20	530
Sandstone, **Naugatuck**, massive	50	580
Sandy shale and concealed, **Dingess**	35	615
Limestone, fossiliferous, Dingess, marine fossils	2	617
Coal blossom, Williamson, (925′ B.)	3	620

The following aneroid section was measured by Krebs, descending from a high summit between Ugly Creek and Mud River, one mile northwest of Estep P. O., southwestward to Ugly Creek at mouth of Grassy Fork, Washington District:

PLATE VII.—View of Big Coal River at mouth of Tony Creek, illustrating topography of Kanawha Series.

Section 1 Mile Northwest of Estep P. O., Washington District.

	Thickness Feet.	Total Feet.	
Pennsylvanian (655')			
Allegheny Series (90')			
Sandstone and concealed	90	90	
Bench, **No. 5 Block Coal horizon**	0	90	
Kanawha Series (565')			
Sandstone, **Homewood**, coarse grained, to bench	80	170	
Sandstone and concealed, to large bench	95	265	
Sandstone and concealed to bench	65	330	
Sandstone and concealed	70	400	
Large bench, **Winifrede Coal horizon**	0	400	310'
Sandstone and concealed, to bench	140	540	
Sandstone and concealed, **marine fossil horizon**	96	636	
Coal blossom, Williamson	4	640	240'
Sandy shale and concealed to creek, (1030' B.)	15	655	

The following aneroid section was measured by Krebs, descending from a high summit on watershed between Ugly Creek and Mud River, one mile northeast of Estep P. O., northward to Bear Camp of Mud River, Washington District:

Section 1 Mile North of Estep P. O., Washington District.

	Thickness Feet.	Total Feet.
Kanawha Series (579')		
Sandstone and concealed, to large bench	110	110
Sandstone and concealed to bench	160	270
Sandstone and concealed	106	376
Sandstone	4	380
Coal, splint.....1' 9" **Coal**, cannel.....3 0 **Coal**, hard.....0 3 } **Chilton**	5	385
Sandstone and concealed	190	575
Coal blossom, Dingess, (1095' B.)	4	579

The following aneroid section was measured by Krebs, descending from a high summit, three miles west of Estep P. O., southwestward to the mouth of Chapman Branch of North Fork of Big Creek, Chapmanville District, Logan County, ½ mile west of the Boone-Logan County Line:

Section 3 Miles West of Estep P. O., Washington District.

	Thickness Feet.	Total Feet.	
Kanawha Series (915′)			
Sandstone, **Homewood**, and concealed	80	80	
Bench, **Stockton Coal horizon**	0	80	
Sandstone and concealed	150	230	
Bench, **Coalburg Coal horizon**	0	230	150′
Sandstone and concealed	100	330	
Bench, **Winifrede Coal horizon**	0	330	100′
Sandy shale and sandstone	30	360	
Sandstone, coarse grained	20	380	
Coal blossom, Chilton	1	381	51′
Sandstone and concealed	69	450	
Concealed, mostly sandstone	30	480	
Concealed	5	485	
Sandstone and concealed to large **bench**	120	605	
Sandstone	3	608	
Concealed	7	615	
Coal blossom, Dingess	2	617	236′
Sandstone, yellowish, massive	48	665	
Concealed	10	675	
Shale	7	682	
Sandstone, massive, yellowish	28	710	
Shale, **coal blossom**, and concealed	4-6″	714-6″	
Coal blossom, Cedar Grove	0-6″	715	98′
Sandstone and concealed	5	720	
Concealed	66	786	
Sandstone	3	789	
Alma: **Coal**, gas 0′ 9″; Slate 0 1; **Coal**, splint 2 0; Concealed by water 1 0	3-10″	792-10″	
Concealed	6-2″	799	
Sandstone, massive	10	809	
Coal, splint	0-10″	809-10″	
Shale and concealed	7-2″	817	
Limestone	1	818	
Sandstone, massive	27	845	
Shale, dark gray, iron ore nodules	4	849	
Campbell Creek (No. 2 Gas): **Coal** 0′ 1″; Shale, gray 0 6; **Coal**, gas 0 6; **Coal**, impure 0 7; **Coal**, gas 0 5; **Coal**, hard, splint 1 9	3-10″	852-10″	
Shale, gray	1-2″	854	
Shale, dark gray	44	898	
Shale, dark	13	911	
Limestone, silicious	2	913	
Dark shale to Big Creek (705′ L.)	2	915	

The foregoing section gives several important coals in the Kanawha Series and indicates that the Kanawha Series is gradually increasing in thickness to the southwest.

In crossing over the divide to Trace Fork of Big Creek the strata rise rapidly towards the crest of the Warfield Anticline, as is shown in the following section, measured by Teets descending from a high point just northwest of Anchor P. O., southwestward along the county road to Anchor P. O.:

Anchor P. O. Section, Washington District.

Kanawha Series (730′)	Thickness Feet.	Total Feet.	
Sandstone, **Coalburg,** massive, coarse grained, pinnacle rock	45	45	
Concealed to bench	35	80	
Sandstone, **Winifrede,** massive, medium coarse grained, buff color	60	140	
Concealed	20	160	
Sandstone, **Lower Winifrede,** massive	25	185	
Concealed	45	230	
Sandy shale and sandstone	27	257	
Coal blossom, Chilton	3	260	260′
Sandstone, **Lower Chilton,** massive, medium coarse grained, friable	60	320	
Sandy shale	10	330	
Sandstone, **Hernshaw,** massive, fine grained	40	370	
Sandy shale	25	395	
Sandstone, **Williamson,** massive, coarse grained, friable	21	416	
Coal 0′ 4″ Slate, gray 0 2 **Coal,** block 0 6 Slate 0 1 **Coal,** block 1 2 Slate 0 0½ **Coal,** gray splint 1 11½ } **Williamson**	4-3″	420-3″	160′
Sandy shale	9-9″	430	
Coal blossom, and fire clay	3	433	
Sandstone, fine grained	57	490	
Coal blossom, Cedar Grove	2	492	72′
Sandstone, massive, fine grained	15	507	
Sandy shale	55	562	
Sandstone, massive, friable, buff color	48	610	
Coal blossom, Campbell Creek (No. 2 Gas)	2	612	120′
Sandy shale	13	625	
Sandstone, **Brownstown,** massive, fine grained	20	645	
Sandy shale	20	665	
Slaty shale	10	675	
Sandstone, massive, gray buff, medium coarse grained	25	700	
Sandy shale to bed of Trace Fork of Big Creek, (920′ B.)	30	730	

The foregoing section was measured to the rise near the crest of the Warfield Anticline, so that the intervals are shown thinner in the section than they really are.

Another section was measured by Teets from a point descending northward along a trail to head of Dog Fork of Trace Fork of Big Creek of the Guyandot River, 1.5 miles southwest from Anchor P. O., which gives the following succession of the strata:

Section 1.5 Miles Southwest from Anchor P. O., Washington District.

		Thickness Feet.	Total Feet.	
Kanawha Series (820')				
Sandy shale and concealed		40	40	
Sandstone, **Winifrede**, massive, medium coarse grained		50	90	
Sandy shale and concealed		55	145	
Sandstone, **Lower Winifrede**, massive		70	215	
Sandy shale		15	230	
Sandstone, massive		25	255	
Sandy shale and concealed		60	315	
Sandstone and concealed		80	395	
Coal blossom, Little Chilton		2	397	397'
Concealed and sandstone		118	515	
Coal blossom, Williamson		4	519	122'
Concealed and sandstone		76	595	
Coal, block ....0' 10"	**Alma**	2-11"	597-11"	79'
Slate 0 1				
Coal, glossy, block....1 2				
Coal, gray splint......0 10				
Slate		1-1"	599	
Sandstone and concealed		101	700	
Coal, abandoned opening fallen shut, **Campbell Creek (No. 2 Gas)**, about		3	703	105'
Sandstone and concealed (955' B.)		117	820	

The following aneroid section was measured by Krebs, descending along road to Garrett Fork of Big Creek, 1.5 miles east of Curry P. O., Chapmanville District, Logan County:

Section 1.5 Miles East of Curry P. O., Chapmanville District.

	Thickness Feet.	Total Feet.
Kanawha Series (462')		
Sandstone and concealed	30	30
Sandstone, shaly	5	35
Slate	0-2"	35-2"
Sandstone	0-8"	35-10"

		Thickness Feet.	Total Feet.	
Coal, gas............ 0′ 6″ Coal, impure......... 0 6 Coal, block, glossy.. 2 2 Fire clay............. ..	Chilton......	3-2″	39	
Sandstone and concealed..................		230	269	
Coal blossom, Alma........................		2	271	232′
Sandstone and concealed..................		110	381	
Coal blossom, Campbell Creek (No. 2 Gas)..		1	382	111′
Sandstone and concealed (1000′ B.).........		80	462	

The following aneroid section was measured by Krebs and Teets, descending from the summit of a high point in the southwestern corner of Washington District, on the Boone-Logan County Line, three miles southwest of Hewett P. O., northeastward to Meadow Fork of Hewett Creek:

Section 3 Miles Southwest of Hewett P. O., Washington District.

	Thickness Feet.	Total Feet.	
Kanawha Series (894′)			
Sandstone, **Homewood**, massive............	35	35	
Sandy shale..............................	20	55	
Sandstone, **Coalburg**, massive, coarse grained	70	125	
Sandy shale and concealed to bench........	40	165	
Sandstone, **Winifrede**, and concealed........	60	225	
Sandy shale and concealed to bench........	20	245	
Sandstone and concealed..................	50	295	
Sandy shale and concealed to bench........	10	305	
Sandstone and concealed..................	85	390	
Sandstone	6	396	
Slate	0-2″	396-2″	
Sandstone	0-8″	396-10″	
Coal, gas..............0′ 7″ Coal, impure...........0 6 Coal, block, glossy.....2 2 } Chilton?.....	3-3″	400-1″	
Sandstone and concealed..................	54-11″	455	
Coal blossom, Hernshaw....................	1	456	56′
Sandstone and concealed..................	19	475	
Coal blossom..............................	1	476	
Sandstone and concealed..................	54	530	
Coal blossom, Dingess.....................	1	531	75′
Sandstone and concealed..................	34	565	
Coal blossom, Williamson..................	1	566	
Sandstone and concealed..................	40	606	
Sandstone, massive.......................	60	666	
Coal blossom, Alma........................	1	667	136′
Sandstone and concealed..................	87	754	
Coal blossom, Campbell Creek (No. 2 Gas), Upper Bench..........................	1	755	88′
Fire clay................................	2	757	

	Thickness Feet.	Total Feet.	
Sandstone and concealed	20	777	
Coal blossom, Campbell Creek (No. 2 Gas), Lower Bench	3	780	25'
Sandstone, **Brownstown,** and concealed	68	848	
Coal blossom, Powellton	1	849	69'
Sandstone, massive	20	869	
Sandstone and concealed to creek (935' B.)	25	894	

The foregoing section was taken about on the strike of the strata, so that the intervals given between the different coals are approximately correct.

On Spruce Fork of Coal River near Low Gap, the following section was measured by Krebs, descending from high hill, westward to Spruce Fork of Coal River at Low Gap P. O.:

Section at Low Gap P. O., Washington District.

	Thickness Feet.	Total Feet.	
Kanawha Series (670')			
Sandy shale and concealed, to bench	60	60	
Sandstone and concealed	40	100	
Shale, gray	20	120	
Slate	1	121	
Alma — **Coal, impure** 0' 3"; **Coal, splint** 0 7; Slate, gray 0 1; **Coal, splint** 0 2; **Coal, impure** 0 3; **Coal, hard, splint** 2 3; **Coal, impure** 0 3; Slate 0 3; **Coal, gas** 0 4	4-5"	125-5"	
Slate	1-7"	127	
Slate and concealed	163	290	
Sandstone, massive	100	390	
Concealed, mostly black slate	60	450	
Dark shale, Eagle, marine fossils	1	451	325'
Shale, dark	29	480	
Sandstone, massive	160	640	
Concealed	10	650	
Sandstone, current bedded, to Spruce River, (705' B.)	20	670	

The above section gives the interval between the Alma Coal and Eagle fossils as 325.5 feet.

The following aneroid section was measured by Krebs and Teets, descending from a high summit, one mile south of Madison, westward to Spruce Fork of Coal River, Washington District, near the crest of the Warfield Anticline:

Section 1 Mile South of Madison, Washington District.

Kanawha Series (950')	Thickness Feet.	Total Feet.	
Sandstone, **Coalburg**, grayish buff, to bench..	40	40	
Sandstone, **Winifrede**, massive	60	100	
Bench, **Winifrede Coal horizon**	0	100	100'
Sandstone and concealed	100	200	
Coal blossom, Little Chilton	1	201	101'
Sandstone, massive, to bench	69	270	
Sandstone and concealed	60	330	
Large bench, **Hernshaw Coal horizon**	0	330	129'
Sandstone and concealed, to bench	140	470	140'
Sandstone and concealed	65	535	
Coal blossom, Alma	5	540	70'
Sandstone and concealed	60	600	
Bench, **Campbell Creek (No. 2 Gas) Coal horizon**	0	600	60'
Sandstone and concealed	65	665	
Bench, **Powellton Coal horizon**	0	665	
Sandstone and concealed	100	765	165'
Bench, **Eagle Coal horizon**	0	765	
Sandstone and concealed	60	825	
Sandy shale and concealed, **Eagle fossil horizon**	10	835	
Sandstone and concealed, (660' B.)	115	950	

The following aneroid section was measured by Krebs and Teets, descending from a high point northward to Low Gap Creek, 2.5 miles southwest of Low Gap P. O., Washington District:

Section 2.5 Miles Southwest of Low Gap P. O., Washington District.

Kanawha Series (450')	Thickness Feet.	Total Feet.	
Sandstone and concealed	178	178	
Coal blossom, Cedar Grove	2	180	
Sandstone, massive	9	189	
Limestone	1	190	
Sandstone and concealed	110	300	
Coal blossom, Campbell Creek (No. 2 Gas)	1	301	121'
Sandstone and concealed	145	446	
Coal blossom, Eagle	1	447	146'
Concealed to creek (955' B.)	3	450	

The following aneroid section was measured by Krebs and Teets, descending from a high knob, ½ mile west of Ottawa, westward to Spruce Fork of Coal River at Ottawa:

Section at Ottawa, Washington District.

	Thickness Feet.	Total Feet.	
Kanawha Series (340′)			
Sandstone, **Coalburg**, massive	60	60	
Bench, **Coalburg Coal horizon**	0	60	
Sandstone and concealed	10	70	
Sandstone, **Winifrede**, massive	120	190	
Bench, **Winifrede Coal horizon**	10	200	
Sandstone, massive, to large bench	80	280	
Sandstone, massive, coarse grained, to bench	120	400	
Sandstone, massive	85	485	
Coal, Chilton	5	490	290′
Sandstone and concealed	124	614	
Coal blossom	1	615	
Sandstone and concealed, to bench	120	735	
Sandstone and concealed	80	815	
Coal, Alma	5	820	330′
Concealed to creek (780′ B.)	20	840	

The Chilton and Alma Coals are both mined at this place by the Coal River Coal Company.

On Bias Branch, about 2 miles west of Ottawa, the following aneroid section was measured by Krebs, descending along road westward to Bias Branch, two miles northeast of Jeffery P. O.:

Section 2 Miles Northeast of Jeffery P. O., Washington District.

	Thickness Feet.	Total Feet.	
Kanawha Series (782′)			
Sandy shale and concealed	105	105	
Bench, **Winifrede Coal horizon**	0	105	105′
Sandy shale and concealed	290	395	
Bench, **Chilton Coal horizon**	0	395	290′
Sandy shale and concealed	98	493	
Coal blossom, Hernshaw	2	495	100′
Sandstone and concealed	163	658	
Coal blossom, Cedar Grove	2	660	
Sandstone and concealed	58	718	
Coal blossom, Alma	2	720	225′
Sandstone and concealed	30	750	
Coal blossom, Little Alma	2	752	
Sandstone and concealed (900′ L.)	30	782	

South of Ottawa, toward Clothier, the measures dip very rapidly into the Coalburg Syncline, as is shown by a large number of diamond core drill holes sunk in prospecting for coal by the Boone County Coal Corporation, which owns a

PLATE VIII.—Looking up West Fork from mouth of Little Ugly Branch; Kanawha Series.

large area of coal land in Boone and Logan Counties. Through the courtesy of Mr. J. C. Blair, Vice President of that Company, the records of these diamond core test holes are permitted to be published in this Report. These records have been of great assistance in making up the structure map, and in the study of the rock strata and coals in this region.

Diamond Core Test No. 2 (109) of the Boone County Coal Corporation.

Located 0.3 mile N. 30° W. of Clothier, Washington District, Boone County; authority, Boone County Coal Corporation; elevation, 797.81' L.

	Thickness Ft.	In.	Total Ft.	In.	
Kanawha Series (184' 1")					
Surface	15	0	15	0	
Slate, sandy	6	3	21	3	
Coal and bone, Williamson	2	4	23	7	
Fire clay	0	3	23	10	
Shale, sandy	6	2	30	0	
Shale	14	1	44	1	
Slate	7	8	51	9	
Sandstone	0	4	52	1	
Slate	3	4	55	5	
Slate, black	12	10	68	3	
Coal 1' 3½"; Bone 0 3½; **Coal** and bone 0 1; **Coal** 0 2 — **Cedar Grove, Upper Bench**	1	10	70	1	46' 6"
Shale, sandy	1	9	71	10	
Sandstone with shale streaks	3	10	75	8	
Sandstone, **Middle Cedar Grove**	34	9	110	5	
Shale	2	0	112	5	
Sandstone with shaly streaks	27	5	139	10	
Coal 1' 3"; **Coal**, core lost 0 7; Fire clay 0 3; Bone 0 1; **Coal** 1 4 — **Alma**	3	6	143	4	73' 3"
Fire clay	1	1	144	5	
Slate, shaly	14	0	158	5	
Shale, with sandstone streaks	16	2	174	7	
Slate, shaly, to bottom	9	6	184	1	

The Alma Coal encountered at 139' 10" in the hole is mined at Ottawa, less than one-half mile north of this core test hole, at an elevation of 800 feet above tide, and this drilling shows the coal to have dipped about 140 feet in this distance, or a fall of more than 5 feet per 100 feet.

Coal Test No. 4 (110) of Boone County Coal Corporation.

Located 0.2 mile N. 75° W. of Clothier, Washington District, Boone County; authority, Boone County Coal Corporation; elevation, 840' L.

	Thickness Ft.	Thickness In.	Total Ft.	Total In.	
Kanawha Series (225' 8")					
Surface	10	0	10	0	
Sandstone	6	6	16	6	
Shale	2	2	18	8	
Sandstone	43	7	62	3	
Shale, sandy	4	0	66	3	
Slate, shaly	22	0	88	3	
Shale, sandy	2	1	90	4	
Coal and bone, Williamson, Upper Bench	1	6	91	10	
Fire clay	2	0	93	10	
Shale, sandy	8	0	101	10	
Coal and bone, Williamson, Lower Bench	0	10	102	8	
Sandstone	4	0	106	8	
Sandstone, with shaly streaks	8	1	114	9	
Shale, sandy	19	10	134	7	
Slate, black	13	2	147	9	
Sulphur	0	1	147	10	
Coal 1' 2"; Slate 0 2½; **Coal** 0 9 — **Cedar Grove, Upper Bench**	2	1½	149	11½	
Fire clay	0	8½	150	8	
Shale, sandy	1	6	152	2	
Sandstone	32	8	184	10	
Slate	1	0	185	10	
Coal, Cedar Grove	1	10	187	8	
Shale	2	4	190	0	
Sandstone, with shale streaks	15	1	205	1	
Shale, with sandstone streaks	9	2	214	3	
Slate	1	0	215	3	
Shale	0	7	215	10	
Coal 3' 4"; Fire clay 0 2½; **Coal and bone** 0 1½; **Coal** 1 7½; **Coal, core lost** 0 3½ — **Alma**	5	7	221	5	71' 5½"
Fire clay to bottom	4	3	225	8	

The following important section was obtained by Ray V. Hennen by combining measurements taken descending from a high knob to diamond core test hole No. 7 (145), located on divide between Left Fork of Beech Creek and Beech Creek, S. 37° W. 3.7 miles from Clothier, 0.3 mile south of the Boone-Logan County Line, and then joining this result to diamond core test No. 1 (142) and Boone County Coal Corporation well No. 1 (65), located S. 10° E. 0.4 mile from Clothier, and 0.2 mile west of the Boone-Logan County Line:

Section Near Clothier, Logan District, Logan County.

	Thickness Ft.	In.	Total Ft.	In.	
Pennsylvanian (2405')					
Conemaugh Series (40')					
Concealed from top of knob	5	0	5	0	
Sandstone, massive, grayish white, coarse, pebbly, making prominent cliffs	35	0	40	0	
Allégheny Series (165')					
Concealed along steep slope, probably mostly shale	25	0	65	0	
Concealed along bench	10	0	75	0	
Concealed along steep slope	15	0	90	0	
Concealed along bench	15	0	105	0	105' 0"
Sandstone, buff, medium coarse, broken, blocky	65	0	170	0	
Concealed along steep slope	20	0	190	0	
Concealed along bench	15	0	205	0	
Kanawha Series (1405')					
Sandstone, **Homewood**, medium grained, partly concealed	110	0	315	0	
Concealed along steep bluff	25	0	340	0	
Concealed with sandstone	35	0	375	0	
Concealed to top of coal test No. 7	15	0	390	0	
Coal Test No. 7 (145) 1290' L.					
Surface	14	0	404	0	
Shale	2	0	406	0	
Sandstone	39	0	445	0	
Coal, Coalburg "Rider"	0	4	445	4	
Shale	17	8	463	0	
Slate	1	0	464	0	
Coal 1' 11"					
Bone 0 1					
Coal 0 3					
Bone 0 1					
Coal 0 7 — **Coalburg** (1210' L.)	6	3	470	3	
Coal and bone 0 2					
Slate 0 4					
Coal and bone, interlaminated 1 9					
Coal 1 1					
Fire clay	0	10	471	1	
Shale, sandy	16	6	487	7	
Slate	0	10	488	5	
Coal 1' 4"					
Slate 0 6 — **Little Coalburg.**	2	10	491	3	21' 0"
Coal 1 0					
Fire clay	1	9	493	0	
Sandy shale	5	0	498	0	
Slate	1	4	499	4	
Fire clay	1	0	500	4	
Shale, sandy	17	8	518	0	
Slate	1	0	519	0	
Coal 0' 5"					
Slate 0 1	2	0	521	0	
Coal 1 6					
Sandstone, sandy	9	4	530	4	

			Thickness Ft.	In.	Total Ft.	In.		
Shale			17	0	547	4		
Coal	1′ 8″							
Slate	0 3		2	5	549	9		
Coal and bone	0 6							
Fire clay			3	2	552	11		
Shale			7	8	560	7		
Coal and bone	1′ 0″							
Shale	11 0	**Buffalo Creek**	12	8	573	3		
Coal and bone	0 8							
Shale			3	0	576	3		
Fire clay			2	0	578	3		
Sandstone, **Upper Winifrede**			74	0	652	3		
Shale			2	6	654	9		
Coal	1′ 9″							
Shale, sandy	17 7	**Winifrede**	19	11	674	8	198′	5″
Coal	0 7							
Sandstone			7	11	682	7		
Shale			22	0	704	7		
Coal			0	5	705	0		
Fire clay			3	0	708	0		
Slate			7	0	715	0		
Shale, sandy			6	1	721	1		
Coal and bone	0′ 8″							
Coal	1 0							
Shale	0 6	**Chilton "A"**	3	0	724	1	49′	5″
Coal	0 3							
Coal and bone	0 7							
Shale			8	0	732	1		
Sandstone	8′ 3″							
Sandstone, hard	12 0							
Sandstone, with hard streaks	9 6	**Upper Chilton Sandstone**	64	9	796	10		
Shale, sandy	11 0							
Sandstone	24 0							
Shale			4	8	801	6		
Slate			0	2	801	8		
Coal	2′ 5″							
Bone	0 0¼	**Chilton**	6	5¼	808	1¼	74′	0¼″
Coal	4 0							
Fire clay, shale and concealed, by hand-level, to top of core test No. 1 (142), located ½ mile south of Clothier, north of Bend Branch, elevation. 804.79′ L.			52	10¾	861	0		
Core Test No. 1 (142).								
Surface			14	0	875	0		
Shale			0	6	875	6		
Coal, Hernshaw			0	4	875	10		
Shale			3	0	878	10		
Shale, sandy			6	0	884	10		
Fire clay			2	2	887			
Shale			58	8	945			
Slate			0	4	946	0		
Sandstone, **Williamson**			19	4	965	0		
Slate, shaly, **Dingess marine fossil horizon**			21	8	987	0		

	Thickness Ft.	Thickness In.	Total Ft.	Total In.	
Coal ....0′ 4″					
Bone 0 1½					
Coal 0 1½					
Shale, sandy......0 7					
Coal 1 2½					
Bone 0 0½					
Coal 0 3					
Fire clay.........1 7					
Shale 5 9					
Coal 0 3					
Bone 0 1					
Coal 0 11					
} Williamson	11	3	998	3	190′ 2″
Fire clay..........................	0	5	998	8	
Shale, sandy.......................	4	4	1003	0	
Sandstone, with shaly streaks, Upper Cedar Grove....................	21	1	1024	1	
Slate, shaly.......................	11	8	1035	9	
Slate, black.......................	12	8	1048	5	
Bone ....0′ 0½″					
Coal 1 3					
Bone 0 0½					
Slate 0 4¼					
Coal 0 7					
} Cedar Grove.....	2	3¼	1050	8¼	
Fire clay..........................	0	4	1051	0¼	
Shale, sandy.......................	3	2¾	1054	3	
Sandstone	0	6	1054	9	
Shale, sandy.......................	0	8	1055	5	
Sandstone	11	6	1066	11	
Sandy shale........................	0	8	1067	7	
Sandstone	3	2	1070	9	
Shale, sandy.......................	0	9	1071	6	
Sandstone	15	2	1086	8	
Coal, Cedar Grove, Lower Bench......	0	8	1087	4	
Shale	4	3	1091	7	
Shale, with sandstone streaks........	12	5	1104	0	
Shale	4	11	1108	11	
Coal and bone, interlaminated ......0′ 2″					
Coal 4 0					
} Alma.....	4	2	1113	1	
Fire clay..........................	0	6	1113	7	
Unrecorded	102	5	1216	0	
Coal, Campbell Creek (No. 2 Gas)....	2	0	1218	0	
Continued with record of well (65).					
Slate	7	0	1225	0	
Sand	175	0	1400	0	
Slate and shells...................	210	0	1610	0	
Lower Pottsville (795′)					
Salt Sand, 1st.....................	270	0	1880	0	
Slate	60	0	1940	0	
Salt Sand, 2nd.....................	435	0	2375	0	
Slate	30	0	2405	0	
Mississippian (1093′)					
Mauch Chunk (345′)					
Limestone, black...................	5	0	2410	0	
Black slate........................	17	0	2427	0	
Lime, black........................	10	0	2437	0	

	Thickness Ft.	In.	Total Ft.	In.
Red rock	3	0	2440	0
Slate and shells	20	0	2460	0
Sand	20	0	2480	0
Lime, black	25	0	2505	0
Slate, white	26	0	2531	0
Sand, black, Maxton	94	0	2625	0
Red rock	10	0	2635	0
Black, limy formation	95	0	2730	0
Red, limy formation	10	0	2740	0
Slate, black	10	0	2750	0
Greenbrier Limestone (230')				
Black limy formation..60' / Gray lime.............45 / Black and gray lime...55 / Red lime..............70 } **Big Lime.**	230	0	2980	0
Pocono Sandstones (518')				
Big Injun, red	40	0	3020	0
Slate and shells	425	0	3445	0
Brown shale	24	0	3469	0
Berea Sand	29	0	3498	0
Devonian (3')				
Slate	3	0	3501	0

It is interesting to note in the foregoing section that fifteen different beds of coal of the Kanawha Series are exposed, making a total thickness of coal of 38.59 feet. This, however, does not include the coals in the No. 5 Block and Stockton-Lewiston horizons, which seams are concealed in the section.

The thickness of the Pottsville Series shown in this section is 2200 feet.

The following section is near Sharples Station, where the Coal River Land Company made its diamond core test No. 1 (143 Lo.):

Coal River Land Company Diamond Core Test No. 1 (143 Lo.).

Located in Logan District, Logan County, at Sharples, S. 23° W. 1.7 miles from Clothier, and 1.6 miles south of the Boone-Logan County Line; authority, Coal River Land Company; elevation, 845' B.

	Thickness Ft.	In.	Total Ft.	In.
Kanawha Series (346' 9")				
Surface	15	6	15	6
Sandstone	17	0	32	6
Coal, Hernshaw	0	7½	33	1½
Slate	2	7½	35	9
Sandstone	44	0	79	9
Slate, sandy	1	8	81	5
Sandstone	13	3	9[illegible]	8

	Thickness Ft.	In.	Total Ft.	In.		
Slate	8	9	103	5		
Sandstone	2	1	105	6		
Slate	23	1½	128	7½		
Coal, Williamson	0	0½	128	8		
Slate, sandy	2	6	131	2		
Sandstone	21	11	153	1		
Slate, sandy	2	2	155	3		
Coal, bony	0	2	155	5		
Sandstone	6	9	162	2		
Slate, sandy, to sandstone	26	2	188	4		
Slate, shaly	26	3	214	7		
Slate, black	15	7½	230	2½		
Coal ....1' 3½" } **Cedar Grove,**						
Slate ....0 4½ } **Upper**	3	1½	233	4		
Coal ....1 5½ } **Bench**						
Slate	0	9½	234	1½		
Sandstone	4	6½	238	8		
Bone and slate	0	1½	238	9½		
Slate and sandstone	13	10½	252	8		
Coal, Cedar Grove, Lower Bench	1	3	253	11		
Shale	9	10	263	9		
Slate ....0' 3 " }						
Coal ....1 2 }						
Slate ....1 4½ } **Alma**	6	1	269	10	34'	6"
Coal ....2 9 }						
Slate ....0 3 }						
Coal ....0 3½ }						
Slate	2	5	272	3		
Sandstone	7	0	279	3		
Slate, sandy and sandstone	4	9	284			
Coal, Little Alma	1	5	285			
Fire clay and slate	5	1	290	0		
Sandstone	9	6	300	6		
Sandstone and sandy shale	35	8	335	8		
Coal	0	7	336	3	66'	5"
Fire clay and sandy shale to bottom	10	6	346	9		

It is unfortunate that this diamond core test was not continued for 20 or 25 feet, until it passed through the Campbell Creek (No. 2 Gas) Coal.

Boone County Coal Corporation Coal Test No. 6 (144 Lo.).

Located in Logan District, Logan County, on Beech Creek, 2½ miles S. 80° W. of Sharples; S. 52° W. 4 miles from Clothier; and 2.7 miles south of the Boone-Logan County Line; elevation, 1018' L.

	Thickness Ft.	In.	Total Ft.	In.
Kanawha Series (103')				
Surface	21	0	21	0
Sandstone	32	10	53	10
Shale	0	6	54	4
Slate	0	9	55	1
Sandstone	1	6	56	7

	Thickness Ft. In.	Total Ft. In.
Shale, sandy	3 0	59 7
Slate	1 8	61 3
Coal, Chilton "A"	0 2	61 5
Fire clay	1 0	62 5
Slate, sandy	8 5	70 10
Slate	4 9	75 7
Slate, sandy	4 3	79 10
Sandstone	9 7	89 5
Slate, sandy	3 9	93 2
Slate	0 6	93 8
Slate, core lost	1 8½	95 4½
Coal 4' 2"; Slate 0 1; **Coal** 0 10 } **Chilton**	5 1	100 5½
Shale to bottom	2 6½	103 0

Boone County Coal Corporation Coal Test No. 3 (146 Lo.)

Located in Logan District, Logan County, 1.7 miles S. 45° E. of Sharples, and S. 11° 30′ E. 3 miles from Clothier, and 1.9 miles southeast of the Boone-Logan County Line; authority, Boone County Coal Corporation; well started 90 feet below Dingess Run Coal; elevation, 880″ B.

	Thickness Fet. In.	Total Ft. In.
Kanawha Series (258′)		
Surface	13 0	13 0
Sandstone, **Williamson**	34 5	47 5
Shale	1 0	48 5
Sandstone	8 1	56 6
Shale	0 6	57 0
Sandstone	7 6	64 6
Shale, sandy	7 0	71 6
Slate, sandy	22 6	94 0
Coal 0′ 8″; Fire clay 1 0; **Coal** 0 6 } **Williamson**	2 2	96 2
Fire clay	0 6	96 8
Shale, sandy	6 8	103 4
Slate	0 3	103 7
Fire clay	1 0	104 7
Sandstone	2 0	106 7
Sandstone, with shaly streaks	16 5	123 0
Shale, dark	3 0	126 0
Shale, sandy	5 10	131 10
Sandstone	2 1	133 11
Shale	0 1	134 0
Sandstone	2 0	136 0
Shale, sandy	30 7	166 7
Slate, black	10 10	177 5
Coal, Cedar Grove, Upper Bench	3 9	181 2
Shale	0 9	181 11
Shale, sandy	3 10	185 9
Slate, shaly	4 10	190 7
Sandstone, **Middle Cedar Grove**	10 9	201 4
Slate, shaly	2 1	203 5

	Thickness Ft.	In.	Total Ft.	In.	
Coal, Cedar Grove, Lower Bench.....	1	3	204	8	
Shale, sandy........................	8	9	213	5	
Sandstone	1	10	215	3	
Shale	2	6	217	9	
Sandstone	0	7	218	4	
Shale, sandy.......................	2	4	220	8	
Shale	4	0	224	8	
Coal and bone, interlaminated ......0′ 4″; **Coal** ....1 1; Sandstone ....0 2; **Coal** and bone......0 1 } **Alma**.....	1	8	226	4	130′ 2″
Fire clay..........................	1	0	227	4	
Shale, sandy.......................	4	0	231	4	
Sandstone, **Logan**...................	20	2	251	6	
Shale	1	0	252	6	
Sandstone	0	3	252	9	
Shale	0	3	253	0	
Slate	0	2	253	2	
Coal, Little Alma....................	2	1	255	3	
Shale, sandy, to bottom...............	2	9	258	0	

Coal River Land Company Coal Test No. 2 (147 Lo.).

Located in Logan District, Logan County, 2½ miles S. 45° E. of Seng, and S. 70° E. 6.8 miles from Clothier, and 1.6 miles southwest of the Boone-Logan County Line; authority, Coal River Land Company; well started 90 feet below the Chilton Coal; elevation, 1120′ B.

	Thickness Ft.	In.	Total Ft.	In.
Kanawha Series (260′ 10″)				
Sand and cobblestone................	14	7	14	7
Sandstone	47	8	62	3
Slate, shaly........................	35	2	97	5
Sandstone	37	0	134	5
Shale, slaty........................	1	5	135	10
Sandstone, **Cedar Grove,** Upper Bench	29	4	165	2
Sandstone, with streaks of sandy shale	10	7	175	9
Slate, sandy........................	12	2	187	11
Coal, Cedar Grove..................	5	8	193	7
Shale, slaty........................	6	7	200	2
Sandstone, with streaks of sandy shale	34	6	234	8
Sandstone	2	2	236	10
Slate	0	7½	237	5½
Coal ....1′ 2½″; Fire clay.........1 7; **Coal** and slate....0 2 } **Alma, Upper Bench**....	2	11½	240	5
Slate, with streaks of sandstone......	9	7	250	0
Sandstone	6	7	256	7
Coal ....0′ 1″; Slate ....0 2; **Coal** ....0 8 } **Alma, Lower Bench**....	0	11	257	6
Slate, black........................	2	0	259	6
Fire clay, sandy, to bottom............	1	4	260	10

It is unfortunate that this core test was not sunk to the Campbell Creek (No. 2 Gas) Coal, as it is probable that it will prove of workable thickness here.

Boone County Coal Corporation Coal Test No. 6 (148 Lo.).

Located in Logan District, Logan County, 3 miles S. 35° E. of Seng, and S. 6° E. 7.3 miles from Clothier, and 1.9 miles southwest of the Boone-Logan County Line; authority, Boone County Coal Corporation; elevation, 1880′ B.

				Thickness Ft.	Thickness In.	Total Ft.	Total In.	
Pennsylvanian (369′ 9″)								
Allegheny Series (31′ 6″)								
Surface				8	0	8	0	
Sandstone				20	0	28	0	
Coal, Clarion				1	6	29	6	
Shale				2	0	31	6	
Kanawha Series (338′ 3″)								
Sandstone, **Homewood**				116	11	148	5	
Coal	2′	0″	**Stockton.** (1785′ B.)	7	0	155	5	
Fire clay	0	7						
Shale	1	1						
Coal and bone	0	2						
Shale	2	8						
Coal and bone	0	6						
Shale				3	3	158	8	
Sandstone				2	1	160	9	
Shale, with sandstone streaks				1	10	162	7	
Sandstone, broken, **Coalburg**				55	2	217	9	
Shale, broken				20	10	238	7	
Shale				10	6	249	1	
Sandstone				9	0	258	1	
Coal and bone, interlaminated	2′	7″	**Coalburg.**	12	3	270	4	115
Slate	0	9						
Coal and bone, interlaminated	2	2						
Shale	0	2						
Coal	0	1						
Fire clay	0	5						
Coal	0	1						
Fire clay	0	7						
Coal	0	1						
Fire clay	0	2						
Sandstone	3	6						
Coal and bone	1	8						
Sandstone				55	3	325	7	
Coal, Little Coalburg				0	3	325	10	
Fire clay				0	7	326	5	
Sandstone				4	3	330	8	
Shale, with sandstone streaks				6	5	337	1	
Sandstone				4	7	341	8	
Sandstone, with shale streaks				7	6	349	2	
Sandstone				18	4	367	6	
Shale, to bottom				2	3	369	9	99′ 5″

Boone County Coal Corporation Coal Test No. 5 (149 Lo.).

Located in Logan District, Logan County, on Laurel Fork, 4½ miles S. 58° E. of Seng, and ½ mile west of the Boone-Logan County Line; authority, Boone County Coal Corporation; well starts flush in the Chilton Coal; elevation, 1385' B.

			Thickness Ft.	Thickness In.	Total Ft.	Total In.
Kanawha Series (303')						
Surface			15	0	15	0
Sandstone, **Lower Chilton**			62	6	77	6
Coal	0' 7"	**Little Chilton**				
Bone	0 2		0	11	78	5
Coal	0 2					
Fire clay			1	0	79	5
Shale, sandy			6	0	85	5
Sandstone			16	10	102	3
Shale			2	2	104	5
Coal	1' 0"	**Hernshaw**				
Shale	8 0		9	6	113	11
Coal and bone	0 6					
Shale			1	1	115	0
Sandstone			45	0	160	0
Shale			50	4	210	4
Coal	0' 6"	**Williamson**				
Shale	7 0					
Coal	0 7					
Fire clay	1 6					
Shale	9 10					
Coal	0 7		25	3	235	7
Sandstone	1 4					
Shale	1 0					
Coal	0 5					
Shale	2 0					
Coal and bone	0 6					
Shale			1	3	236	10
Fire clay			1	0	237	10
Shale			17	0	254	10
Sandstone, with shale streaks			16	0	270	10
Shale, sandy			2	6	273	4
Sandstone			5	4	278	8
Shale			13	10	292	6
Coal	1' 5"	**Cedar Grove**				
Sulphur	0 0½					
Coal	0 2					
Shale	0 1½					
Coal	2 5½		6	1	298	7
Bone	0 0¼					
Coal	1 7¼					
Coal and sulphur	0 3					
Fire clay			1	1	299	8
Shale, sandy			1	9	301	5
Sandstone, with shale streaks, to bottom			1	7	303	0

The following diamond core test hole was drilled the latter part of 1914 by the Boone County Coal Corporation:

Record of Boone County Coal Corporation Diamond Core Test Hole No. 21 (157).

Located in Washington District, Boone County, on a branch of Spruce Fork flowing from the west, 0.7 mile northwest from Clothier; authority, A. R. Montgomery; completed, 1914; elevation, 760' B.

	Thickness Ft.	Thickness In.	Total Ft.	Total In.	
Surface	13	3	13	3	
Fire clay	4	9	18	0	
Soft shale	1	0	19	0	
Slate, black	17	8	36	8	
Cedar Grove: Coal 1' 8"; Slate and bone 0 3; Coal 0 7	2	6	39	2	
Shale	1	11	41	1	
Sand rock (**Middle Cedar Grove**)	33	8	74	9	
Shale	2	10	77	7	
Coal, Lower Cedar Grove	1	10	79	5	
Fire clay	1	1	80	6	
Shale with sand rock streaks	27	0	107	6	
Alma: Coal 3' 11½"; Slate 0 5; Coal 1 9½	6	2	113	8	74' 6"
Fire clay to bottom	1	4	115	0	

Several diamond core test holes were drilled the latter part of 1914 by the Boone County Coal Corporation, on Spruce Fork of Coal River in Logan District, Logan County, the records of which were received too late to be included in the Logan-Mingo Detailed County Report by Messrs. Hennen and Reger. As these core test holes are located just west of the Boone-Logan County Line, and furnish a great deal of important data, the records are published on the following pages, and Mr. Ray V. Hennen has added an interesting commentary on the same, as follows:

"A careful analysis of the logs of boring Nos. 158 to 169, inclusive, the tests which they represent all being located on the waters of Spruce Fork in Logan County, and scattered from the northern edge of Logan District to its southern boundary, shows that the approximate minable areas assigned for the Chilton, Cedar Grove, Alma and Campbell Creek (No. 2 Gas) Coals in Figures 11, 14, 16 and 17, respectively, of the Report for Logan and Mingo Counties are in close harmony with the facts for this portion of the former county. It also shows that the Williamson and Lower Cedar Grove beds are very unreliable in this region, and that the remarks in the Authors' Preface in the latter Report are fully justified, in which it is suggested that there are probably numerous patches within the shaded areas on these Figures where the coal designated is too thin and impure to be minable. Only one boring; viz, No. 161 on Seng Camp Creek, penetrated deep enough to test the Campbell Creek (No. 2 Gas) Coal, and its record exhibits this bed in fine development. In this locality the latter coal should be one of the most important deposits of fuel as regards thickness, purity and minable area, especially above the mouth of Seng Camp Creek, since, southward along the valley of Buffalo Creek, it is mined successfully on a commercial scale. Nos. 167 to 169, inclusive, also failed to penetrate to a sufficient depth to test the Alma Coal. The latter should attain minable dimensions in the locality of these borings."—Ray V. Hennen. (February 16, 1915).

Record of Boone County Coal Corporation Diamond Core Test Hole No. 9 (158Lo.).

Located on Bend Branch, 1.0 mile from its mouth and 1.5 miles from Clothier; authority, A. R. Montgomery; completed, 1914; elevation, 940.6′ L.

		Thickness Ft.	In.	Total Ft.	In.	
Surface		20	5	20	5	
Coal, slate and coal, no core		2	4	22	9	
Shale		11	0	33	9	
Coal 0′ 4″	Chilton					
Shale 0 6						
No core 1 4						
Slate 0 9						
Coal 0 1						
Slate 0 10		10	4	44	1	44′ 1″
Shale 2 8						
Coal 1 4						
Bone 0 4						
Coal 1 4½						
Coal, no core 0 9½						
Slate		2	7	46	8	
Shale		4	1	50	9	
Coal		0	1	50	10	
Shale		4	4	55	2	
Sand rock and shale		6	0	61	2	
Sand rock		51	8	112	10	
Shale		5	4	118	2	
Sand rock		7	3	125	5	
Shale		0	7	126	0	
Slate		0	4	126	4	
Coal		0	4	126	8	
Sand rock		6	4	133	0	
Coal		0	1	133	1	
Shale		5	0	138	1	
Sand rock (Williamson)		57	9	195	10	
Shale 20′ 6″	Dingess Limestone					
Shaly slate 29 11	Marine horizon.	50	5	246	3	

			Thickness Ft. In.	Total Ft. In.	
Coal	0′ 9″	Williamson	6 5½	252 8½	208′ 7½″
Bone	0 5				
Coal	0 2				
Bone	0 5				
Coal	0 3				
Shale	2 10½				
Coal	0 3½				
Slate	0 2½				
Coal	0 5½				
Bone	0 3½				
Coal	0 4				
Fire clay			1 10	254 6½	
Shale			7 0½	261 7	
Shale with sand rock streaks	15′ 5″	(Upper Cedar Grove)	22 6	284 1	
Sand rock with shale streaks	7 1				
Shaly slate			12 2	296 3	
Black slate			11 8	307 11	
Coal	1′ 2″	Cedar Grove	2 7	310 6	57′ 9½″
Bone	0 2				
Coal, no core, 6″	1 3				
Fire clay			1 7	312 1	
Sandy shale			2 6	314 7	
Sand rock			24 11	339 6	
Sand rock and shale			0 8	340 2	
Sand rock			8 11	349 1	
Shale			2 8	351 9	
Coal (Lower Cedar Grove)			1 10	353 7	
Shale			6 4	359 11	
Coal	0′ 11″	Alma	5 9½	365 8½	55′ 2½″
Bone	0 10				
Coal and bone	0 4				
Coal	3 8½				
Fire clay to bottom			1 11½	367 8	

Record of Boone County Coal Corporation Diamond Core Test Hole No. 12 (159Lo.).

Located on Spruce Fork, 0.5 mile south of mouth of Rockhouse Creek, and 1.0 mile southeast of Clothier, Logan District, Logan County; authority, A. R. Montgomery; completed, 1914; elevation, 835′ B.

	Thickness Ft. In.	Total Ft. In.
Surface	22 0	22 0
Shale	6 6	28 6
Sandstone	15 9	44 3
Shale	0 7	44 10
Sandstone	99 0	143 10
Shale	4 9	148 7
Sandstone	2 4	150 11
Shale	20 0	170 11

	Thickness Ft. In.	Total Ft. In.	
Coal ....0' 11 " / Bone ....0 1½ / Fire clay........1 4½ / Coal ....1 0 } Cedar Grove....	3 5	174 4	
Fire clay..............................	1 3	175 7	
Shale, sandstone streaks.............	15 0	190 7	
Sandstone, shale streaks.............	17 2	207 9	
Gray sandy shale.....................	20 3	228 0	
Dark shale...........................	12 0	240 0	
Coal ....0' 1 " / Bone and sulphur.0 0¼ / Coal ....3 10 / Bone and coal....0 7 } Alma.......	4 6¼	244 6¼	70' 2¼"
Shale, sandstone streaks, to bottom...	5 6	250 0¼	

Record of Boone County Coal Corporation Diamond Core Test Hole No. 14 (160Lo.).

Located on Rockhouse Creek of Spruce Fork, 1.8 miles from its mouth and 1.9 miles from Clothier, Logan District, Logan County; authority, A. R. Montgomery; completed, 1914; elevation, 901.8' L.

	Thickness Ft. In.	Total Ft. In.	
Surface	18 0	18 0	
Shale	3 0	21 0	
Coal, Chilton........................	1 0	22 0	
Fire clay............................	1 6	23 6	
Sand rock and shale..................	8 2	31 8	
Sand rock, Lower Chilton.............	24 6	56 2	
Coal and bone......0' 5" / Coal ....0 5 } Little Chilton...	0 10	57 0	35' 0"
Sandstone	5 2	62 2	
Shale	18 2	80 4	
Coal, Hernshaw.......................	0 10	81 2	24' 2"
Shale	4 7	85 9	
Sand rock............................	54 4	140 1	
Slate	26 4	166 5	
Sand rock............................	20 10	187 3	
Slate	0 3	187 6	
Shale	1 3	188 9	
Slate and coal.......................	0 10	189 7	
Coal ....0' 2 " / Slate ....2 3 / Coal ....0 1 / Slate ....0 1 / Coal ....0 5½ / Coal and bone.0 6 / Coal ....0 7½ } Williamson..	4 2	193 9	112' 7"
Fire clay............................	1 0	194 9	
Shale	1 10	196 7	
Slate	7 2	203 9	
Shale streaks with sand rock.........	24 5	228 2	
Slate, black.........................	14 3	242 5	
Coal, Cedar Grove....................	2 3	244 8	51' 11"

		Thickness Ft. In.	Total Ft. In.	
Bone		0 5	245 1	
Fire clay		1 1	246 2	
Shale		1 7	247 9	
Sand rock		4 10	252 7	
Shale		5 6	258 1	
Sand rock		17 1	275 2	
Shale		3 7	278 9	
Bone		0 2	278 11	
Coal ....0' 11" Slate ....0 2 **Coal** 0 3	Lower Cedar Grove.	1 4	280 3	35' 7"
Wash, lost		1 6	281 9	
Fire clay		1 9	283 6	
Shale and sand rock, mixed		16 10	300 4	
Slate		0 5	300 9	
Coal ....1' 1½" Bone ....0 2 **Coal** ....2 4½ Shale ....0 2½ **Coal** 0 1	Alma	3 11½	304 8½	24' 5½"
Slate with fire clay to bottom		2 2	306 10½	

Record of Boone County Coal Corporation Diamond Core Test Hole No. 15 (161Lo.).

Located on Seng Camp Branch of Spruce Fork, 1.5 miles from its mouth and 3.0 miles southeast from Clothier, Logan District, Logan County; authority, A. R. Montgomery; completed, 1914; elevation, 1020' B.

	Thickness Ft. In.	Total Ft. In.	
Surface	15 0	15 0	
Shale	4 0	19 0	
Coal and bone (**Hernshaw**)	0 4	19 4	
Fire clay	3 0	22 4	
Shale	2 8	25 0	
Sandstone, shale streaks	3 2	28 2	
Sandy shale	3 3	31 5	
Sandstone	35 2	66 7	
Sandstone, **coal** streaks	18 9	85 4	
Sandstone	20 8	106 0	
Shale	0 4	106 4	
Sandstone, shale streaks	1 5	107 9	
Coal (Williamson)	0 11	108 8	89' 4"
Fire clay, sandstone streaks	1 1	109 9	
Sandstone	53 11	163 8	
Sandy shale	1 8	165 4	
Sandstone	0 5	165 9	
Sandy shale	2 2	167 11	
Sandstone	29 6	197 5	
Sandy shale, sandstone streaks	8 4	205 9	
Dark sandy shale	21 0	226 9	
Shale, sandstone, mixed	2 0	228 9	
Shale, sandstone streaks	4 4	233 1	

PLATE IX.—View from head of Whites Branch looking west, illustrating topography of Allegheny Series.

		Thickness Ft. In.	Total Ft. In.	
Sandy shale		6 5	239 6	
Coal ......1′ 8″ **Coal**, core lost......0 7 Bone ......0 3 Sandy shale......2 9 **Coal** and bone......0 7	**Alma**	5 10	245 4	136′ 8″
Sandstone ...... 3′ 2″ Sandstone, shale streaks ......18 0 Sandstone ...... 3 0 Sandstone, shale streaks 18 1	**Logan**	42 3	287 7	
Gray shale		19 9	307 4	
Black slate, **coal** streaks		10 6	317 10	
Coal ......4′ 8½″ **Coal** and bone....0 3 Shale ......1 10½ **Coal** and bone....0 4½ Shale ......0 7½ **Coal** and bone....1 5	**Campbell Creek (No. 2 Gas)**	9 3	327 1	81′9″
Shale to bottom		6 0	333 1	

Record of Boone County Coal Corporation Diamond Core Test Hole No. 17 (162Lo.).

Located on Spruce Fork, 1.0 mile below Pigeonroost Branch and 4.0 miles south of Clothier, Logan District, Logan County; authority, A. R. Montgomery; completed, 1914; elevation, 920′ B.

		Thickness Ft. In.	Total Ft. In.	
Surface		11 6	11 6	
Sand rock		26 6	38 0	
Coal		0 6	38 6	
Sand rock		27 0	65 6	
Shale		19 4	84 10	
Coal		1 0	85 10	
Bone		0 2	86 0	
Fire clay		0 6	86 6	
Sand rock		15 10	102 4	
Shale		1 4	103 8	
Slate		1 6	105 2	
Coal, Williamson		1 2	106 4	
Slate		3 5	109 9	
Sand rock and sandy shale		31 5	141 2	
Shale		17 11	159 1	
Slate, black		9 1	168 2	
Bone		0 2	168 4	
Coal ......2′ 0″ Bone ......0 1 **Coal** 1 3½	**Cedar Grove**	3 4½	171 8½	65′ 4½″
Slate		2 0	173 8½	
Shale		7 1	180 9½	
Sand rock		3 8	184 5½	
Shale		1 10	186 3½	

			Thickness Ft.	In.	Total Ft.	In.	
Coal			1	6½	187	10	
Shale			2	0	189	10	
Sand rock			11	5	201	3	
Shale			0	10	202	1	
Sand rock			8	6	210	7	
Shale			1	9	212	4	
Bone			0	2	212	6	
Coal	0' 5"	**Lower**					
Sulphur	0 1½	**Cedar**					
Coal	0 5½	**Grove**	1	4½	213	10½	42' 2"
Bone	0 2						
Coal	0 2½						
Fire clay			0	8	214	6½	
Shale			1	10	216	4½	
Sand rock			8	0	224	4½	
Shale			2	4	226	8½	
Slate			0	3½	227	0	
Coal	1' 5½"						
Bone	0 1	**Alma**	1	10	228	10	14' 11½"
Coal	0 3½						
Fire clay			0	4	229	2	
Shale			2	7	231	9	
Sand rock			17	5	249	2	
Shale			3	7	252	9	
Slate			0	3½	253	0½	
Coal, Little Alma			0	8½	253	9	24' 11"
Slate, dark			6	6	260	3	
Shale, dark			3	11	264	2	
Sand rock, to bottom			15	4	279	6	

Record of Boone County Coal Corporation Diamond Core Test Hole No. 18 (163Lo.).

Located on Brushy Fork of Spruce Fork, 2.0 miles from its mouth, and 9.5 miles southeast from Clothier, Logan District, Logan County; authority, A. R. Montgomery; completed, 1914; elevation, 1500' B.

			Thickness Ft.	In.	Total Ft.	In.	
Boulders and sand			10	0	10	0	
Hard sand and gravel			6	3	16	3	
Sandstone			26	2	42	5	
Coal	1' 6"						
Fire clay	0 4		2	2	44	7	
Coal	0 4						
Shale and sandstone			9	0	53	7	
Sandstone			3	5	57	0	
Slate and shale			12	0	69	0	
Coal	0' 5"						
Slate and shale	3 0						
Coal	0 2	**Chilton**	15	5	84	4	39' 10"
Slate and shale	9 10						
Coal	2 0						
Fire clay and slate			4	5	88	10	
Sandstone			35	2	124	0	

		Thickness Ft.	In.	Total Ft.	In.	
Slaty shale		3	3	127	3	
Sandstone		32	1	159	4	
Coal		1	2	160	6	
Slate and sandstone		7	2	167	8	
Coal ..... 0' 8" Slate ..... 0 5 **Coal** 1 2	**Hernshaw**	2	3	169	11	85' 6"
Fire clay		1	7	171	6	
Fire clay and sandstone		3	2	174	8	
Sandstone		60	0	234	8	
Slate		42	1	276	9	
Coal, Williamson		0	8	277	5	
Fire clay		0	8	278	1	
Sandstone, streaks of slate		7	2	285	3	
Sandstone		67	1	352	4	
Slate, streaks of sandstone		13	2	365	6	
Slate, black		2	6	368	0	
Coal ..... 1' 9" Slate ..... 0 5 **Coal** ..... 3 5 Slate and bony **coal** . 0 4 **Coal** 0 4	**Cedar Grove**	6	3	374	3	204' 4"
Fire clay to bottom		8	2	382	5	

Record of Boone County Coal Corporation Diamond Core Test Hole No. 19 (164Lo.).

Located on Pigeonroost Branch of Spruce Fork, about 0.5 mile from its mouth and 4.8 miles south from Clothier, Logan District, Logan County; authority, A. R. Montgomery; completed, 1914; elevation, 1090' B.

		Thickness Ft.	In.	Total Ft.	In.	
Surface		13	0	13	0	
Shale wash		9	0	22	0	
Coal ..... 3' 0" Shale ..... 9 5 **Coal** ..... 1 6 Shale ..... 3 11 **Coal** ..... 1 6 Shale ..... 0 4 **Coal** 0 1	**Chilton**	19	9	41	9	
Shale		3	4	45	1	
Sandstone		13	6	58	7	
Sandstone, **coal** streaks		13	9	72	4	
Sandstone, shale streaks		1	3	73	7	
Sandstone, **coal** and shale streaks		10	4	83	11	
Sandstone		1	7	85	6	
Coal, Little Chilton		1	0	86	6	45' 9"
Sandy shale		1	6	88	0	
Shale and sandstone streaks		2	4	90	4	
Sandstone		8	2	98	6	
Coal		0	2	98	8	
Shale		1	8	100	4	

	Thickness Ft.	In.	Total Ft.	In.	
Sandstone	26	7	126	11	
Sandy shale	7	1	134	0	
Sandstone, **coal** streaks	28	7	162	7	
Sandy shale	40	0	202	7	
Coal	0	3	202	10	
Shale	0	4	203	2	
Sandstone	33	9	236	11	
Coal .....0′ 1″					
Sandstone .....0 1 } **Williamson**..	0	3	237	2	150′ 8″
Coal 0 1					
Sandstone	6	0	243	2	
Dark sandy shale	44	1	287	3	
Black slate	7	7½	294	10½	
Coal, Cedar Grove	3	9½	298	8	62′ 6″
Shale	1	5	300	1	
Sandy shale	10	0	310	1	
Sandstone	19	1	329	2	
Coal, Lower Cedar Grove	0	9	329	11	31′ 3″
Dark shale	4	4	334	3	
Sandstone, shale streaks	8	7	342	10	
Sandy shale	8	6	351	4	
Coal, Alma	1	7	352	11	23′ 0″
Shale to bottom	0	4	353	3	

Record of Boone County Coal Corporation Diamond Core Test Hole No. 18 (165Lo.).

Located on Oldhouse Branch of Spruce Fork, 700 feet from its mouth and 5.0 miles southwest of Clothier, Logan District, Logan County; authority, A. R. Montgomery; completed, 1914; elevation, 980′ B.

	Thickness Ft.	In.	Total Ft.	In.
Surface	15	0	15	0
Sand rock	81	11	96	11
Shale	17	8	114	7
Sand rock	44	1	158	8
Sand rock and shale	14	7	173	3
Shale	7	4	180	7
Sand rock and shale	9	11	190	6
Sand rock	14	5	204	11
Shale	8	0	212	11
Slate	1	6	214	5
Coal .....0′ 5″				
Slate, black....6 3				
Coal 0 6				
Bone 0 2				
Coal .....1 5½ } **Cedar Grove**..	11	9	226	2
Coal and bone..0 3½				
Slate 0 8½				
Coal 1 3				
Slate 0 6				
Coal and bone..0 2½				
Shale	8	4	234	6

	Thickness Ft. In.	Total Ft. In.	
Sand rock	4 2	238 8	
Coal, Cedar Grove, Lower Bench	0 10	239 6	
Fire clay	2 8	242 2	
Shale	3 1	245 3	
Sand rock	20 10	266 1	
Shale	1 3	267 4	
Slate	0 7½	267 11½	
Bone	0 2	268 1½	
Coal, Alma "A"	0 5	268 6½	
Bone	0 2	268 8½	
Fire clay	2 0	270 8½	
Shale	5 5	276 1½	
Sand rock and shale	4 10	280 11½	
Shale	2 0	282 11½	
Coal ....0′ 5″ Bone ....0 3½ **Coal** ....0 2½ } **Alma**	0 11	283 10½	57′ 8½″
Sandy shale	6 0	289 10½	
Sand rock to bottom	1 7	291 5½	

Record of Boone County Coal Corporation Diamond Core Test Hole No. 16 (166Lo.).

Located on Whiteoak Branch of Spruce Fork, 1.0 mile from its mouth and 6.0 miles southeast from Clothier, Logan District, Logan County; authority, A. R. Montgomery; completed, 1914; elevation, 1270′ B.

	Thickness Ft. In.	Total Ft. In.	
Boulders and sand	17 0	17 0	
Sandstone	31 0	48 0	
Coal, Chilton "A"	1 3	49 3	
Fire clay	0 9	50 0	
Slate	11 6	61 6	
Coal and slate	0 4	61 10	
Slate, sand streaks	23 2	85 0	
Coal ....1′ 4″ Bone and slate.....0 8 Slate ....0 5 **Coal** ....1 2 Slate and fire clay..2 0 **Coal** ....0 7 } **Chilton**	6 2	91 2	91′ 2″
Slate, sandy	1 10	93 0	
Sandstone, streaks of slate	3 2	96 2	
Sandstone	68 10	165 0	
Coal	0 4	165 4	
Slate and sandstone	12 4	177 8	
Coal and slate, **Hernshaw**	1 0	178 8	87′ 6″
Fire clay and slate	1 4	180 0	
Sandstone	70 6	250 6	
Slate	35 8	286 2	
Coal, Cedar Grove	0 6	286 8	108′ 0″
Sandstone and slate	8 8	295 4	
Slate	7 10	303 2	

	Thickness Ft.	In.	Total Ft.	In.	
Coal	0	1½	303	3½	
Fire clay	2	0	305	3½	
Sandstone	1	8½	307	0	
Slate, sand streaks	22	2	329	2	
Sandstone, streaks of slate	19	10	349	0	
Slate, black	15	1	364	1	
Coal 3' 9" Slate 0 7 **Coal** 1 3 } **Alma**	5	7	369	8	83' 0"
Fire clay to bottom	4	4	374	0	

Record of Boone County Coal Corporation Diamond Core Test Hole No. 20 (167Lo.).

Located on Adkins Branch of Spruce Fork, 0.5 mile up same, and 7.0 miles southwest from Clothier, Logan District, Logan County; authority, A. R. Montgomery; completed, 1914; elevation, 1075' B.

	Thickness Ft.	In.	Total Ft.	In.	
Boulders and sand	14	9	14	9	
Sandstone	18	7	33	4	
Coal, Little Chilton	0	5	33	9	
Sandstone	28	8	62	5	
Slate and sandstone	9	1	71	6	
Sandstone	40	2	111	8	
Shale	1	10	113	6	
Sandstone	0	2	113	8	
Shale	0	8	114	4	
Sandstone	2	4	116	8	
Broken sandstone	19	6	136	2	
Shale	0	6	136	8	
Sandstone	0	2	136	10	
Shale	31	7	168	5	
Sandstone	60	10	229	3	
Sandy shale	5	9	235	0	
Shale, gray	12	0	247	0	
Slate	0	4	247	4	
Coal 0' 9" **Coal and bone** 0 2 Slate, black 2 11 Sandstone, **coal** streaks 0 2½ **Coal** 1 7 Slate 0 1½ **Coal, core lost, very soft** 1 11 } **Cedar Grove**	7	8	255	0	221' 3"
Sandy shale	0	9	255	9	
Sandstone	42	8	298	5	
Coal, Lower Cedar Grove	1	6	299	11	44' 11"
Fire clay	1	4	301	3	
Shale with sandstone streaks to bottom	5	6	306	9	

Record of Boone County Coal Corporation Diamond Core Test Hole No. 13 (168Lo.).

Located on Garland Fork of Spruce Fork, 0.7 mile from its mouth, and 8.0 miles south from Clothier, Logan District, Logan County; authority, A. R. Montgomery; completed, 1914; elevation, 1200' B.

	Thickness Ft.	In.	Total Ft.	In.
Boulders and sand	12	0	12	0
Sandstone	15	0	27	0
Coal, Little Chilton	0	6	27	6
Fire clay	0	7	28	1
Shale, streaks of sandstone	18	5	46	6
Shale, little bony **coal**	0	6	47	0
Coal and bony **coal**	0	11	47	11
Fire clay	0	5	48	4
Sandstone	59	5	107	9
Slate	33	3	141	0
Slate, sandy	13	1	154	1
Coal	0	8	154	9
Slate, sandy	0	11	155	8
Sandstone	58	4	214	0
Slate, sandy	10	6	224	6
Sandstone	12	6	237	0
Sandstone, slate streaks	13	0	250	0
Slate, black	0	10	250	10
Coal 3' 5½" Slate 0 2½ **Coal** 0 10 Slate 0 1 **Coal** 0 10½ Slate and bony **coal** 0 8½ } **Cedar Grove**	6	2	257	0
Slate to bottom	4	0	261	0

Record of Boone County Coal Corporation Diamond Core Test Hole No. 11 (169Lo.).

Located on Brushy Fork of Spruce Fork, 8.5 miles southeast from Clothier, Logan District, Logan County; authority, A. R. Montgomery; completed, 1914; elevation, 1270' B.

	Thickness Ft.	In.	Total Ft.	In.
Boulders and sand	10	0	10	0
Hard sandstone	5	0	15	0
Sandstone	16	0	31	0
Coal, little bony	1	3	32	3
Fire clay	1	0	33	3
Sandstone and slate	9	0	42	3
Coal, bony 0' 6" **Coal** 1 8 } **Hernshaw?**	2	2	44	5
Slate, shaly	6	7	51	0
Sandstone	40	6	91	6
Slate	1	9	93	3
Sandstone	9	9	103	0

		Thickness Ft.	In.	Total Ft.	In.	
Sandstone, mixed		12	5	115	5	
Slate		34	2	149	7	
Coal, Williamson		1	0	150	7	106' 2"
Fire clay, soft		2	5	153	0	
Sandstone, streaks of slate		5	0	158	0	
Sandstone		74	10	232	10	
Slate, black		22	0	254	10	
Coal	0' 5"					
Coal, bony	0 2					
Coal	0 10½					
Slate	0 1½					
Coal	2 7½					
Slate and bony coal	0 4	**Cedar Grove** 7	7½	262	5½	111' 10½"
Coal	1 0					
Slate	0 3½					
Coal	0 10					
Slate and bony coal	0 11½					
Slate, sandy		1	10½	264	4	
Slate and sandstone to bottom		4	8	269	0	

Spruce Laurel Fork flows into Spruce Fork at Clothier. For a distance of more than fifteen miles this stream has very little grade, only twenty to forty feet per mile.

Along this stream occurs one of the richest coal fields in Boone County, if not in West Virginia. A great many diamond core test holes have been drilled to test the lower coal measures by the Laurel Coal and Land Company, owned by Messrs. Chilton, MacCorkle and Chilton and Meany. This firm owns a large area of land in Washington and Crook Districts.

Through the courtesy of Ex-Governor W. A. MacCorkle, the Survey has received a complete copy of the records of these diamond core test holes. This information has been of great assistance to the writer in the study of the rock series and mapping up the structure on the detailed Map II, and has greatly assisted in the computation of the available coal tonnage in Boone County.

PLATE X.—View of Seth, also showing topography of Kanawha Series, capped with the Allegheny.

Boone County Coal Corporation Coal Test No. 5B (111).

Located on Spruce Laurel Fork, 1 mile east of Clothier, Washington District, Boone County; authority, Montgomery, Clothier and Tyler; elevation, 835' B.

Kanawha Series (275')			Thickness Ft.	In.	Total Ft.	In.	
Surface			21	0	21	0	
Shale, sandstone streaks			10	8	31	8	
Sandstone			58	11	90	7	
Shale			0	2	90	9	
Sandstone			18	6	109	3	
Shale			0	8	109	11	
Sandstone			22	11	132	10	
Sandy shale			29	3	162	1	
Coal	0'	5¼"					
Shale	0	5					
Coal	0	10½					
Slate	0	2					
Coal	1	1½					
Fire clay	0	1					
Coal	1	0					
Shale	1	0	Cedar Grove, Upper Bench.... 5	11	168	0	168'
Coal	0	2					
Slate	0	0½					
Coal and slate, interlaminated	0	6					
Shale	0	4					
Coal	0	9					
Sandy shale			2	10	170	10	
Sandstone, shale streaks			7	2	178	0	
Sandy shale			17	6	195	6	
Sandy slate			5	1	200	7	
Black slate			11	4	211	11	
Coal	0'	10½"					
Bone	0	0¼					
Coal	0	1	Cedar Grove Lower Bench..... 1	6¾	213	5¾	
Shale	0	2					
Coal	0	5					
Fire clay			0	11	214	4¾	
Sandy shale			1	6	215	10¾	
Sandstone			37	4¼	253	3	
Coal	1'	5"					
Shale	0	3	Alma..... 2	5	255	8	41' 3¼"
Coal	0	9					
Slate			0	6	256	2	
Shaly slate to bottom			18	10	275	0	

The above record begins 20 feet under the Chilton Coal, so that the coal encountered at 255 feet is the Alma seam.

Spruce Laurel Diamond Core Test Hole No. 9 (112).

Located on Jerry Fork of Spruce Laurel on the D. Bias farm, on land of Chilton, MacCorkle and Chilton, and Meany, S. 60° E. 4 miles from Clothier, Washington District, Boone County; authority, Ex-Gov. W. A. MacCorkle; elevation, 1038′ L.

				Thickness		Total		
Kanawha Series (356′)				Ft.	In.	Ft.	In.	
Surface				4	6	4	6	
Sandstone, very hard, **Williamson**				68	3	72	9	
Slate, **coal** streaks				1	2	73	11	
Fire clay				1	5	75	4	
Sandy shale				2	0	77	4	
Slate				0	6	77	10	
Coal and bone	0′	2″						
Fire clay	0	9						
Sandy shale	5	3						
Bone	0	1	**Williamson**	11	3	89	1	
Coal	1	1						
Fire clay	3	9						
Coal and bone	0	2						
Fire clay				0	10	89	11	
Sandy shale				4	4	94	3	
Sandstone, **Upper Cedar Grove**				34	9	129	0	
Sandy shale				6	3	135	3	
Sandstone				12	9	148	0	
Sandy shale				41	4	189	4	
Coal				0	8	190	0	
Slate				0	6	190	6	
Sandy shale				10	1	200	7	
Coal	0′	1″						
Fire clay	0	5						
Shale	0	9						
Coal	0	1						
Fire clay	1	10	**Alma**	9	1	209	8	120′ 7″
Sandy shale	2	6						
Slate	0	4						
Coal and bone	0	4						
Coal	2	9						
Fire clay				3	5	213	1	
Sandstone				1	4	214	5	
Sandy shale				0	6	214	11	
Sandstone				25	0	239	11	
Shale, sandstone streaks				1	6	241	5	
Sandstone, **coal** streaks				1	10	243	3	
Sandstone, **Malden**				22	6	265	9	
Sandy shale				2	3	268	0	
Black slate				12	11	280	11	
Coal	2′	6″	**Campbell Creek**					
Bone	0	0⅛						
Coal	0	10½	**(No. 2**					
Bone	0	0⅛	**Gas)**	6	5¼	287	4¼	77′8¼″
Coal	1	0½						
Coal and bone, interlaminated	0	2						
Coal	1	10						
Shale				5	1¾	292	6	

			Thickness Ft.	In.	Total Ft.	In.	
Sandstone, **Brownstown**			22	8	315	2	
Sandy shale			4	0	319	2	
Sandstone			14	4	333	6	
Coal	0′ 9″	Powellton					
Coal and bone, interlaminated	0 10	"Rider"	1	7	335	1	
Fire clay			1	3	336	4	
Shale			9	5	345	9	
Coal	1′ 4″						
Fire clay	0 5						
Coal	3 5	Powellton	5	4	351	1	63′ 8¾″
Coal and bone, interlaminated	0 2						
Fire clay			1	1	352	2	
Sandy shale			3	10	356	0	

The following section was obtained by combining drill core test hole No. 4 (136), located on the top of the divide between Casey and Whiteoak Creeks, and core drill hole No. 10 (113), located on the Right Fork of Whiteoak Creek, drilled on the land of Chilton, MacCorkle and Chilton and Meany; elevation, 1790′ B.:

Section 5.2 Miles S. 82° E. from Clothier, Washington District.

			Thickness Ft.	In.	Total Ft.	In.	
Pennsylvanian (1034′ 4″)							
Allegheny Series (84′ 4″)							
Surface			7	0	7	0	
Sandstone			45	6	52	6	
Slate			5	6	58	0	
Sandstone			5	10	63	10	
Slate			3	0	66	10	
Sandstone, pebbly			14	6	81	4	
Coal			0	6	81	10	
Sandy shale			2	6	84	4	
Kanawha Series (950′)							
Sandstone	44′ 2″	**Homewood**	70	6	154	10	
Shale	1 10	**Sandstone**					
Sandstone	24 6						
Slate			5	8	160	6	
Coal	4′ 2″						
Slate	0 1	**Stockton-**					
Coal	0 9	**Lewiston,**	5	3½	165	9½	84′
Bone	0 2½	**Upper Bench**					
Coal	0 1						
Sandy shale			5	11½	171	9	
Coal			0	5	172	2	
Sandy shale			6	0	178	2	
Sandstone			6	0	184	2	
Sandy shale			0	4	184	6	

Stratum	Detail	Bed	Thickness Ft. In.	Total Ft. In.	
Coal	0′ 6″	Stockton-Lewiston Middle Bench	0 11	185 5	20′
Bone	0 3				
Coal	0 2				
Sandy shale			10 0	195 5	10′
Coal	2′ 0″	Stockton-Lewiston Lower Bench	5 5	200 10	
Coal, dirty	1 0				
Coal	2 5				
Sandy shale			6 0	206 1	
Sandstone, Coalburg			51 8	258	
Sandy shale			12 0	270	
Shale			3 3	273 9	
Bone	0′ 3″	Coalburg.	7 9	281 6	80′
Coal	0 9				
Slate	0 1				
Coal	0 3				
Slate	0 1				
Coal	0 6				
Bone	0 6				
Coal and bone	1 0				
Coal	1 6				
Coal and bone	0 4				
Soft shale	1 0				
Coal	1 6				
Slate			2 5	283 11	
Sandy shale			3 0	286 11	
Sandstone			6 6	293 5	
Sandy shale			0 6	293 11	
Coal	1′ 2″	Little Coalburg.	3 9	297 8	16′
Slate	0 9				
Fire clay	0 8				
Shale	1 0				
Coal	0 2				
Sandy shale			5 0	302 8	
Sandstone			13 3	315 11	
Sandy shale			7 3	323 2	
Sandstone, Lower Coalburg			71 4	394 6	
Slate			0 6	395 0	
Coal and bone, Buffalo Creek			0 7	395 7	
Sandy shale			4 0	399 7	
Sandstone			36 4	435 11	
Shale, Upper Winifrede Sandstone			0 11	436 10	
Coal			0 2	437 0	
Sandstone			29 10	466 10	
Slate			2 5	469 3	
Sandstone			8 5	477 8	
Coal	0′ 3″	Winifrede (1308.75′ B.)	3 7	481 3	184′
Slate	0 7				
Coal	2 0				
Bone	0 1				
Coal	0 8				
Fire clay			2 0	483 3	
Sandstone			4 0	487 3	

	Thickness Ft.	In.	Total Ft.	In.
Sandstone and concealed to top of Core Drill Hole No. 10 (113)[1]	90	9	578	0
Surface	19	6	597	6
Sandstone	3	8	601	2
Shale	15	8	616	10
Sandstone, Lower Chilton	13	2	630	0
Shale	1	0	631	0
Sandstone	11	9	642	9
Shale	0	6	643	3
Coal, Little Chilton	0	1	643	4
Fire clay	2	5	645	9
Sandstone, Hernshaw	31	2	676	11
Shale	2	6	679	5
Sandstone	5	0	684	5
Shale	1	3	685	8
Sandstone	2	2	687	10
Coal and bone, Hernshaw	2	2	690	0
Fire clay	1	6	691	6
Shale	6	9	698	3
Fire clay	3	9	702	0
Shale	6	3	708	3
Sandstone, Naugatuck	51	2	759	5
Bastard limestone	2	0	761	5
Sandstone, Williamson	78	4	839	9
Shale	2	1	841	10
Coal, Williamson	1	5	843	3
Sandstone, Upper Cedar Grove	36	8	879	11
Shale	2	0	881	11
Sandstone	13	9	895	8
Shale, sandy	40	5	936	1
Sandstone	5	11	942	0
Coal, Cedar Grove, Lower Bench	0	1	942	1
Shale and sandstone	19	8	961	9
Bone and coal, Little Alma	0	4	962	1
Shale, sandy	9	0	971	1
Black slate	18	9	989	10
Shale	15	0	1004	10
Black slate	7	2	1012	0
Coal 2′ 5¼″; Shale and coal 0 2¾; Shale 14 1; Coal 3 10 — Campbell Creek (No. 2 Gas)	20	7	1032	7
Fire clay	1	9	1034	4

Coal opening located on Right Fork of Whiteoak Creek, S. 45° W. one mile from diamond core test hole No. 4 (136); elevation, 1328′ B., being located a little to the rise of core test hole No. 4 (136), shows the following section:

[1]Core drill hole No. 10 (113) is located S. 45° W. 1000 feet from the above opening in the Winifrede Coal and the interval from the base of the opening to the top of core drill hole No. 10 (113) is 90′ 9″ by hand level.

		Thickness Feet	Inches
Sandstone		6	0
Dark shale..........0′ 2″	Winifrede	4	2
Coal 0 5			
Dark shale..........0 10			
Coal, splint..........0 10			
Coal, impure..........0 4			
Coal, hard, gas..........1 0			
Shale, gray..........0 1			
Coal, hard, splint..........0 8			

The above section correlates with the coal at 481′ 3″ in the section of the core hole No. 4 (136).

Record of Spruce Laurel Diamond Core Test Hole No. 3 (114)

Located on Spruce Laurel Fork, on land of Chilton, MacCorkle and Chilton, and Meany, S. 67° E. 4.2 miles from Clothier, Washington District, Boone County; authority, Ex-Gov. W. A. MacCorkle; elevation, 970′ L.

	Thickness Ft.	In.	Total Ft.	In.
Kanawha Series (217′ 10″)				
Surface	17	1	17	1
Sandstone	4	1	21	2
Sandy shale	3	10	25	0
Sandstone	21	9	46	9
Sandy shale	1	4	48	1
Sandstone	30	8	78	9
Sandy shale	36	10	115	7
Coal and bone, Upper Cedar Grove	0	2	115	9
Shale	0	2	115	11
Fire clay	1	4	117	3
Sandy shale	8	9	126	0
Slate	0	1	126	1
Coal, Lower Cedar Grove	0	1	126	2
Sandy shale	7	6	133	8
Sandstone	8	5	142	1
Sandy shale	2	10	144	11
Coal and bone, Alma	0	2	145	1
Shale	1	0	146	1
Sandstone, with **coal** streaks in lower portion	25	4	171	5
Coal	0	1	171	6
Shale	0	5	171	11
Sandstone	10	5	182	4
Sandy shale	1	6	183	10
Sandstone	1	0	184	10
Sandy shale	14	0	198	10
Black slate	12	1	210	11
Coal, Campbell Creek (No. 2 Gas)	6	5	217	4
Fire clay	0	2	217	6
Sandy shale	0	4	217	10

The following aneroid section was measured by Krebs, descending from a high summit on east side of Spruce Laurel Fork, just north of Bear Hollow, southwesterly, and joined to Spruce Laurel core drill hole No. 1 (115):

Section 4.6 Miles Southeast of Clothier, Washington District.

	Thickness		Total	
Kanawha Series (889' 11")	Ft.	In.	Ft.	In.
Sandstone and concealed	150	0	150	0
Coal, Stockton-Lewiston	4	0	154	0
Sandstone and concealed to large bench	126	0	280	0
Sandstone and concealed	130	0	410	0
Coal blossom, Winifrede	2	0	412	0
Sandstone and concealed	183	0	595	0
Coal and slate, **Hernshaw**	5	0	600	0
Sandstone and concealed to top of Spruce Laurel diamond core test hole No. 1, elevation, 1006' L	45	0	645	0
Core Test Hole No. 1 (115)				
Surface	17	2	662	2
Shale, **Dingess fossil horizon**	11	11	674	1
Coal 1' 3"; Fire clay and **coal** 2 7 } **Williamson**	3	10	677	11
Shale	7	9	685	8
Sandstone	47	6	733	2
Shale	45	1	778	3
Coal 1' 3"; Slate 0 3½; **Coal** 0 3; Shale 0 8½; **Coal** 0 4; Slate 3 7; **Coal** 0 1 } **Alma Coal**	6	6	784	9
Sandy shale	2	11	787	8
Sandstone	13	0	800	8
Coal, Little Alma	0	2	800	10
Fire clay and shale	3	0	803	10
Sandstone, **Logan**	35	5	839	3
Sandy shale	22	3	861	6
Shale, black	12	3	873	9
Coal, Campbell Creek (No. 2 Gas)	7	3	881	0
Shale and sandstone	8	11	889	11

The interval between the Campbell Creek (No. 2 Gas) Coal and the Stockton-Lewiston Coal is 727 feet, and between the Winifrede Coal and the Campbell Creek (No. 2 Gas) is 469 feet, as compared with 781 feet and 552 feet between the same coals in the Bald Knob Sections, and 903 feet and 543 feet at Clothier.

Record of Spruce Laurel Diamond Core

Test Hole No. 16 (116).

Located on Spruce Laurel Fork on land of Chilton, MacCorkle and Chilton, and Meany, near Reed Jarrold's house, S. 53° E. 5 miles from Clothier, Washington District, Boone County; authority, Ex-Gov. W. A. MacCorkle; elevation, 1025′ L.

	Thickness Ft.	Thickness In.	Total Ft.	Total In.
Kanawha Series (236′ 7″)				
Surface	11	4	11	4
Sandstone	2	10	14	2
Williamson: Coal 0′ 8″; Sandstone 6 8; Coal 1 0	8	4	22	6
Fire clay	1	6	24	0
Shale	1	10	25	10
Sandstone, Upper Cedar Grove	26	3	52	1
Shale	1	0	53	1
Sandstone	0	5	53	6
Shale	0	9	54	3
Sandstone	7	0	61	3
Sandstone, coal streaks, Upper Cedar Grove	2	10	64	1
Sandstone, Middle Cedar Grove	20	6	84	7
Sandy shale	47	5	132	0
Alma: Coal 1′ 0″; Coal and bone 0 4; Coal 0 4; Shale 1 2; Coal 0 4; Sandy shale 6 7; Coal and bone 0 5	10	2	142	2
Sandy shale	3	4	145	6
Sandstone and shale	6	2	151	8
Sandstone, Logan	36	5	188	1
Sandy shale	25	10	213	11
Black slate	12	5	226	4
Campbell Creek (No. 2 Gas): Coal 0′ 3″; Slate 0 1½; Coal 2 6½; Shale 0 0¼; Coal 1 8; Bone 0 0¾; Coal 0 6; Coal and bone, interlaminated 0 4½; Shale 0 1½; Coal 2 3	7	11	234	3
Fire clay	1	0	235	3
Shale	1	4	236	7

The foregoing section shows the Campbell Creek (No. 2 Gas) Coal to be split up considerably with shale and impure coal. The Alma bed is also broken up with shale and bone.

Record of Spruce Laurel Diamond Core Test Hole No. 4 (117)

Located on Sycamore Fork of Spruce Laurel Fork, on land of Chilton, MacCorkle and Chilton, and Meany, S. 54° E. 5.1 miles from Clothier, Washington District, Boone County; authority, Ex-Gov. W. A. MacCorkle; elevation, 1065' B.

		Thickness Ft.	Thickness In.	Total Ft.	Total In.
Kanawha Series (276')					
Surface		9	2	9	2
Sandstone, **Williamson**		58	2	67	4
Coal 1' 2"; Fire clay 1 2; **Coal** 0 2	**Williamson**	2	6	69	10
Sandy shale		2	0	71	10
Sandstone		16	0	87	10
Sandstone, hard		5	2	93	0
Sandstone		10	10	103	10
Sandy shale		1	4	105	2
Sandstone		25	0	130	2
Sandy shale		43	6	173	8
Slate		1	8	175	4
Coal 0' 1"; Shale 0 9; **Coal** 1 2; Shale 0 3; **Coal** 0 4; Shale 0 6; **Coal** 0 3	**Alma**	8	1	183	5
Sandy shale		5	0	188	5
Sandstone		1	6	189	11
Sandy shale		2	9	192	8
Sandstone		0	5	193	1
Sandy shale		1	0	194	1
Fire clay		0	10	194	11
Sandy shale		2	0	196	11
Sandstone		36	0	232	11
Sandstone and sulphur		2	6	235	5
Sandy shale		19	2	254	7
Black slate		11	10	266	5
Coal 0' 5"; Slate 0 1; **Coal** 5 4; Slate 0 2; **Coal** 2 2	**Campbell Creek (No. 2 Gas)**	8	2	274	7
Fire clay		1	5	276	0

The above section shows the Campbell Creek (No. 2 Gas) Coal of excellent thickness and almost free from impurities.

Record of Spruce Laurel Diamond Core Test Hole No. 5 (118)

Located on Spruce Laurel Fork, on land of Chilton, MacCorkle and Chilton, and Meany, S. 49° 30' E. 5.2 miles from Clothier, Washington District, Boone County; authority, Ex-Gov. W. A. MacCorkle; elevation 1042' L.

				Thickness Ft.	In.	Total Ft.	In.	
Kanawha Series (228′ 8″)								
Surface				14	0	14	0	
Sandstone				10	8	24	8	
Coal	0′	2″	**Williamson**	3	8	28	4	28′ 4″
Fire clay	3	0						
Coal	0	6						
Fire clay				0	9	29	1	
Shale				3	0	32	1	
Sandstone, hard, **Upper Cedar Grove**				45	9	77	10	
Shale				0	5	78	3	
Black slate				0	11	79	2	
Sandy shale				49	8	128	10	
Coal	0′	6″	**Alma**	7	6	136	4	108′
Laminated shale	0	1						
Coal	0	7						
Laminated shale	0	3						
Coal	0	3						
Shale	0	1						
Coal	0	2						
Coal and slate	0	3						
Shale	0	11						
Coal	0	5						
Shale	2	0						
Coal	0	2						
Shale	1	8						
Coal and bone	0	2						
Slate				0	5	136	9	
Shale				0	9	137	6	
Sandstone				4	9	142	3	
Sandy shale				3	0	145	3	
Fire clay				1	6	146	9	
Sandy shale				4	9	151	6	
Sandstone, **Logan**				52	0	203	6	
Sandy shale				3	6	207	0	
Black slate				12	8	219	8	
Coal	3′	5¾″	**Campbell Creek (No. 2 Gas)**	7	7 15/16	227	3 15/16	
Shale	0	0⅛						
Coal	0	0⅞						
Slate	0	0 1/16						
Coal	0	1½						
Laminated shale	0	4½						
Slate	0	0⅜						
Coal	0	7½						
Slate and coal streaks	0	2						
Coal and bone, interlaminated	0	2¾						
Slate	0	3						
Coal	0	0⅜						
Slate	0	0⅛						
Coal	2	3						
Fire clay to bottom				1	4 1/16	228	8	91′

Record of Spruce Laurel Diamond Core Test Hole No. 11 (119).

Located on Bear Hollow of Spruce Laurel Fork, on lands of Chilton, MacCorkle and Chilton, and Meany, S. 47° E. 5 miles from Clothier, Washington District, Boone County; authority, Ex-Gov. W. A. MacCorkle. Well completed, January 31, 1912; elevation, 1125' B.

	Thickness Ft.	In.	Total Ft.	In.	
Kanawha Series (316' 6")					
Surface	2	0	2	0	
Fire clay	3	0	5	0	
Shale	1	11	6	11	
Sandstone	2	0	8	11	
Shale	1	5	10	4	
Sandstone ......93' 1"					
Shale 0 4					
Sandstone 2 6					
} **Naugatuck Sandstone**	95	11	106	3	
Coal, Williamson	1	2	107	5	
Fire clay	2	6	109	11	
Shale	1	6	111	5	
Sandstone, shale streaks	7	6	118	11	
Shale	12	6	131	5	
Sandstone	36	7	168	0	
Sandy shale	51	8	219	8	
Coal ....1' 0"					
Shale 0 3					
Coal 0 5					
Coal and bone......0 2					
Coal 0 7					
Shale 2 2					
Coal 0 3					
Shale 0 6					
Coal 0 2					
Shale 1 3					
Coal 0 2					
Shale 0 2					
Coal and bone......0 3					
} **Alma**	7	4	227	0	119' 7"
Shale	2		229	6	
Fire clay	1		231	0	
Sandstone, Logan	45		276	6	
Shale, sandstone streaks	10		286	6	
Shale	7		293	6	
Black slate	13	6	307	3	
Coal ....0' 3½"					
Slate 0 1					
Coal 0 1					
Coal and bone, interlaminated..0 1½					
Coal 4 9					
Coal and bone, interlaminated..0 5					
Slate 0 2½					
Coal 2 0½					
} **Campbell Creek (No. 2 Gas)**	8	0	315	3	88' 3"
Shale to bottom	1	3	316	6	

The above core test hole is located one-half mile west of core test hole No. 4 (118), and shows a total thickness of the Campbell Creek (No. 2 Gas) Coal, 8′ 0″, compared with 7′ 7 15/16″ in hole No. 118.

Record of Spruce Laurel Diamond Core Test Hole No. 6 (120)

Located on Spruce Laurel Fork, one-half mile southeast of diamond core test hole No. 118, and S. 49° E. 5.5 miles from Clothier, Washington District, Boone County, on land of Chilton, MacCorkle and Chilton, and Meany; authority, Ex-Gov. W. A. MacCorkle; elevation, 1067′ B.

		Thickness Ft.	In.	Total Ft.	In.
Kanawha Series (453′ 9″)					
Surface		15	5	15	5
Soft shale, wash		2	7	18	0
Sandy shale		7	8	25	8
Sandstone, **Upper Cedar Grove**		26	8	52	4
Sandy shale		0	9	53	1
Sandstone		0	5	53	6
Sandy shale		0	1	54	4
Sandstone, hard		15	0	69	4
Broken sandy shale		13	4	82	8
Sandy shale		39	4	122	0
Coal 0′ 10″ Shale 0 9 **Coal** 0 10 Shale 2 0 **Coal** 0 3	**Alma**	4	8	126	8
Sandy shale		8	0	134	8
Fire clay		2	0	136	8
Sandstone, **Logan**		47	3	183	11
Sandy shale		15	11	199	10
Black slate		5	9	205	7
Coal 0′ 6″ **Coal and bone, interlaminated** 0 3 **Coal** 2 5 Shale 0 0¼ **Coal** 2 2½ Bone and coal streaks 0 2½ Slate 0 2½ **Coal** 0 2	**Campbell Creek (No. 2 Gas) Upper Bench**	5	11¾	211	6¾
Fire clay		0	3¼	211	10
Shale		1	2	213	0
Sandstone		8	6	221	6
Shale		0	8	222	2
Coal and bone, Campbell Creek (No. 2 Gas) Lower Bench		2	0	224	2
Shale		6	6	230	8
Sandstone		20	4	251	0
Black slate		1	0	252	0
Fire clay		1	4	253	4

	Thickness Ft. In.	Total Ft. In.	
Shale	6 8	260 0	
Coal	1 10	261 10	
Sandy shale	13 9	275 7	
Slate	0 6	276 1	
Coal, Powellton	3 1	279 2	
Fire clay	0 6	279 8	
Sandy shale	48 9	328 5	
Coal ..... 1' 3"; Fire clay ..... 1 6; **Coal and bone, interlaminated** ..... 1 2 } **Matewan**	3 1	332 4	120' 9¼"
Sandy shale	5 3	337 7	
Slate	0 10	338 5	
Coal, Eagle "A"	0 3	338 8	
Fire clay	1 0	339 8	
Sandy shale	4 9	344 5	
Sandstone, **Eagle**	24 3	368 8	
Coal ..... 0' 3"; Sandy shale ..... 2 6; **Coal and bone** ..... 0 6 } **Eagle**	3 3	371 11	39' 7"
Fire clay	1 0	372 11	
Sandy shale	4 10	377 9	
Sandstone, **Decota**	30 9	408 6	
Sandy shale	1 3	409 9	
Sandstone	16 2	425 11	
Sandy shale to bottom	27 10	453 9	

The Powellton Coal shows a thickness of 3' 1", while the Eagle Coal appears to be broken up with shale, fire clay and bone.

Record of Spruce Laurel Core Test Hole No. 7 (121).

Located at the mouth of Skin Poplar Branch of Spruce Laurel Fork, on land of Chilton, MacCorkle, and Chilton, and Meany, S. 42° 30' E. 5.5 miles from Clothier, Washington District, Boone County; authority, Ex-Gov. W. A. MacCorkle; elevation, 1091' L.

Kanawha Series (286' 9")	Thickness Ft. In.	Total Ft. In.	
Surface	13 0	13 0	
Sandy shale	20 9	33 9	
Sandstone, **Upper Cedar Grove**	24 9	58 6	
Sandy shale	0 6	59 0	
Sandstone	0 6	59 6	
Coal and bone, Upper Cedar Grove	0 1	59 7	59' 7"
Sandstone, hard, **Middle Cedar Grove**	16 10	76 5	
Sandy shale	47 3	123 8	

			Thickness Ft.	In.	Total Ft.	In.	
Coal	1′ 1″						
Slate	0 6						
Coal	0 9						
Coal and bone......	0 1						
Coal	0 3						
Shale	1 6	**Alma**.....	6	10	130	6	70′ 11″
Coal	0 5						
Shale	1 1						
Coal and bone......	0 2						
Shale	0 11						
Coal and bone......	0 1						
Sandstone			6	9	137	3	
Sandy shale........................			13	4	150	7	
Sandstone, **Logan**....................			33	7	184	2	
Sandy shale........................			20	3	204	5	
Black slate........................			5	4	209	9	
Coal	0′ 8¼″						
Bone	0 0¼	**Campbell**					
Coal	0 1½	**Creek**					
Bone	0 0¼	**(No. 2 Gas)**					
Coal	2 11¼	**Upper**					
Bone	0 0¾	**Bench**....	7	0	216	9	86′ 3″
Coal	1 4						
Bone	0 0¼						
Coal	1 3						
Bone	0 3						
Coal and bone....	0 3½						
Sandy shale........................			9	0½	225	9½	
Coal and bone....	0′ 3″						
Coal	0 5	**Campbell**					
Bone	0 0½	**Creek**					
Coal	0 8	**(No. 2 Gas)**					
Coal and bone....	0 1	**Lower**					
Shale	0 7	**Bench**....	2	2½	228	0	
Coal and bone.....	0 2						
Sandy shale........................			5	1	233	1	
Sandstone, **Brownstown**..............			21	2	254	3	
Slate			0	8	254	11	
Fire clay........................			0	10	255	9	
Coal and bone......	1′ 1″						
Shale	4 1	**Powellton**					
Slate	0 4	**"Rider"**...	7	4	263	1	46′ 4″
Coal	1 10						
Sandy shale........................			5	7	268	8	
Sandstone			9	5	278	1	
Sandy shale........................			2	7	280	8	
Coal and bone, **Powellton**...........			2	10	283	6	20′ 5″
Sandy shale to bottom...............			3	3	286	9	

The Campbell Creek (No. 2 Gas) Coal is divided into two benches in the above section, the Upper being the better bed, but containing some impurities. The total thickness of the coal makes it an important seam at this point.

Record of Spruce Laurel Diamond Core Test Hole No. 15 (122).

Located on Skin Poplar Branch of Spruce Laurel Fork, just below the mouth of Jigly Branch, on land of Chilton, MacCorkle and Chilton, and Meany, S. 40° 30′ E. 6.3 miles from Clothier, Washington District, Boone County; authority, Ex-Gov. W. A. MacCorkle; elevation, 1160′ B.

			Thickness		Total		
Kanawha Series (230′ 9″)			Ft.	In.	Ft.	In.	
Surface			17	0	17	0	
Sandstone, **Williamson**			9	9	26	9	
Coal	0′ 10″						
Sandy shale	4 8	**Williamson**	5	10	32	7	32′ 7″
Coal	0 4						
Shale			10	5	43	0	
Sandstone, **Upper Cedar Grove**			55	2	98	2	
Sandy shale			46	9	144	11	
Coal	1′ 0″						
Shale	1 6						
Coal	0 4						
Shale	0 2	**Alma**	6	10	151	9	119′ 2″
Coal	1 0						
Shale	0 10						
Slate	1 8						
Coal	0 4						
Shale			1	0	152	9	
Fire clay			1	0	153	9	
Sandstone			4	6	158	3	
Shale and sandstone streaks			55	6	213	9	
Sandstone			2	0	215	9	
Shale			3	0	218	9	
Black slate			4	0	222	9	
Coal	1′ 1″						
Slate	0 1½						
Coal and bone, interlaminated	0 2⅛						
Coal	0 8½	**Campbell Creek (No. 2 Gas)**	6	8	229	5	77′ 8″
Sulphur, coal streaks	0 0½						
Coal	3 6						
Shale, **coal** streaks	1 0						
Shale to bottom			1	4	230	9	

The above section shows the Alma and the Campbell Creek (No. 2 Gas) Coals to have commercial thickness. The impurities in the Campbell Creek (No. 2 Gas) bed are thin and will not lessen its value materially.

Record of Cassingham Diamond Core Test Hole No. 3 (123)

Located on Right Fork of Jigly Branch of Skin Poplar Branch of Spruce Laurel, on land of Chilton, MacCorkle and Chilton, and Cassingham, S. 44° E. 7 miles from Clothier, Washington District, Boone County; authority, Crawford and Ashby; elevation, 1410′ B.

	Thickness Ft.	Thickness In.	Total Ft.	Total In.
Kanawha Series (383′ 6″)				
Surface	3	0	3	0
Sandstone, broken	12	8	15	8
Sandstone, hard	20	7	36	3
Sandstone, dark	3	2	39	5
Sandstone, coal streaks	8	8	48	1
Sandy shale	0	5	48	6
Sandstone	34	2	82	8
Sandstone, coal streaks	10	6	93	2
Sandstone	6	3	99	5
Sandy shale, sandstone streaks	1	9	101	2
Sandstone	6	10	108	0
Sandstone, coal streaks	4	3	112	3
Shale, dark	1	5	113	8
Coal 0′ 8″; **Coal** and bone, interlaminated 0 6	1	2	114	10
Sandy shale, limestone nodules	4	9	119	7
Sandstone	30	4	149	11
Sandstone, coal streaks	13	8	163	7
Sandstone, shale, and coal streaks	8	5	172	0
Sandy shale	20	2	192	2
Sandy shale, **marine fossils, Upper Dingess**	0	2	192	4
Sandy shale	11	2	203	6
Sandy shale, dark, **marine fossils**	10	0	213	6
Sandy shale and impure limestone, **Dingess fossils**	0	7	214	1
Sandstone, coal streaks	0	2	214	3
Coal	0	6	214	9
Sandy shale	10	0	224	9
Williamson: Bone 0′ 2″; **Coal** 0 7; **Coal** and bone 0 3; **Coal** 0 6; Sandy shale 7 4; **Coal** and bone 0 6; Shale 0 3; **Coal** 0 1; Shale 0 3; **Coal** 0 1	10	0	234	9
Sandy shale	1	5	236	2
Sandy shale, sandstone streaks	25	11	262	1
Sandstone, shale streaks	16	0	278	1
Sandy shale, sandstone streaks	28	2	306	3
Black slate	3	5	309	8

PLATE XI.—View at mouth of Laurel Creek, looking west, at Seth, Kanawha Series.

		Thickness Ft. In.		Total Ft. In.	
Coal and bone, interlaminated ..0′ 2½″ Coal ....0 11 Shale ....0 3 Coal and bone, interlaminated ..0 4½ Coal ....1 4 Coal, hard..........0 8 Coal ....0 5 Sandstone ....0 1½ Coal ....2 4 Slate ....0 6 Coal and bone......0 1½ Slate ....0 5 Coal and bone, interlaminated ..0 5½	Cedar Grove....	8	1½	317	9½
Sandstone, shale streaks.............		7	7½	325	5
Sandstone, coal streaks...............		0	4	325	9
Coal and bone, interlaminated.......		0	1	325	10
Sandstone		12	7	328	5
Sandy shale and sandstone streaks...		4	6	342	11
Coal, cannel..........0′ 3″ Coal ....1 3 Coal and bone, interlaminated ...0 4 Shale ....3 4 Coal 0 2	Alma.....	5	4	348	3
Fire clay.............................		1	5	349	8
Shale		8	5	358	1
Coal and bone, interlaminated........		0	4	358	5
Fire clay.............................		1	7	360	0
Sandy shale..........................		11	0	371	0
Coal and bone, interlaminated........		0	8	371	8
Shale, dark..........................		2	0	373	8
Coal and bone....0′ 5 ″ Shale, black......1 6 Shale ....2 6½ Bone ....0 0⅛ Coal ....1 2½ Bone ....0 1½ Coal ....1 4½ Shale ....0 10 Coal 0 1	Little Alma.....	8	1½	381	9½
Fire clay to bottom..................		1	8½	383	6

The foregoing core test hole stopped about 65 feet above the Campbell Creek (No. 2 Gas) Coal horizon.

Record of Cassingham Diamond Core Test Hole No. 4 (155).

Located on Skin Poplar Branch of Spruce Laurel Fork, 5.8 miles southeast from Clothier, on land of Chilton, MacCorkle, Chilton, and Cassingham; Washington District, Boone County; authority, Crawford and Ashby, Charleston, W. Va.; elevation, 1245′ B.

Kanawha Series (253' 3")		Thickness Ft.	Thickness In.	Total Ft.	Total In.
Surface		15	0	15	0
Shale and wash		6	0	21	0
Sandstone		24	2	45	2
Coal, Williamson		0	2	45	4
Fire clay		2	8	48	0
Shale		7	3	55	3
Sandstone, shale streaks		4	0	59	3
Sandstone, **coal** streaks		10	0	69	3
Sandstone		28	6	97	9
Sandstone, **coal** streaks		13	4	111	1
Sandstone		7	0	118	1
Sandy shale		13	4	131	5
Sandy shale, dark		0	2	131	7
Sandy shale		23	11	155	6
Coal		0	6	156	0
Sandy shale		6	9	162	9
Bone ....0' 2"	**Alma Cedar Grove? (I. C. W.)**	1	5	164	2
Coal 0 4					
Coal and bone, interlaminated 0 3					
Slate 0 3					
Coal 0 5					
Sandstone, shale streaks		7	7	171	9
Coal ....0' 2"	**Little Alma**	0	7	172	4
Coal and bone, interlaminated 0 3					
Coal 0 2					
Fire clay		1	4	173	8
Shale, sandstone streaks		7	11	181	7
Coal and bone		0	7	182	2
Fire clay		1	5	183	7
Sandy shale		8	0	191	7
Sandstone, shale streaks		46	5	238	0
Slate, black		2	8	240	8
Coal ....0' 5½"	**Campbell Creek (No. 2 Gas)**	6	6½	247	2½
Coal and bone, interlaminated 0 0½					
Coal 0 5½					
Coal and bone, interlaminated 0 1					
Coal 0 7					
Shale and **coal** 0 4½					
Coal 0 1					
Sulphur 0 0½					
Coal 4 4					
Shale		3	10½	251	1
Sandstone to bottom		2	2	253	3

The Campbell Creek (No. 2 Gas) Coal encountered at 247′ 2½″ shows a good core and is of commercial value.

Record of Spruce Laurel Diamond Core Test Hole No. 18 (124).

Located on Spruce Laurel Fork at the mouth of Trough Fork, S. 37° 30′ E. 5.9 miles from Clothier, Washington District, Boone County, on land of Chilton, MacCorkle, Chilton, and Meany; authority, Ex-Gov. W. A. MacCorkle, elevation, 1123′ L.

			Thickness Ft.	Thickness In.	Total Ft.	Total In.	
Kanawha Series (310′ 9″)							
Surface			12	9	12	9	
Hard, broken sandstone, Williamson			30	3	43	0	
Fire clay, Williamson Coal horizon			0	8	43	8	43′ 8″
Sandy shale			3	11	47	7	
Sandstone	9′ 6″	Upper Cedar Grove Sandstone	69	2	116	9	
Sandy shale	0 4						
Sandstone, hard	22 6						
Sandy shale	1 8						
Sandstone, hard	35 2						
Sandy shale			40	4	157	1	
Coal and bone	0′ 4″	Alma	4	8	161	9	118′ 1″
Shale	0 3						
Fire clay	0 4						
Coal and bone	0 4						
Coal	0 7						
Coal and bone	0 3						
Fire clay	2 3						
Coal	0 4						
Fire clay			1	3	163	0	
Sandy shale			5	1	168	1	
Sandstone			6	0	174	1	
Sandy shale			9	5	183	6	
Sandy shale, sandstone streaks			19	0	202	6	
Sandstone			13	6	216	0	
Sandy shale			17	8	233	8	
Coal			0	1	233	9	
Black slate			5	2	238	11	
Coal and bone, interlaminated	0′ 6″	Campbell Creek (No. 2 Gas)	6	6	245	5	83′ 8″
Coal	0 7½						
Shale	0 3						
Coal	1 2						
Coal and bone	0 4						
Coal	0 6						
Bone	0 1½						
Coal	0 3½						
Bone	0 0½						
Coal	0 7						
Sulphur	0 1						
Coal	1 0						
Shale	0 1½						
Coal and bone	0 2						
Coal	0 0¾						
Shale	0 3						
Bone and coal	0 5¾						

	Thickness Ft.	In.	Total Ft.	In.	
Sandstone	15	11	261	4	
Shale	4	8	266	0	
Coal, Powellton "A"	1	2	267	2	21' 9"
Fire clay	1	2	268	4	
Shale	0	6	268	10	
Sandstone	18	3	287	1	
Shale	1	0	288	1	
Fire clay	1	0	289	1	
Sandy shale	2	0	291	1	
Sandstone	8	10	299	11	
Sandy shale	1	3	301	2	
Shale, coal streaks	0	4	301	6	
Coal, Powellton	1	3	302	9	35' 7"
Fire clay	1	8	304	5	
Sandy shale to bottom	6	4	310	9	

The above section shows two seams of coal of commercial thickness and purity—the Alma and the Campbell Creek (No. 2 Gas) Coals. The impurities shown in the Campbell Creek (No. 2 Gas) bed can be easily eliminated from the coal when mined.

Record of Spruce Laurel Diamond Core Test Hole No. 13 (125).

Located on Trough Fork of Spruce Laurel Fork, about one-half mile from its mouth, on land of Chilton, MacCorkle and Chilton, and Meany, S. 32° E. 5.8 miles from Clothier, Washington District, Boone County; authority, Ex-Gov. W. A. MacCorkle; elevation, 1245' B.

	Thickness Ft.	In.	Total Ft.	In	
Kanawha Series(355')					
Surface	9	6	9	6	
Shale	9	7	19	1	
Sandstone, **Naugatuck**	35	8	54	9	
Shale	1	6	56	3	
Sandstone, **Williamson**	65	3	121	6	
Bone	0	1	121	7	
Sandy shale	8	5	130	0	
Fire clay	2	0	132	0	
Sandy shale	13	2	145	2	
Coal, Williamson	1	0	146	2	146' 2"
Fire clay	0	11	147	1	
Sandy shale	6	0	153	1	
Sandstone, **Upper Cedar Grove**	76	9	229	10	
Sandy shale	40	8	270	6	

		Thickness Ft. In.	Total Ft. In.	
Coal ....1' 0" Shale ....0 6 Coal ....1 0 Coal and bone......0 4 Shale ....2 7 Coal ....0 8 Shale ....1 8 Coal 0 2	Alma....	7 11	278 5	132' 3"
Shale		0 4	278 9	
Sandstone		3 10	282 7	
Sandy shale..............................		5 4	287 11	
Sandstone		4 7	292 6	
Shale		5 4	297 10	
Sandstone, Malden..............................		32 2	330 0	
Shale, sandstone streaks..............................		9 9	339 9	
Shale		1 1	340 10	
Black slate..............................		5 6	346 4	
Coal ....1' 0 " Fire clay.........0 1 Coal ....2 6½ Bone ....0 0¼ Coal 2 1½	Campbell Creek (No. 2 Gas)	5 9¼	352 1¼	73' 8¼"
Fire clay..............................		0 3¾	352 5	
Shale to bottom..............................		2 7	355 0	

The Alma and the Campbell Creek (No. 2 Gas) Coal beds both appear to be of commercial thickness and purity in the above section.

Record of Spruce Laurel Diamond Core Test Hole No. 12 (126).

Located on Spruce Laurel Fork, S. 34° E. 6.3 miles from Clothier, Washington District, Boone County, on land of Chilton, MacCorkle and Chilton, and Meany; authority, Ex-Gov. W. A. MacCorkle; elevation, 1190' B.

	Thickness Ft. In.	Total Ft. In.
Kanawha Series (225')		
Surface	7 0	7 0
Sandstone, Williamson..............................	9 9	16 9
Sandy shale..............................	4 0	20 9
Coal, Williamson..............................	1 0	21 9
Fire clay..............................	0 6	22 3
Sandstone, Upper Cedar Grove..............................	50 4	72 7
Sandstone coal streaks, Upper Cedar Grove	1 6	74 1
Sandstone	20 2	94 3
Sandy shale..............................	31 0	125 3
Coal	0 6	125 9
Fire clay..............................	1 5	127 2
Sandy shale..............................	11 7	138 9

		Thickness Ft. In.	Total Ft. In.
Coal ....0′ 10″ Shale ....1 1 **Coal** and bone.....0 8	Alma.....	2 7	141 4
Shale		2 2	143 6
Sandstone, **Logan**....................		13 4	156 10
Sandy shale, sandstone streaks.......		23 0	179 10
Sandstone, **Malden**..................		28 7	208 5
Sandy shale.........................		5 1	213 6
Black slate.........................		3 0	216 6
Coal ....0′ 4 ″ Bone ....0 0½ **Coal** ....1 0 Shale ....0 0¼ **Coal** ....2 5½ Bone ....0 0¼ **Coal** 1 9½	**Campbell Creek (No. 2 Gas)**	5 8	222 2
Shale		2 10	225 0

The Campbell Creek (No. 2 Gas) Coal appears of sufficient thickness and purity for mining purposes, but the Alma bed appears to have thinned and to be of little value at this point.

Record of Spruce Laurel Diamond Core Test Hole No. 14 (127).

Located on Spruce Laurel Fork, on land of Chilton, MacCorkle and Chilton, and Meany, S. 31° 30′ E. 6.9 miles from Clothier, Washington District, Boone County; authority, Ex-Gov. W. A. MacCorkle; elevation, 1225′ B.

		Thickness Ft. In.	Total Ft. In.	
Kanawha Series (245′)				
Surface		8 0	8 0	
Sandstone, **Williamson**		23 9	31 9	
Coal ....0′ 6″ Fire clay...........1 4 Shale ....5 6 **Coal** 1 0	**Williamson**	8 4	40 1	40′ 1″
Fire clay...........................		1 2	41 3	
Shale		2 0	43 3	
Sandstone, **Upper Cedar Grove**.......		47 4	90 7	
Shale		1 0	91 7	
Sandstone		12 2	103 9	
Sandy shale.........................		47 1	150 10	
Coal ....0′ 10″ Shale ....2 2 **Coal** 1 0	Alma.....	4 0	154 10	114′ 9″
Fire clay...........................		1 6	156 4	
Shale		9 0	165 4	
Sandstone, **Logan**....................		14 4	179 8	
Sandstone, shale streaks, **Malden**.....		38 4	218 0	

			Thickness Ft. In.	Total Ft. In.	
Sandy shale			8 0	226 0	
Sandstone			3 6	229 6	
Sandy shale			5 0	234 6	
Black slate			1 10	236 4	
Coal	0' 1"	Campbell Creek (No. 2 Gas)	5 10	242 2	87' 4"
Sulphur	0 1½				
Bone	0 1				
Coal	0 5½				
Sulphur	0 0½				
Coal	0 3½				
Shale	0 2				
Coal	2 4½				
Shale	0 0½				
Coal	2 2				
Shale, coal streaks			1 6	243 8	
Shale to bottom			1 4	245 0	

The above section shows the Campbell Creek (No. 2 Gas) Coal to be of sufficient thickness and purity to make it of commercial value, while the Alma bed has thinned so as to be of but little value.

Record of Spruce Laurel Diamond Core Test Hole No. 15 (128).

Located on Spruce Laurel Fork on land of Chilton, MacCorkle and Chilton, and Meany, S. 39° E. 7.7 miles from Clothier; authority, Ex-Gov. W. A. MacCorkle; elevation, 1233' B.

			Thickness Ft. In.	Total Ft. In.	
Kanawha Series (201' 9")					
Boulders and sand			9 11	9 11	
Sandstone, **Upper Cedar Grove**			44 1	54 0	
Shale			42 10	96 10	
Coal, Lower Cedar Grove			0 4	97 2	97' 2"
Fire clay, sandstone and shale			18 8	115 10	
Coal	0' 11"	Alma	4 6	120 4	23' 2"
Shale	0 5				
Fire clay	1 2				
Coal	0 10				
Shale	0 6				
Coal	0 3				
Shale	0 3				
Coal	0 2				
Fire clay and sandstone			6 11	127 3	
Shale and sandstone			7 9	135 0	
Shale			9 0	144 0	
Sandstone, streaks of shale			46 7	190 7	
Shale, black			2 7	193 2	
Coal, Campbell Creek (No. 2 Gas)			6 9	199 11	
Shale, sandy			1 10	201 9	

Record of Boone County Coal Corporation Diamond Core Test Hole No. 3 (129).

Located on the Boone-Logan County Line, 2.9 miles north of Stow P. O., and S. 26° E. 7.6 miles from Clothier, Washington District, Boone County; authority J. C. Blair, Vice President, Boone County Coal Corporation; elevation, 1920′ B.

			Thickness Ft.	In.	Total Ft.	In.
Pennsylvanian (540′)						
Allegheny Series (30′ 6″)						
Surface			5	0	5	0
Sandstone, reddish			22	0	27	0
Coal and bone			2	6	29	6
Fire clay			1	0	30	6
Kanawha Series (509′ 6″)						
Sandstone, **Homewood**			71	3	101	9
Shale			0	4	102	1
Coal and bone, **Stockton**			1	6	103	7
Fire clay			0	8	104	3
Sandstone			34	7	138	10
Shale			0	1	138	11
Coal			0	2	139	1
Fire clay			1	6	140	7
Sandstone			12	0	152	7
Shale			2	9	155	4
Sandstone			50	11	206	3
Shale			1	0	207	3
Sandstone			0	9	208	0
Shale with **coal** streaks, **Coalburg**			2	3	210	3
Sandstone			48	9	259	0
Shale			0	3	260	0
Sandstone			13	10	273	1
Coal	0′ 1″	**Little Coalburg**				
Shale	1 1					
Coal	0 7		11	11	285	0
Shale	9 6					
Coal	0 8					
Fire clay			1	0	286	0
Sandstone			54	8	340	8
Coal	0′ 3″	**Buffalo Creek**				
Sandstone	0 8		1	1	341	9
Coal	0 2					
Fire clay			2	0	343	9
Sandstone			37	9	381	6
Sandstone with **coal** streaks	2′ 3″	**Winifrede**				
Slate	0 10		3	5	384	11
Coal	0 4					
Shale, sandy			3	6	388	5
Slate			14	1	402	6
Sandstone			3	6	406	0
Shale			2	10	408	10
Coal	0′ 3″		0	8	409	6
Coal and bone	0 5					
Sandstone			21	2	430	8
Coal and bone			0	10	431	6

PLATE XII.—Old fashioned water-power mill on head of Pond Fork.

		Thickness Ft. In.	Total Ft. In.
Shale, sandy		1 8	433 2
Sandstone		33 4	466 6
Coal 0' 2"	Chilton "A"	7 9	474 3
Shale 3 6			
Coal 0 10			
Fire clay 0 6			
Shale 1 9			
Coal 1 0			
Shale		3 5	477 8
Sandstone		1 5	479 1
Shale		4 9	483 10
Fire clay		5 2	489 0
Shale, sandy		15 10	504 10
Fire clay		1 6	506 4
Shale, sandy		13 3	519 7
Sandstone		4 3	523 10
Coal, local		0 3	524 1
Shale		3 0	527 1
Sandstone to bottom		12 11	540 0

It is regrettable that the foregoing diamond core test hole was not completed through to the Chilton bed of coal at this point, so as to have given a complete section from the top of the Kanawha Series to that bed.

Record of Cassingham Diamond Core Test Hole No. 2 (130).

Located on Spruce Laurel Fork, about ⅓ mile below mouth of Dennison Fork, S. 33° E. 9 miles from Clothier, on land of Chilton, MacCorkle and Chilton, and Cassingham; authority, Crawford and Ashby; elevation, 1400' B.

		Thickness Ft. In.	Total Ft. In.	
Kanawha Series (339' 3")				
Surface		12 0	12 0	
Sandstone		7 0	19 0	
Sandy shale, dark		38 8	57 8	
Dark limy shale, **Dingess, marine fossils**		0 3	57 11	
Sandy shale		3 4	61 3	
Coal, Upper Williamson		0 10	62 1	62' 1"
Fire clay		2 4	64 5	
Sandy shale		2 1	66 6	
Shale, sandstone streaks		7 4	73 10	
Coal and bone, interlaminated 0' 4"	**Williamson**	3 5	77 3	19' 4"
Shale, sandstone streaks 1 0				
Black slate 1 3				
Coal and bone, interlaminated 0 10				
Fire clay		1 0	78 3	
Sandstone, shale streaks		10 6	88 9	

	Thickness Ft.	In.	Total Ft.	In.	
Sandy shale, sandstone streaks	22	1	110	10	
Sandstone	12	11	123	9	
Sandstone, shale streaks	10	4	134	1	
Coal, Cedar Grove, Upper Bench	1	10	135	11	58′ 8″
Shale	2	0	137	11	
Shale, sandstone streaks	6	6	144	5	
Sandy shale	3	8	148	1	
Dark shale	0	4½	148	5½	
Coal 1′ 0″; **Coal** and bone, interlaminated 0 3; **Coal** 0 5; Slate 0 0¼; **Coal** 0 4½; Sulphur 0 0¼; **Coal** 0 10½ — **Cedar Grove, Lower Bench**	2	11½	151	5	15′ 7″
Fire clay	0	2	151	7	
Sandy shale	1	5	153	0	
Sandstone	30	6	183	6	
Coal 1′ 5″; **Coal** and bone, interlaminated 0 2; Dark shale 3 4; **Coal** and bone 0 3 — **Alma?**	5	2	188	8	37′ 8″
Fire clay	1	6	190	2	
Sandy shale	2	8	192	10	
Sandstone	3	5	196	3	
Sandy shale	1	2	197	5	
Sandy shale, sandstone streaks	11	2	208	7	
Sandstone, **coal** streaks	0	7	209	2	
Shale	0	4	209	6	
Sandstone, **coal** streaks	2	2	211	8	
Dark shale	0	2	211	10	
Sandstone, **coal** streaks	10	0	221	10	
Sandstone, reddish cast	0	10	222	8	
Sandstone	9	6	232	2	
Sandstone, **coal** streaks	29	11	262	1	
Shale	0	10	262	11	
Sandstone	32	6	295	5	
Sandstone, **coal** streaks	1	9	297	2	
Sandstone	4	3	301	5	
Shale, sandstone streaks	1	4	302	9	
Sandstone, **coal** streaks	2	0	304	9	
Shale	0	6	305	3	
Sandstone, shale streaks, very pronounced shale	1	8	306	11	
Sandstone, shale and **coal** streaks	17	8	324	7	
Dark shale	3	3	327	10	
Coal and bone, interlaminated 0′ 3″; **Coal** 1 2 — **Campbell Creek (No. 2 Gas) Lower Bench**	1	5	329	3	
Sandy shale to bottom	10	0	339	3	

The Upper Bench of the Campbell Creek (No. 2 Gas) Coal appears to be entirely absent from the foregoing section.

The overlying sandstones carry coal streaks, which evidently indicate that considerable erosion has occurred in the strata at this point.

Record of Cassingham Diamond Core Test Hole No. 1 (131).

Located on Spruce Laurel Fork, on land owned by Chilton, MacCorkle and Chilton, and Cassingham, 3¼ miles southwest of Bald Knob P. O., and S. 39° E. 10 miles from Clothier, Washington District, Boone County; drilling began December, 1913, completed January, 1914; authority, Crawford and Ashby; well begins 90 feet under Winifrede Coal; elevation 1570′ B.

		Thickness Ft.	In.	Total Ft.	In.	
Kanawha Series (642′ 3″)						
Surface		6	0	6	0	
Sandstone		37	1	43	1	
Shale		8	10	51	11	
Coal ... 1′ 8″; Shale ... 1 2; Slate ... 0 6; **Coal** and bone, interlaminated ... 0 9	**Chilton**	4	1	56	0	56′ 0″
Fire clay		1	0	57	0	
Sandstone		37	3	94	3	
Sandy shale		45	6	139	9	
Coal ... 0′ 1″; Sandy shale ... 1 10; **Coal** ... 0 10	**Hernshaw Upper Bench**	2	9	142	6	86′ 6″
Fire clay		2	6	145	0	
Sandy shale with sandstone streaks		9	3	154	3	
Black slate		0	10	155	1	
Coal and bone, interlaminated ... 0′ 3″; **Coal** ... 0 4; **Coal** and bone, interlaminated ... 0 5; **Coal** ... 0 2; **Coal** and bone, interlaminated ... 0 3	**Hernshaw Lower Bench**	1	5	156	6	
Fire clay		1	5	157	11	
Shale and sandstone streaks		31	5	189	4	
Sandstone, shale streaks		14	4	203	8	
Sandy shale, **marine fossils, Dingess**		1	0	204	8	
Sandy shale		6	2	210	10	
Shale, black		1	4	212	2	
Coal, Williamson		2	9½	214	11½	
Sandy shale		13	6	228	5½	
Coal, impure		0	10½	229	4	
Fire clay		1	10	231	2	
Sandstone		27	11	259	1	
Sandstone, **coal** streaks		1	0	260	1	
Coal, Upper Cedar Grove		1	11	262	0	
Fire clay		1	5	263	5	
Sandstone		5	0	268	5	
Sandy shale, dark		3	1	271	6	
Coal		0	4	271	10	

	Thickness Ft.	Thickness In.	Total Ft.	Total In.
Shale, dark	0	5	272	3
Shale	0	10	273	1
Sandstone	12	11	286	0
Coal and bone, Lower Cedar Grove	0	2	286	2
Sandstone	22	0	308	2
Sandstone, **coal** streaks	27	4	335	6
Coal	0	3	335	9
Slate, dark	0	9	336	6
Fire clay	0	8	337	2
Sandy shale	12	2	349	4
Coal ... 3′ 1″; **Coal and bone, interlaminated** ... 0 3 } **Alma**	3	4	352	8
Sandy shale	1	10	354	6
Sandstone, dark	1	6	356	0
Sandstone with shales	8	0	364	0
Sandstone	18	6	382	6
Sandstone, **coal** streaks	21	0	403	6
Sandstone	8	10	412	4
Slate, **coal** streaks	1	0	413	4
Coal ... 1′ 1″; Bone and **coal** ... 0 1; Bone ... 0 6; **Coal** and bone, interlaminated ... 1 0 } **Little Alma**	2	8	416	0
Fire clay ... 1′ 2″; Sandstone ... 8 1; Sandy shale ... 0 4 }	9	7	425	7
Coal, Campbell Creek, "Rider"	1	7	427	2
Sandy shale with **coal** streaks	8	0	435	2
Sandstone	6	1	441	3
Sandstone with lime	7	4	448	7
Slate	0	5	449	0
Bone ... 0′ 1″; **Coal** ... 1 0; Fire clay ... 1 0; Shale, dark and sandy ... 2 2; **Coal** and bone, interlaminated ... 1 0 } **Campbell Creek (No. 2 Gas)**	5	3	454	3
Sandstone	2	0	456	3
Sandstone, hard, **Brownstown**	58	5	514	8
Shale	0	3	514	11
Sandstone, hard	12	2	527	1
Sandy shale	9	11	537	0
Sandstone with shale streaks	44	0	581	0
Sandy shale with sandstone streaks	10	0	591	0
Sandstone with shale streaks	11	6	602	6
Sandstone, shale streaks	7	7	610	1
Sandstone, hard	0	4	610	5
Dark shale	0	2	610	7
Sandy shale, sandstone streaks	9	6	620	1
Sandy shale	0	7	620	8
Sandstone, coarse	8	0	628	8
Coarse conglomerate sandstone, **Decota**, to bottom	13	7	642	3

The Alma and the Campbell Creek (No. 2 Gas) Coal beds appear to be thin and the latter contains considerable impurities. It will be noted that the sandstone over the lower bed contains streaks of coal, showing that there evidently was erosion in the strata at this point during Pennsylvanian time.

Sherman District.

Sherman District lies east of Peytona, Scott and Crook Districts, and just south of the Warfield Anticline. The Coalburg Syncline crosses in a northeast and southwest direction through the entire district, north of the center of the same.

The following aneroid section was measured by Krebs, descending southwestward to Tony Creek from a high point in the Boone-Kanawha County Line, 3.8 miles north of Comfort P. O., Sherman District:

Section 3.8 Miles North of Comfort P. O., Sherman District.

	Thickness Feet.	Total Feet.	
Pennsylvanian (810')			
Allegheny Series (55')			
Sandstone and concealed, **East Lynn**, to bench	55	55	
Kanawha Series (755')			
Sandstone and concealed	60	115	
Sandstone, massive, coarse grained	60	175	
Coal blossom, Stockton, (1432' L.)	5	180	
Sandy shale	43	223	
Coal blossom	2	225	
Sandstone and concealed	35	260	
Sandstone, massive	5	265	
Coalburg: **Coal blossom** 2' 6"; Shale, dark 0 4; **Coal**, block 0 8	3.5	268.5	86' 6"
Slate	1.5	270	
Sandstone and concealed	32	302	
Coal blossom, Buffalo Creek	2	304	
Sandstone, massive	22	326	
Sandy shale	40	366	
Bench, **Winifrede Coal horizon**	0	366	97' 6"
Sandstone and concealed	110	476	
Bench, **Chilton Coal horizon**	0	476	
Sandstone and concealed	60	536	
Coal blossom, Hernshaw	2	538	
Sandstone and concealed	108	646	
Sandy shale	40	686	
Coal blossom, Alma	1	687	
Sandy shale	60	747	
Sandstone, massive	20	767	

	Thickness Feet.	Total Feet.
Coal blossom, Campbell Creek (No. 2 Gas), Upper Bench	4	771
Sandstone	30	801
Coal blossom, Campbell Creek (No. 2 Gas), Lower Bench	2	803
Concealed to creek	7	810

The foregoing section shows the interval between the base of the Stockton Coal and the Campbell Creek (No. 2 Gas), Upper Bench, to be 591 feet, and that of the Lower Bench of the Campbell Creek (No. 2 Gas) Coal, 623 feet, compared with 579 feet in the Hernshaw Section, 4 miles north of this section, showing that the thickening occurs in the Lower Bench of the Campbell Creek (No. 2 Gas) Coal.

The following aneroid section was measured by Krebs, descending from the summit of a hill located on the Boone-Kanawha County Line, southward to Trace Fork of Joes Creek, 1.8 miles northeast from Orange P. O., Sherman District, and combined with the Winifrede well No. 1 (56), drilled by the Carter Oil Company:

Section 1.8 Miles Northeast of Orange P. O.

	Thickness Feet.	Total Feet.
Pennsylvanian (2127')		
Allegheny Series (205')		
Sandstone	18	18
Fire clay	2	20
Sandy shale and sandstone	130	150
Fire clay and slate	2	152
Sandy shale and sandstone	51	203
Fire clay, **No. 5 Block Coal horizon**	2	205
Kanawha Series (1175')		
Sandstone, **Homewood**, massive	88	293
Coal blossom, Stockton	2	295
Sandstone and concealed	65	360
Coal blossom, Coalburg	2	362
Sandstone and concealed	88	450
Coal, Winifrede	5	455
Sandstone and concealed to top of Winifrede well No. 1 (56), elevation 900' B	145	600
And thence with well:		
Soil	30	630
Sandstone	50	680
Coal, Cedar Grove, Upper Bench	2	682
Slate	18	700
Coal, Cedar Grove, Lower Bench	4	704
Slate and lime	146	850

	Thickness Feet.	Total Feet.
Coal, Campbell Creek (No. 2 Gas)	5	855
Lime	35	890
Sand	30	920
Slate and shells	20	940
Sand	8	948
Slate and shells	37	985
Coal, Eagle	5	990
Slate	25	1015
Sand	35	1050
Lime	45	1095
Sand	25	1120
Slate and shells	48	1168
Lime	27	1195
Sand	35	1230
Slate and shells	150	1380
Middle and Lower Pottsville Series (753')		
Sand, hard, **Nuttall**	130	1510
Sand and shells	105	1615
Sand	10	1625
Slate and shells	30	1655
Sand, 1st Salt	145	1800
Slate	36	1836
Sand and shale	16	1852
Slate	180	2032
Sand	95	2127
Mississippian (837')		
Mauch Chunk (173')		
Unrecorded	73	2200
Sand, Maxton	35	2235
Little Lime	50	2285
Unrecorded	10	2295
Pencil cave	5	2300
Greenbrier Limestone (218')		
Big Lime	218	2518
Pocono Sandstones (453')		
Slate	30	2548
Sand, **Big Injun**	22	2570
Unrecorded	15	2585
Sand, **Squaw**	20	2605
Slate and shells	360	2965
Sand, **Berea**	6	2971
Devonian (579')		
Slate and shells to bottom	579	3550

The Winifrede Coal is mined just to the south of the boring by the Winifrede Coal Company. The interval between the Campbell Creek (No. 2 Gas) Coal and top of the Big Lime is 1445 feet, showing a rapid increase to the southeast.

The following aneroid section was measured by Krebs, descending from a high knob on the Boone-Kanawha County Line, 3 miles northeast of Racine, southward to Short Creek, and joined to J. E. Toney well No. 1 (49):

Section 3 Miles Northwest of Racine, Sherman District.

	Thickness Feet.	Total Feet.
Pennsylvanian (1375')		
Kanawha Series (715')		
Sandstone and concealed	70	70
Sandstone, hard, grayish	7	77
Sandy shale and concealed	10	87
Concealed, mostly sandstone	88	175
Coal blossom, Hernshaw	5	180
Concealed, mostly sandstone	85	265
Coal blossom, Williamson	2	267
Concealed, mostly sandstone	60	327
Coal blossom, Cedar Grove	3	330
Sandstone and concealed	20	350
Coal blossom, Alma	2	352
Sandstone, massive, micaceous	60	412
Dark shale	8	420
Sandstone and concealed	4	424
Shale, gray	1	425
Coal 0' 3"; Shale, gray 1 2; **Coal**, gas 1 0; Shale, dark 0 3; **Coal**, semi-splint 1 3; **Coal**, splint 1 3 — **Campbell Creek (No. 2 Gas)**	5-2"	430-2"
Slate and concealed	4-10"	435
Sandstone, massive	21	456
Shale	1	457
Sandstone, massive	43	500
Concealed to top of **J. E. Toney Well No. 1 (49)**, elevation, 730' B	25	525
Gravel	30	555
Sand, white	130	685
Slate, black	30	715
Middle and Lower Pottsville (675')		
Sand, white	120	835
Slate, black	20	855
Sand, white	90	945
Slate, black	20	965
Sand, white, 1st Salt	205	1170
Coal, Sewell?	3	1173
Lime, black	22	1195
Sand, white, 2nd Salt	180	1375
Mississippian (1069')		
Mauch Chunk Series (360')		
Lime, black	180	1555
Red rock	10	1565
Sand, white, **Maxton**	135	1700
Little Lime, black	25	1725
Pencil cave	10	1735
Greenbrier Limestone (229')		
Lime, black 37'; Lime, white 192 — **Big Lime**	229	1964
Pocono Sandstones (480')		
Sand, red 30'; Sand, white 31 — **Big Injun**	61	2025

	Thickness Feet.	Total Feet.
Slate, black	407	2432
Sand, gray, **Berea**	12	2444
Devonian (16')		
Slate, black, to bottom	16	2460

The interval between the Campbell Creek (No. 2 Gas) Coal and top of Big Lime is 1305 feet in this well.

Near Seth, the following aneroid section was obtained by Krebs, descending northward from a high point on divide between Sandlick Creek and Laurel Creek, two miles west of Nelson P. O., to Sandlick Creek, and joined to well No. 2 (51), drilled by Lackawanna Coal & Lumber Company:

Section 2 Miles West of Nelson P. O., Sherman District.

	Thickness Feet.	Total Feet.
Pennsylvanian (2052')		
Allegheny and Kanawha Series (1147')		
Sandstone and concealed	150	150
Sandstone, coarse grained, full of coal streaks	10	160
Coal, block, hard.......3' 4" Dark shale, with **coal** streaks ....2 0 **Coal**, splint............1 0 } **Coalburg?**	6-4"	166-4"
Sandstone and concealed	90-8"	257
Sandstone	10	267
Coal, splint............2' 0" Slate ....0 4 **Coal**, visible...........1 2 } **Winifrede**	3-6"	270-6"
Sandstone and concealed, to bench	90-6"	361
Sandstone and concealed, to bench	45	406
Coal, splint, **Chilton**	1	407
Fire clay	2	409
Sandstone and concealed	13	422
Sandstone, sandy shale and concealed to top of No. 2 well (51), elevation, 875' B.	25	447
and thence with well:		
Loose rock and gravel	50	497
Slate	145	642
Coal, Alma	4	646
Slate	31	677
Shale and sandstone	130	807
Sandstone	85	892
Slate and shells	65	957
Sand and lime	42	999
Slate	148	1147
Lower Pottsville Series (905')		
Sandstone, **Nuttall**	120	1267
Coal	3	1270
Sandstone	97	1367
Coal	4	1371
Sandstone	316	1687

	Thickness Feet.	Total Feet.
Coal, Fire Creek	5	1692
Sandstone	235	1927
Slate	60	1987
Slate and shells	5	1992
Sandstone	60	2052
Mississippian (822')		
Mauch Chunk (125')		
Slate	43	2095
Little Lime	33	2128
Sand and lime	45	2173
Pencil cave	4	2177
Greenbrier Limestone (205')		
Big Lime	205	2382
Pocono Sandstones (486')		
Slate	10	2392
Sand, **Big Injun**	30	2422
Slate	75	2497
Sandy lime and shale	75	2572
Slate	295	2867
Sand, **Berea**	1	2868
Devonian (6')		
Slate	6	2874

The above section shows the interval between the Alma Coal and top of Big Lime to be 1531 feet, and that from the Big Lime to the Berea Sand, 691 feet. The coals encountered at 1267, 1371, and 1692 feet are representatives of New River Measures, but it is more than probable that the beds are not pure, being mixed with slate and fire clay.

The following aneroid section was measured by Teets, descending from a point just west of Seth eastward to Coal River, and combined with the Lackawanna Coal & Lumber Company's well No. 1 (54):

Section at Seth, Sherman District.

	Thickness Feet.	Total Feet.
Pennsylvanian (1638')		
Kanawha Series (790')		
Sandstone, massive	10	10
Coal blossom, Winifrede	2	12
Concealed	28	40
Sandstone and concealed	132	172
Coal, gray splint......0' 8" / Gray shale............0 7 / **Coal**, block, visible.....1 0 } **Little Chilton**	2-3"	174-3"
Sandstone and concealed to bench	105-9"	280
Gray shale	10	290
Limestone, Dingess, impure, marine fossils	0-6"	290-6"
Gray shale	20-6"	311

	Thickness Feet.	Total Feet.
Coal, Williamson	1-10"	312-10"
Gray shale	18-2"	331
Limestone, (Seth), sandy, impure, marine fossils	1-10"	332-10"
Concealed to top of well at 690' B	17-2"	350
Loose rock and gravel	50	400
Unrecorded	34	434
Slate	40	474
Coal, Campbell Creek (No. 2 Gas), Upper Bench	3-6"	477-6"
Slate	32-6"	510
Coal, Campbell Creek (No. 2 Gas), Lower Bench	3-6"	513-6"
Slate	216-6"	730
Shale	60	790
Middle and Lower Pottsville (858')		
Sandstone	230	1020
Lime	88	1108
Sandstone	302	1410
Sand and lime	210	1620
Sandstone	18	1638
Mississippian (532')		
Mauch Chunk (172')		
Slate	32	1670
Sand and lime	140	1810
Sand and lime, **Big Lime, Greenbrier**	275	2085
Pocono Sandstones (85')		
Sand, **Big Injun**	55	2140
Slate	20	2160
Sand, **Squaw**	7	2167
Slate	3	2170

The interval between the Lower Bench of the Campbell Creek (No. 2 Gas) Coal and the top of the Big Lime is 1296.5 feet at this point.

The section shows the Dingess Fossiliferous Limestone and another fossiliferous ore above the top of the well which may be called the **Seth Limestone** from this locality, since it would appear to be too high above the Campbell Creek Coal for the Campbell Creek Limestone, which is only 91 feet above that coal at Kayford.

The following aneroid section was measured by Teets, descending from a high point north of the mouth of Haggle Branch of Coal River, 1.5 miles northwest of Orange P. O., southward to the LaFollette-Robson et al. core test hole No. 136 near the mouth of Haggle Branch:

Section 1.5 miles Northwest of Orange P. O., Sherman District

	Thickness Feet.	Total Feet.
Pennsylvanian (990')		
Allegheny Series (75')		
Sandstone, **East Lynn**, and concealed, to bench, **No. 5 Block Coal horizon**	75	75
Kanawha Series (915' 5")		
Concealed	30	105
Sandy shale and concealed	50	155
Sandstone, **Homewood**	30	185
Sandy shale, **Stockton-Lewiston Coal horizon**	20	205
Sandstone, massive, **Coalburg**	85	290
Sandy shale and concealed	45	335
Sandstone, massive, micaceous, medium coarse grained, **Winifrede**	110	445
Sandy shale and concealed	25	470
Sandstone, medium grained, buff, micaceous	40	510
Concealed, bench, **Chilton Coal horizon**	15	525
Sandy shale	40	565
Concealed	19	584
Coal blossom, Little Chilton	1	585
Sandstone, massive	60	645
Concealed, bench	10	655
Sandstone and concealed	45	700
Coal, (Cedar Grove?)	3	703
Sandy shale and concealed	97	800
Limestone, impure, (Campbell Creek) marine fossils	0-8"	800-8"
Gray shale to top of core test, 750' B.	4-4"	805
(Core Test Hole No. 136)		
Surface	7	812
Dark shale	19	831
Coal (No. 2 Gas "Rider")	1-2"	832-2"
Fire clay	1-4"	833-6"
Sandy shale	5-6"	839
Sand rock	17	856
Coal 1' 0"; Fire clay 0 5; **Coal** 0 4; Bone 0 8; Clay 0 6; **Coal** 0 2 — **Peerless (U. C. C.)**	3-1"	859-1"
Dark shale	18-11"	878
Soft clay	4	882
Dark shale	13	895
Coal 2' 11"; Fire clay 0 9; **Coal** 2 9; Slate 0 1; **Coal** 0 8 — **Campbell Creek (No. 2 Gas)**	7-2"	902-2"
Fire clay	0-10"	903
Sand rock, to bottom	87-5"	990-5"

The interval between the No. 5 Block and the Campbell Creek (No. 2 Gas) Coals is 825 feet at this point.

The following aneroid section was measured by Krebs, descending northward from the summit of a high point, 4 miles southwest of Nelson P. O., Sherman District, to Cold Fork of Laurel Creek:

Section 4 Miles Southwest of Nelson P. O., Sherman District.

	Thickness Feet.	Total Feet.	
Pennsylvanian (955')			
Allegheny Series (226')			
Sandy shale and concealed	70	70	
Sandstone, massive	58	128	
Fire clay	2	130	
Sandstone, massive, **East Lynn**	88	218	
Slate	1	219	
Coal, block 1' 0"; Shale, gray 0 4; **Coal**, splint 3 0; Shale, gray 0 8; **Coal**, splint 1 0 } **No. 5 Block**	6	225	225'
Slate	1	226	
Kanawha Series (729')			
Sandstone, **Homewood**	117	343	
Coal blossom, Stockton	2	345	
Sandy shale and concealed	178	523	
Coal blossom, Buffalo Creek	2	525	
Shale and sandstone	43	568	
Coal blossom, Winifrede	2	570	345'
Sandstone and concealed	105	675	
Bench, **Chilton Coal horizon**	0	675	105'
Sandstone and concealed	55	730	
Bench, **Little Chilton Coal horizon**	0	730	
Sandstone and concealed	160	890	
Sandstone, massive	19	909	
Coal, splint, Cedar Grove, Upper Bench	2	911	236
Sandstone and concealed	6	917	
Coal, Cedar Grove, Upper Bench	1	918	
Sandy shale	10	928	
Coal blossom, Cedar Grove, Lower Bench	2	930	
Sandy shale and concealed to 810' L	25	955	

The following aneroid section was measured by Krebs, descending from a high summit between Robinson and Laurel Creeks, northward to Laurel Creek, 5 miles southwest of Nelson P. O., Sherman District:

Section 5 Miles Southwest of Nelson P. O., Sherman District.

	Thickness Feet.	Total Feet.
Pennsylvanian (450')		
Allegheny Series (147')		
Sandstone, massive, coarse grained, to bench	65	65
Sandstone and concealed, **East Lynn**	80	145
Coal blossom, No. 5 Block	2	147
Kanawha Series (303')		
Sandstone and concealed, **Homewood**	110	257
Large bench, **Stockton Coal horizon**	0	257
Sandstone and concealed	83	340
Bench, **Coalburg Coal horizon**	0	340
Sandstone and concealed	85	425
Gray shale	1	426
Coal, splint 2' 0" / Slate 0 8 / **Coal**, splint, visible 1 4 } **Buffalo Creek?**	4	430
Sandstone and concealed to creek, (1120' B.)	20	450

The Buffalo Creek Coal shown in the above section has been opened and mined for local fuel use.

The following aneroid section was measured by Krebs, descending southwestward to Mudlick Fork of Laurel Creek from a summit, 4 miles southwest of Nelson P. O., Sherman District:

Section 4 Miles Southwest of Nelson P. O., Sherman District.

	Thickness Feet.	Total Feet.	
Pennsylvanian (576')			
Conemaugh Series (115')			
Sandstone and concealed, **Mahoning**	115	115	
Allegheny Series (190')			
Bench, **Upper Freeport Coal horizon**	0	115	
Sandstone and concealed, to large bench	80	195	
Sandstone, massive, conglomerate, **East Lynn**	110	305	
Bench, **No. 5 Block Coal horizon**	0	305	190'
Kanawha Series (271')			
Sandstone and concealed	185	490	
Coal blossom, Coalburg	5	495	190'
Sandstone and concealed	30	525	
Concealed	50	575	
Coal blossom, Winifrede, to Mudlick Fork, elevation, 1155' B.	1	576	

The above section begins in the lower part of the Conemaugh Series and the coal found at the base of the section is undoubtedly the Winifrede bed, as the strata dip rapidly in going down **Mudlick** Fork to mouth of Stolling Fork, where the **Winifrede** was once mined at an elevation of 1085′ B.

Near the head of Lavinia Fork of Hopkins, considerable development in prospecting for the coals has been done by the Lackawanna Coal & Lumber Company, and there the following aneroid section was measured by Teets, descending from a high point eastward along the trail to Lavinia Fork of Hopkins Fork, at a point two miles up Lavinia, and three miles south from Seth:

Section 3 Miles South from Seth, Sherman District.

	Thickness Feet.	Total Feet.	
Pennsylvanian (580′)			
Allegheny Series (115′)			
Sandy shale and concealed	20	20	
Sandstone, massive, full of large pebbles	90	110	
Prospect **coal** opening, fallen shut, **No. 5 Block**, about	5	115	
Kanawha Series (465′)			
Sandstone and concealed	95	210	
Bench, concealed	10	220	
Sandstone and concealed	80	300	
Concealed bench	10	310	
Sandy shale and concealed	85	395	
Concealed bench	10	405	
Sandy shale and concealed	47	452	
Coal blossom, Winifrede	3	455	340′
Sandstone and concealed	100	555	
Coal 0′ 4″ / **Slate** 0 7 / **Coal** 2 1 } **Chilton**	3	558	
Sandstone, massive, elevation, 1060′ B	22	580	

In passing across the divide between Lavinia and Hopkins Forks, the following hand level section was measured by Teets, descending from a high summit southward to Hopkins Fork, and joined to Lackawanna Coal & Lumber Company well No. 4 (59) at the mouth of Jarrolds Fork, 1 mile northeast of the new mining town of Griffith:

Section 1 Mile Northeast of Griffith, Sherman District.

	Thickness Feet.	Total Feet.
Pennsylvanian (2181′)		
Sandstone, **Homewood**, massive, coarse grained	90	90
Coal, gray splint 3′ 11″ / **Bone** 0 1 / **Coal, gray splint** 1 10 } **Stockton-Lewiston**	5-10″	95-10″

	Thickness Feet.	Total Feet.
Sandstone, massive, and concealed, **Coalburg**	98-2″	194
Coal blossom, Coalburg	2	196
Sandstone and concealed, **Upper Winifrede.**	93	289
Coal blossom, Winifrede, prospect opening fallen shut	7	296
Sandstone, massive, **Lower Winifrede**	53	349
Coal blossom, Chilton "A"	3	352
Sandstone, massive, **Upper Chilton**	34	386
Coal blossom, Chilton	1	387
Sandstone, massive, **Lower Chilton**	74	461
Coal blossom, Hernshaw	3	464
Sandstone and concealed	128	592
Sandstone, massive	44	636
Coal, block 0′ 10″; Slate, gray 0 0½; **Coal,** gray splint 0 7½; **Coal,** block 1 7 — **Cedar Grove Upper Bench.**	3-1″	639-1″
Sandy shale, slaty	2-11″	642
Coal 1′ 0″; **Slate** 0 4; **Coal** 0 2 — **Cedar Grove, Lower Bench.**	1-6″	643-6″
Slate, gray	3-6″	647
Sandstone to top of well, 935′ L	9	656
Thence with Well No. 59:		
Gravel	10	666
Slate and shells	113	779
Coal, Campbell Creek (No. 2 Gas)	2	781
Sandstone	60	841
Slate and shells	15	856
Sandstone	50	906
Slate and lime shells	400	1306
Coal, Sewell?	3	1309
Lime	82	1391
Sand	130	1521
Sand and lime	120	1641
Shells and slate	115	1756
Sand	150	1906
Salt Sand	220	2126
Lime and sand to base of Pottsville Series	55	2181
Mississippian (1165′)		
Mauch Chunk Series (470′)		
Lime	30	2211
Red rock	25	2236
Lime and sand	25	2261
Red rock	120	2381
Sand	15	2396
Red rock	25	2421
Sand	15	2436
Lime and shells	70	2506
Red rock	30	2536
Slate and lime shells	115	2651
Greenbrier Limestone (280′)		
Big Lime	280	2931
Pocono Sandstones (415′)		
Big Injun	75	3006

PLATE XIII.—View at head of Whites Branch, looking east over Kanawha Series.

	Thickness Feet.	Total Feet.
Slate and shells	20	3026
Sand, Squaw	30	3056
Sand and shells	160	3216
Slate and shells	98	3314
Sand, **Berea**	32	3346
Devonian (160')		
Slate and shells	64	3410
Sand, Gordon?	15	3425
Sand and shells	25	3450
Sand	16	3466
Slate to bottom of boring	40	3506

The interval between the Campbell Creek (No. 2 Gas) Coal and the top of the Greenbrier Limestone is 1870 feet, showing a gradual thickening of the lower portion of the Pottsville Series to the south.

The following aneroid section was measured by Teets, descending from a point between Lots Branch and Jarrolds Fork, 5.8 miles due west of Whitesville, descending to Lots Branch:

Section 5.8 Miles Due West of Whitesville, Sherman District.

			Thickness Ft.	Thickness In.	Total Ft.	Total In.
Pennsylvanian (450')						
Allegheny Series (138')						
Sandy shale, sandstone and concealed			130	0	130	0
Coal, impure	4' 2"	**No. 5 Block**	6	3	136	3
Slate, black	0 3					
Coal, gray splint	1 0					
Slate, gray	0 4					
Coal, gray	0 6					
Slate, visible			1	9	138	0
Kanawha Series (312')						
Concealed, **Homewood Sandstone**			66	0	204	0
Coal, gray splint	1' 1"	**Stockton**	6	3	210	3
Slate, dark	0 1					
Coal, dark	3 4					
Slate, gray	0 7					
Coal, gray splint	0 8					
Coal, gas	0 6					
Slate, visible			0	9	211	0
Sandstone and concealed			159	0	370	0
Coal, gray splint	1' 10"	**Coalburg**	6	3	376	3
Bony slate, dark	0 1					
Coal, block	0 9					
Slate, dark	0 4					
Coal, impure	1 0					
Coal, semi-splint	2 3					
Slate			1	9	378	0
Concealed			57	0	435	0

		Thickness Feet.	Total Feet.
Coal, hard splint....1' 9" Niggerhead, dark...0 5 Coal, block.........1 5 Slate, gray.........0 1 Coal, block.........0 10	Winifrede	4 6	439 6
Concealed, elevation, 1255' B.........		10 6	450 0

The above section ends about 450 feet above the Campbell Creek (No. 2 Gas) Coal.

Another aneroid section was measured by Teets, descending from a point in the divide between the head of Seng Fork of Logan Fork and Hopkins Fork at the Three Forks:

Section at Three Forks of Hopkins Fork, Sherman District.

		Thickness Ft. In.	Total Ft. In.	
Kanawha Series (505')				
Sandy shale and concealed...........		25 0	25 0	
Sandstone, Homewood, massive, coarse grained		39 0	64 0	
Coal, impure........4' 6" Slate ....0 8 Coal, block, visible. 0 10	Stockton-Lewiston	6 0	70 0	70'
Concealed and sandstone.............		163 0	233 0	
Coal, weathered....2' 2" Slate, gray.........0 3 Coal ....0 3 Slate ....0 7 Coal, weathered....3 3	Coalburg.	6 6	239 6	169'
Slate, visible........................		1 6	241 0	
Sandstone and concealed............		64 0	305 0	
Coal, weathered.....2' 5" Slate, gray.........0 4 Coal, block, visible..2 8	Winifrede	5 5	310 5	71'
Sandstone and concealed to bench...		99 7	410 0	
Sandstone and concealed to bench....		50 0	460 0	
Sandstone, massive, to bottom, (1205' B.)......................		45 0	505 0	

The following section was measured by Teets descending from the divide between Little Jarrolds and Jarrolds Forks of Hopkins Fork, about 2½ miles up Jarrolds Fork, eastward to Jarrolds Fork, 5.7 miles northwest of Jarrolds Valley:

Section 5.7 Miles Northwest from Jarrolds Valley.

Kanawha Series (525')			Thickness Feet.	Total Feet.
Sandstone and concealed	40'	Homewood	95	95
Sandstone, massive	55			
Coal, impure	1' 4"	Stockton-Lewiston	4-6"	99-6"
Slate, gray	0 4			
Coal, block, visible	2 10			
Sandstone, massive, medium coarse grained			168-6"	268
Prospect **coal** opening fallen shut, about			5	273
Sandstone and concealed			52	325
Coal, weathered	1' 3 "	Winifrede	6	331
Slate, black	0 1			
Coal, block	2 1			
Slate, gray	0 7			
Coal, gray splint	1 6			
Slate	0 0½			
Coal, visible	0 5½			
Sandstone and concealed to bench			94	425
Sandstone and concealed			60	485
Sandstone, massive, to bed of stream, (1155' B.)			40	525

The following section was obtained by Teets on the east side of Hopkins Fork, at a point 1¾ miles southeast from mouth of Logan Fork of Hopkins Fork, 4½ miles northwest from Jarrolds Valley:

Section 4½ Miles Northwest of Jarrolds Valley.

Kanawha Series (570')	Thickness Feet.	Total Feet.
Sandstone, massive	75	75
Coal, impure, **Stockton**, blossom in prospect opening	5	80
Sandstone, massive and concealed	63	143
Coal blossom	2	145
Sandstone and concealed to bench	85	230
Sandstone and concealed	99	329
Coal blossom, **Winifrede**	1	330
Sandstone and concealed to bench	100	430
Sandstone and concealed to bench	60	490
Sandstone, massive, to Hopkins Fork (1130' B.)	80	570

Beyond the divide between Logan Fork of Laurel Creek, near the head of Elk Run of Coal River, the following section was measured by Teets from a high point on the divide, descending along the public road to Elk Run, 3.2 miles southwest of Jarrolds Valley:

Section 3.2 Miles Southwest of Jarrolds Valley.

	Thickness Feet.	Total Feet.
Pennsylvanian (1165')		
Allegheny Series (150')		
Sandstone and concealed	40	40
Sandstone, **East Lynn**, massive, full of large pebbles	40	80
Sandy shale and sandstone to bench	70	150
Kanawha Series (1015')		
Sandy shale and sandstone to bench	90	240
Sandy shale and sandstone to bench	80	320
Sandy shale and sandstone	50	370
Coal blossom	1	371
Sandstone, massive, to bench	54	425
Sandy shale and sandstone	55	480
Coal blossom, Coalburg	1	481
Sandy shale and concealed to bench	79	560
Sandy shale and concealed	65	625
Coal blossom, Chilton "A"	1	626
Sandy shale and concealed to bench	44	670
Sandy shale and concealed	60	730
Coal blossom in bench, Chilton	0	730
Sandy shale and concealed	60	790
Concealed to **Campbell Creek (No. 2 Gas) Coal horizon**, (1390' B.)	375	1165

North of Big Coal River, in the northern part of Sherman District, the following aneroid section was measured by Krebs, descending from a high hill on the Kanawha-Boone County Line into Spicelick Fork of Left Fork of Joes Creek, Sherman District, Boone County, 4 miles north of Orange P. O.:

Section 4 Miles North of Orange P. O., Sherman District.

	Thickness Feet.	Total Feet.
Pennsylvanian (555')		
Conemaugh-Allegheny Series (300')		
Sandy shale and concealed	100	100
Sandy shale and sandstone	100	200
Sandstone, conglomerate, **East Lynn**	96	296
Coal, No. 5 Block	4	300
Kanawha Series (255')		
Sandstone and concealed	250	550
Coal, Winifrede, (1040' B.)	5	555

The above section was measured to the rise of the strata, so that the interval between the No. 5 Block Coal and the Winifrede is about 60 feet too small.

Another section was measured by Krebs, descending from a high point on the Boone-Kanawha County Line along the road into Whiteoak, ½ mile north of Orange P. O., Sherman District:

Section ½ Mile North of Orange P. O., Sherman District.

	Thickness Feet.	Total Feet.	
Pennsylvanian (900')			
Allegheny Series (162')			
Sandy shale and concealed	44	44	
Coal blossom	1	45	
Sandy shale and concealed	30	75	
Fire clay	2	77	
Sandstone and concealed	83	160	
Fire clay and **coal blossom, No. 5 Block**	2	162	117'
Kanawha Series (738')			
Sandstone, massive, **Homewood**	128	290	
Coal blossom, Stockton	5	295	
Sandstone and concealed	50	345	
Coal blossom, Coalburg	2	347	
Sandstone and concealed	113	460	
Coal blossom, Winifrede	5	465	
Sandstone and concealed	60	525	
Coal blossom, Chilton "A"	2	527	
Sandstone and concealed	103	630	
Coal blossom, Hernshaw	1	631	
Sandstone and concealed	144	775	
Coal 1' 0" / Slate 0 2 / **Coal** 0 8 — **Cedar Grove, Lower Bench**	1-10"	776-10"	
Sandstone and concealed	18-2"	795	
Coal, Alma, Lower Bench	3	798	
Sandy shale	22	820	
Concealed	78	898	
Coal, Campbell Creek (No. 2 Gas), (720' B.)	2	900	

The base of the above section ends in the Campbell Creek (No. 2 Gas) Coal, found in Big Coal River just above the mouth of Whiteoak Creek where it was once mined for local fuel use.

The following aneroid section was measured by Teets, descending from a high point at the head of Pack Branch of Coal River, southwestward to Coal River at mouth of Pack Branch, 1.4 miles northwest of Whitesville:

Section 1.4 Miles Northwest of Whitesville, Sherman District.

		Thickness Ft.	In.	Total Ft.	In.
Kanawha Series (905')					
Sandstone and concealed		125	0	125	0
Slate, gray		0	2	125	2
Coal, splint 1' 7"	Stockton	5	5	130	7
Slate, gray 0 3					
Coal, splint 2 1					
Slate, gray 0 4					
Coal, gray splint 1 2					
Sandstone, massive		34	5	165	0
Coal blossom, Coalburg		1	0	166	0
Sandstone, massive		84	0	250	0
Coal, Winifrede, impure		2	0	252	0
Sandstone and concealed		318	0	570	0
Coal, Cedar Grove		6	0	576	0
Sandstone and concealed		149	0	725	0
Coal 0' 6"	**Campbell Creek (No. 2 Gas)**	4	9	729	9
Slate 0 5					
Coal, block, glossy 2 4					
Slate 0 4					
Coal, reported 1 2					
Sandstone, massive		42	3	772	0
Coal, Powellton		2	7	774	7
Sandstone, to bed of Coal River (760' B.)		130	5	905	0

Near the head of Seng Creek, the following aneroid section was measured by Krebs and Teets, descending from the summit of a high hill southwestward to Seng at High Coal P. O., and thence along the grade of the Cabin Creek Branch of the C. & O. Ry. to mouth of Seng Creek:

Section at High Coal, Sherman District.

		Thickness Feet.	Total Feet.
Pennsylvanian (1330')			
Allegheny Series (99')			
Sandy shale and concealed		40	40
Sandstone, massive		50	90
Slate		1	91
Coal, hard, block 6' 9"	**No. 5 Block.**	8	99
Slate 0 1			
Coal, hard, block 1 2			
Kanawha Series (1231')			
Sandstone and concealed		156	255
Slate		0-4"	255-4"
Coal, block 2' 8"	**Stockton-Lewiston**	6-2"	261-6"
Slate 1 2			
Coal, block 2 4			
Slate		1-6"	263
Sandstone, top portion massive, **Coalburg**		97	360

		Thickness Feet.	Total Feet.	
Coal, Coalburg		2-11"	362-11"	
Slate		1-1"	364	
Sandstone and concealed		157	521	
Coal, Winifrede		8	529	430'
Sandstone and concealed		90	619	
Coal blossom, Chilton		1	620	
Sandstone and concealed		135	755	
Coal blossom, Hernshaw		1	756	
Sandstone and concealed		86	842	
Limestone, impure, fossiliferous, Dingess		1	843	
Sandy shale and concealed		86	929	
Limestone, dark gray, fossiliferous, Campbell Creek		1	930	
Gray shale		44	974	
Coal 0' 1"	**Peerless**			
Slate, gray 0 1				
Coal, gray splint 1 5		3	977	
Slate 0 1				
Coal, gray splint 1 4				
Shale, gray		7	984	
Coal, gas, Peerless		1-8"	985-8"	
Fire clay		1-4"	987	
Sandstone, massive		29	1016	
Gray shale		10	1026	
Coal, gas, Campbell Creek (No. 2 Gas)		3	1029	
Sandstone, massive		31	1060	
Coal, Powellton		2	1062	
Sandstone and concealed		90	1152	
Sandstone, massive		60	1212	
Shale, dark gray		0-4"	1212-4"	
Coal, gray splint 0' 4"	**Eagle**			
Slate 0 0½				
Coal, splint, gray 0 10				
Slate 0 0½		3-3"	1215-7"	
Coal, gas 0 8				
Shale 1 0				
Coal, gas 0 4				
Shale, gray		1-5"	1217	
Sandstone, massive, **Decota**		60	1277	
Shale, gray		24	1301	
Limestone, Eagle		1	1302	
Shale, dark		8	1310	
Limestone, Eagle		1	1311	
Concealed to Coal River, 795' B		19	1330	

The foregoing section shows the succession of the various coals in the Kanawha Series at this point, and contains three of the impure, fossiliferous limestones that have been found in the Kanawha Series. The interval between the Dingess and the Eagle Limestone is 468 feet, and that between the Campbell Creek and the Eagle, 382 feet.

The following aneroid section was measured by Teets, descending from a high point at Coal River Siding, one mile north of the Boone-Raleigh County Line, westward to Coal River:

Section at Coal River Siding, Sherman District.

	Thickness Feet.	Total Feet.
Pennsylvanian (1320')		
Allegheny Series (125')		
Sandy shale and concealed	20	20
Sandstone, massive, coarse grained, **East Lynn**	90	110
Concealed bench, **No. 5 Block Coal horizon**	15	125
Kanawha Series (1195')		
Sandstone, massive, medium coarse grained	70	195
Sandy shale, **Stockton Coal horizon**	10	205
Sandstone, **Coalburg**, massive cliff rock	135	340
Sandy shale	30	370
Sandstone	30	400
Coal blossom, Coalburg	1	401
Sandstone, massive	99	500
Concealed bench	15	515
Sandstone	60	575
Winifrede: **Coal**, block 0' 5"; **Coal**, gray splint 0 5; Bone 0 1; **Coal**, gray splint 1 1; **Coal**, block 2 1; Slate 0 2; **Coal** 0 2	4-5"	579-5"
Slate	1-7"	581
Sandstone, massive	109	690
Concealed	20	710
Sandy shale	24	734
Coal blossom, Little Chilton, plant fossils	1	735
Sandstone	40	775
Sandy shale and concealed bench	35	810
Sandstone	75	885
Sandy shale, iron ore nodules	35	920
Sandstone, massive	68	988
Sandstone, massive, full of iron ore nodules	2	990
Gray shale and bench	5	995
Sandstone and concealed to bench	75	1070
Bench, **Campbell Creek (No. 2 Gas) Coal horizon**	0	1070
Sandstone and concealed to bench	90	1160
Sandstone and concealed to bench	105	1265
Sandstone and sandy shale to railroad level, (850' B.)	55	1320

The Winifrede Coal is mined at this point by the Seng Creek Coal Company.

The following aneroid section was measured by Teets,

PLATE XIV.—View at mouth of Robinson Creek, showing topography of Kanawha Series.

descending from a point just east of Clear Fork, westward to Clear Fork Creek, ½ mile south of the Boone-Raleigh County Line:

Section at Clear Fork, Marsh Fork District, Raleigh County.

		Thickness Feet.	Total Feet.
Kanawha Series (1102')			
Sandstone, **Coalburg**, massive		80	80
Concealed, bench		15	95
Sandstone, massive, medium coarse grained, micaceous		85	180
Concealed bench		10	190
Sandstone, massive		47	237
Coal 0' 3"	**Winifrede**	8-5"	245-5"
Slate, gray 0 2			
Coal, block 1 3			
Slate, gray 0 2			
Coal 0 6			
Slate, gray 0 2			
Coal, gray splint 4 1			
Slate 0 1			
Coal 0 6			
Slate, gray 0 2			
Coal, block 1 1			
Slate		1-7"	247
Sandy shale and concealed		33	280
Sandstone, massive, to bench		100	380
Sandstone and concealed to bench		75	455
Sandstone, massive		80	535
Concealed bench		10	545
Sandstone, massive, coarse grained		135	680
Bench, **Alma Coal horizon**		0	680
Sandstone and concealed		45	725
Sandy shale		8	733
Coal blossom, Campbell Creek (No. 2 Gas)		2	735
Sandstone. massive		40	775
Sandy shale		7	782
Coal 0' 8"	**Powellton**	3-7½"	785-7½"
Slate, dark 0 2			
Coal, gray splint 1 11			
Slate 0 11½			
Coal 0 11			
Slate		1-4½"	787
Sandstone and concealed		95	882
Bench, **Eagle Coal horizon**		0	882
Concealed and sandstone		65	947
Bench, **Little Eagle Coal horizon**		0	947
Concealed, mostly sandstone, to bench		95	1042
Sandstone, massive, to railroad grade, (845' B.)		60	1102

The Winifrede Coal is mined here by the Clear Fork Coal Company.

Crook District.

Crook District occupies the central and southeastern portion of Boone County, being a strip about 25 miles in length and 5.5 miles in width, running in a southeast direction from Madison to Walnut Gap on the Boone-Wyoming County Line. It is the largest District in Boone County and represents practically the entire drainage area of Pond Fork and its tributaries.

The Warfield Anticline passes through the northeastern part of the district, running in a northeast and southwest direction. The Coalburg Syncline passes through the district almost parallel to the Warfield Anticline, about two miles northwest of the mouth of West Fork.

The following aneroid section was measured by Teets, descending from the top of Workman Knob, southwestward to mouth of Price Fork of Pond Fork, 1.5 miles southwest of Madison:

Workman Knob Section, Crook District.

	Thickness Feet.	Total Feet.
Pennsylvanian (1275')		
Allegheny Series (55')		
Sandstone, massive, **East Lynn**	55	55
Bench, **No. 5 Block Coal horizon**	0	55
Kanawha Series (1220')		
Sandstone and sandy shale ..... 95' } **Homewood Sandstone** Sandstone and concealed . 60 }	155	210
Bench, **Stockton Coal horizon**	0	210
Sandstone and concealed, **Coalburg**	90	300
Bench; **Coalburg Coal horizon**	0	300
Sandstone, **Winifrede**, massive	100	400
Bench, **Winifrede Coal horizon**	0	400
Sandstone, massive	60	460
Sandstone, massive and concealed	55	515
Concealed, **Chilton bench**	5	520
Sandstone, massive	90	610
Coal blossom, Little Chilton	2	612
Sandstone, massive	73	685
Bench, **Hernshaw Coal horizon**	5	690
Sandstone and concealed	75	765
Bench, **Cedar Grove Coal horizon**	5	770
Sandstone and concealed	100	870
Bench, **Little Alma Coal horizon**	0	870
Sandstone and concealed	55	925
Bench, **Campbell Creek (No. 2 Gas) Coal horizon**	5	930
Sandstone and concealed	55	985
Sandstone and concealed to mouth of creek, (690' B.)	290	1275

The foregoing section was taken just south of the Warfield Anticline.

The following aneroid section was measured by Teets, descending from a high summit southward to Pond Fork at the mouth of Robinson Creek and joined to Arbogast well No. 1 (63), two miles northwest of Lantie P. O., Crook District:

Section 2 Miles North of Lantie P. O., Crook District.

	Thickness Feet.	Total Feet.
Pennsylvanian (2404')		
Conemaugh and Allegheny Series (250')		
Sandstone, massive, coarse grained	95	95
Sandy shale	15	110
Sandstone, **East Lynn**, massive, ferriferous	40	150
Sandy shale	10	160
Sandstone, massive, coarse grained	30	190
Sandy shale	8	198
Slate, **No. 5 Block Coal horizon**	2	200
Sandy shale	50	250
Kanawha Series (1289')		
Sandstone, massive	30	280
Sandy shale	28	308
Coal blossom, Stockton	2	310
Sandstone, massive, cliff rock	195	505
Bench, **Winifrede Coal horizon**	5	510
Sandstone, massive	25	535
Bench	5	540
Sandstone, massive, cliff rock	60	600
Sandy shale	10	610
Sandstone, massive medium coarse	69	679
Coal blossom, Chilton	1	680
Shale, gray, sandy	68	748
Shale, gray, plant fossils	1	749
Coal, Hernshaw	1	750
Sandstone, massive	60	810
Sandy shale and concealed	40	850
Sandy shale and concealed and sandstone	40	890
Limestone, Dingess, silicious, marine fossils.	1	891
Shale with coal partings	3	894
Shale, gray	6	900
Sandstone, massive, micaceous	23	923
Coal, soft 0' 6"; Fire clay 0 4; Coal, soft 1 3; Fire clay, plant fossils 1 6; Coal, soft 0 6; Fire clay, plant fossils 1 6; Coal, cannel 3 4 } **Williamson**	8-11"	931-11"
Sandstone, massive	30-1"	962
Shale, gray, full of plant fossils	10	972
Coal, gray splint 0' 2"; Coal, block, glossy 1 10; Coal, gray splint 0 8 } **Cedar Grove**	2-8"	974-8"

	Thickness Feet.	Total Feet.
Soft fire clay	1-4"	976
Concealed to top of **Arbogast well No. 1 (63),** (734' L.)	12	988
Thence continuing with well:		
Gravel	23	1013
Slate	60	1073
Sand	16	1089
Slate	10	1099
Coal, Campbell Creek (No. 2 Gas)	3	1102
Slate	12	1114
Sand	60	1174
Slate	15	1189
Lime	10	1199
Sand	90	1289
Slate	15	1304
Coal, Eagle	5	1309
Slate	145	1454
Sand	40	1494
Lime	45	1539
Middle and Lower Pottsville Series (865')		
Sand	140	1679
Slate	70	1749
Sand	30	1779
Slate	2	1781
Sand, Salt	108	1889
Coal, Sewell?	3	1892
Slate and lime	17	1909
Sandstone, **Raleigh**	275	2184
Slate	20	2204
Lime	15	2219
Sand	185	2404
Mississippian (904')		
Mauch Chunk (210')		
Red rock	35	2439
Lime shells	123	2562
Slate	7	2569
Lime	10	2579
Shells	35	2614
Greenbrier Limestone (183')		
Big Lime	183	2797
Pocono Sandstones (511')		
Red rock	22	2819
Slate	20	2839
Shells	125	2964
Black slate	75	3039
Slate and shells, brown shale	242	3281
Sand, **Berea**	27	3308
Devonian (2')		
Slate	2	3310

The foregoing section shows the thickness of the Pottsville Series at this point to be 2154 feet. The **Dingess Limestone** with marine fossils appears 211 feet above the **Campbell Creek (No. 2 Gas) Coal.**

Another aneroid section was measured from a point in the head of Jacks Branch, along the county road to a point in Jacks Branch, 2.8 miles southwest of Lantie P. O.:

Section 2.8 Miles Southwest of Lantie P. O., Crook District.

Kanawha Series (490')				Thickness Feet.	Total Feet.
Sandstone, Homewood, massive, large pebbles				25	25
Bench				0	25
Sandstone and concealed, Homewood				95	120
Coal, gas	0'	6"	Stockton-Lewiston	5-3"	125-3"
Slate, gray	0	4			
Coal, gas	1	4			
Coal, gray splint	3	1			
Concealed				0-9"	126
Sandstone and concealed				203	329
Coal, block	1'	11"	Winifrede	6	335
Slate, gray	2	0			
Coal, block	2	1			
Sandstone and sandy shale				40	375
Sandstone, massive				30	405
Coal, slaty	0'	3"	Chilton "A".	1-9"	406-9"
Coal, gas	1	6			
Gray shale with coal streaks				18-3"	425
Blue slate				1-6"	426-6"
Coal	0'	6"	Chilton	5-8"	432-2"
Slate, gray, plant fossils	1	5			
Coal, block	1	3			
Bone	0	2			
Coal, block	1	1			
Slate, black	0	3			
Coal, gray splint	1	0			
Sandstone and concealed				55-10'	488
Coal blossom (1025' B.)				2	490

The foregoing section was taken toward the greatest rise, and in order to get the true interval between the Stockton and Chilton Coals, it will be necessary to add about 190 feet to the interval given in the section, and the interval between the other coals should be apportioned on this basis.

The following aneroid section was measured by Krebs along road descending westward to Jacks Branch, three miles west of Lantie P. O., Crook District:

Section 3 Miles West of Lantie P. O., Crook District.

Kanawha Series (446′)	Thickness Feet.	Total Feet.
Sandstone, **Coalburg**, massive	50	50
Bench, **Coalburg Coal horizon**	0	50
Sandstone and concealed	45	95
Bench	10	105
Sandstone and concealed, to bench	60	165
Sandstone and concealed	40	205
Fire clay, **Winifrede Coal horizon**	2	207
Sandstone and concealed	128	335
Shale, gray	1	336
Coal, splint ... 1′ 0″ / **Coal**, hard splint ... 1 5 / **Coal**, impure ... 0 2 / **Coal**, soft ... 0 6 } **Chilton**	3-1″	339-1″
Slate	0-11″	340
Sandstone and concealed	100	440
Coal blossom, Dingess	1	441
Concealed to creek, (1003′ B.)	5	446

Bull Creek flows into Pond Fork from the southwest, three-fourths mile south of Jacks Branch, and the following aneroid section was measured by Teets, descending eastward to Bull Creek, one-fourth mile southwest of Lantie P. O.:

Section ¼ Mile Southwest of Lantie P. O., Crook District.

Kanawha Series (730′)	Thickness Feet.	Total Feet.
Sandstone, massive	75	75
Concealed bench	15	90
Sandstone, **Homewood**	60	150
Concealed bench, **Stockton Coal horizon**	15	165
Sandstone, massive, pebbly	65	230
Sandy shale and sandstone	70	300
Large bench, **Coalburg Coal horizon**	10	310
Sandstone and concealed	52	362
Coal blossom, Buffalo Creek	3	365
Sandstone and sandy shale	91	456
Coal blossom, Winifrede?	4	460
Sandstone and sandy shale	130	590
Bench	5	595
Sandy shale and concealed	45	640
Bench	5	645
Sandy shale and concealed	53	698
Coal blossom, Hernshaw	2	700
Sandstone, massive, to head of Bull Creek, (780′ B.)	30	730

The base of the foregoing section ends about 80 feet above the Dingess Limestone, so that the coal encountered at 700 feet correlates with the Hernshaw Coal.

On Pond Fork one mile above the mouth of West Fork, the following section was measured by Krebs, descending eastward to Pond Fork on the land of Betsey Polley:

Betsey Polley Section 2 Miles South of Van P. O.

				Thickness Feet.	Total Feet.	
Pennsylvanian (970')						
Allegheny Series (190')						
Sandstone, massive, pebbly				70	70	
Concealed				10	80	
Sandstone, **East Lynn**, massive				90	170	
Concealed, **No. 5 Block Coal horizon**				20	190	
Kanawha Series (780')						
Sandstone, **Homewood**, massive, grayish white				60	250	
Concealed				.70	320	
Coal	3'	6"	**Stockton-Lewiston**	5	325	
Shale	0	2				
Coal	1	4				
Concealed				70	395	
Coal blossom				0	395	
Concealed				60	455	
Coal	4'	10"	**Coalburg**	9-7"	464-7"	139-7"
Fire clay shale	0	7				
Coal and shale	0	6				
Coal, slaty	0	10				
Fire clay and shale	0	3				
Coal, fair	2	7				
Concealed to bench				150-5"	615	
Coal blossom, Winifrede				0	615	
Concealed and sandstone				160	775	
Coal	0'	6"	**Chilton**	13	788	
Clay	0	5				
Coal	0	9				
Shale, gray	0	2				
Coal, bony	1	2				
Fire clay and shale	7	6				
Coal, bony	0	3				
Coal, good	2	3				
Concealed and sandstone				130	918	
Coal, Dingess				2	920	
Concealed to Pond Fork, (820' B.)				50	970	

The above section ends about 20 feet above the Dingess fossil horizon, which rises out of Pond Fork about one-half mile south from where this section was measured.

Dr. White once measured a section at this point and published same in Volume II(A), pages 292 and 293, West Virginia Geological Survey, 1908. The writer has made some changes in the correlation from that given by Dr. White.

The coal encountered at 468′ 9″ is classed by Dr. White

as the same bed encountered at 515′ 7″ in the Bald Knob Section, on a previous page of this Report, known as the William Price opening at that point. This correlation is correct according to the sections, and the coal bed is the **Coalburg,** instead of the **Winifrede,** as formerly supposed.

Casey Creek flows into Pond Fork from the south, about three miles south of mouth of West Fork. The following aneroid section was measured by Krebs, descending along a trail eastward to Casey Creek, about one-fourth mile from its mouth, and four miles south of Lantie P. O.:

Section 4 Miles South of Lantie P. O., Crook District.

			Thickness Feet.	Total Feet.
Kanawha Series (836′)				
Sandstone and concealed			65	65
Sandy shale			15	80
Sandstone and concealed			80	160
Coal blossom, Stockton			5	165
Sandy shale and concealed			40	205
Sandstone, **Coalburg,** massive			50	255
Sandy shale			20	275
Coal blossom, Coalburg			4	279
Sandstone and concealed			191	470
Coal blossom, Winifrede			4	474
Sandy shale			6	480
Sandstone and concealed			220	700
Coal blossom, Hernshaw			3	703
Sandstone and concealed			72	775
Sandstone, massive			20	795
Shells, gray			17	812
Limestone, Dingess, marine fossils			2	814
Shells, gray			0-8″	814-8″
Shale, sandy			10-4″	825
Shale, gray			1-6″	826-6″
Coal	0′	3″		
Shale	0	2		
Coal	0	5		
Gray shale, iron nodules	1	4		
Sandstone	2	0	**Williamson** 9	835-6″
Shale, gray	4	0		
Coal	0	3		
Slate	0	2		
Coal	0	2		
Slate	0	2		
Coal	0	1		
Shale, gray, full of iron nodules			2-6″	838

Three diamond core test holes were sunk on Casey Creek, on the lands of Chilton, MacCorkle and Chilton, and Meany.

Through the courtesy of Ex-Gov. MacCorkle, the Survey received copies of the records of these borings, which have given much valuable information in the study of the different strata.

Record of Casey Diamond Core Test Hole No. 2 (137).

Located on Casey Creek, 1.3 miles from its mouth, on land of Chilton, MacCorkle and Chilton, and Meany, N. 47° W. 5.7 miles from Bald Knob P. O.; authority, Ex-Gov. W. A. MacCorkle; elevation, 973' B.

	Thickness Ft.	Thickness In.	Total Ft.	Total In.
Kanawha Series (353')				
Surface	1	0	1	0
Sandstone, Naugatuck	43	0	44	0
Shale	54	6	98	6
Shale and coal	2	8	101	2
Shale	4	0	105	2
Shale and coal	1	1	106	3
Shale	2	11	109	2
Fire clay	2	10	112	0
Shale	46	4	158	4
Coal	0	8	159	0
Shale	12	1	171	1
Alma: Bone and coal, interlaminated 2' 2½"; Shale 0 4; Coal 2 0; Shale 0 5; Coal 2 1	7	0½	178	1½
Sandstone	19	11½	198	1
Shale and coal, Little Alma	1	1	199	2
Fire clay	2	4	201	6
Sandstone	35	4	336	10
Coal	1	2	338	0
Shale	7	2	345	2
Coal, Campbell Creek (No. 2 Gas)	4	9	349	11
Shale	0	3½	350	2½
Sandstone, to bottom	2	9½	353	0

Record of Casey Diamond Core Test Hole No. 1 (138).

Located on Casey Creek, one mile south of core test No. 2 (137), on land of Chilton, MacCorkle and Chilton, and Meany, N. 55° W. 5.2 miles from Bald Knob P. O.; authority, Ex-Gov. W. A. MacCorkle; elevation, 1035' L.

	Thickness Ft.	Thickness In.	Total Ft.	Total In.
Kanawha Series (291' 6")				
Sandstone	19	6	19	6
Shale	2	9	22	3
Coal, Hernshaw	2	1	24	4
Fire clay	0	2	24	6
Shale	2	0	26	6
Sandstone	28	5	54	11
Shale	1	2	56	1
Sandstone	19	2	75	3

				Thickness		Total	
				Ft.	In.	Ft.	In.
Shale				45	3	120	6
Coal	0′	10″					
Coal and shale	0	5					
Coal	0	4	**Williamson**				
Coal and shale	0	5		3	3	123	9
Shale	0	10					
Soft clay	0	5					
Shale				6	4	130	1
Coal				0	5	130	6
Shale				12	6	143	0
Sandstone				3	6	146	3
Shale				2	0	148	3
Coal and shale, **Upper Cedar Grove**				1	9	150	0
Shale				1	1	151	1
Sandstone, **Middle Cedar Grove**				10	8	161	9
Shale				4	3	166	0
Sandstone, **Middle Cedar Grove**				19	2	185	2
Black slate				3	7	188	9
Coal, Lower Cedar Grove				0	5	189	2
Shale				15	6	204	8
Coal and bone, interlaminated	0′	8½″					
Slate	0	1½					
Coal and bone, interlaminated	0	9					
Shale	0	2					
Coal and bone, interlaminated	0	5½	**Alma**	6	5	211	1
Coal, good	1	10					
Shale	0	5					
Coal	0	2½					
Shale	0	1½					
Coal	1	2½					
Shale	0	0½					
Coal	0	4½					
Fire clay				2	7	213	8
Shale				3	3	216	11
Sandstone, **Logan and Malden**				68	3	285	2
Shale				0	6	285	8
Coal	0′	1″					
Shale	0	0¼	**Campbell**				
Coal	0	7	**Creek**				
Shale	0	1¼	**(No. 2**				
Coal	0	3½	**Gas)**	4	1½	289	9½
Shale	0	2½					
Coal	2	10					
Shale				0	2½	290	0
Fire clay to bottom				1	6	291	6

The above section shows two seams of coal of commercial thickness. The Campbell Creek (No. 2 Gas) Coal is thinner than on Spruce Laurel, as shown in the core drill holes on that stream, and it is possible that the section at the base of the diamond core test hole is the Upper Bench of this bed.

Record of Casey Diamond Core Test Hole No. 3 (139).

Located on Casey Creek, 0.6 mile south of core test No. 2 (138) on lands of Chilton, MacCorkle and Chilton, and Meany, N. 59° W. 4.8 miles from Bald Knob P. O.; authority, Ex-Gov. W. A. MacCorkle; elevation, 1190′ B.

				Thickness.		Total.		
Kanawha Series (347′)				Ft.	In.	Ft.	In.	
Surface				3	0	3	0	
Sandstone, **Hernshaw**				58	0	61	0	
Coal	0′	2″	Hernshaw..					
Fire clay	1	0						
Shale	7	8		10	11	71	11	
Coal	2	1						
Fire clay				1	4	73	3	
Shale				4	10	78	1	
Sandstone, **Naugatuck**				30	1	108	2	
Shale				2	6	110	8	
Sandstone	16′	1″	Williamson Sandstone.					
Shale	2	11						
Sandstone	10	6		29	6	140	2	
Shale				34	11	175	1	
Coal	0′	7″	Williamson....					
Shale	0	3						
Coal	0	9						
Shale	0	2		10	3	185	4	
Coal	0	7						
Shale	6	10						
Coal and shale	1	1						
Shale				21	4	206	8	
Sandstone				25	1	231	9	
Shale				6	9	238	6	
Slate				2	5	240	11	
Coal, Lower Cedar Grove				0	8	241	7	
Sandstone				9	4	250	11	
Shale				5	2	255	1	
Slate				1	4	256	5	
Coal	0′	6″	Alma.....					
Slate	0	0½						
Coal	3	9						
Shale	2	3		9	2	265	7	
Coal	0	1						
Shale	0	9						
Coal	1	9½						
Fire clay				1	8	267	3	
Shale				15	6	282	9	
Sandstone, **Logan**				18	0	300	9	
Shale				2	3	303	0	
Sandstone, **Malden**				33	0	336	0	
Shale				2	8	338	8	
Coal	1′	10″	Campbell Creek (No. 2 Gas)					
Shale	0	1						
Coal	0	4						
Shale	0	8		6	8½	345	4½	80′
Coal	3	9½						
Fire clay				0	2½	345	7	
Sandstone to bottom				1	5	347	0	

The foregoing section shows the Alma and Campbell Creek (No. 2 Gas) Coals of sufficient thickness to be of commercial value.

Record of Cassingham Diamond Core Test Hole No. 5 (156).

Located on the land of Chilton, MacCorkle, Chilton and Cassingham, on Cow Creek, three miles southwest from Pond P. O.; authority, Crawford & Ashby, Charleston, W. Va.; completed, June 5, 1914; elevation, 1335' B.

	Thickness Ft.	Thickness In.	Total Ft.	Total In.
Kanawha Series (483' 4")				
Surface	5	0	5	0
Shale	3	0	8	0
Coal, Hernshaw	2	3	10	3
Fire clay	0	8	10	11
Sandstone	5	7	16	6
Shale	10	0	26	6
Shale, **coal** and sandstone streaks	11	0	37	6
Sandstone, shale streaks	4	9	42	3
Sandy shale, sandstone streaks	22	6	64	9
Williamson: **Coal** 0' 10"; Shale, sandstone streaks 14' 11"; Shale 1' 2"; **Coal** 1' 4"; Fire clay 2' 1"; Shale, sandstone streaks 3' 0"; **Coal** 0' 10"	24	2	88	11
Sandstone, shale streaks	15	7	104	6
Coal, interlaminated with slate	0	9	105	3
Sandy shale	2	2	107	5
Coal 0' 10"; Bone 0' 2"; **Coal** 0' 6"	1	6	108	11
Sandstone, shale streaks	15	7	124	6
Coal	0	2	124	8
Sandstone, shale streaks	11	8	136	4
Shale	0	5	136	9
Cedar Grove: **Coal** 1' 6"; Dark sandstone 1' 5"; **Coal** 0' 6"	3	5	140	2
Dark sandstone, **coal** streaks	1	4	141	6
Sandstone	43	3	184	9
Shale	1	7	186	4
Coal, Alma, (Peerless? I. C. W.)	2	7	188	11
Slate	2	6	191	5
Sandstone	0	4	191	9
Shale	4	2	195	11
Sandstone, **coal** streaks	41	1	237	0
Sandstone, **coal** and shale streaks	11	2	248	2
Sandy shale	1	3	249	5

		Thickness Ft.	Thickness In.	Total Ft.	Total In.
Black slate		1	1	250	6
Coal ....2′ 0″ Slate ....0 2½ Fire clay....0 11½ Dark shale....3 1 Shale, sandstone streaks....11 2 Coal .... 1 1 Bone and coal... 0 3 Coal 0 3	Campbell Creek (No. 2 Gas)	19	0	269	6
Fire clay		2	7	272	1
Shale, sandstone streaks		5	3	277	4
Shale		4	0	281	4
Coal		0	7	281	11
Fire clay		0	11	282	10
Shale, sandstone streaks		7	4	290	2
Shale, coal streaks		0	2	290	4
Coal		0	10	291	2
Sandy shale		2	0	293	2
Hard sandstone		21	2	314	4
Hard sandstone, coal streaks		13	0	324	4
Hard sandstone		7	8	335	0
Hard sandstone, shale and coal streaks		8	7	343	7
Hard sandstone		8	8	352	3
Coal and bone		0	2	352	5
Hard sandstone		4	3	356	8
Shale		0	2	356	10
Sandstone		1	2	358	0
Shale, coal and sandstone streaks		3	6	361	6
Shale, sandstone streaks		52	0	413	6
Sandy shale		19	2	432	8
Gray and dark shale		1	8	434	4
Black slate, Eagle Coal horizon		1	0	435	4
Gray and dark sandstone		3	5	438	9
Sandstone		21	9	460	6
Sandstone, coarse grained		17	11	478	5
Sandstone, hard, dark, to bottom		4	11	483	4

The above core test hole shows the Alma Coal thin and the Campbell Creek (No. 2 Gas) Coal broken up with "splits" so as to be unavailable commercially at present.

The following aneroid section was measured by Teets, down the dip from a point in head of Left Hand Fork of Coal Branch of Pond Fork to the mouth of Coal Branch, 1.2 miles northwest of Bald Knob P. O.:

Section 1.2 Miles Northwest of Bald Knob P. O.

			Thickness Ft.	In.	Total Ft.	In.
Kanawha Series (665')						
Sandstone and concealed			110	0	110	0
Coal, block	1' 9"	Chilton "Rider"	5	4	115	4
Slate, gray	0 0½					
Coal, block	0 5					
Slate, gray	0 1					
Coal, gray splint	1 10					
Slate	0 0¼					
Coal, gas	0 4					
Slate	0 1¼					
Coal, gas	0 9					
Sandstone			8	8	124	0
Slate			2	0	126	0
Coal	0' 3"	Chilton	4	3	130	3
Slate, gray	0 0½					
Coal, gray splint	2 6½					
Slate, gray	0 1					
Coal, gray splint	1 4					
Concealed and sandstone			164	9	295	0
Coal, block	1' 3"	Hernshaw	5	6	300	6
Coal, splint	0 6					
Coal, gas	0 3					
Coal, gray splint	2 3					
Coal, gas	1 3					
Sandstone and concealed			259	6	560	0
Coal, block	1' 3"	Alma, Upper Bench	2	3	562	3
Fire clay	1 0					
Sandstone			9	0	571	3
Coal, gas	1' 0"	Alma, Lower Bench	1	9	573	0
Coal, gray splint	0 9					
Concealed			87	0	660	0
Slate			3	0	663	0
Coal, block, Powellton? (1005' B.)			2	0	665	0

The coal at the base of the section is apparently the Powellton Coal, the Campbell Creek (No. 2 Gas) being concealed in the interval below 573 feet.

The following hand leveled section was measured by Teets, descending from a high point on the eastern side of Pond Fork, one mile due north of Bald Knob P. O., to the mouth of Workman Branch of Pond Fork:

Section 1 Mile Due North of Bald Knob P. O., Crook District.

	Thickness Ft.	In.	Total Ft.	In.
Pennsylvanian (1136')				
Allegheny Series (52')				
Sandstone, massive cliff capping the hill, ferriferous, micaceous	50	0	50	0
Fire clay, **No. 5 Block Coal horizon**	2	0	52	0

Kanawha Series (1084′)		Thickness Ft.	In.	Total Ft.	In.
Sandstone, massive		18	0	70	0
Concealed to bench		70	0	140	0
Sandstone and concealed to bench		70	0	210	0
Sandy shale and concealed, **Stockton**		20	0	230	0
Sandstone and concealed		110	0	340	0
Sandy shale and concealed to bench, **Coalburg Coal horizon**		30	0	370	0
Sandstone, massive		173	0	543	0
Slate, gray		1	1	544	1
Coal, gas..........0′ 11″ **Coal**, block........3 0	**Winifrede**	3	11	548	0
Gray shale		24	0	572	0
Coal		0	8	572	8
Gray shale		10	0	582	8
Slate, gray		0	4	583	0
Coal............0′ 11½″ Slate, gray......0 0½ **Coal**............0 4 Slate, gray......1 0 **Coal**, gray splint.2 1 Slate, gray.......0 2 **Coal**............0 8 Slate, gray......0 1 **Coal**............0 8	**Chilton "A"**	6	0	589	0
Slate, gray		1	0	590	0
Sandstone, massive		9	0	599	0
Coal, gas........0′ 6½″ Slate, gray.......0 0½ **Coal**, gray splint.2 0 Slate, gray......0 2 **Coal**, gray splint.1 3	**Chilton "Rider"**	4	0	603	0
Slate, gray		1	0	604	0
Sandstone		93	8	697	8
Coal, block, **Chilton**		2	4	700	0
Sandstone, massive		11	0	711	0
Concealed		3	3	714	3
Slate, gray		1	4	715	7
Slate, black		0	4	715	11
Coal..............0′ 6″ Slate, gray.........0 10 **Coal**...............0 9	**Little Chilton**	2	1	718	0
Gray slate		9	7	727	7
Coal		0	5	728	0
Gray slate		1	0	729	0
Sandstone, massive		28	2	757	2
Coal, gas...........0′ 9″ **Coal**, gray splint....3 1	**Hernshaw**	3	10	761	0
Sandstone, massive		44	0	805	0
Gray slaty shale		23	0	828	0
Limestone, Dingess, gray, marine fossils		1	0	829	0
Gray shale		35	10	864	10
Coal. Williamson		1	2	866	0
Sandy shale		36	0	902	0
Sandstone		22	6	924	6
Limestone, hard, gray		1	6	926	0

	Thickness Ft.	In.	Total Ft.	In.
Sandstone and sandy shale..........	26	0	952	0
Limestone, bluish gray..............	1	0	953	0
Sandstone, massive..................	28	0	981	0
Coal, gas, Campbell Creek (No. 2 Gas), Upper Bench....................	2	0	983	0
Sandstone and concealed...........	33	0	1016	0
Coal, visible, Campbell Creek (No. 2 Gas), Lower Bench..............	1	0	1017	0
Sandstone and concealed............	72	0	1089	0
Slaty limestone, marine fossils at base, Bald Knob, (Stockton?).....	7	0	1096	0
Coal ....1′ 0″ / Slate, gray.........0 4 / **Coal, gas**...........1 6 } **Matewan**..	2	10	1098	10
Sandstone and sandy shale...........	36	2	1135	0
Coal, visible, Eagle, (1014′ L.)........	1	0	1136	0

The foregoing interesting section gives a large number of the coals in the Kanawha Series, and also two very important limestone horizons carrying **marine fossils**, the **Dingess** and the **Bald Knob (Stockton?)**.

The following section was measured with hand level by Krebs and Teets, descending from Bald Knob Mountain southwestward along the road to Bald Knob P. O., Crook District:

Bald Knob Section Near Bald Knob P. O., Crook District.

	Thickness Ft.	In.	Total Ft.	In.
Pennsylvanian (1247′)				
Allegheny Series (156′)				
Sandy shale and concealed...........	31	0	31	0
Yellow shale.........................	3	0	34	0
Sandy shale and concealed..........	36	0	70	0
Concealed and sandstone, mostly sandstone	70	0	140	0
Sandstone, massive, coarse grained...	8	0	148	0
Coal, gas...........1′ 0″ / **Coal, splint**.........3 1 / **Coal, splint, harder**..2 5 } **No. 5 Block**.	6	6	154	6
Slate	1	6	156	0
Kanawha Series (1091′)				
Concealed, mostly sandstone, to bench	149	0	305	0
Sandy shale and concealed..........	121	0	426	0
Coal blossom.......................	1	0	427	0
Sandy shale and concealed...........	62	0	489	0
Shale, gray..........................	19	0	508	0

PLATE XV.—View where Dingess Limestone Fossil Horizon goes under Pond Fork, 1 mile north of Lantie.

		Thickness Ft.	In.	Total Ft.	In.
Coal, cannel........0' 2" Shale, gray.........1 6 Coal, gas...........2 0 Shale, gray.........0 1 Coal, block..........0 6 Fire clay...........0 4 Coal, splint.........2 0 Coal, hard splint....1 0	Coalburg.	7	7	515	7
Slate		3	5	519	0
Sandstone, Winifrede, massive.......		97	0	616	0
Dark sandy shale....................		2	0	618	0
Sandstone, massive..................		18	0	636	0
Shale, gray.........................		4	0	640	0
Shale, dark.........................		0	2	640	2
Coal, splint.........1' 4" Shale, gray.........0 2 Coal, splint.........1 10 Coal, cannel........0 10 Coal, splint.........0 10	Winifrede	5	0	645	2
Shale, gray.........................		2	10	648	0
Concealed, mostly sandstone.........		57	0	705	0
Coal blossom........................		5	0	710	0
Shale and concealed.................		6	6	716	6
Coal, visible, Chilton "A"..........		1	6	718	0
Shale, gray.........................		1	0	719	0
Sandstone, Upper Chilton, massive, micaceous		82	6	801	6
Coal, visible, Chilton..............		1	6	803	0
Shale, gray.........................		3	0	806	0
Sandy shale and concealed...........		44	0	850	0
Coal, visible, Hernshaw.............		3	0	853	0
Shale, gray.........................		3	0	856	0
Sandstone		5	0	861	0
Sandy shale.........................		11	0	872	0
Sandstone, massive..................		19	0	891	0
Shale, gray.........................		4	0	895	0
Concealed, mostly sandstone.........		32	0	927	0
Sandstone, dark.....................		8	0	935	0
Sandy shale and concealed...........		10	0	945	0
Fire clay...........................		3	0	948	0
Shale, gray.........................		13	0	961	0
Shale, gray and sandy...............		6	0	967	0
Concealed, mostly shale.............		36	0	1003	0
Shale, gray.........................		2	0	1005	0
Coal, Alma..........................		0	6	1005	6
Fire clay...........................		0	6	1006	0
Sandstone		4	0	1010	0
Concealed, mostly sandstone.........		14	0	1024	0
Concealed, mostly sandy shale.......		49	6	1073	6
Coal, visible, Campbell Creek (No. 2 Gas)		0	6	1074	0
Sandy shale.........................		4	0	1078	0
Concealed, mostly sandstone.........		35	0	1113	0
Sandstone		4	0	1117	0
Shale and concealed.................		9	0	1126	0

	Thickness Ft.	In.	Total Ft.	In.
Coal, Powellton	1	0	1127	0
Shale, gray	5	0	1132	0
Sandstone, massive	38	0	1170	0
Shale, gray and sandy	23	0	1193	0
Shale, dark gray	2	10	1195	10
Shale, dark gray, **Bald Knob (Stockton?) Limestone horizon, marine fossils**	1	0	1196	10
Coal, gas, **Matewan**	2	2	1199	0
Sandstone	0	6	1199	6
Fire clay	0	6	1200	0
Concealed	5	0	1205	0
Sandstone, massive	20	0	1225	0
Coal, gas, **Eagle, Upper Bench**	2	0	1227	0
Sandstone, massive	6	0	1233	0
Shale, dark, full of plant fossils	7	0	1240	0
Coal 0′ 2″; Fire clay 2 0; **Coal** 1 0; Shale, gray 0 4; **Coal**, gas 2 0 } **Eagle**	5	6	1245	6
Slate	0	6	1246	0
Sandstone to creek at 1085′ L.	1	0	1247	0

This interesting section exhibits nearly all the coals in the Kanawha Series, and also the **Bald Knob (Stockton?) fossil horizon.**

In passing up Pond Fork from Bald Knob, the Campbell Creek (No. 2 Gas) Coal rises faster than the bed of the stream, as is shown in the following aneroid section, measured by Teets, descending from a high summit westward to Pond Fork, 0.9 mile north of Echart P. O.:

Section 0.9 Mile North of Echart P. O., Crook District.

	Thickness Feet.	Total Feet.	
Kanawha Series (1100′)			
Sandstone, massive, coarse grained	110	110	
Coal blossom, Coalburg	2	112	112′
Sandstone, massive, coarse grained, buff color	66	178	
Coal, good, **Buffalo Creek**	2	180	68′
Sandstone, **Upper Winifrede**, massive, medium grained, micaceous	69	249	
Coal blossom, Winifrede	1	250	70′
Sandstone, fine grained	100	350	
Sandy shale	20	370	
Sandstone, medium grained	58	428	
Coal blossom, Chilton	2	430	180′
Sandstone, flaggy, micaceous	70	500	
Coal blossom, Little Chilton	2	502	72′

	Thickness Feet.	Total Feet.	
Sandstone, flaggy	36	538	
Coal blossom, Hernshaw	2	540	38′
Sandstone, massive	87	627	
Coal blossom, Upper Cedar Grove	3	630	90′
Sandy shale and sandstone	160	790	
Coal blossom, Campbell Creek (No. 2 Gas)	1	791	161′
Sandstone and concealed, to bench	139	930	
Sandy shale and concealed	100	1030	
Bench, Little Eagle Coal horizon	0	1030	
Sandstone and concealed to Pond Fork, (1280′ B.)	70	1100	

The following aneroid section was measured by Teets in descending from a high point west of Pond Fork and opposite the mouth of Burnt Camp Branch, 0.8 mile southeast of Echart P. O., eastward to the mouth of Burnt Camp Branch:

Section 0.8 Mile Southeast of Echart P. O., Crook District.

			Thickness Ft.	In.	Total Ft.	In.
Kanawha Series (935′)						
Sandstone and concealed			200	0	200	0
Coal	1′ 3″	Chilton "A"	4	9	204	9
Slate	0 1					
Coal	0 9					
Slate	0 1					
Coal	2 7					
Sandstone and concealed			220	3	425	0
Coal	0′ 2″	Williamson	6	0	431	0
Slate, gray	0 2					
Coal	0 3					
Slate	0 4					
Coal, block	1 11					
Coal, gray splint	1 8					
Coal, block	0 11					
Coal, gray splint	0 7					
Sandy shale and concealed			241	9	672	9
Coal, Campbell Creek (No. 2 Gas)			2	3	675	0
Concealed			40	0	715	0
Coal blossom, Powellton "A"?			1	6	716	6
Sandstone and concealed			18	6	735	0
Coal blossom, Powellton?			2	0	737	0
Sandstone and concealed			63	0	800	0
Coal, block	1′ 6″	Matewan?	3	8	803	8
Slate	0 4					
Coal, gray splint	1 10					
Sandstone and concealed			60	4	864	0

		Thickness Ft. In.	Total Ft. In.
Coal ....0′ 4″	Eagle?...		
Slate, gray.........0 3			
Coal 0 4			
Slate, dark.........0 1			
Coal 0 1			
Slate 0 1		7 2	871 2
Coal, block.........1 10			
Slate, gray, soft.....1 0			
Coal, gray splint....1 2			
Slate 0 3			
Coal, gray splint....1 9			
Concealed to creek bed, (1380′ B.)...		63 10	935 0

The following aneroid section was measured by Teets, descending from a high point in Guyandot Mountain, just west of Walnut Gap at the Wyoming-Boone County Line, and in the extreme southern part of Boone County, descending northward along slope to Walnut Gap, thence along county road to head of Skin Fork of Pond Fork:

Walnut Gap Section, Crook District.

	Thickness Feet.	Total Feet.
Pennsylvanian (1520′)		
Allegheny Series (190′)		
Sandstone, **East Lynn**, massive, coarse grained, buff color, to bench	160	160
Concealed bench	30	190
Kanawha Series (1330′)		
Sandstone, massive, ferriferous, to bench...	80	270
Sandstone	80	350
Concealed to Walnut Gap	140	490
Sandy shale, **Coalburg Coal horizon**, and concealed	35	525
Sandstone, **Winifrede**, massive, micaceous, ferriferous	180	705
Bench, **Winifrede Coal horizon**	0	705
Sandstone, massive, micaceous, coarse grained	110	815
Coal blossom, Chilton "A"	2	817
Sandstone and concealed	108	925
Coal blossom, Chilton?	1	926
Sandstone, massive, to bench	119	1045
Sandy shale	10	1055
Fire clay	3	1058
Sandy shale and concealed	47	1105
Sandstone	55	1160
Fire clay	2	1162
Sandy shale	63	1225
Sandstone, massive, fine grained	100	1325
Sandy shale	17	1342
Coal blossom, Campbell Creek (No. 2 Gas)	3	1345
Sandstone	150	1495
Slate, concealed	8	1503

		Thickness Feet.	Total Feet.
Coal		1	1504
Slaty shale, marine fossils, Bald Knob (Stockton?)		4	1508
Coal ... 0' 4"	Matewan	2-6"	1510-6"
Slate, gray ... 0 6			
Coal ... 1 8			
Slaty shale to 1710' B.		9-6"	1520

This section was measured to the dip, and the intervals given between the different coals are too great. The section begins with the East Lynn Sandstone and ends at the base near the **Eagle Coal**. Very few exposures of the several coal beds were found, owing to a lack of development, so it is really difficult properly to classify the different coal horizons.

The following aneroid section was measured from a point in the southern end of Cook Mountain at head of West Fork of Pond Fork, descending southward along trail to the mouth of Burgess Branch of Pond Fork, one-half mile above mouth of Lacey Branch, and three and one-fourth miles due north of Walnut Gap:

Section 3¼ Miles Due North of Walnut Gap, Crook District

		Thickness Feet.	Total Feet.
Kanawha Series (1105')			
Sandstone and sandy shale, **Coalburg**		90	90
Bench, **Coalburg Coal horizon**		0	90
Sandstone and sandy shale, to bench		70	160
Sandstone and sandy shale to low gap		55	215
Bench, **Winifrede Coal horizon**		0	215
Sandy shale and concealed		165	380
Bench, **Chilton Coal horizon**		0	380
Sandy shale and sandstone		165	545
Bench, **Hernshaw Coal horizon**		0	545
Sandstone and sandy shale, to bench		85	630
Sandstone, **Williamson**, massive, to bench		60	690
Sandy shale and concealed		15	705
Sandstone, **Logan and Lower Cedar Grove**		130	835
Sandy shale and sandstone, to bench		125	960
Sandy shale and sandstone, to bench		75	1035
Sandstone and concealed		61	1096
Coal ... 0' 1"	**Eagle**	4-10"	1100-10"
Slate, gray ... 0 1			
Coal, block ... 0 2			
Slate ... 0 3			
Coal, gray splint ... 0 5			
Slate ... 0 1			
Coal, gas ... 3 9			
Fire clay and concealed to creek level, (1610' B.)		4-2"	1105

West Fork flows into Pond Fork about nine miles southeast of **Madison**, and almost parallels Pond Fork for its entire length, being separated from the latter stream by Cook **Mountain**.

Whites Branch flows from the northeast into West Fork about one mile northeast from the latter's junction with Pond Fork.

The following aneroid section was measured by Krebs, descending from a high summit on the Sherman-Crook District Line, southwestward along county road to Whites Branch, four miles northeast of Gordon P. O.:

Section 4 Miles Northeast of Gordon P. O., Crook District.

		Thickness Ft.	In.	Total Ft.	In.	
Pennsylvanian (550')						
Allegheny Series (145')						
Sandstone and concealed		65	0	65	0	
Fire clay		2	0	67	0	
Sandstone and concealed		75	0	142	0	
Coal blossom, No. 5 Block		3	0	145	0	
Kanawha Series (405')						
Sandstone and concealed		100	0	245	0	
Bench, **Stockton Coal horizon**		0	0	245	0	
Sandstone and concealed		125	0	370	0	
Shale, gray		1	0	371	0	
Coal, impure........1' 0"						
Coal, splint.........2 0						
Coal, block.........1 0	**Coalburg "A"**	4	10	375	10	230'
Coal, splint.........0 4						
Coal..............0 6						
Sandstone and concealed		40	2	416	0	
Sandstone		2	0	418	0	
Coal...............1' 0"						
Shale, dark.........1 0						
Coal, gas...........1 2	**Coalburg**	3	7	421	7	
Slate 0 1						
Coal, visible........0 4						
Sandy shale and concealed		25	0	446	7	
Sandstone, massive, coarse grained		10	5	457	0	
Coal blossom, Buffalo Creek, (1097' L.)		1	6	458	6	
Sandstone		74	6	533	0	
Slate		0	8	533	8	
Coal, gas...........0' 1"						
Shale, gray.........0 2						
Coal, splint.........0 10						
Slate 0 2	**Winifrede**	3	7	537	3	
Coal...............0 1						
Slate, dark.........0 3						
Fire clay...........0 6						
Coal, gas...........1 6						
Sandstone and concealed to creek, (1015' B.)		12	9	550	0	

Another aneroid section was measured by Krebs, descending westward from a point in the Sherman-Crook District Line, 2.3 miles northeast of Gordon P. O., Crook District, to Lick Branch of Whites Branch:

Section 2.3 Miles Northeast of Gordon P. O., Crook District.

Kanawha Series (655′)				Thickness Ft.	In.	Total Ft.	In.
Sandy shale				20	0	20	0
Sandstone, massive				90	0	110	0
Coal blossom				4	0	114	0
Sandstone and concealed				30	0	144	0
Slate				1	0	145	0
Coal, block	4′	2″					
Slate	0	1					
Coal, hard splint	0	6	Stockton-				
Slate	0	4	Lewiston,				
Coal, hard splint	1	2	Upper				
Slate	0	6	Bench	9	2	154	2
Coal	0	5					
Slate	1	0					
Coal	1	0					
Sandstone and concealed				58	0	212	2
Slate				1	10	214	0
Coal, splint	0′	2″	Stockton-				
Slate	0	1	Lewiston,				
Coal, splint	1	8	Lower				
Coal, impure	1	10	Bench	4	3	218	3
Coal, hard	0	6					
Sandstone and concealed				130	9	349	0
Slate				1	0	350	0
Coal, splint	2′	8″					
Slate, dark	0	4	Coalburg	6	0	356	0
Coal, splint	1	6					
Coal, impure	1	6					
Slate				1	0	357	0
Sandstone and concealed				270	0	627	0
Coal, block	1′	0″					
Slate, full of plant fossils	3	0	Chilton	6	4	633	4
Coal, block	2	4					
Shells, gray				1	8	635	0
Sandstone and concealed, (1015′ B.)				20	0	655	0

The following aneroid section was measured by Krebs, descending from a high summit on Cook Mountain, westward to Greens Branch, three miles northwest of Chap P. O.:

Section 3 Miles Northwest of Chap P. O., Crook District.

	Thickness Feet.	Total Feet.
Pennsylvanian (1035')		
Allegheny Series (160')		
Sandstone and concealed	160	160
Bench, No. 5 Block Coal horizon	0	160
Kanawha Series (875')		
Sandstone and concealed, to bench	100	260
Sandstone, massive	140	400
Large bench, Stockton Coal horizon	0	400
Sandstone and concealed	125	525
Bench, Coalburg Coal horizon	0	525
Sandstone and concealed	120	645
Large bench, Winifrede Coal horizon	0	645
Sandstone and concealed	114	759
Coal blossom, Chilton "A"	1	760
Shale, gray	4	764
Sandstone, massive	8	772
Dark shale	2	774
Sandstone and concealed	50	824
Sandstone, massive	6	830
Shale, gray	5	835
Coal, splint 1' 8" / Coal, block 1 10 } Chilton	3-6"	838-6"
Slate	1-6"	840
Sandstone and concealed	50	890
Sandstone, massive	10	900
Coal, gas, Little Chilton	3	903
Sandstone, massive	10	913
Sandy shale	85	998
Sandstone, massive, to creek, (991' L.)	37	1035

Messrs. Crawford and Ashby of Charleston have drilled five diamond core test holes on West Fork, the records of which they have kindly furnished the Survey, and these will now be given:

The following aneroid section was measured by Teets, descending from a point just northwest of the mouth of Jarrolds Branch of West Fork, southwestward, and combined with diamond core test hole No. 5 (132) of Crawford and Ashby, located at the mouth of Jarrolds Branch:

Section at Mouth of Jarrolds Branch, 1.3 Miles Northwest of Chap P. O., Crook District.

	Thickness Ft.	Thickness In.	Total Ft.	Total In.
Kanawha Series (453' 5")				
Sandstone and concealed	90	0	90	0
Coal 2' 3" / Bone 0 1 / Coal, gray splint 0 10 } Hernshaw	3	2	93	2

PLATE XVI.—View looking up West Fork from point ¾ mile south of mouth of Browns Branch, and topography of Kanawha Series.

	Thickness Ft. In.	Total Ft. In.
Slate	1 10	95 0
Concealed and sandstone	100 0	195 0
Sandstone	25 0	220 0
Sandy shale	20 0	240 0
Gray shale full of concretions	1 0	241 0
Sandstone to top of well (1041' L.)	3 0	244 0
Thence with core test hole:		
Shale	25 9	269 9
Slate	4 0	273 9
Coal 0' 9" / Slate 0 3 / **Coal** 0 9 / Slate 0 1 / **Coal** 1 1 / Slate 0 6 / **Coal** 0 4½ / Slate 0 1½ / **Coal** 2 1 / Slate 0 4 / **Coal** 0 4 — **Campbell Creek (No. 2 Gas) Upper Bench**	6 8	280 5
Shale	6 8	287 1
Sand rock	5 9	292 10
Shale	3 7	296 5
Black slate	1 10	298 3
Sand rock	15 3	313 6
Coal, Campbell Creek (No. 2 Gas) Lower Bench	1 0	314 6
Sand rock	3 4	317 10
Shale	18 8	336 6
Coal, Powellton	2 0	338 6
Sand rock	50 8	389 2
Shale	10 7	399 9
Coal, Matewan	0 8	400 5
Sand rock	42 10½	443 3½
Slate	0 7½	443 11
Coal 0' 11½" / Slate 0 2½ / **Coal** 4 3½ — **Eagle**	5 5½	449 4½
Shale	2 0½	451 5
Sand rock to bottom	2 0	453 5

The coal at the base of the section correlates with the Eagle.

James Creek flows into West Fork three-fourths mile southeast from Chap P. O., and the following aneroid section was measured by Teets, descending from a high point about one mile southeast to James Creek:

Section 1.4 Miles Southeast of Chap P. O., Crook District.

		Thickness Ft.	In.	Total Ft.	In.
Kanawha Series (455')					
Sandstone and concealed		100	0	100	0
Coal, soft.........1' 9"	Coalburg?				
Slate 0 0½					
Coal 0 10					
Niggerhead and impure coal.....1 6					
Coal, gray splint..1 10					
Slate, gray.......0 1					
Coal, gray splint..1 10½		7	11	107	11
Sandstone and concealed		102	1	210	0
Coal, block.........1' 1"	Winifrede?				
Slate, gray.........0 1					
Coal, block.........1 9					
Slate, grayish black.0 7					
Coal 0 4					
Slate 0 1					
Coal, block.........0 7					
Slate, black.........0 4					
Coal, gray splint....1 9		6	7	216	7
Sandstone and concealed		224	8	441	3
Coal, block.........2' 4"	Chilton?..				
Bone 0 2					
Coal 1 3		3	9	445	0
Concealed to creek, (1290' B.)		10	0	455	0

At the mouth of Little Ugly Branch, the following aneroid section was measured by Krebs, descending from top of Bald Knob northwestward along road to West Fork at mouth of Little Ugly, and combined with Crawford and Ashby diamond core test hole No. 4 (133), one mile southeast of Chap P. O.:

Section 1 Mile Southeast of Chap P. O., Crook District.

	Thickness Ft.	In.	Total Ft.	In.
Pennsylvanian (1392')				
Allegheny Series (200')				
Sandy shale	70	0	70	0
Sandstone, fine grained	30	0	100	0
Sandy shale	10	0	110	0
Sandstone and concealed, **East Lynn**	90	0	200	0
Bench, **No. 5 Block Coal horizon**	0	0	200	0
Kanawha Series (1192')				
Sandy shale and concealed, to bench..	80	0	280	
Sandstone and concealed	240	0	520	
Bench, **Stockton-Lewiston Coal horizon**	0	0	520	
Sandstone and concealed	130	0	650	
Bench, **Coalburg Coal horizon**	0	0	650	0
Sandstone and concealed	70	0	720	0

				Thickness Ft.	In.	Total Ft.	In.
Bench, **Winifrede Coal** horizon				0	0	720	0
Sandstone and concealed				90	0	810	0
Bench, **Chilton Coal** horizon				0	0	810	0
Sandstone and concealed				156	0	966	0
Coal blossom, Hernshaw				4	0	970	0
Sandstone and concealed to top of diamond core test hole No. 4 (133), 1208′ L				156	0	1126	0
Continuing with core test hole:							
Sand rock				4	8	1130	8
Shale				3	5	1134	1
Coal	0′	11″					
Shale	8	4	**Alma**	10	2	1144	3
Coal	0	11					
Slate				2	6	1146	9
Shale				3	8	1150	5
Coal				0	5	1150	10
Fire clay				2	6	1153	4
Shale				9	6	1162	10
Coal	1′	0″	**Campbell Creek**				
Shale	0	10	**(No. 2 Gas)**	3	0	1165	10
Coal	1	2	**Upper Bench**				
Shale				20	6½	1186	4½
Slate				0	6½	1186	11
Coal	0′	1″					
Slate	0	1					
Coal, no core	0	8	**Campbell**				
Slate	0	1	**Creek**				
Coal	2	2	**(No. 2 Gas),**				
Slate	0	0½	**Lower**				
Coal	0	3	**Bench**	4	4½	1191	3½
Slate	0	2					
Coal	0	10					
Shale				0	6½	1191	10
Slate				0	8	1192	6
Sand rock				31	1	1223	7
Shale				0	7	1224	2
Sand rock				0	6	1224	8
Coal				0	8	1225	4
Slate				0	2	1225	6
Shale				4	9	1230	3
Sand rock				10	4	1240	7
Coal and slate	1′	0″					
Slate	1	7					
Shale	0	4					
Soft clay	0	4	**Powellton**	8	8	1249	3
Slate	3	5					
Coal	2	0					
Fire clay				3	0	1252	3
Shale				12	7	1264	10
Sand rock				28	5	1293	3
Shale				16	5	1309	8
Coal, Matewan				1	0	1310	8
Fire clay				2	5	1313	1
Sand rock				53	4	1366	5
Sand slate				3	2	1369	7

		Thickness Ft. In.	Total Ft. In.
Slate		1 10½	1371 5½
Coal ....5′ 8½″ Soft clay.........1 0 Shale ....7 9 **Coal**, no core 2″..1 0	**Eagle**....	15 5½	1386 11
Shale		1 0	1387 11
Sand rock to bottom		4 1	1392 0

The foregoing interesting section was taken to the dip, so that the intervals between the several coals and their horizons are too large. This fact is especially true in regard to the interval between the No. 5 Block and the Stockton-Lewiston Coals.

Another aneroid section was measured by Teets, descending from a high point westward to West Fork, two-tenths mile south of Little Ugly Branch, and combined with diamond core test hole No. 2 (134), drilled by Crawford and Ashby, one and two-tenths miles southwest of Chap P. O.:

Section 1.2 Miles South of Chap P. O., Crook District.

		Thickness Ft. In.	Total Ft. In.
Pennsylvania (1237′)			
Allegheny Series (149′ 8″)			
Sandstone and sandy shale		70 0	70 0
Coal blossom, Upper Kittanning		5 0	75 0
Sandstone and sandy shale		45 0	120 0
Coal ....0′ 5″ Slate, black.......0 2 **Coal, block**.......3 0 **Coal, gray splint**..4 0 Shale, gray.......2 5 **Coal, soft**........1 0 **Coal, gray splint**..1 6 **Coal, block**.......1 4 Slate, dark........0 2 **Coal, gray splint**..1 1 Slate, gray........0 8 **Coal** ....0 3 Slate, gray........1 0 Sandstone ....4 0 Slate, gray........0 3 **Coal** ....0 1 Slate, gray........1 1 **Coal, block**........1 11 Slate, dark.......0 0½ **Coal, block**.......1 5 **Coal, gray splint**..3 10½	**No. 5 Block**	29 8	149 8
Kanawha Series (1187′ 4″)			
Sandstone and concealed		170 4	320 0

	Thickness Ft.	Thickness In.	Total Ft.	Total In.
Bench, Stöckton-Lewiston Coal horizon	0	0	320	0
Sandstone and concealed	152	0	472	0
Coal, block........0' 10"				
Slate, dark........0 2				
Coal, block........1 6				
Niggerhead ....1 2 } Coalburg..	7	9	479	9
Coal, gray splint...2 0				
Slate, gray........0 3				
Coal, gray splint...1 10				
Slate	1	3	481	0
Sandstone and concealed	119	0	600	0
Coal blossom, Winifrede	5	0	605	0
Sandstone, massive	140	0	745	0
Sandy shale	20	0	765	0
Sandstone, massive	71	0	836	0
Coal, impure, Little Chilton	4	0	840	0
Sandstone and concealed	145	0	985	0
Coal, Alma, top of core drill hole No. 2 (134), drilled by Crawford and Ashby at mouth of Little Ugly Branch, elevation, 1203' L.	2	0	987	0
Thence continuing with core drill hole.				
Shale	13	6	1000	6
Coal, no core	0	3	1000	9
Slate	0	4	1001	1
Soft clay	1	5	1002	6
Shale	13	5	1015	11
Coal and slate...1' 6" } Campbell Creek				
Slate ....0 8 } (No. 2 Gas).	3	3	1019	2
Coal, no core....1 1 } Upper Bench				
Shale	17	5	1036	7
Slate	0	5	1037	0
Coal ....0' 1 " } Campbell				
Slate ....0 1 } Creek				
Coal ....0 1½ } (No. 2 Gas)	2	6	1039	6
Slate ....0 4 } Lower Bench				
Coal, no core 11"..1 10½				
Slate	0	8	1040	2
Fire clay	0	11	1041	1
Sand rock	37	1	1078	2
Coal	0	7	1078	9
Fire clay	1	3	1080	0
Shale	14	3	1094	3
Slate	0	2	1094	5
Coal and slate.....0' 5"				
Coal, no core......0 6				
Slate ....2 7 } Powellton	8	11	1103	4
Fire clay..........2 0				
Slate 1 1				
Coal, no core......2 4				
Slate	1	3	1104	7
Sand rock	41	6	1146	1
Shale	17	11	1164	0
Coal, Matewan	1	0½	1165	0½
Fire clay	2	4½	1167	5
Sand rock	57	9	1225	2

		Thickness Ft. In.	Total Ft. In.
Slate		2 3	1227 5
Coal ....0′ 11″ Slate ....0 2 **Coal**, no core 2′ 6½″ 5 3	**Eagle**.....	6 4	1233 9
Slate		1 4	1235 1
Sand rock to bottom		1 11	1237 0

It is interesting to note in the above section that the total thickness of coal exposed and measured is 57 feet 11 inches. The great local thickness of coal at 120′ to 149′ 8″ is correlated as the No. 5 Block.

Mats Creek flows into West Fork from the southeast two miles southeast of Chap P. O., and a core test hole by Crawford and Ashby, No. 3 (135), has been drilled at the mouth of Pettry Fork of Mats Creek, one and one-fourth miles from its mouth.

The record of this boring is connected with a section measured by Teets, descending from a high point on the east side of West Fork northeastward to the mouth of Pettry Fork of Mats Creek, and as thus combined reads as follows:

Section 2.6 Miles Southeast of Chap P. O., Crook District.

		Thickness Ft. In.	Total Ft. In
Allegheny Series (106′)			
Sandstone and concealed		75 0	75 0
Coal blossom, No. 5 Block, Upper Bench		3 0	78 0
Sandy shale and concealed		22 0	100 0
Coal ....1′ 2 ″ Slate ....0 0½ **Coal**, gray splint..4 9½	**No. 5 Block Lower Bench**	6 0	106 0
Kanawha Series (1194′)			
Sandstone and sandy shale		164 0	270 0
Coal, impure......2′ 1 ″ **Coal**, gray splint..3 7 Slate, gray.......0 0¾ **Coal**, gray splint..0 4¼ Slate, gray.......0 5 **Coal**, gray splint..2 7	**Stockton-Lewiston**	9 1	279 1
Sandstone and concealed		130 11	410 0

			Thickness Ft.	In.	Total Ft.	In.
Coal, splint	1'	10"				
Slate, gray	0	1				
Coal	0	6				
Slate, gray	0	2				
Coal	0	1				
Slate, gray	0	1				
Coal	0	2	Coalburg.	8	7	418 7
Slate, dark	0	5				
Coal	0	1				
Slate	0	7				
Coal, gray splint	2	2				
Slate, gray	0	4				
Coal, gray splint	2	1				
Sandstone and concealed				91	5	510 0
Coal	1'	0"				
Slate, dark	1	7				
Coal	0	4				
Slate	5	0				
Coal, block	0	11	Winifrede.	12	6	522 6
Slate, gray	0	1				
Coal, block	1	11				
Coal, cannel	0	10				
Coal, gas	0	10				
Sandstone and concealed				278	9	801 3
Coal, Hernshaw					9	805 0
Sandstone and sandy shale				33	0	838 0
Limestone, fossiliferous, Dingess				2	0	840 0
Sandstone and concealed				80	0	920 0
Coal, Cedar Grove horizon				2	0	922 0
Concealed to top of core test hole No. 3 (135), 1350' B., and thence with core test				3	0	925 0
Sand rock				20	7	945 7
Slate				3	8	949 3
Coal, Alma				1	0	950 3
Fire clay				0	9	951 0
Sand rock				14	9	965 9
Shale				4	5	970 2
Slate				1	6	971 8
Coal				0	2	971 10
Fire clay				2	0	973 10
Sand rock				41	2	1015 0
Coal, no core 8"	2'	7"	Campbell Creek			
Fire clay	0	4	(No. 2 Gas)	5	3	1020 3
Coal, no core	2	4				
Slate				0	1	1020 4
Shale				2	7	1022 11
Sand rock				27	3½	1050 2½
Coal, Powellton				0	4½	1050 7
Fire clay				2	6	1053 1
Shale				3	2	1056 3
Sand rock				81	5	1137 8
Shale				21	4	1159 0
Coal, Matewan				0	11	1159 11
Shale				1	8	1161 7
Sand rock				3	9	1165 4

		Thickness Ft. In.	Total Ft. In.
Shale		1 1	1166 5
Sand rock		35 1	1201 6
Shale		15 4	1216 10
Slate		0 1	1216 11
Coal 0′ 9½″	Eagle		
Bone 0 2			
Coal 2 11¾			
Slate 0 1			
Coal 0 9¼	**Eagle**....	5 11½	1222 10½
Slate 0 0½			
Coal 0 1			
Slate 0 3½			
Coal and slate 0 3			
Coal 0 6			
Shale		8 0½	1230 11
Coal, Little Eagle?		1 0	1231 11
Shale		1 7	1233 6
Sand rock		30 8	1264 2
Shale to bottom		35 10	1300 0

The foregoing section was taken a little towards the rise, so that the interval between the base of the Coalburg Coal and the Campbell Creek (No. 2 Gas) will be a little too small. The total amount of coal given in the above section is 50 feet 6 inches.

The following hand leveled section, measured by Teets on the west side of the West Fork of Pond Fork, three and three-tenths miles south of Chap P. O., and combined with the diamond core test hole No. 1 (140) of Crawford and Ashby, reads as follows:

Section 3.3 Miles South of Chap P. O., Crook District.

		Thickness Ft. In.	Total Ft. In.
Kanawha Series (355′ 0″)			
Gray shale		20 0	20 0
Limestone, Seth, marine fossils		0 4	20 4
Slate		0 2	20 6
Coal		0 3	20 9
Shale and sandstone to core test, 1425′ B		10 3	31 0
Thence with core test:			
Slate		0 3½	31 3½
Coal, no core 0′ 5 ″			
Slate 0 3½	**Cedar**		
Coal 0 5½	**Grove**....	1 10½	33 2
Slate 0 6½			
Coal 0 2			
Slate		0 4½	33 6½
Shale		44 8½	78 3

PLATE XVII(a)—View on Coal River at Dartmont.

PLATE XVII(b).—The twin trees, birch and maple, on Haggle Branch of Coal River.

		Thickness		Total	
		Ft.	In.	Ft.	In.
Coal, no core...... 1' 5"					
Shale 12 5	**Alma**.....	14	5	92	8
Coal, no core...... 0 7					
Fire clay..........................		2	4	95	0
Shale		7	1	102	1
Sand rock..........................		20	0	122	1
Shale		9	6	131	7
Sandy shale..........................		11	2	142	9
Coal ....2' 5 "					
Slate 1 4½	**Campbell Creek**				
Coal 0 11	(**No. 2 Gas**)	6	2	148	11
Slate 0 0½					
Coal, no core..1 5					
Sand rock..........................		1	11	150	10
Shale		6	4	157	2
Sand rock..........................		64	6	221	8
Coal, no core, **Powellton**............		2	0	223	8
Sand rock..........................		40	5	264	1
Slate		24	2	288	3
Coal, no core 4"....1' 1"					
Slate 3 5	**Matewan**..	5	9	294	0
Coal, no core 8"....1 0					
Coal, bone.........0 3					
Fire clay..........................		5	6	299	6
Shale		17	1	316	7
Sand rock..........................		5	4	321	11
Shale		19	7	341	6
Slate		3	0	344	6
Coal, no core 3"..0' 11 "					
Slate 0 2½					
Coal 4 4	**Eagle**....	6	8	351	2
Coal and slate...1 0					
Coal 0 2½					
Slate to bottom..........................		3	10	355	0

The above core test begins with the Cedar Grove bed, so that the coal at the base of this section is the Eagle seam. This proves to be a valuable coal on a portion of West Fork, as shown by the foregoing core test hole, sunk by Messrs. Crawford and Ashby.

On the waters of Marsh Fork of Coal River, the following aneroid section was measured, descending from the Boone-Raleigh County Line, five and one-half miles northeast of Bald Knob P. O., northeastwardly to the mouth of Centley Branch of Marsh Fork, three and two-tenths miles southwest of Jarrolds Valley P. O., Marsh Fork District, Raleigh County:

Section 3.2 Miles Southwest of Jarrolds Valley P. O., Marsh Fork District, Raleigh County.

	Thickness Ft.	In.	Total Ft.	In.
Kanawha Series (1380′)				
Sandy shale and concealed..........	160	0	160	0
Sandstone	10	0	170	0
Coal, hard, block 3′ 10″ } **Stockton-Lewiston**				
Coal, splint......2 8 } **Upper Bench**	6	6	176	6
Slate	3	6	180	0
Sandy shale and concealed..........	48	0	228	0
Coal blossom, Stockton-Lewiston, Lower Bench........................	3	0	231	0
Sandstone	127	0	358	0
Coal blossom (reported) **Coalburg**....	9	11	367	11
Sandy shale........................	1	1	369	0
Sandy shale and concealed...........	140	0	509	0
Coal blossom (reported) **Winifrede**...	5	10	514	10
Sandy shale and concealed...........	133	2	648	0
Coal blossom (reported) **Chilton "A"**..	3	8	651	8
Sandy shale and concealed...........	184	4	836	0
Coal blossom, Hernshaw.............	4	0	840	0
Sandy shale and sandstone...........	203	0	1043	0
Coal, splint, **Alma**..................	2	6	1045	6
Sandy shale and sandstone...........	98	0	1143	8
Slate	4	6	1148	0
Coal, splint.......2′ 5″ } **Campbell Creek (No. 2 Gas)**				
Slate, gray........0 2				
Coal, hard, gnarly.1 1				
Slate, dark........0 0½	4	8½	1152	8½
Coal, splint.......1 0				
Shale	1	3½	1154	0
Sandy shale and concealed...........	196	0	1350	0
Coal ....0′ 11″ } **Eagle**				
Slate 0 2½				
Coal ....2 3 ...	5	7	1355	7
Slate 0 8				
Coal 1 6½				
Sandy shale........................	2	5	1358	0
Sandstone and concealed to 920′ B....	22	0	1380	0

The above section was taken to the dip, so that the intervals given between the different coals are too great.

SUMMARY.

The following table gives the intervals between the different coals in the Allegheny and Kanawha Series, with reference to the Campbell Creek (No. 2 Gas) Coal, taken at different points in Kanawha, Boone and Logan Counties:

Intervals of the Coals in the Kanawha Series Above and Below the Campbell Creek (No. 2 Gas) Coal in and Adjacent to Boone County.

NAME OF COAL SEAMS	KANAWHA COUNTY								BOONE COUNTY													LOGAN CO.	
	Hernshaw	Winifrede	Shrewsbury	Coalburg	North Carbon	South Carbon	Republic	Weveco	High Coal	Coal River Siding	Griffith	Cold Fork (Head of)	Robinson Creek (Mouth of)	Elk Run (Head of)	Bald Knob	Skin Poplar Branch (Mouth of)	Clothier	Dennison Fork (Head of)	Manila	Toney Creek (Head of)	Big Creek Station	Curry	Workman Branch (Near Mouth)
No. 5 Block Coal	707	815	745	785	....	....	806	924	930	945	....	855	902	1015	1042	....	1113	....	....	....	1047	....	....
Stockton-Lewiston Coal	571	....	....	595	687	748	711	730	768	....	705	735	792	....	....	....	903	....	....	....	....	....	745
Coalburg Coal	499	642	516	505	625	655	....	660	666	669	600	....	....	684	688	....	748	....	....	....	....	....	655
Little Coalburg Coal	....	....	....	....	....	....	....	....	....	....	....	....	....	....	....	....	727	....	....	....	....	....	....
Buffalo Creek Coal	....	....	....	....	....	543	....	....	....	....	....	555	....	....	....	....	645	....	....	....	....	....	....
Winifrede Coal	414	462	439	455	509	493	533	522	500	491	495	510	592	....	553	....	543	....	....	....	551	....	560
Chilton "A" Coal	361	....	380	....	....	386	375	....	....	....	....	....	....	539	472	....	494	....	....	....	496	....	....
Chilton Coal	300	....	....	....	339	....	332	....	409	....	....	405	422	435	388	....	410	....	....	....	399	....	380
Little Chilton Coal	261	....	....	....	....	....	....	....	....	335	....	350	....	....	....	....	....	....	....	....	381	....	....
Hernshaw Coal	236	....	....	....	254	286	296	....	....	....	....	....	352	....	341	....	342	....	....	....	338	....	280
Dingess Coal	....	....	....	....	....	....	265	....	....	....	....	....	....	....	....	....	....	....	....	....	....	....	220
Dingess Limestone	....	....	....	....	....	....	....	....	186	....	....	....	211	....	....	....	231	260	....	....	270	....	185
Williamson Coal	161	....	....	....	224	....	248	225	182	....	....	....	178	....	189	....	220	....	....	....	233	....	180
Cedar Grove Coal, Upper	117	150	150	145	152	180	219	....	....	....	156	156	127	....	121	....	149	....	....	....	....	....	130
Cedar Grove Coal, Lower	....	....	....	....	....	123	145	....	....	....	153	....	....	....	....	....	131	....	....	....	116	....	....
Alma Coal	85	102	....	....	....	....	96	63	52	....	....	....	....	....	69	86	105	80	85	95	82	95	85
CAMPBELL CREEK (NO. 2 GAS)	0	0	0	0	0	0	0	0	0	0	0	0	0	0	0	0	0	0	0	0	0	0	0
[illegible]lton Coal	....	....	....	....	38	22	73	....	....	....	....	....	....	....	....	....	....	....	....	....	37	....	....
Eagle Coal	....	....	....	....	196	195	168	....	187	....	....	....	....	....	....	....	....	....	....	....	....	....	....
Little Eagle Coal	....	....	....	....	227	....	....	....	....	....	....	....	....	....	....	....	....	....	....	....	....	....	....

CHAPTER V.

STRATIGRAPHY—THE CONEMAUGH SERIES.

The Conemaugh Series is limited to that division of the rock column that extends from the base of the Pittsburgh Coal to the top of the Upper Freeport Coal.

Dr. I. C. White gives an interesting account of this series in West Virginia in Volume II, pages 225-226, of the State Geological Survey Reports.

In Boone County, the Dunkard, the Monongahela, and the top portion of the Conemaugh Series have been entirely eroded, if they were ever present over this area, so that only the basal members of the Conemaugh Series remain in the northwestern part of the county, throughout a portion of Peytona, Scott and Washington Districts.

DESCRIPTION OF THE CONEMAUGH FORMATIONS.

The following are the principal formations included in the Conemaugh Series in descending order whose geological horizons exist in Boone County:

Pittsburgh Red Shale.
Saltsburg Sandstone.
Bakerstown (Barton) Coal.
Pine Creek (Cambridge) Limestone.
Buffalo Sandstone.
Brush Creek Limestone.
Brush Creek Coal.
Upper Mahoning Sandstone.
Mahoning Coal.
Lower Mahoning Sandstone.

THE PITTSBURGH RED SHALE.

The Pittsburgh Red Shale is the first formation in descending order in the Conemaugh Series that was observed in the rock strata in Boone County.

This Red Shale occurs just below the Ames Limestone horizon and consists of a series of red and brown shales frequently with lime nuggets scattered through them. It crumbles easily and especially when it comes in contact with water, is easily disintegrated and converted into mud. The lime nuggets also dissolve and enrich the soil, so that the latter brings forth excellent grass and finely flavored fruits, like apples, plums and peaches.

Peytona District.—In Peytona District the Pittsburgh Red Shale caps the tops of the hills between Fork and Dicks Creeks, and ranges in thickness from five to twenty feet, containing some lime nuggets.

Scott District.—The Pittsburgh Red Shale occurs in the tops of the hills between Little Hewitt and Dicks Creeks, also on the tops of the range of hills between Peters Cave Fork and main Horse Creek west of Horse Creek at the Boone-Lincoln County Line. It ranges in thickness from five to fifty feet, and is very red and full of limestone nodules. It also occurs on the tops of the high hills between the head of Wash Hill Fork of Horse Creek and Mud River. Its section is as follows:

	Thickness Feet.	Total Feet.
Red limy shale	5	5
Sandstone, buff	29	34
Red lime shale	20	54
Sandstone, Saltsburg	..	..

The Red Shale is frequently almost entirely replaced with sandy shale and sandstone.

Washington District.—The Pittsburgh Red Shale caps a few high knobs between Ballard Fork of Mud River and Jule Webb Fork of Horse Creek, occurring in traces from five to ten feet in thickness.

THE SALTSBURG SANDSTONE.

Underneath the Pittsburgh Red Shale occurs the Saltsburg Sandstone. In Boone County it occurs in the hills under the Pittsburgh Red Shale, and ranges in thickness from twenty to thirty feet, and is massive and buff in color. Very few cliffs of this formation are visible in Boone.

THE BAKERSTOWN (BARTON) COAL.

The Bakerstown (Barton) Coal occurs beneath the Saltsburg Sandstone, and from sixty to ninety feet under the Ames Limestone horizon. It is reported with dark shale and coal, ranging in thickness from one to three feet, but no clean section of the coal was observed.

THE BUFFALO SANDSTONE.

From five to twenty feet underneath the Saltsburg Sandstone occurs another bed of sandstone that has been named the Buffalo Sandstone. It is a coarse, friable sandstone, and frequently makes massive cliffs ranging in thickness from thirty to sixty feet. In Boone County this formation extends over a very small area in the northwestern part of the county.

THE BRUSH CREEK COAL.

Neither the Brush Creek Coal nor its overlying Limestone was observed in Boone, although the coal appears to have been drilled through in the core test of the Standard Fuel Company, 2.2 miles northwest of MacCorkle, as shown on page 67, where it had a thickness of 3′ 9″, including 5″ of slate and bone.

THE MAHONING SANDSTONE.

From one to twenty feet below the Brush Creek Coal horizon occurs a massive sandstone from 30 to 60 feet thick that correlates with the Mahoning Sandstone. This sandstone is often separated by a thin layer of dark slate and fire clay, from the overlying Sandstone (Buffalo) and the two beds form cliffs 60 to 75 feet thick along the tops of the ridges in the northeastern and eastern parts of Boone County. However, it covers a small area of Boone County, and it is only found on the tops of the high ridges.

CHAPTER VI.

STRATIGRAPHY—THE ALLEGHENY SERIES.

The Allegheny Series begins at the top of the Upper Freeport Coal and extends down the rock column to the Homewood Sandstone. Dr. I. C. White gives an interesting account of the same in West Virginia on pages 333-341 of Volume II of the State Survey Reports. The crop of these rocks in Boone County, as shown in detail on Map II, is confined to the northern and eastern portions of the county, and only on top of the high ridges and summits.

The general section of the Allegheny Series is given on page 44 of this volume.

DESCRIPTION OF THE ALLEGHENY FORMATIONS.

THE UPPER FREEPORT COAL.

The topmost formation of the Allegheny Series is the **Upper Freeport Coal** bed which was so designated by the First Geological Survey of Pennsylvania, from the town of Freeport in western Pennsylvania, where it is a multiple-bedded seam, a characteristic feature that accompanies it everywhere in West Virginia.

The following section located on Peters Cave Fork of Horse Creek, Lincoln County, three miles northwest of Mistletoe P. O., and one-half mile west of the Boone-Lincoln County Line, in the local fuel mine of C. Wilkinson, illustrates its structure and character in the area discussed in this volume:

Section of C. Wilkinson's Coal Opening.

			Ft.	In.
Sandstone, massive			30	0
Coal, impure	0'	8"		
Coal, hard splint	1	6		
Slate, dark	0	8		
Coal, with streaks of slate	1	0		
Coal, hard block, (slate floor)	0	8	4	6

In Boone County, the Upper Freeport Coal is near the tops of the hills, and, therefore, has not been opened for local fuel use, and since the area underlain by this seam is so small compared with that of the Kanawha Coals, the bed has been practically ignored, and hence it was not possible to get measurements showing the true thickness of the coal in Boone.

THE UPPER FREEPORT SANDSTONE.

Underneath the Upper Freeport Coal, there comes a buff sandstone from 20 to 50 feet thick, medium coarse and often massive, that would correspond with the **Upper Freeport Sandstone** of Pennsylvania. The character of this bed is revealed in the vertical sections already given on previous pages.

THE LOWER FREEPORT COAL.

The **Lower Freeport Coal,** occurring under the Upper Freeport Sandstone, was not recognized in the detailed study of Boone County.

THE EAST LYNN SANDSTONE.

From 60 to 100 feet below the Upper Freeport Coal is a grayish buff, massive, nearly always pebbly sandstone that has been designated the **East Lynn Sandstone.** This sandstone is often full of iron ore nodules. It is hard and does not erode easily, and thus usually forms massive, rugged, projecting cliffs.

Peytona District.

In Peytona District, the **East** Lynn **Sandstone** occurs near the Boone-Kanawha County Line, at the mouth of Fork Creek, 275 to 300 feet above the level of the creek, at an elevation of 915′, and shows a thickness of 30 to 40 feet. Near the mouth of River Fork, the East Lynn Sandstone forms rugged cliffs along the tops of the hills at an elevation of 1150′ to 1175′ B., showing a rapid rise of the strata to the south.

Continuing up Fork Creek to the mouth of Jimmy Fork, the East Lynn Sandstone forms rugged cliffs 40 feet high, 475 to 500 feet above the bed of the creek, at an elevation of 1250′ B.

On the Boone-Kanawha County Line, north of Dartmont, the East Lynn Sandstone caps the highest points and varies in thickness from 50 to 60 feet, containing iron ore nuggets, and is pebbly.

On the head of Roundbottom Creek, this sandstone sometimes appears in the tops of the highest points at an elevation of 1600 feet, or about 710 feet above the Campbell Creek (No. 2 Gas) Coal. Going to the southwest, and crossing Coal River to the head of Whiteoak Creek in the southern portion of Peytona District, the East Lynn Sandstone is found capping the highest summits in massive, coarse grained sandstone cliffs, from 30 to 50 feet high, at a tidal elevation of 1620 to 1650 feet.

At the head of Drawdy Creek, on the crest of the great Warfield Anticline, this sandstone caps the highest knobs in a coarse grained ledge, 30 to 40 feet thick, at an elevation of 1650 feet; and this ledge is found on the highest peaks in the Scott-Peytona District Line, as far north as Cabell P. O., ranging in thickness from 40 to 50 feet.

Scott District.

The East Lynn Sandstone occurs in the northwestern portion of Scott District. Just south of Pinnacle Tunnel on Little Coal River, this sandstone is in massive cliffs, about 50 feet high, and about 280 feet above the bed of the river, at an elevation of 910′ B.

Passing farther south and west of Little Coal River, at Altman, the East Lynn Sandstone is almost a continuous rugged cliff as far south as Julian, where it forms massive, pebbly cliffs, 60 feet high, at an elevation of 950 feet.

At Van Camp Station, about 3 miles southeast of Julian, the East Lynn Sandstone occurs on the hills at an elevation of 1400′, 85 feet in thickness, coarse grained and massive. It occurs here 620 feet above the Alma Coal, 1795 feet above the Greenbrier Limestone, and 2490 feet above the Berea Sand, as shown in the Van Camp Section, page 72 of this Report.

At the head of Wash Hill Fork of Horse Creek, in the northwestern part of Scott District, near the Boone-Lincoln County Line, the East Lynn Sandstone occurs in a massive, rugged cliff, 40 feet high, at an elevation of 1020′ B.

Washington District.

In the extreme northwestern part of Washington District, the East Lynn Sandstone occurs in great cliffs, 30 to 40 feet high, at the head of Connelly Branch of Mud River, near the tops of the hills. It is coarse grained and often pebbly. Projecting cliffs are also formed by this sandstone near the head of Lukey Fork of Mud River on the Boone-Lincoln County Line. This sandstone just caps the highest peaks on the head of Whiteoak Branch of Spruce Laurel Fork of Spruce Fork, 5.5 miles southeast of Clothier P. O., just south of the Coalburg Syncline.

Sherman District.

The East Lynn Sandstone occurs in the top of the ridge between Coal River and its tributaries and Kanawha River and its tributaries, along the dividing line of the Boone-Kanawha County Line. At the head of Tony Creek, it is 40 feet thick at an elevation of 1550′. At the head of Joes Creek, the sandstone becomes pebbly and is 60 to 80 feet thick, forming bold cliffs. Massive cliffs also occur on the head of Whiteoak Creek of Big Coal River, and vertical cliffs 60 feet thick occur in this sandstone near the Boone-Raleigh County

Line at an elevation of 2130 feet, just southeast of High Coal.

In passing up Laurel Creek from Seth to Gordon, the East Lynn Sandstone is found on the dividing ridge between the waters of Laurel Creek of Big Coal River and the waters of West Fork of Pond Fork. It is the cliff rock along this dividing ridge, capping the highest points to the southeast until southern extremity of Sherman District on Cherry Pond Mountain.

Crook District.

In Crook District the East Lynn Sandstone occurs on Cook Mountain, between West and Pond Forks, capping the highest hills and often 60 to 75 feet thick, frequently pebbly, coarse grained and brownish-buff.

It is this sandstone that caps the highest summits of Guyandot Mountain from Walnut Gap to Pilot Knob, in the extreme southern part of Boone County.

THE UPPER KITTANNING (NORTH COALBURG) COAL.

Under the East Lynn Sandstone, there often occurs a bed of coal which has been opened at several localities, being from 2 to 6 feet in thickness in Boone County, and possibly representing the North Coalburg or Upper Kittanning bed. It is separated from the next lower bed by 2 to 8 feet of shale and slate.

MIDDLE AND LOWER KITTANNING COAL (NO. 5 BLOCK).

Underlying the East Lynn Sandstone and often separated from it by 2 to 20 feet of shale, slate and coal, there occurs one of the most persistent coal beds of the entire Allegheny Series. This bed is mined extensively in the Kanawha Coal Field and is there known as the **No. 5 Block Coal.**

This bed appears to be almost universally present at its proper geological horizon and nearly always in commercial thickness.

This coal appears to be identical with the Lower Kittanning of the Pennsylvania column, or possibly with the Lower and Middle Kittanning combined, as correlated by I. C.

White[1], since its bed is always a multiple one, being separated into two or more benches by partings of shale and bone.

This bed occurs in the northern part of Boone County in Peytona, Scott and Washington Districts, and in eastern Boone, in Sherman and Crook Districts. The bed rises rapidly to the southeast and in the southern part of the county it appears near the tops of the highest summits where it forms a thick multiple-bedded coal.

The crop of the No. 5 Block Coal is shown in detail on Map II by an appropriate symbol.

The openings numbered on the map correspond with the numbers of the openings in the text, showing the approximate location in the county.

The sections of the various openings in this coal bed will be discussed by magisterial districts.

Peytona District.

The Forks Creek Coal Company owns a large tract of coal on the waters of Fork Creek south of Brounland. E. B. Snider of Charleston has recently made several prospect coal openings on this property. The following opening was fallen shut when visited by Teets, but Snider reports the measurement as given below:

Forks Creek Coal Company Opening—No. 1 on Map II.

Located on the east side of Fork Creek, opposite the mouth of Jimmy Fork, 3.8 miles south of Brounland; **No. 5 Block Coal**; elevation, 1295' B.

	Ft.	In.
Shale roof		
Coal, cannel, (slate floor)	4	6

Forks Creek Coal Company Opening—No. 2 on Map II.

Located on the south side of River Fork of Fork Creek, 1.0 mile south of Emmons; **No. 5 Block Coal**; elevation, 1270' B.

			Ft.	In.
Shale roof				
Coal, impure	2'	6"		
Coal, block	2	0		
Slate	0	7		
Coal (slate floor)	1	0	6	1

[1]See Volume II(A), page 495, W. Va. Geol. Survey; 1908.

Forks Creek Coal Company Opening—No. 3 on Map II.

Located on the head of Jimmy Fork of Fork Creek, 3.0 miles southeast of Emmons P. O.; section furnished by Snider; **No. 5 Block Coal**; elevation, 1365' B.

				Ft.	In.
Shale, gray	0'	6"			
Coal (slate floor)	1	4	....	1	10

Scott District.

Mohler Lumber Company Local Mine Opening. No. 4 on Map II.

Located south of the mouth of Little Hewitt Creek and 0.25 mile east of Altman; section by Teets; **No. 5 Block Coal**; elevation, 945' B.

				Ft.	In.
Sandstone, massive					
Coal, gray splint	0'	6"			
Shale, gray	0	0½			
Coal, gray splint	2	3			
Slate, dark	0	2			
Coal, block, (slate floor)	0	7	....	3	6½

Butts, N. 50° W.; faces, N. 40° E.

Mohler Lumber Company Opening—No. 5 on Map II.

Located on east side of Little Coal River, 0.1 mile north of Altman P. O.; section taken by Teets at the local mine of **Columbus Hill**; **No. 5 Block Coal**; elevation, 905' B.

					Ft.	In.
1.	Sandstone, massive				20	0
2.	**Coal**, gray splint	1'	8"			
3.	**Coal**, bony	0	8			
4.	**Coal**, gray splint, visible	1	2	....	3	6

Butts, N. 50° W.; faces, N. 40° E.

The analysis of a sample collected from Nos. 2 and 4, as reported by Messrs. Hite and Krak, is published in the table of coal analyses under **No. 13.**

Tuncil Price Mine—No. 6 on Map II.

Located 0.75 mile north of Julian P. O.; section by Teets; **No. 5 Block Coal**; elevation, 905' B.

					Ft.	In.
1.	Sandy shale and sandstone roof					
2.	**Coal**, bony	0'	8"			
3.	**Coal**, gray splint	1	8			
4.	**Coal**, bony	0	4			
5.	**Coal**, gray splint (fire clay floor)	1	4	....	4	0

The analysis of a sample collected from Nos. 3 and 5, as reported by Messrs. Hite and Krak, is published in the table of coal analyses under **No. 1.**

Tuncil Price is mining the coal for local fuel, and usually mines about 400 bushels of coal per annum.

D. G. Courtney Mine—No. 7 on Map II.

Located on the south side of Wash Hill Fork of Horse Creek, 2.8 miles southwest of Mistletoe P. O.; section secured by Krebs; **No. 5 Block Coal**; elevation, 980' B.

				Ft.	In.
1.	Slate roof				
2.	**Coal**, splint	0'	8"		
3.	Slate	1	6		
4.	**Coal**, impure	1	4		
5.	Slate	0	2		
6.	**Coal**, hard, gray splint	1	6		
7.	**Coal**, soft (slate floor)	1	2	6	4

The coal was formerly mined here for locomotive use in hauling timber by the Leatherwood Lumber Company. A sample was collected for analysis from Nos. 2, 6 and 7, and the analysis, as reported by Messrs. Hite and Krak, is published in the table of coal analyses under **No. 2.**

C. A. Croft Local Mine—No. 8 on Map II.

Located on head of Wash Hill Fork of Horse Creek on north side of same; section taken by Krebs; **No. 5 Block Coal**; elevation, 850' B.

			Ft.	In.
Slate roof				
Coal	1'	0"		
Shale, gray	2	6		
Coal, impure	0	8		
Coal, bony	0	8		
Slate	0	2		
Coal, hard, splint (slate floor)	2	8	7	8

Horse Creek Coal Land Co. Opening—No. 9 on Map II.

Located on the north side of Wash Hill Fork of Horse Creek, 1.5 miles northwest of Mistletoe P. O., on the land of **Ephraim Griffith**; section taken by Krebs; **No. 5 Block Coal**; elevation, 950' B.

			Ft.	In.
Shale, gray				
Coal, semi-cannel	0'	3"		
Slate, gray	0	2		
Coal, semi-cannel	0	5		
Slate	0	2		
Coal, cannel	0	1		
Fire clay	0	8		
Coal, block	0	2		
Coal, impure	0	8		
Coal, block	1	4		
Coal, gray splint	1	8		
Slate	0	1		
Coal, soft (slate floor)	1	4	7	0

C. A. Croft Prospect Opening—No. 10 on Map II.

Located on Wash Hill Fork of Horse Creek, 3.0 miles southwest of Mistletoe P. O.; section taken by Krebs; **No. 5 Block Coal**; elevation, 950' B.

			Ft.	In.
Sandstone				
Shale, bluish			0	8
Coal, semi-cannel	0'	2"		
Slate	0	4		
Coal, block	0	5		
Slate, gray	0	3		
Coal, impure	0	1		
Fire clay	0	7		
Coal	0	2		
Niggerhead	0	7		
Coal, gray splint	2	10		
Coal, block (slate floor)	1	5	6	10

Washington District.

Several prospect openings have been made on the property of the Little Coal Land Company in the northwestern part of Washington District:

Little Coal Land Company Opening—No. 11 on Map II.

Located on Parson Branch of Mud River; 0.7 mile southeast of Mud P. O.; section taken by Krebs; **No. 5 Block Coal**; elevation, 1100' B.

			Ft.	In.
1. Slate roof				
2. Coal, splint	1'	2"		
3. Shale, gray	0	3		
4. Coal, splint, (slate floor)	3	4	4	9

The analysis of a sample collected from Nos. 2 and 4, as reported by **Messrs.** Hite and Krak, is published in the table of coal analyses under **No. 3.**

Little Coal Land Company **Opening—No. 12 on Map II.**

Located on west side of Mud River, 0.25 mile north of mouth of Ballard Fork; section by Krebs; **No. 5 Block Coal;** elevation, 1100' B.

				Ft.	In.
1.	Sandstone, massive				
2.	Slate			0	8
3.	**Coal**	1'	0"		
4.	Slate	0	4		
5.	**Coal,** hard, block, (slate floor)	4	0	5	4

The analysis of a sample collected from Nos. 3 and 5, as reported by **Messrs.** Hite and Krak, is published in the table of coal analyses under **No. 4.**

Little Coal Land Company **Opening—No. 13 on Map II.**

Located on Coal Hollow of Connelly Branch of Mud River, 1.0 mile south of Mud P. O., in Harts Creek District, Lincoln County, 0.25 mile west of the Boone-Lincoln County Line; section taken by Krebs; **No. 5 Block Coal;** elevation, 1125' B.

				Ft.	In.
1.	Sandstone, massive				
2.	Slate			1	6
3.	**Coal,** splint	1'	2"		
4.	Slate	0	1		
5.	**Coal**	0	2		
6.	Slate	0	1		
7.	**Coal** (slate floor)	3	7	5	1

The analysis of a sample collected from Nos. 3, 5 and 7, as reported by **Messrs.** Hite and Krak, is published in the table of coal analyses under **No. 5.**

Little Coal Land Company **Opening—No. 14 on Map II.**

Located on Lukey Fork of Mud River, 1.0 mile south of Mud P. O.; section measured by Krebs; **No. 5 Block Coal;** elevation, 1100' B.

				Ft.	In.
1.	Sandstone, massive				
2.	**Coal,** soft	0'	10"		
3.	**Coal,** hard splint	0	10		
4.	Slate, gray	0	2		
5.	**Coal,** hard, splint	2	0		
6.	**Coal,** gas (slate floor)	1	2	5	0

PLATE XVIII(a)—Irwin Green coal opening, Campbell Creek (No. 2 Gas) Coal, on Drawdy Creek.

PLATE XVIII(b)—"Turtleback" Limestone concretions on Turtle Creek.

The analysis of a sample collected from Nos. 2, 3, 5 and 6, as reported by Messrs. Hite and Krak, is published in the table of coal analyses under No. 6.

Samples collected from the coal sections in openings Nos. 11, 12, 13 and 14 by Clark and Krebs in June, 1912, for the Little Coal Land Company, and analyses made by the Charleston Testing Laboratory, gave the following results:

Analyses of No. 5 Block Coal, Little Coal Land Company.

Coal Opening No. on Map II.	Moisture.	Volatile Matter.	Fixed Carbon.	Ash.	Sulphur.	Phosphorus.	B. T. U.
11	2.56	35.51	56.01	5.92	0.506	.007	13,013
12	2.91	35.92	57.37	3.80	1.305	.012	13,263
13	2.34	33.25	54.33	10.08	0.515	.014	12,575
14	2.78	37.26	57.42	2.54	0.655	.004	13,793
Average	2.65	35.48	56.29	5.58	0.745	.009	13,161

The excessive ash in No. 13 is possibly due to some impurity in the sample that should have been eliminated.

D. G. Courtney Opening—No. 15 on Map II.

Located on the east side of Mud River, 1.4 miles northwest of Mud P. O., Jefferson District, Lincoln County, and 0.5 mile north of the Boone-Lincoln County Line; section taken by Krebs; **No. 5 Block Coal**; elevation, 970′ B.

			Ft.	In.
Slate roof..				
Coal	1′	0″		
Shale, gray..........................	0	2		
Coal, hard splint, (slate floor)........	3	4	4	6

At the head of Dennison Fork of Spruce Laurel Fork, the following openings have been made by A. R. Montgomery:

Dennison Fork Prospect Opening—No. 16 on Map II.

	Thickness Ft. In.	Total Ft. In.
Sandstone, **East Lynn**, visible	20 0	20 0
Coal, Upper Kittanning	4 1	24 1
Sandstone and concealed	45 11	70 0
Slate	0 6	70 6
Coal, hard.......2' 8" / Slate, dark......0 6 / **Coal**............0 5 } **No. 5 Block.** 2660' B.	3 7	74 1

Sherman District.

The No. 5 Block Coal occurs in the dividing ridge between Kanawha and Boone Counties, also in the southern and western parts of Sherman District on the highest hills between Big Coal River and West Fork.

Asa Williams Coal Mine—No. 17 on Map II.

Located on head of Cold Fork of Laurel Creek, 3.5 miles southwest of Nelson P. O.; section taken by Teets near crop; **No. 5 Block Coal**; elevation, 1550' B.

				Ft.	In.
1.	Slate roof				
2.	**Coal**, splint	1'	1"		
3.	Bone	0	4		
4.	**Coal**, hard, block	2	0		
5.	**Coal**, gray splint	1	2		
6.	Shale and fire clay	1	0		
7.	**Coal**, gray splint (slate floor)	1	1	6	8

A sample was collected for analysis from Nos. 2, 4, 5 and 7, and the results as reported by Messrs. Hite and Krak are given in the table of coal analyses under No. 7.

Another measurement was made by Krebs, at the face of the above mine, about 200 feet from the crop, where the following section was obtained:

				Ft.	In.
1.	Slate				
2.	**Coal**, splint	1'	0"		
3.	**Coal**, impure	0	6		
4.	**Coal**, block	1	3		
5.	**Coal**, impure	1	0		
6.	**Coal**, splint	1	6		
7.	Fire clay	0	4		
8.	Slate	0	5		

			Ft.	In.
9. **Coal**, block	0′	6″		
10. Slate	0	1		
11. **Coal**, splint, (slate floor)	1	6	8	1

The analysis of a sample collected from Nos. 2, 3, 4, 6, 9 and 11, as reported by Messrs. Hite and Krak, is published in the table of coal analyses under No. 8.

Lackawanna Coal & Lumber Company Prospect Opening. No. 18 on Map II.

Located on the west side of the Right Fork of Lavinia Fork of Hopkins Fork of Laurel Creek, 3.0 miles south of Nelson P. O.; section taken by Teets; **No. 5 Block Coal**; elevation, 1530′ B.

			Ft.	In.
Shale roof				
Coal, impure	0′	8 ″		
Coal, block	1	1		
Slate, gray	0	1		
Coal, splint	1	7		
Slate, gray	0	8		
Coal	0	0½		
Slate, dark	0	5		
Coal, gray splint (slate floor)	2	11	[illegible]	5½

Winifrede Coal Company Opening—No. 19 on Map II.

Located on the head of Spruce Fork of Joes Creek, 3.2 miles northeast of Comfort P. O.; section taken by Krebs; **No. 5 Block Coal**; elevation, 1345′ B.

			Ft.	In.
Slate roof				
Coal, splint, hard	2′	6″		
Shale, dark	0	2		
Coal, block (slate floor)	2	0	4	8

The Anchor Coal Company has an operating coal plant on the head of Seng Creek, southwest of Kayford. The No. 5 Block Coal occurs near the tops of the high hills. The following is a section taken in a prospect opening made by this company:

Anchor Coal Company Opening—No. 19A on Map II.

No. 5 Block Coal; elevation, 1990′ B.

			Ft.	In.
Sandstone, massive, **East Lynn**				
Slate			0	2
Coal, splint	6′	9″		
Slate	0	1		
Coal, splint, (slate floor)	1	2	8	0

The coal appears very pure, black and glossy in the above opening.

Crook District.

The No. 5 Block Coal occurs on top of the highest hills in Cook and Cherry Pond Mountains.

E. J. Berwind Prospect Opening—No. 20 on Map II.

Located on the west side of West Fork, 0.5 mile south of Chap P. O.; section taken by Teets; **No. 5 Block Coal**; elevation, 2020′ B.

				Ft.	In.
1.	Slate roof				
2.	**Coal**, gray splint	7′	7″		
3.	Shale, gray	0	10		
4.	**Coal**, gray splint	1	5		
5.	**Coal**, block	1	8		
6.	Bone **coal**	0	2		
7.	**Coal**, gray splint	1	9		
8.	Slate, gray	1	9		
9.	**Coal**, gray splint	0	7		
10.	Slate, gray	0	4		
11.	**Coal**	0	2		
12.	Slate, gray	0	8		
13.	**Coal**, block	1	9		
14.	Bone	0	1		
15.	**Coal**, gray splint (slate floor)	3	5	22	2

A sample was collected for analysis from Nos. 2, 4, 5, 7 and 9 and the results, as reported by Messrs. Hite and Krak, are given in the table of coal analyses under **No. 12.**

Another sample was collected for analysis from Nos. 11, 13 and 15, and the results, as reported by Messrs. Hite and Krak, are given in the table of coal analyses under **No. 9.**

Rowland Land Company Opening—No. 21 on Map II.

Located on Cherry Pond Mountain, on the west side of Marsh Fork, 1.3 miles northwest of Launa P. O., Marsh Fork District, Raleigh County, 0.5 mile east of the Boone-Raleigh County Line; section taken by Teets; **No. 5 Block Coal**; elevation, 2470' B.

			Ft.	In.
Sandstone				
Coal	0'	8"		
Fire clay	1	9		
Coal, block	1	1		
Shale, gray	1	5		
Coal, block (slate floor)	4	8	9	7

Lackawanna Coal and Lumber Company Opening. No. 22 on Map II.

Located on Lick Branch of Whites Branch, 2.1 miles northeast of Gordon P. O.; section by Krebs; **No. 5 Block Coal**; elevation, 1560' B.

				Ft.	In.
1.	Slate roof				
2.	**Coal, block**	4'	2"		
3.	Slate	0	1		
4.	**Coal, hard splint**	0	6		
5.	Slate	0	5		
6.	**Coal, hard splint**	1	2		
7.	Slate	0	6		
8.	**Coal, splint**	0	5		
9.	Slate	1	0		
10.	**Coal, splint (slate floor)**	1	0	9	3

The analysis of a sample collected from Nos. 2, 4, 6, 8 and 10, as reported by **Messrs.** Hite and Krak, is published in the table of coal analyses under **No. 10.**

The No. 5 Block Coal is mined by W. C. Cook, just east of Bald Knob, where the following section was obtained by Krebs:

W. C. Cook Mine—No. 23 on Map II.

				Ft.	In.
1.	Sandstone, massive			8	0
2.	**Coal**	1'	0"		
3.	**Coal, splint**	3	0		
4.	**Coal, splint, hard (slate floor)** 2180' B.	2	5	6	5

The analysis of a sample collected from Nos. 2, 3 and 4, as reported by **Messrs.** Hite and Krak, is published in the table of coal analyses under **No. 11.**

Mr. J. S. Cunningham, of Charleston, W. Va., collected a sample from the entire section of the above mine and the analysis made by McCreath gives the following results:

	Per cent.
Moisture	2.12
Volatile Matter	33.74
Fixed Carbon	55.23
Ash	8.11
Sulphur	0.80
Total	100.00
B. T. U.	13,537

E. J. Berwind Prospect Opening—No. 66 on Map II.

Located on north side of James Creek, 1.6 miles southeast of Chap P. O.; section taken by Teets; **No. 5 Block Coal**; elevation, 1930′ B.

			Ft.	In.
Slate roof				
Coal, hard splint	2′	10″		
Shale	0	8		
Coal, hard splint (slate floor)	2	10	6	4

E. J. Berwind Prospect Opening—No. 67 on Map II.

Located on east side of West Fork, 2.8 miles southeast of Chap P. O.; section taken by Teets; **No. 5 Block Coal**; elevation, 2160′ B.

			Ft.	In.
Slate roof				
Coal, block	1′	2 ″		
Slate, gray	0	0½		
Coal, gray splint	4	10	6	0½

E. J. Berwind Local Mine Opening—No. 68 on Map II.

Located on the north side of James Creek, 1.5 miles southeast of Chap P. O.; section taken by Teets; **No. 5 Block Coal**; elevation, Lower Bench, 2060′ B.

						Ft.	In.
Slate roof							
Coal	0′	5″	Upper Bench	7′	7″		
Slate, black	0	2					
Coal, block	3	0					
Coal, gray splint	4	0					
Slaty shale				2′	6″		
Coal, soft	1′	0″	Middle Bench	6′	5″		
Coal, gray splint	1	6					
Coal, block	1	4					
Slate, black	0	2					
Coal, gray splint	1	1					
Slate, gray	0	8				29	3
Coal	0	8					
Sandstone and concealed				4′	0″		
Slate, gray	0′	3″	Lower Bench	8′	9″		
Coal	0	1					
Slate, gray	1	1					
Coal, block	1	11					
Slate, black	0	0½					
Coal, block	1	5½					
Coal, gray splint (slate floor)	3	11					

The lower bench is being mined by **Lewis Jarrell** for local fuel use.

Rowland Land Co. Local Mine Opening—No. 69 on Map II.

Located on eastern slope of Cherry Pond Mountain, on waters of Hazy Creek, 1.5 miles northeast of Launa P. O., Marsh Fork District, Raleigh County, and just east of the Boone-Raleigh County Line; measurement by Teets; **No. 5 Block Coal**; elevation, 2340′ B.

	Ft.	In.
Sandstone roof		
Coal, block (slate floor)	6	8

Rowland Land Co. Prospect Opening—No. 70 on Map II.

Located on eastern slope of Cherry Pond Mountain, 1.8 miles southwest of Hecla P. O., Marsh Fork District, Raleigh County, and just east of the Boone-Raleigh County Line, on head of Big Branch; **No. 5 Block Coal**; elevation, Lower Bench, 2200′ B.

					Ft.	In.
1.	Sandstone					
2.	**Coal, Upper Bench**				8	0
3.	Shale				7	0
4.	**Coal**, splint	2′	0″	**Middle Bench**	3	0
5.	**Coal**, cannel	1	0			
6.	Sandy shale and concealed				65	0
7.	**Coal**, hard splint	3′	6″	**Lower Bench**	7	0
8.	**Coal**, harder splint (slate floor)	3	6			

Here at this prospect opening the parting shales have greatly thickened and separated the great bed into three well defined coal seams, totaling 18 feet of coal and 72 feet of rock partings.

A sample collected from Nos. 7 and 8 by J. S. Cunningham, and analysis made by the chemist for the New River and Pocahontas Consolidated Coal Company, Berwind, W. Va., gave the following results:

	Per cent.
Moisture	2.74
Volatile Matter	33.98
Fixed Carbon	56.01
Ash	6.612
Sulphur	0.658
Total	100.000

Mr. J. B. Dilworth also collected a sample from the above opening for E. V. d'Invilliers, and analysis made by A. S. McCreath, Harrisburg, Pa., gave the following results:

	Per cent.
Moisture	1.700
Volatile Matter	33.800
Fixed Carbon	55.872
Ash	7.900
Sulphur	0.728
Total	100.000

The foregoing analyses agree very closely, and indicate that the bed is an excellent fuel, steam, and domestic coal.

A further discussion of the character and quality and probable available area and tonnage of the No. 5 Block Coal will be given on subsequent pages in the Chapter on Coal.

FLORA OF THE ALLEGHENY SERIES.

Dr. David White, the eminent paleobotanist of the U. S. Geological Survey, has studied the collection of fossil plants made by himself, Mr. M. R. Campbell and others from shales in the Allegheny Series at Mason, Clendenin, Pleasant Retreat and other localities in Clay and Kanawha Counties, and from apparently the same horizon at Furnace Hollow, Wayne County. The list of plants identified from this horizon is given in a paper published by Dr. David White, March, 1900, in the Bulletin of the Geological Society of America, pages 170-172, inclusive, the same list being reproduced in the Detailed County Report of Cabell, Wayne and Lincoln Counties, pages 223-4, it having been previously republished in Volume II, West Virginia Geological Survey, pages 283-4.

PLATE XIX.—Workman coal opening, No. 5 Block Coal, 1 mile north of Bald Knob, Crook District.

CHAPTER VII.

STRATIGRAPHY—THE POTTSVILLE SERIES. NO. XII OF ROGERS.

The Pottsville Series, as agreed upon by geologists, begins with the top of the Homewood Sandstone and extends down through a series of rocks to the Mauch Chunk Red Shale, having a thickness of 300 feet in the northern portion of the State, and 2,500 feet or more in the southwestern part of the same.

"Near the eastern part of the present coal fields was the edge of a great basin extending northeastward to the Anthracite district of Pennsylvania and southwestward to Alabama, into which the rivers from the mountain regions to the southeast poured their load of detrital material until it was filled to a depth of 2,400 feet or more with Carboniferous sediments before the peat marshes could spread westward and northward into western Pennsylvania, southeastern Ohio and northeastern Kentucky, thus making the Pottsville deposits and coal beds of the New River and Pocahontas regions distinctly older than the Pottsville of northern West Virginia, western Pennsylvania and southeastern Ohio.

"According to this view, the most of the Kanawha Series of coals and sediments would belong in the Pottsville of western Pennsylvania, and principally in the Mercer and Connoquenessing stages of the Beaver Group."[1]

In the territory covered by this Report, the Pottsville Series has increased in thickness from 1,500 feet in the northern part to 2,400 feet in the southern part of the county.

[1]I. C. White, Vol. II(A), W. Va. Geol. Survey, pp. 12-13.

Dr. I. C. White has sub-divided the Pottsville Series into three great groups, named, respectively, Upper, Middle and Lower Pottsville, as expressed in the following scheme of classification:

Pottsville Series	Upper	Beaver (Kanawha) Group
	Middle	New River Group
	Lower	Pocahontas Group

DESCRIPTION OF THE KANAWHA SERIES OR UPPER POTTSVILLE FORMATIONS.

The Kanawha Series is the only group of the Pottsville that is exposed above the surface in Boone County. This group has been still further sub-divided by Dr. White into two well marked groups, the Upper and Lower Kanawha.

The Upper Kanawha beds extended from the top of the Homewood Sandstone to the base of a grayish white sandstone (the Lower Winifrede). This division includes those coals that are usually of a blocky or splinty type, being hard and glossy. The Lower Kanawha Series extends from the base of this sandstone to the top of the Nuttall Sandstone, and usually includes the coals that are generally of a softer and more gaseous character, being good coking and by-product coals. The list and names of the formations are given in the generalized section on pages 44-7 of this Report.

THE HOMEWOOD SANDSTONE.

The Homewood Sandstone of I. C. White, which forms the top of the Kanawha Series, is hard and forms massive cliffs and steep bluffs, extending entirely across the State in a southwest direction from Monongalia County on the north to Mingo at the Kentucky State Line.

Peytona District.

In Boone County, Peytona District, the Homewood Sandstone varies in thickness from 60 to 100 feet, as shown in the sections already given. The bed rises out of Coal River between Forks of Coal and Sproul, and passing up Coal River to the mouth of Fork Creek, the sandstone forms cliffs 30 to 40 feet thick. In passing up Fork Creek to the mouth of Jimmy Fork the base of the sandstone rises much faster than the bed of the creek, as is shown in the following section:

	Thickness Feet.	Total Feet.
Sandstone, massive, coarse.................	40	40
Sandy shale..............................	25	65
Sandstone, massive, quartz pebbles, 1175' B.	30	95

At Dartmont, the Homewood Sandstone caps the highest points, at an elevation of 1330 feet, or about 640 feet above the bed of Coal River.

The Homewood Sandstone forms cliffs near the top of the divide between Bull and Lick Creeks, and at the head of Roundbottom, just west of the Peytona-Sherman District Line, near the crest of the Warfield Anticline, this sandstone forms massive cliffs 60 feet high, medium coarse grained, and of a grayish buff color, at an elevation of 1450 feet B., or 575 feet above the Upper Bench of the Campbell Creek (No. 2 Gas) Coal. Crossing Coal River to the southwest near the head of Whiteoak Creek, this sandstone forms bold cliffs 30 to 50 feet high, at an elevation of 1350 feet B., or 575 feet above the Upper Bench of the Campbell Creek (No. 2 Gas) Coal.

The great Warfield Anticline crosses through the divide between Dawdry Creek and Hubbard Fork of Rock Creek, where the Homewood Sandstone is found near the top of the highest summit, forming a massive cliff 40 feet high, coarse grained, at an elevation of 600 feet above the Campbell Creek (No. 2 Gas) Coal.

Scott District.

The Homewood Sandstone rises above the level of Little Coal River, about one mile northeast of the Boone-Kanawha County Line, and rises faster than the bed of the stream in passing up Little Coal River to the south. At 0.75 mile south of Pinnacle Tunnel, 0.5 mile north of Altman P. O., the following section is exposed on the east side of the river:

	Ft.	In.
Sandstone, massive, **East Lynn**	50	0
Coal, No. 5 Block	3	5
Sandstone, Homewood, massive, coarse grained, 835′ B.	71	0

Passing southward along Coal River to Julian, this bed forms great massive cliffs, ranging in thickness from 50 to 60 feet. At Julian, the section shows 62 feet of this bed, at an elevation of 855′ B.

Passing south to the head of Little Horse Creek, the following section is exposed:

	Ft.	In.
Sandstone, Homewood, massive, medium coarse grained	63	0
Coal blossom, **Stockton,** 970′ B.	3	0

Eastward across Little Coal River, near the head of Camp Creek, 1.3 miles west of Cabell P. O., the Homewood Sandstone occurs in two benches, as shown in the following section:

	Feet.	
Sandy shale and sandstone	35	**Homewood**.. 95′
Sandstone, medium coarse	40	
Sandy shale	20	

Elevation, 1430′ B., and about 600 feet above the Campbell Creek (No. 2 Gas) Coal.

Another section near Van Camp Station on Little Coal River was as follows:

	Thickness Feet.	Total Feet.
Sandy shale	10	10
Sandstone, Homewood, massive, cliff rock	90	100
Coal blossom, **Stockton-Lewiston**	0	100

Elevation, 1265′ B., and 515 feet above the Campbell Creek (No. 2 Gas) Coal.

The Van Camp Section, published on page 72 of this Report shows the Homewood Sandstone at this point 1665 feet above the Greenbrier Limestone and 2360 feet above the top of the Berea Sand. Southwest from Van Camp, near the head of Price Branch, the following exhibits a section of the Homewod Sandstone:

	Thickness Feet.	Total Feet.
Sandy shale	25	25
Sandstone, Homewood, massive, 1525′ B.	70	95

About 2 miles east of Madison, near the crest of the Warfield Anticline, the Homewood Sandstone appears, massive, coarse grained, current bedded, buff colored, 75 feet thick, capping the highest points, at an elevation of 1815′ B., and 550 feet above the Hernshaw Coal.

Washington District.

In the northern part of Washington District, the Homewood Sandstone is more or less broken up into sandy shale, and thin ledges of cliff rock, 20 to 30 feet thick, as is shown on the head of Stanley Fork of Mud River, where it forms a cliff 20 feet high, being the lower portion of the bed, at an elevation of 930′ B. However, continuing to the southwest, this sandstone becomes massive again and forms rugged cliffs, as is exhibited in the following section at the head of Mud River:

	Thickness Feet.	Total Feet.
Sandstone, **East Lynn,** coarse grained	60	60
Sandy shale	15	75
Sandstone, Homewood, coarse grained, massive, 1510′ B.	75	150

Almost due south 3.5 miles from the last locality, near the head of Trace Fork of Big Creek, the following section is exposed just northeast of Anchor P. O.:

	Thickness Feet.	Total Feet.
Sandstone, Homewood, massive, forming pinnacle rocks	45	45
Sandy shale and concealed	35	80
Sandstone, **Coalburg,** buff color, massive, medium coarse, 1605′ B.	60	140

The Homewood Sandstone occurs on the dividing ridges between Spruce Laurel Fork and Casey Creek, and Spruce Laurel and Spruce Forks, often forming massive cliffs, projecting on the tops of the short spur ridges, and ranging in thickness from 60 to 80 feet.

Sherman District.

The Homewood Sandstone occurs along the northeastern edge of Sherman District, where it ranges in thickness from 40 to 60 feet. It also occurs on the high ridges between Big Coal River and Pond and West Forks.

The following section on the head of Cold Fork of Laurel Creek exhibits the general character of this bed of sandstone:

	Thickness Feet.	Total Feet.
Sandstone, **East Lynn**, massive	55	55
Coal, No. 5 Block	5	60
Sandstone, Homewood, massive, current bedded, 1430′ B	122	182

Near the mouth of Jarrolds Fork of Hopkins Fork, the Homewood Sandstone appears, as shown in the following section, where its base comes 1430′ B.:

	Thickness Feet.	Total Feet.
Sandstone, Homewood, massive, current bedded, coarse, buffish-gray	90	90
Coal, Stockton-Lewiston, with several slate partings	6	96
Sandy shale	9	105
Sandstone, **Coalburg**	110	215

The Homewood Sandstone appears on the dividing ridge between West Fork and Marsh Fork of Coal River, ranging in thickness from 60 to 90 feet, to the southern boundary of Sherman District.

East of Coal River, on the head of Little Whiteoak Creek, the Homewood Sandstone forms cliffs 60 to 90 feet in thickness, and is buffish-gray and medium coarse grained.

Southward, near the head of Pack Branch of Big Coal River, the following section is exposed by the Homewood Sandstone:

	Feet.
Sandy shale	10
Sandstone, Homewood, buffish-gray, 1540′ B	60
Coal, Stockston-Lewiston	5

Crook District.

The Homewood Sandstone occurs on the dividing ridges in the northern, eastern, southern and western boundary lines of Crook District; also on Cook Mountain between West and Pond Forks.

Just northeast of the mouth of Robinson Creek, the Homewood Sandstone was measured in the following section:

	Thickness Feet.	Total Feet.
Draw slate, No. 5 Block Coal horizon		
Sandy shale	50	50
Sandstone, Homewood, massive	58	108
Coal blossom, Stockton-Lewiston, 1420′ B	2	110

Southeastward, near the mouth of Bull Creek, on the hill just west of Lantie P. O., the following section is exposed:

	Thickness Feet.	Total Feet.
Sandstone, Homewood, massive, large quartz pebbles	65	65
Sandy shale, 1320′ B	30	95

This sandstone bed forms massive cliffs along Bull Creek to its head, ranging in thickness from 60 to 90 feet, and usually carries some large quartz pebbles in the upper portion through a thickness ranging from 2 to 10 feet.

THE STOCKTON-LEWISTON-BELMONT COAL.

From 5 to 20 feet under the Homewood Sandstone, and from 100 to 200 feet under the No. 5 Block Coal, there occurs a great multiple-bedded coal of widely extended distribution, and which is generally present wherever the Kanawha Series has any considerable development. This bed was first mined as a cannel coal opposite Montgomery on the Kanawha River at Cannelton by Mr. Stockton, and was therefore named the **Stockton Coal.** It was also called **Lewiston,** from Lewiston P. O., Winifrede Junction, where it was once mined. The name Lewiston has often been erroneously applied to the Winifrede and Coalburg seams. It has also been called the **Belmont** seam where it is being mined near a village of that name, 23.5 miles southeast of Charleston.

It is a multiple bed, separated by shales, slate and fire clay, and in Boone County there are often two benches of this bed that are of sufficient thickness and purity to be of commercial value.

The sections of the various openings in this coal will be discussed by magisterial districts:

Peytona District.

In Peytona District, the **Stockton-Lewiston Coal** has been uncovered in several prospect openings on Fork Creek by Mr. E. B. Snider for the Forks Creek Coal Company. This bed rises above the railroad grade just south of Sproul Tunnel, and rises more rapidly than the bed of Coal River, until at the mouth of Fork Creek it is 310 feet above the bed of the stream.

Forks Creek Coal Company Opening—No. 23A on Map II.

Located on the west side of Coal River, 0.5 mile southwest of Brounland, Washington District, Kanawha County, and 0.25 mile north of the Boone-Kanawha County Line; section by E. B. Snider; **Stockton-Lewiston Coal**; elevation, 922' B.

			Ft.	In.
Shale roof				
Coal	1'	9"		
Sandstone	5	0		
Coal (slate floor)	1	6	8	3

At the mouth of Locust Fork of Fork Creek another prospect opening was measured by Teets:

Forks Creek Coal Company Opening—No. 24 on Map II.

Located on the east side of Locust Fork of Fork Creek, 0.5 mile above its mouth; **Stockton-Lewiston Coal**; elevation, 935′ B.

			Ft.	In.
Shale roof				
Coal, soft, weathered	2′	6″		
Fire clay	2	8		
Coal, weathered	4	0	9	2

Forks Creek Coal Company Opening—No. 25 on Map II.

Located near the head of Locust Fork, 0.5 mile east of Kanawha-Boone County Line; **Stockton-Lewiston Coal**; elevation, 980′ B.

			Ft.	In.
Sandstone roof				
Coal, gray splint	2′	1″		
Fire clay	0	6		
Coal, splint	0	7		
Fire clay	0	5		
Coal, block (fire clay floor)	0	6	4	1

Staunton Grailey Opening—No. 25A on Map II.

Located on the east side of Fork Creek, 0.25 mile north of the mouth of Jimmy Fork; section by Teets; **Stockton-Lewiston Coal**; elevation, 990′ B.

			Ft.	In.
1. Shale roof				
2. **Coal**, gas	0	6		
3. Fire clay	0	6		
4. **Coal**, splint (fire clay floor)	1	6	2	4

Butts, N. 54° E.; Faces, N. 26° W.

The analysis of a sample collected from Nos. 2 and 4 of the above section, as reported by Messrs. Hite and Krak, is given in the table of coal analyses under **No. 14A**.

In the southern part of Peytona District, this bed appears in the highest points of the hills. An old abandoned opening, located on the divide between Brush Creek and Big Coal River, one mile south of Brushton, showed 2 feet of coal at an elevation of 1450′ B.

Another opening located on the divide between Brush Creek and Wilderness Fork of Fork Creek, at an elevation of 1420′ B., was fallen shut. The coal was formerly mined here by **Mack Walker** and reported to be 3 feet thick and of excellent quality as a domestic coal.

Scott District.

The Stockton-Lewiston Coal rises out of Little Coal River just south of Blue Tom Tunnel of the Coal River Branch of the C. & O. Ry., and from this point to MacCorkle the river has a meandering course, the general direction of which is nearly on the line of strike, so that at the mouth of Cobb Creek the coal is about 20 feet above the railroad grade. From MacCorkle, the course of the river is almost due south, and the Stockton-Lewiston Coal rises faster than the bed of the stream.

Horse Creek Coal Land Co. Local Mine—No. 26 on Map II.

Located on the south side of Little Horse Creek, 0.25 mile west of Julian; section measured by Teets; **Stockton-Lewiston Coal**; elevation, 895' B.

			Ft.	In.
Sandstone roof				
Coal, splint	1'	0"		
Slate, gray	0	1		
Coal, gray splint	0	5		
Slate, gray	0	2		
Coal, gray splint	0	10		
Slate, dark	0	1		
Coal, splint	0	7		
Bone	0	2		
Coal, (slate floor)	0	4	3	8

Butts, N. 39° W.; faces, N. 51° E.

Farther up Horse Creek this seam of coal develops into a thicker bed, and is being mined locally at several places for domestic fuel. Also two mining operations are being installed for the purpose of mining and shipping this seam of coal.

Price Heirs' Local Mine Opening—No. 27 on Map II.

Located on Trace Branch of Horse Creek, 1.0 mile north of Altman P. O.; section taken by Krebs; **Stockton-Lewiston Coal**; elevation, 850' B.

			Ft.	In.
Shale roof				
Coal, splint	4'	0"		
Fire clay	0	9		
Coal, block	3	5		
Slate	0	1		
Coal, block (slate floor)	0	6	8	9

The fire clay parting shown as 9 inches in the above section increases in thickness to the west going up Horse Creek, until it reaches a thickness of 8 to 10 feet.

Horse Creek Land and Mining Company Opening. No. 28 on Map II.

Located on west side of Horse Creek, on Bear Branch, 1.0 mile southwest of Woodville P. O., Duval District, Lincoln County, and 0.75 mile west of the Lincoln-Boone County Line; measured by Krebs; **Stockton-Lewiston Coal**; elevation, 810′ B.

			Ft.	In.
Shale roof				
Coal, splint (slate floor)			4	4

Horse Creek Coal Land Co. Opening—No. 29 on Map II.

Located on the south side of Horse Creek, 0.25 mile west of mouth of Old House Branch, and 1.0 mile south of Woodville P. O., Duval District, Lincoln County, and 0.25 mile north of the Lincoln-Boone County Line; section taken by Krebs; **Stockton-Lewiston Coal**; elevation, 830′ B.

			Ft.	In.
Sandstone, massive				
Coal, impure	1′	0″		
Sandstone	3	0		
Coal, hard splint	3	4		
Slate	0	4		
Coal, gas	0	8	8	4

The coal was mined here for boiler fuel by the railroad contractor when the tunnel on the Horse Creek Branch of the C. & O. Ry. was constructed.

Horse Creek Block Coal Co. Opening—No. 30 on Map II.

Located on Horse Creek, 0.5 mile northwest of Mistletoe P. O., on land of Horse Creek Coal Land Company; **Stockton-Lewiston Coal**; elevation 900′ B.

				Ft.	In.
1.	Slate roof				
2.	**Coal**, splint	0′	8″		
3.	**Coal**, bone	0	2		
4.	**Coal**, large block	3	8		
5.	**Coal**, splint	1	6		
6.	Shale	1	0	7	0
7.	Sandstone and concealed			6	6
8.	**Coal**, block (slate floor)			3	3

The composition and calorific value of the sample from Nos. 2, 4 and 5 only of section, as reported by Messrs. Hite and Krak, are given under **No. 15** in the table of coal analyses.

On Wash Hill Fork, 1.0 miles northwest of Mistletoe, the coal is mined for local fuel use by **Augustus Miller,** on the land of the Horse Creek Coal and Land Company, where the following section was obtained by Krebs:

Horse Creek Coal Land Company Opening—No. 31 on Map II.

			Ft.	In.
Sandy shale				
Coal, semi-splint			1	3
Sandy shale and sandstone			8	9
Coal	0′	6″		
Niggerhead	0	8	5	8
Coal, block	4	6		
Shale			1	0
Sandstone and concealed			7	0
Coal, block			2	0
Concealed to creek, 880′ B.			1	0

The above section shows three main divisions of the Stockton-Lewiston bed.

Horse Creek Coal Land Company Opening—No. 32 on Map II.

Located on a branch of Wash Hill Fork of Horse Creek, 2.2 miles southwest of Woodville; section taken by Teets; **Stockton-Lewiston Coal**; elevation, 865′ B.

			Ft.	In.
1. Slaty shale roof				
2. **Coal**, gas	0′	8″		
3. Slate, gray	2	0		
4. **Coal**, block	0	3		
5. Slate, gray	0	5		
6. **Coal**, block	1	1		
7. Niggerhead	0	3		
8. **Coal**, gray splint	1	7		
9. **Coal**, block, (slate floor)	1	0	7	3

The analysis of a sample collected from Nos. 4, 6, 8 and 9 of the above section, as reported by Messrs. Hite and Krak, is given in the table of coal analyses under **No. 14.**

D. G. Courtney Opening—No. 33 on Map II.

Located on Charley Branch of Mud River, 0.25 mile from its mouth and 1.8 miles northwest of Mud P. O., Jefferson District, Lincoln County, and 1.5 miles west of the Boone-Lincoln County Line; measurements taken by Krebs; **Stockton-Lewiston Coal**; elevation, 800' B.

			Ft.	In.
Slate roof				
Coal, hard	0'	5"		
Slate	0	3		
Coal, gray splint	2	0		
Shale	0	0½		
Coal, hard block	0	5		
Coal, block, softer	0	10		
Niggerhead	0	3		
Coal, hard block, (slate floor)	1	6	5	8½

The coal is mined for local fuel use by **S. V. Mullens** at this place.

D. G. Courtney Opening—No. 34 on Map II.

Located on the north side of Mud River, 0.25 mile west of the mouth of Sugartree Branch, 1.9 miles northeast of Mud P. O., and just west of the Boone-Lincoln County Line; measurements taken by Krebs; **Stockton-Lewiston Coal**; elevation, 900' B.

			Ft.	In.
Sandstone, massive				
Slate	1'	0"		
Coal, splint	1	0		
Slate	0	2		
Coal, splint	1	10		
Slate	0	1		
Coal, splint (slate floor)	2	0	5	1

The coal is mined in several places for local fuel use up Mud River as far as Connelly Branch.

Mary Ann Thompson Opening—No. 35 on Map II.

Located on Connelly Branch of Mud River, 0.25 mile east of Mud P. O., Jefferson District, Lincoln County, and 1.0 mile west of the Boone-Lincoln County Line; section taken by Teets; **Stockton-Lewiston Coal**; elevation, 1000' B.

			Ft.	In.
Sandstone				
Slate	0'	2"		
Coal	0	2		
Coal, impure	0	2		
Coal, good	0	4		

			Ft.	In.
Shale	0′	6″		
Coal, block	0	5		
Slate	0	2		
Coal, splint, visible	0	8		
Slate	0	3		
Coal, block	1	0	3	10

On the east side of Little Coal River, near Van Camp Station, several openings have been made in this seam, as follows:

Pryor and Allen Local Mine Opening—No. 36 on Map II.

Located on the north side of Browns Branch of Little Coal River, 0.3 mile northeast of Van Camp Station; section measured by Teets; **Stockton-Lewiston Coal**; elevation, 1215′ B.

			Ft.	In.
Sandstone, massive, Homewood				
Coal, gray splint	2′	11″		
Shale, gray	1	0		
Coal, block	0	6		
Shale, gray	0	2		
Coal, hard block	0	9		
Slate, dark	0	10		
Coal, splint, visible	1	5	7	7

Pryor and Allen Prospect Opening—No. 37 on Map II.

Located on south side of Browns Branch of Little Coal River, 0.5 mile southeast of Van Camp Station; section taken by Teets; **Stockton-Lewiston Coal**; elevation, 1220′ B.

			Ft.	In.
Sandstone				
Coal, block	0′	11″		
Coal, impure	0	7		
Coal, splint	0	4		
Shale, gray	0	1		
Coal, block	0	10		
Slate, gray	0	1		
Coal, gray splint (slate floor)	1	2	4	0

Pryor and Allen Prospect Opening—No. 38 on Map II.

Located on the south side of Browns Branch, 0.1 mile southeast of Van Camp Station; section taken by Teets; **Stockton-Lewiston Coal**; elevation, 1260′ B.

			Ft.	In.
Sandstone roof				
Coal, impure	0'	7"		
Slate, dark	0	1		
Coal, gray splint	2	10		
Shale, gray	1	0		
Coal, hard	0	4		
Slate, dark	0	6		
Coal, block	0	8		
Shale, gray	0	11		
Coal, gray splint	2	2	9	1

Washington District.

Washington District lies south of Scott District, and there is possibly a considerable area of Stockton-Lewiston Coal of commercial thickness and purity in the northwestern part of the district. However, in the central and southern part the seam is high on the hills, so that only a small area is underlain by this bed.

Yawkey and Freeman Opening—No. 39 on Map II.

Located on the head of Stollings Branch of Spruce Fork, 0.6 mile east of Jeffery P. O.; section taken by Teets; **Stockton-Lewiston Coal**; elevation, 1675' B.

			Ft.	In.
Sandstone, massive, **Homewood**				
Coal, splint	1'	3"		
Coal, cannel (slate floor)	1	0	2	3

The coal has been mined here for local fuel use by D. R. Mullins.

This bed of coal occurs in the divide between Spruce Laurel Fork and Casey Creek.

Yawkey and Freeman Opening—No. 40 on Map II.

Located on the head of Whiteoak Branch of Spruce Laurel Fork, 5.0 miles east of Clothier; section taken by Krebs; **Stockton-Lewiston Coal**; elevation, 1590' B.

			Ft.	In.
Shale, gray				
Coal, splint	0'	4"		
Coal, impure	0	3		
Coal, soft	2	0		
Slate	0	1		
Coal, hard splint	2	0		
Bone	0	1		
Coal, hard splint (slate floor)	1	6	6	3

The coal has been mined here to use in firing boiler for sinking diamond core test hole.

Spruce Laurel Diamond Core Test Hole (136). No. 41 on Map II.

Section of coal reported in core test hole (136) drilled by Boone County Coal Corporation. For section of coal, see pages 123-4 of this Report. The diamond core test hole shows three benches in this bed, making a total section of 24 feet 11 inches.

Cassingham Opening—No. 42 on Map II.

Located on north side of Spruce Laurel, 1.0 mile northeast of mouth of Dennison Fork; prospect opening fallen shut; section by A. R. Montgomery, of Clothier; **Stockton-Lewiston Coal**; elevation, 2020′ B.

	Ft.	In.
Sandstone roof		
Coal (with 13″ parting in center)	5	10

Sherman District.

The Stockton-Lewiston Coal occurs in the hill from 400 to 600 feet above the valleys in Sherman District, so that only a small area in the entire district carries this seam of coal. However, it attains such a thickness and purity that give it considerable commercial value in this area. This seam appears in the road on the divide, in the Boone-Kanawha County Line, between Tony Creek and Lefthand Fork of Lens Creek, at an elevation of 1430 feet. The coal has never been faced up, but looks to be about 3 feet thick.

Allen Foster Local Mine Opening—No. 43 on Map II.

Located in head of Joes Creek, 2.4 miles northeast of Orange P. O.; section taken by Krebs; **Stockton-Lewiston Coal**; elevation, 1425′ B.

			Ft.	In.
1. Slate roof				
2. **Coal**, splint	1′	0″		
3. Slate	0	2		
4. **Coal**, splint (slate floor)	5	11	(	1

PLATE XX.—Band Saw Mill of Lackawanna Coal & Lumber Co., showing mill pond, logs and conveyor line, at

The analysis of a sample collected from Nos. 2 and 4, as reported by **Messrs.** Hite and Krak, is published in the table of coal analyses under **No. 16.**

Thomas Foster Local Mine—No. 44 on Map II.

Located on the head of Joes Creek, 2.0 miles northeast of Orange P. O., and about 0.5 mile south of Opening No. 43 above; section taken by Krebs; **Stockton-Lewiston Coal**; elevation, 1430′ B.

			Ft.	In.
Slate roof				
Coal, splint, hard	2′	6″		
Coal, splint, soft (slate floor)	3	5	5	11

LaFollette-Robson et al. Prospect Opening. No. 45 on Map II.

Located on head of Coal Fork of Cabin Creek, 2.8 miles northeast of Orange P. O., and 0.25 mile east of the Boone-Kanawha County Line; section taken by Krebs; **Stockton-Lewiston Coal**; elevation, 1465′ B.

			Ft.	In.
Sandstone, massive, **Homewood**				
Coal, splint	2′	2″		
Slate	0	1		
Coal, gray splint	0	9		
Slate	0	2		
Coal	0	6		
Slate	0	1		
Coal, splint (slate floor)	0	9	4	6

LaFollette-Robson et al. Opening—No. 46 on Map II.

Located on the head of Left Fork of Whiteoak Creek, 3.5 miles east of Orange P. O.; section taken by Teets; **Stockton-Lewiston Coal**; elevation, 1480′ B.

			Ft.	In.
Sandstone, massive, **Homewood**				
Shale, gray			0	4
Coal, block	0′	7″		
Shale, gray	0	1		
Coal, splint	1	8		
Shale, gray	0	2		
Coal, gray splint	1	0		
Coal, semi-splint	0	3		
Coal, block	0	10		
Slate	0	1		
Coal, block, glossy (slate floor)	0	4	5	0

LaFollette-Robson et al. Prospect Opening.
No. 47 on Map II.

Located on the head of Pack Branch of Coal River, 1.4 miles north of Whitesville P. O.; section taken by Teets; **Stockton-Lewiston Coal**; elevation, 1535' B.

			Ft.	In.
Sandstone, massive, **Homewood**				
Slate, dark			0	2
Coal, splint	1'	7"		
Shale, gray	0	3		
Coal, hard, block	2	1		
Shale, gray	0	4		
Coal, gray splint (slate floor)	1	2	5	5

On Seng Creek, about 4 miles east of No. 47 Prospect, this bed has been opened by the Anchor Coal Company in testing for coal, as follows:

Anchor Coal Company Prospect Opening—No. 48 on Map II.

Located at High Coal, on the south side of Seng Creek, just south of the Anchor Mine; section taken by Krebs, not fully under cover; **Stockton-Lewiston Coal**; elevation, 1970' B.

			Ft.	In.
Sandstone, massive, **Homewood**				
Coal, impure	0'	4"		
Coal, splint	2	8		
Slate	0	2		
Coal, splint	2	4	5	6
Slate (to sandstone)			0	2

On Hopkins Fork of Laurel Creek considerable prospecting for the different coals has been done by the Lackawanna Coal and Lumber Company, which owns a large area of land on Laurel Creek and its tributaries. The Stockton-Lewiston bed appears to be an important seam on this property, and frequently occurs in two benches, both of which are of commercial thickness and purity.

Lackawanna Coal and Lumber Co. Prospect Opening.
No. 49 on Map II.

Located on the west side of Hopkins Fork, 3.5 miles south of Nelson P. O.; section taken by Teets; **Stockton-Lewiston Coal**; elevation, 1495' B.

			Ft.	In.
Sandstone roof				
Coal, gray splint	3′	11″		
Bone	0	1		
Coal, gray splint (slate floor)	1	10	5	10

Lackawanna Coal and Lumber Co. Opening. No. 50 on Map II.

Located on east side of Lots Branch of Little Jarrolds Fork of Hopkins Fork, 5.5 miles south of Nelson P. O.; section taken by Teets; **Stockton-Lewiston Coal, Lower Bench**; elevation, 1475′ B.

			Ft.	In.
1. Slate roof				
2. **Coal**, gray splint	1′	1″		
3. Shale, dark	0	1		
4. **Coal**, splint	3	4		
5. Slate, gray	0	7		
6. **Coal**, gray splint	0	8		
7. **Coal**, gas (slate floor)	0	6	6	3

The analysis of a sample collected from Nos. 2, 4, 6 and 7, as reported by Messrs. Hite and Krak, is published in the table of coal analyses under No. 17.

Lackawanna Coal and Lumber Company Prospect Opening. No. 51 on Map II.

Located on east side of Lots Branch of Little Jarrolds Fork of Hopkins Fork, 5.5 miles south from Nelson P. O., and just south of Opening No. 49; measurements by Teets; **Stockton-Lewiston Coal, Upper Bench**; elevation, 1545′ B.

			Ft.	In.
Slate roof				
Coal, soft, weathered	4′	2″		
Slate, black	0	3		
Coal, gray splint	1	0		
Shale, gray	0	4		
Coal, gas (slate floor)	0	6	6	3

Lackawanna Coal and Lumber Company Opening. No. 52 on Map II.

Located on the west side of Little Jarrolds Fork of Hopkins Fork, 5.7 miles south of Nelson P. O.; section taken by Teets; **Stockton-Lewiston Coal, Lower Bench**; elevation, 1580′ B.

			Ft.	In.
Sandstone roof				
Coal, soft	4′	1″		
Shale, gray	0	5		
Coal, block, visible	1	0	5	6

Butts, N. 58° W.; faces, N. 32° E.; greatest rise to the southeast; opening partly fallen shut.

Lackawanna Coal and Lumber Company Prospect Opening. No. 53 on Map II.

Located on the west side of Little Jarrolds Fork of Hopkins Fork, just west of No. 52; section taken by Teets; **Stockton-Lewiston Coal, Upper Bench**; elevation, 1635′ B.

			Ft.	In.
Slate roof				
Coal, soft	3′	0″		
Coal, impure	2	10	5	10

Lackawanna Coal and Lumber Company Prospect Opening. No. 54 on Map II.

Located on the east side of Jarrolds Fork of Hopkins Fork, 6.9 miles south of Nelson P. O.; section taken by Teets; **Stockton-Lewiston Coal, Upper Bench**; elevation, 1630′ B.

			Ft.	In.
Slate roof				
Coal, block	0′	4″		
Shale, dark	0	3		
Coal, block	1	2		
Slate, gray	0	1		
Coal, splint	3	9		
Niggerhead	0	7		
Coal, splint (slate floor)	0	9	6	11

Lackawanna Coal and Lumber Company Prospect Opening. No. 55 on Map II.

Located on the east side of Hopkins Fork, 7.0 miles southeast of Nelson P. O.; section taken by Teets; **Stockton-Lewiston Coal, Lower Bench**; elevation, 1655′ B.

			Ft.	In.
Slate roof				
Coal, soft	3′	1″		
Fire clay	1	4		
Coal, block, visible	1	0	5	5

Lackawanna Coal and Lumber Company Prospect Opening. No. 56 on Map II.

Located on the east side of Hopkins Fork of Laurel Creek, 6.0 miles south from Nelson P. O.; section taken by Teets; **Stockton-Lewiston Coal**; elevation, 1620' B.

			Ft.	In.
Sandstone roof				
Slate, dark			0	5
Sandstone			0	3
Coal, soft	4'	4"		
Slate, gray	0	1		
Coal, block, visible	0	10	5	3

Lackawanna Coal and Lumber Company Prospect Opening. No. 57 on Map II.

Located on the east side of Hopkins Fork, 6.6 miles southeast from Nelson P. O.; section taken by Teets; **Stockton-Lewiston Coal**; elevation, 1640' B.

			Ft.	In.
Shale, sandy				
Coal, soft, weathered	4'	6"		
Shale, gray	0	8		
Coal, block, visible	0	10	6	0

Crook District.

Crook District lies south of Sherman District and contains considerable of the Stockton-Lewiston seam in the hills at an elevation of 500 to 800 feet above the beds of the streams, and the sections show it to be of commercial thickness and purity.

Lackawanna Coal and Lumber Company Opening. No. 58 on Map II.

Located on the south side of Robinson Creek, 2.0 miles northeast of Lantie P. O.; section taken by Krebs; **Stockton-Lewiston Coal**; elevation, Lower Bench, 1300' B.

	Thickness Ft.	In.	Total Ft.	In.
Sandstone, massive, **Homewood**	6	0	6	0
Shale, gray	1	0	7	0
Coal, splint, **Upper Bench**	3	6	10	6
Concealed, mostly sandstone	29	6	40	0

				Thickness Ft. In.	Total Ft. In.
Coal, impure	0′	6″	Lower Bench		
Coal, splint	2	0			
Shale, dark	0	6		5 0	45 0
Coal, impure	2	0			

The coal appears in two benches at this point, as shown in the above section.

Presumably on the Upper Bench, a sample was collected by J. B. Dilworth, E. V. and I. Nicholas, and analyzed by A. S. McCreath, which gave the following results:

	Per cent.
Moisture	1.422
Volatile Matter	35.128
Fixed Carbon	51.855
Ash	10.873
Sulphur	0.722
Total	100.00

Mr. Dilworth had correlated this bed as the Coalburg Coal.

Lackawanna Coal and Lumber Company Prospect Opening. No. 59 on Map II.

Located on Lick Branch of Whites Branch, 2.1 miles northeast of Gordon P. O.; measurements taken by Krebs; **Stockton-Lewiston Coal**; elevation, 1515′ B.

			Ft.	In.
Slate roof				
Coal, splint	0′	2″		
Slate	0	1		
Coal, splint	1	8		
Coal, impure	1	10		
Coal, splint	0	6	4	3

Lackawanna Coal and Lumber Company Prospect Opening. No. 60 on Map II.

Located on Camps Branch of Whites Branch, 2.6 miles northeast of Gordon P. O.; section measured by Krebs; **Stockton-Lewiston Coal**; elevation, 1480′ B.

			Ft.	In.
Slate roof				
Coal, block	2′	10″		
Slate	0	1		
Coal, splint	0	4		

			Ft.	In.
Shale, gray	0′	10″		
Coal, splint	0	1		
Slate	0	10		
Coal, block	0	10		
Coal, splint	0	4		
Coal, block	1	0		
Slate	0	2		
Coal, block	0	7	[illegible]	11

Westward across Pond Fork to the head of Jacks Branch, this bed of coal is opened in several places.

Yawkey and Freeman Opening—No. 61 on Map II.

Located on head of Jacks Branch, 2.5 miles southwest of Lantie P. O.; section measured by Teets; **Stockton-Lewiston Coal**; elevation, 1390′ B.

				Ft.	In.
1.	Slate roof				
2.	**Coal**, gas	0′	6″		
3.	Shale, gray	0	4		
4.	**Coal**, block	1	4		
5.	**Coal**, gray splint	3	1		
6.	**Coal**, to water	0	6	5	9

The analysis of a sample collected from Nos. 2, 4 and 5, as reported by Messrs. Hite and Krak, is published in the table of coal analyses under **No. 18.** The coal is mined at this point by **Samuel Hunter** and used for local fuel.

Yawkey and Freeman Opening—No. 62 on Map II.

Located on the head of Bull Creek, 3.7 miles south from Lantie P. O.; section measured by Teets; **Stockton-Lewiston Coal**; elevation, 1600′ B.

			Ft.	In.
Slate roof				
Coal, gas	1′	0″		
Shale, gray	0	1		
Coal, block	1	11		
Bone	0	1		
Coal, block	0	10		
Shale	0	1		
Coal, splint, (slate floor)	0	10	4	10

Butts, N. 47° W.; faces, N. 43° E.

Coal mined for local fuel use by **Mrs. Mary Smoot.**

Yawkey and Freeman Opening—No. 63 on Map II.

Located on the west side of Bull Creek, 3.3 miles south of Lantie P. O.; section taken by Teets; **Stockton-Lewiston Coal**; elevation, 1530′ B.

			Ft.	In.
Shale, gray				
Coal, block	1′	4″		
Slate, gray	0	2		
Coal	0	1		
Slate, gray	0	1		
Niggerhead	0	2		
Coal, block	1	5		
Slate, dark	0	2		
Coal, visible	0	8	4	1

Mined for local fuel use by Doughton Heirs.

On Old Camp Branch, some prospecting in the coal seams has been done on the lands of the Wharton estate.

Wharton Estate Opening—No. 64 on Map II.

Located on the south side of Old Camp Branch, 1.0 mile northeast of Pond P. O., just back of R. Price's field; section taken by Krebs; **Stockton-Lewiston Coal**; elevation, 1800″ B.

			Ft.	In.
Sandstone roof				
Coal	3′	6″		
Shale, gray	1	6		
Coal	0	5		
Slate	0	5		
Coal (slate floor)	1	8	7	6

E. J. Berwind Prospect Opening—No. 65 on Map II.

Located on the west side of West Fork, 0.5 mile south of Chap P. O.; section taken by Teets; **Stockton-Lewiston Coal**; elevation, 1850′ B.

			Ft.	In.
Slate roof				
Coal, splint	2′	6″		
Shale	0	1		
Coal	0	4		
Shale	0	9		
Coal, splint	1	9		
Slate	0	3		
Coal (slate floor)	2	2	7	10

PLATE XXI.—Winifrede Coal opening on hill south of Bald Knob.

E. J. Berwind Opening—No. 71 on Map II.

Located in the head of James Creek of West Fork, 3.5 miles southeast of Chap P. O.; section taken by Teets; **Stockton-Lewiston Coal**; elevation, 2275′ B.

			Ft.	In.
Sandstone roof				
Coal, splint (slate floor)			2	7

Pocahontas Coal and Coke Co. Opening—No. 72 on Map II.

Located on the head of a west branch of West Fork, 2.2 miles northeast from Echart P. O.; section by Teets; **Stockton-Lewiston Coal**; elevation, 2595′ B.

			Ft.	In.
Slate roof				
Coal, soft	1′	11″		
Coal, block (slate floor)	2	3	4	2

Coal mined for local fuel use by **George Daniels.**

Near the mouth of West Fork, the Stockton-Lewiston Coal has been prospected back of **Mrs. Betsey Polley's** house, on the Wharton estate:

Wharton Estate Opening—No. 72A on Map II.

Located on west side of Pond Fork, 1.5 miles south of Van P. O.; section taken by Krebs; **Stockton-Lewiston Coal**; elevation, 1485′ B.

			Ft.	In.
Sandy shale				
Coal, splint	3′	6″		
Shale, gray	0	2		
Coal, splint (slate floor)	1	4	5	0

Wharton Estate Opening—No. 73 on Map II.

Located in first hollow below Pond P. O. putting into Pond Fork from the east; 0.6 mile north of Pond P. O.; section taken by A. R. Montgomery; **Stockton-Lewiston Coal**; elevation, 1780′ B.

			Ft.	In.
Sandstone, massive				
Coal	2′	6½″		
Slate	1	4		
Coal	0	5		
Slate	0	3½		
Coal	1	5	6	0

A further discussion of the Stockton-Lewiston Coal, with the estimates of its probable tonnage, will be given in the Chapter on Coal.

THE COALBURG SANDSTONE.

At an interval of 5 to 10 feet under the Lower Bench of the Stockton-Lewiston Coal, and separated from the same by shale and fire clay, there usually occurs a massive, coarse grained, bluish gray sandstone, from 50 to 80 feet in thickness, named the **Coalburg Sandstone** by Dr. I. C. White, from its occurrence over the coal of that name. This sandstone often weathers into "Chimney Towers" and "Table Rocks", where exposed on the summits and spurs of ridges. In Boone County, this bed of sandstone is frequently divided into two benches by a seam of coal named the Buffalo Creek by Ray V. Hennen, in which cases the sandstones are named the **Upper Coalburg** and **Lower Coalburg**.

In Boone County the base of this bed rises out of Big Coal River between Sproul and Brounland, and on Little Coal River the sandstone rises out of the bed of the river between MacCorkle and Altman.

A short discussion of this sandstone will now be taken up by magisterial districts.

Peytona District.

In Peytona District, the base of the **Upper Coalburg Sandstone** appears from 100 to 150 feet above the bed of Coal River, at the mouth of Fork Creek, where the Boone-Kanawha County Line crosses, and it rises rapidly to the southeast toward the Warfield Anticline, on the crest of which this sandstone appears from 650 to 700 feet above the bed of Coal River.

Near the head of Falling Rock Fork of Lick Creek, the following section was measured:

	Ft.	In.
Sandstone, Coalburg, massive, coarse grained, bluish gray, 1365′ B......................	65	0

Another section in the head of Roundbottom Creek shows the following:

	Thickness Feet.	Total Feet.
Sandstone, massive, **Homewood**.............	60	60
Sandy shale, **Stockton-Lewiston Coal horizon**	30	90
Sandstone, Coalburg, massive, buffish-gray, coarse grained.........................	60	150

Elevation of base of sandstone, 1360′ B., and 485′ above the Campbell Creek (No. 2 Gas) Coal at this point.

Near the head of Drawdy Creek, the following section was measured:

	Thickness Feet.	Total Feet.
Sandstone, Coalburg, massive, capping a spur of ridge...........................	55	55
Sandy shale and concealed, **Coalburg Coal horizon**	15	70
Sandstone, **Upper Winifrede,** massive.......	85	155

The base of the Coalburg Sandstone comes 1480′ B., and 530 feet above the Campbell Creek (No. 2 Gas) Coal at this point.

Scott District.

The Coalburg Sandstone appears above Coal River in the northern part of Scott District at the mouth of Dicks Creek, and rising gradually out of Little Coal River as far up as Altman, and from thence the course of the river swings to the southeast, and farther up the river, the Coalburg Sandstone gets higher and higher until at Danville the base of the sandstone is more than 600 feet above the bed of Coal River.

The following measurement taken 0.8 mile south of Pinnacle Tunnel exhibits the structure and character of the sandstone:

	Thickness Feet.	Total Feet.
Coal blossom, Stockton-Lewiston...........	2	2
Sandy shale and concealed.................	6	8
Sandstone, Coalburg, massive, buffish-gray, 770′ B.................................	67	75

On Little Coal River at Julian, the Coalburg Sandstone is 40 feet thick, and has an elevation of 790′ B.

About 2.5 miles up Camp Creek, the following section was taken:

	Ft.	In.
Sandstone, Coalburg, massive, coarse grained....	35	0
Sandy shale..................................	30	0

Elevation of base, 1375′ B., and 475 feet above the Campbell Creek (No. 2 Gas) Coal.

Almost due west from this point, at Van Camp Station, the following section is exposed:

	Thickness Feet.	Total Feet.
Coal blossom, Stockton-Lewiston............	2	2
Sandstone, Coalburg, massive..............	43	45
Coal blossom, Coalburg....................	2	47

Elevation of sandstone, 1175′ B., and 400 feet above the Alma Coal.

On Little Coal River, at Danville, the following section is exposed on the east side of the river:

	Ft.	In.
Sandy shale, Stockton-Lewiston Coal horizon....		
Sandstone, Coalburg, massive, medium coarse grained, buffish-gray, 1320′ B................	95	0

This sandstone forms cliffs, projecting along the ridges and spurs of ridges, on the east side of Coal River, east of Madison, towards Workman Branch. The section on Workman Branch shows at the crest of the Warfield Anticline as follows:

	Ft.	In.
Sandy shale..................................	10	0
Sandstone, Coalburg, massive, forming abrupt cliffs, 1615′ B..............................	65	0
Coal blossom, Coalburg........................	0	2

Washington District.

The Coalburg Sandstone rises out of Mud River, just north of Spurlocksville, and is a massive cliff-maker along Mud River to its head, ranging in thickness from 40 to 90 feet, as shown in the following sections:

	Thickness Feet.	Total Feet.
Sandstone, Homewood, massive.............	75	75
Sandy shale and concealed, Stockton-Lewiston Coal horizon......................	10	85
Sandstone, Coalburg, forming massive cliffs, 1400′ B................................	90	175

The base at this point is 360 feet above the Dingess Limestone.

Passing to the head of Trace Fork of Big Creek, this sandstone is 50 to 60 feet thick and a cliff-maker.

It is this bed of sandstone that forms the "Pinnacle Rocks" between the water-sheds of Mud River and head of Big Creek, and along the ridges and spurs, on each side of Spruce Fork, between Madison and Clothier. The sandstone is usually hard and does not erode easily.

Sherman District.

The Coalburg Sandstone occurs near the tops of the hills, on the head of Tony Creek, with elevation of its base 1434′ L., in the northern part of Sherman District, and on up Coal River, this sandstone forms high projecting cliffs 60 to 75 feet in thickness.

On the head of Little Whiteoak Creek just south of the Boone-Kanawha County Line, the Coalburg Sandstone is 60 feet thick and the elevation of its base is 1370′ B.

On Coal River, at the mouth of Pack Branch, the Coalburg Sandstone shows a thickness of 35 feet at an elevation of 1500′ B. This sandstone forms cliffs along Laurel Creek and its tributaries in the central part of Sherman District.

The following section shows the character of the sandstone 0.5 mile northeast of the mouth of Jarrolds Fork of Hopkins Fork:

	Thickness Feet.	Total Feet.
Sandstone, Homewood, massive	90	90
Coal blossom, Stockton-Lewiston	6	96
Sandstone, Coalburg, cliff-maker, medium coarse, buffish-gray	110	206
Coal blossom, Coalburg, 1395′ B.	2	208

Crook District.

The Coalburg Sandstone occurs from 300 to 800 feet above the bed of the streams in Crook District. It retains its hardness and is a "cliff-maker" in the hills throughout the entire district.

A section at a point 1.0 mile north of Echart P. O. shows the following:

	Feet.
Sandstone, Coalburg, massive, buffish-gray	110
Coal blossom, Coalburg	2
Sandstone, Lower Coalburg, 2270′ B	66

The Coalburg Sandstone comes here about 680 feet above the Campbell Creek (No. 2 Gas) Coal.

Another section at the extreme southwestern part of the district shows as follows:

		Feet.
Sandstone and sandy shale	70′	
Sandstone, massive, 2495′ B	55	125

THE COALBURG COAL.

Underlying the Coalburg Sandstone from 0 to 10 feet is the next seam of coal which has been named the **Coalburg Coal** from its occurrence near a small town of that name on the Kanawha River in Kanawha County, where the coal was first mined on a commercial scale, and where mining operations in this seam first established the character and reputation of the Kanawha "Splint" Coals in the commercial markets of the country.

This bed contains much splint coal as well as alternate layers of soft, or "gas", coal, and one or more partings of shale. Frequently the layers of shale will thicken into several feet of rock material.

The following section was measured at Coalburg in the mine of the **Coalburg-Kanawha Coal Company,** and represents a typical section of the seam:

				Ft.	In.
1.	Sandstone, massive, visible			6	0
2.	**Coal,** hard splint	0′	8″		
3.	Shale, dark	0	5		
4.	**Coal,** splint	0	3		
5.	Shale, dark, soft, iron ore nodules	1	2		
6.	Slate, dark, hard	0	7		
7.	Shale, dark gray, iron ore nodules	4	0		
8.	Sandstone, hard	0	5		
9.	Shale, gray, with iron ore nodules	1	3		
10.	**Coal,** splint	0	10		
11.	Niggerhead	0	3		
12.	**Coal,** hard splint	2	11	12	9

The shale separating Nos. 2, 3 and 4 from Nos. 9, 10 and 11 often thins as the coal increases in thickness. Sometimes it is only represented by a small parting from 0 to 4 feet, as will be shown in the sections given on succeeding pages.

An analysis of a sample from No. 11 (niggerhead) from the above mine, as reported by Messrs. Hite and Krak, shows as follows:

	Per cent.
Moisture	0.88
Volatile Matter	22.37
Fixed Carbon	31.68
Ash	45.07
Total	100.00
Sulphur	0.38
Phosphorus	0.008

In several parts of Boone County, the **Coalburg Coal** is valuable, and the several coal openings on this bed will now be discussed by magisterial districts:

Peytona District.

The Coalburg Coal rises out of Coal River, just west of Brounland, where it is mined for local fuel use, and, passing up Coal River, it rises rapidly until at the mouth of Fork Creek, it is 80 feet above the bed of the river.

The Forks Creek Coal Company made several openings in this seam on its holdings along Fork Creek, and there it ranges from 2 to 3 feet in thickness.

Forks **Creek Coal Company Opening—No. 74 on Map II.**

Located on west side of Coal River, just opposite Brounland Station, Washington District, Kanawha County, and 0.5 mile north of the Boone-Kanawha County Line; section taken by Teets; **Coalburg Coal**; elevation, 620' B.

	Ft.	In.
Sandstone roof		
Coal, hard, block (slate floor, 620' B.)	2	7

Mined for local fuel use by Jacob **Chandler.**

Forks Creek Coal Company Opening—No. 75 on Map II.

Located on west side of Coal River, 0.5 mile south of Brounland Station, just north of the Boone-Kanawha County Line; section measured by Teets; **Coalburg Coal**; elevation, 670′ B.

			Ft.	In.
Sandstone roof, visible			20	0
Coal	0′	6″		
Slate, gray	0	3		
Coal, impure, visible	1	4	2	1
Concealed to Little **Coalburg Coal**			15	0

Forks Creek Coal Company Opening—No. 76 on Map II.

Located on Rock Branch of Fork Creek, 1.0 mile south of Brounland Station; section taken by Teets; **Coalburg Coal**; elevation, 720′ B.

			Ft.	In.
Sandstone, massive				
Coal	0′	6″		
Slate, gray	0	5		
Coal	0	7		
Coal, impure	1	7		
Coal, visible	1	5	4	6

Forks Creek Coal Company Opening—No. 77 on Map II.

Located on west side of Fork Creek at mouth of Locust Fork; section by E. B. Snider; **Coalburg Coal**; elevation, 920′ B. (Teets).

			Ft.	In.
Shale roof				
Coal	2′	0″		
Shale, gray	11	5		
Coal (slate floor)	1	2	14	7

Forks Creek Coal Company Opening—No. 78 on Map II.

Located on Lynn Hollow of Locust Fork; section taken by E. B. Snider; **Coalburg Coal**; elevation, 960′ B.

			Ft.	In.
Coal	2′	8″		
Sandstone	1	2		
Coal	2	6		
Bone	0	3		
Slate	1	3		
Coal	0	7	8	5

Forks Creek Coal Company Opening—No. 79 on Map II.

Located on Pigeon Roost Branch of Locust Fork; section taken by E. B. Snider; **Coalburg Coal**; elevation, 840' B.

			Ft.	In.
Sandstone roof				
Coal	2'	6"		
Bone	0	3		
Slate	1	3		
Coal	0	7	4	[illegible]

Forks Creek Coal Company Opening—No. 80 on Map II.

Located on Pigeon Roost Branch of Locust Fork; section taken by E. B. Snider; **Coalburg Coal**; elevation, 850' B.

			Ft.	In.
Sandstone roof				
Coal	2'	5"		
Bone	0	3		
Coal	0	6		
Slate	0	6		
Coal	0	9	4	5

Forks Creek Coal Company Opening—No. 81 on Map II.

Located on west side of Locust Fork, near head of same; section by E. B. Snider; **Coalburg Coal**.

			Ft.	In.
Sandstone roof				
Coal	2'	10"		
Slate	0	4		
Coal	0	9		
Slate	0	6		
Coal	1	0	5	5

Forks Creek Coal Company Opening—No. 82 on Map II.

Located on north side of Locust Fork, 1.4 miles from its mouth; section taken by E. B. Snider; **Coalburg Coal**; elevation, 875' B.

			Ft.	In.
Slate roof				
Coal	3'	5 "		
Slate and rock	2	5		
Coal	3	2		
Slate	0	2½		
Coal	1	11	11	1½

Forks Creek Coal Company Opening—No. 83 on Map II.

Located on east side of Locust Fork, 1.5 miles from its mouth; section taken by E. B. Snider; **Coalburg Coal**; elevation, 970' B.

			Ft.	In.
Sandstone roof				
Coal	1'	7"		
Slate	0	5		
Coal	1	4	3	4

Forks Creek Coal Company Opening—No. 84 on Map II.

Located in the head of Locust Fork of Fork Creek; section measured by E. B. Snider; **Coalburg Coal**; elevation, 990' B.

			Ft.	In.
Sandstone roof				
Coal	2'	0"		
Slate	0	6		
Coal	2	3	4	9

Forks Creek Coal Company Opening—No. 85 on Map II.

Located on head of Locust Branch near the Boone-Kanawha County Line; section taken by E. B. Snider; **Coalburg Coal**; elevation,

			Ft.	In.
Sandstone roof				
Coal	1'	8 "		
Slate	0	4		
Coal	0	5		
Slate	0	6		
Coal	0	7½	3	6½

Forks Creek Coal Company Opening—No. 86 on Map II.

Located on the east side of Fork Creek, at mouth of Jimmy Fork; section by E. B. Snider; **Coalburg Coal**; elevation, 995' B.

			Ft.	In.
Coal	0'	11"		
Slate	0	3		
Coal	1	1	2	3

Scott District.

Very little development has been made in the Coalburg Coal in Scott District. In the northwestern part of the district, where it rises out of Coal River and Horse Creek, it is usually, as far as it has been proved, thin and of little commercial value, and in the southern part of the district the seam occurs from 500 to 650 feet above the bed of Coal River, so that very little work has been done to prove the thickness of this seam.

Pryor and Allen Coal Prospect Opening—No. 87 on Map II.

Located on the south side of Browns Branch, 0.5 mile southeast of Van Camp Station; section measured by Teets; **Coalburg Coal**; elevation, 1180′ B.

			Ft.	In.
Sandstone roof				
Coal, cannel	1′	6″		
Slate, cannelly	0	6		
Coal, cannel	0	4		
Slate, semi-cannel	1	1		
Coal, impure, visible	1	0	4	5

Washington District.

Washington District does not carry much commercial coal in the Coalburg seam as far as it has been proved. In the northern part, no openings have been made to prove the coal, while in the central and southern parts, the coal was encountered in several diamond core test holes drilled near the tops of the hills.

Core Test Hole No. 145 (Logan), published on page 99 of this Report, shows the Coalburg Coal as follows:

Boone County Coal Corporation Core Test. No. 145 (Logan) on Map II.

			Ft.	In.
Slate			1	0
Coal	1′	11″		
Bone	0	1		
Coal	0	3		
Bone	0	1		
Coal	0	7		
Coal and bone	0	2		
Slate	0	4		
Coal and bone, interlaminated	1	9		
Coal, 1210′ L	1	1	6	3

The bed shows considerable impurities. Another section of the coal is shown in core test hole No. 148 (Logan) published on page 106 of this Report. The coal there presents the following section:

Boone County Coal Corporation Core Test. No. 148 (Logan) on Map II.

			Ft.	In.
Coal and bone, interlaminated	2′	7″		
Slate	0	9		
Coal and bone, interlaminated	2	2		
Shale	0	2		
Coal	0	1		
Fire clay	0	5		
Coal	0	1		
Fire clay	0	7		
Coal	0	1		
Fire clay	0	2		
Sandstone	3	6		
Coal and bone, 1610′ B.	1	8	12	3

The above record reveals much impurity and is a typical section of the bed.

Diamond core test hole No. 136, published on pages 123-4 of this Report, gives another section of this seam, as follows:

Chilton, MacCorkle and Chilton and Meany Core Test Hole No. 136—No. 136 on Map II.

			Ft.	In.
Bone	0′	3″		
Coal	0	9		
Slate	0	1		
Coal	0	3		
Slate	0	1		
Coal	0	6		
Bone	0	6		
Coal and bone	1	0		
Coal	1	6		
Coal and bone	0	4		
Soft shale	1	0		
Coal, (slate floor, 1510′ L.)	1	6	7	9

Another section of the Coalburg Coal is shown in diamond core test hole No. 129, published on page 144 of this Report, where the coal is almost entirely absent.

Cassingham Opening—No. 88 on Map II.

Located on the second hollow from the south above the mouth of Dennison Fork; measurement by A. R. Montgomery; **Coalburg Coal**; elevation, 1920' B.

	Ft.	In.
Sandstone roof		
Coal	2	2

Cassingham Opening—No. 89 on Map II.

Located on main Dennison Fork on the point between the main branches of Dennison, 1.3 miles above its mouth; section by A. R. Montgomery; **Coalburg Coal.**

	Ft.	In.
Sandstone roof		
Coal (slate floor)	2	10½

This is possibly the top member of this seam, the lower portion being doubtless mixed with slate and shale, and of very little commercial value.

Sherman District.

The Coalburg Coal attains its best development in Sherman District of Boone County. Considerable prospect work has been done by the Lackawanna Coal and Lumber Company on their lands in this seam, and it has proved to be a valuable bed of coal in the southern part of the district. A few openings have been made in the northern part of the district, where it is of commercial thickness and purity.

Wm. Price Opening—No. 90 on Map II.

Located on the head of Tony Creek, 3.8 miles north of Comfort P. O.; section measured by Krebs; **Coalburg Coal**; elevation, 1300' B.

				Ft.	In.
1.	Sandstone, **Coalburg**, buffish-gray			10	0
2.	**Coal**, hard splint	2'	6"		
3.	Slate, dark	0	4		
4.	**Coal**, splint (slate floor)	0	8	3	6

The analysis of a sample collected from Nos. 2 and 4, as reported by Messrs. Hite and Krak, is published in the table of coal analyses under **No. 19.** The coal is mined for local fuel use by Mr. Price at this point.

On the south side of Sandlick Creek of Laurel, two openings have been made in this seam:

Lackawanna Coal and Lumber Company Local Mine Opening. No. 91 on Map II.

Located on the south side of Sandlick Creek, 2.0 miles west of Nelson P. O.; section taken by Krebs; **Coalburg Coal**; elevation, 1225′ B.

				Ft.	In.
Sandstone, **Coalburg**, coarse, full of **coal** streaks..				10	0
Coal, splint..........................	3′	4″			
Shale, dark, with **coal** streaks..........	2	0			
Coal, splint, visible...................	1	0	....	6	4

One mile and a half farther up Sandlick Creek, this seam has been opened again, as follows:

Lackawanna Coal and Lumber Company Opening. No. 92 on Map II.

Located on north side of Sandlick Creek, 3.0 miles west of Nelson P. O.; section measured by Krebs; **Coalburg Coal**; elevation, 1255′ B.

				Ft.	In.
Sandstone, **Coalburg**, massive, full of fossil plants, visible				8	0
Shale, gray..................................				0	4
Coal, splint........................	0′	2 ″			
Slate	0	0½			
Coal, splint........................	0	5			
Shale, gray........................	0	3			
Coal, splint........................	0	2			
Coal, impure........................	2	0			
Slate	0	1			
Coal, splint, good..................	1	0			
Slate, gray........................	0	1			
Coal, block........................	0	11			
Slate	0	1			
Coal, splint........................	0	4			
Slate	0	0½			
Coal, splint........................	0	5			
Shale, gray........................	0	7			
Coal, block........................	1	8			
Coal, splint........................	0	1			
Coal, semi-splint, (slate floor).......	0	8	....	9	0

The coal has been mined at this place by **Cyrus Green** for local fuel purposes.

Lackawanna Coal and Lumber Company Local Mine Opening. No. 93 on Map II.

Located on a branch of Cold Fork, 3.0 miles southwest of Nelson P. O.; section measured by Teets; **Coalburg Coal**; elevation, 1285′ B.

			Ft.	In.
Slate roof				
Coal, impure	0′	4″		
Coal, gray splint	2	4		
Niggerhead	0	7		
Coal, gray splint, visible	1	5	4	8

Lackawanna Coal and Lumber Company Opening. No. 94 on Map II.

Located in the head of Right Fork of Lavinia Fork of Hopkins Fork, 4.0 miles south of Nelson P. O.; measurements by Teets; **Coalburg Coal**; elevation, 1240′ B.

				Ft.	In.
1.	Sandstone, **Coalburg**, massive, buffish-gray			15	0
2.	Shale, dark	0′	8″		
3.	**Coal**, gas	0	3		
4.	**Coal**, gray splint	1	11		
5.	Bone	0	3		
6.	**Coal**, hard splint	1	0		
7.	Shale, dark	1	6		
8.	**Coal**, gray splint (slate floor)	1	6	7	1

The analyses of two samples collected from Nos. 3, 4, 6 and 8, as reported by **Messrs**. Hite and Krak, are published in the table of coal analyses under **No. 20.**

Several openings have been made on Lots Branch, as follows:

Lackawanna Coal and Lumber Company Opening. No. 95 on Map II.

Located on the east side of Lots Branch, 5.5 miles south of Nelson P. O.; measurement by Teets; **Coalburg Coal**; elevation, 1255′ B.

			Ft.	In.
Slate roof				
Coal, splint	1′	9″		
Niggerhead	0	5		
Coal, block	1	5		
Slate, gray	0	1		
Coal, splint (slate floor)	0	10	4	6

On the east side of Lots Branch, another opening shows the following section as measured by Teets:

Lackawanna Coal and Lumber Company Prospect Opening. No. 96 on Map II.

				Ft.	In.
1.	Slate roof				
2.	**Coal**, gray splint	1′	10″		
3.	Shale, dark	0	1		
4.	**Coal**, block	0	9		
5.	Shale, dark	0	4		
6.	**Coal**, gnarly	1	10		
7.	**Coal**, semi-splint, 1310′ B.	2	3	7	1

Butts, N. 53° W.; faces, N. 37° E.

The analysis of a sample collected from Nos. 2, 4, 6 and 7, as reported by Messrs. Hite and Krak, is published in the table of coal analyses under **No. 21.**

Lackawanna Coal and Lumber Company Opening. No. 97 on Map II.

Located on Lease No. 43 on Little Jarrolds Fork of Hopkins Fork, 5.7 miles south from Nelson P. O.; section taken by Teets; **Coalburg Coal**; elevation, 1370′ B.

			Ft.	In.
Slate roof				
Coal	0′	8″		
Shale	0	5		
Coal, splint	1	6		
Slate	1	2		
Coal, splint	1	4		
Slate	0	1		
Coal, visible	0	9	5	11

This opening had fallen shut and it was impossible to get measurements to the bottom.

Lackawanna Coal and Lumber Company Opening. No. 98 on Map II.

Located on Little Jarrolds Fork of Jarrolds Fork on Lease No. 43, south of Opening No. 97; section taken by Teets; **Coalburg Coal**; elevation, 1400′ B.

			Ft.	In.
Slate roof				
Coal	2′	6″		
Shale, gray	0	4		
Coal	4	1	6	11

PLATE XXII.—Upper Winifrede Sandstone on hill to south of Bald Knob.

The opening was not fully driven under cover, so the coal appeared soft and weathered.

Lackawanna Coal and Lumber Company Opening. No. 99 on Map II.

Located on the east side of Jarrolds Fork of Hopkins Fork, on Lease No. 32, 7.8 miles south of Nelson P. O.; section measured by Teets; **Coalburg Coal**; elevation, 1450′ B.

				Ft.	In.
1.	Slate, roof				
2.	**Coal**, splint	1′	9″		
3.	Shale, gray	0	1		
4.	**Coal**, block	2	4		
5.	Niggerhead	1	1		
6.	**Coal**, splint	1	9		
7.	Shale, gray	0	3		
8.	**Coal**, splint	0	9		
9.	Shale, dark	0	1		
10.	**Coal**, gray splint	0	4		
11.	Shale, dark	0	1		
12.	**Coal**, gray splint	1	1		
13.	Shale, dark	0	3		
14.	**Coal**, gray splint	2	1		
15.	Slate, dark	0	1		
16.	**Coal**, block, visible	1	6	13	6

Butts, N. 34° W.; faces, N. 56° E.

The analysis of a sample collected from Nos. 2, 4, 6, 8, 10, 12, 14 and 16, as reported by Messrs. Hite and Krak, is published in the table of coal analyses under **No. 22**.

Lackawanna Coal and Lumber Company Prospect Opening. No. 100 on Map II.

Located on the west side and near the three forks of Hopkins Fork, 6.8 miles south from Nelson P. O.; section taken by Teets; **Coalburg Coal**; elevation, 1400′ B.

			Ft.	In.
Slate				
Coal, soft	2′	5″		
Shale, gray	0	4		
Coal, splint	2	0	4	9

Lackawanna Coal and Lumber Company Prospect Opening. No. 101 on Map II.

Located on west side of Hopkins Fork, about 7.2 miles southeast of Nelson P. O.; section measured by Teets; **Coalburg Coal**; elevation, 1472′ B.

			Ft.	In.
Slate roof				
Coal, soft, weathered	2′	2″		
Shale, gray	0	3		
Coal	0	3		
Slate	0	7		
Coal, splint (slate floor)	3	3	6	6

Lackawanna Coal and Lumber Company Prospect Opening. No. 102 on Map II.

Located on the west side of Jarrolds Fork of Hopkins Fork, about 2.0 miles from its mouth; section taken by Teets; **Coalburg Coal**; elevation, 1335′ B.

			Ft.	In.
Shale, gray, roof				
Coal, soft	1′	3″		
Shale, dark	0	1		
Coal, splint	2	1		
Shale, gray	0	7		
Coal, gray splint	1	6		
Shale, gray	0	0½		
Coal, splint, visible	0	8	6	2½

Lackawanna Coal and Lumber Company Prospect Opening. No. 103 on Map II.

Located on the east side of Logan Fork of Hopkins Fork, about 2.0 miles from its mouth; section taken by Teets; **Coalburg Coal**; elevation, 1395′ B.

			Ft.	In.
Slate roof				
Coal	0′	6″		
Shale, gray	0	2		
Coal, soft	2	1		
Niggerhead	0	8		
Shale, dark	0	4		
Coal, splint	1	9		
Shale, gray	0	6		
Coal, visible	0	8	6	0

Lackawanna Coal and Lumber Company Opening. No. 104 on Map II.

Located on the east side of Logan Fork of Hopkins Fork, about 0.5 mile south of Opening No. 103, on Lease No. 21; section taken by Teets; **Coalburg Coal**; elevation, 1420′ B.

			Ft.	In.
Sandstone				
Coal, soft	2′	6″		
Shale, dark	1	0		
Coal, splint	1	6	5	0

The opening had fallen in, so that it was not possible to obtain a full section.

Another opening located on the east side of Logan Fork of Hopkins Fork, about 0.5 mile south of Opening No. 104, gives the following section as measured by Teets, at an elevation of 1450′ B.:

Lackawanna Coal and Lumber Company Prospect Opening. No. 105 on Map II.

			Ft.	In.
Sandstone roof				
Coal	1′	9″		
Shale, gray	0	2		
Coal, splint, visible	1	8	3	[illegible]

The foregoing gives a fair average as to thickness and purity of the Coalburg seam in the southern and central parts of Sherman District, and from these sections it can readily be seen that this is an important coal bed in this part of Boone County.

Near the head of Whiteoak Creek, about 4.0 miles east of Orange P. O., this seam of coal is mined on the property of LaFollette-Robson et al., where Teets measured the following section:

LaFollette-Robson et al. Coal Opening—No. 106 on Map II.

			Ft.	In.
Sandstone, massive, **Coalburg**				
Shale, dark			0	4
Coal, splint	0′	7″		
Shale, gray	0	1		
Coal, black, glossy	1	8		

			Ft.	In.
Slate	0′	2″		
Coal, gray splint	1	0		
Coal, bony	0	3		
Coal, splint	0	10		
Slate	0	1		
Coal, glossy, (slate floor, 1410′ B.)	0	7	5	3

The coal is mined for local fuel by **Silas Massy.**

At High Coal, the Anchor Coal Company has made a section from their mine to the top of the hill. The following is a section measured by Krebs at this point:

Anchor Coal Company Opening—No. 107 on Map II.

	Ft.	In.
Sandstone, massive		
Coal, splint, (slate floor, 1865′ B.)	2	11

This coal is 165 feet above the Winifrede or Dorothy seam mined by the Anchor Coal Company.

Crook District.

From the developments and openings made in the Coalburg seam and the measurements taken, which will now be given in detail, it is evident that this seam is an important bed of coal in Crook District, and that it has considerable commercial value there.

Lackawanna Coal and Lumber Company Opening. No. 108 on Map II.

Located on Lick Branch of Whites Branch, 2.0 miles northeast of Gordon P. O.; section taken by Krebs; **Coalburg Coal**; elevation, 1370′ B.

				Ft.	In.
1.	Dark shale			1	2
2.	Coal, splint	2′	8″		
3.	Dark shale	0	4		
4.	Coal, impure	1	6		
5.	Coal, splint, (slate floor)	1	6	6	0

The analysis of a sample collected from Nos. 2 and 5, as reported by Messrs. Hite and Krak, is published in the table of coal analyses under **No. 23.**

Betsey Polley Prospect Opening—No. 109 on Map II.

Located on west side of Pond Fork, 1.0 mile south of Lantie P. O.; section measured by Krebs; **Coalburg Coal**; elevation, 1325' B.

			Ft.	In.
Shale roof, dark				
Coal, splint	4'	10"		
Dark shale	0	7		
Shale, with layers of **coal**	0	6		
Coal, slaty	0	10		
Shale, gray	0	3		
Coal, splint, (slate floor)	2	7	9	7

Old Camp Branch flows into Pond Fork from the southeast, 4.0 miles above the mouth of West Fork and several test openings have been made on this branch by the Boone County Coal Corporation, under the direction of A. R. Montgomery, who has kindly placed the coal sections of these tests at the disposal of the Survey:

Wharton Estate Prospect Opening—No. 110 on Map II.

Located on south side of Old Camp Branch, 0.75 mile from its mouth; section by A. R. Montgomery; **Coalburg Coal**; elevation, 1600' B. (Krebs).

			Ft.	In.
Sandstone				
Slate			1	1
Coal	2'	4 "		
Slate	0	2½		
Coal	2	3½		
Slate	0	5		
Coal	1	3	6	6

Marion Ferrell Opening—No. 110A on Map II.

Located on west side of Bull Creek, about 0.5 mile west of Lantie P. O.; section by Teets; **Coalburg Coal**; elevation, 1160' B.

			Ft.	In.
1. Sandstone roof				
2. **Coal**, gray splint	1'	8"		
3. Bone	0	2		
4. **Coal**, gray splint	1	1		
5. Slate, gray	0	1		
6. **Coal**, visible, hard block	0	6	3	6

The analysis of a sample collected from Nos. 2, 4 and 6, as reported by Messrs. Hite and Krak, is published in the table of coal analyses under **No. 22A.**

Wharton Estate Opening—No. 111 on Map II.

Located on south side of Old Camp Branch about 1.5 miles from Pond P. O.; section measured by Krebs; **Coalburg Coal**; elevation, 1680′ B.

				Ft.	In.
1.	Slate			1	10
2.	**Coal**, gas	0′	1″		
3.	**Coal**, block	0	1		
4.	**Coal**, gas	0	4		
5.	Slate	0	0½		
6.	**Coal**, gas	0	2		
7.	**Coal**, splint	0	4		
8.	Slate	0	2		
9.	**Coal**, splint	2	1		
10.	Shale, gray	0	2		
11.	**Coal**	0	1		
12.	Shale, gray	0	1		
13.	**Coal**, gas	0	2		
14.	**Coal**, splint	1	0		
15.	**Coal**, splint, hard, (slate floor)	0	6	5	3½

The analysis of a sample collected from Nos. 2, 3, 4, 6, 7, 9, 11, 13, 14 and 15, as reported by Messrs. Hite and Krak, is published in the table of coal analyses under **No. 24.**

Wharton Estate Prospect Opening—No. 112 on Map II.

Located in head of a branch of Pond Fork, 1.0 mile east of Pond P. O.; section taken by A. R. Montgomery; **Coalburg Coal**; elevation, 1650′ B. (Krebs).

			Ft.	In.
Slate roof				
Coal	1′	1″		
Slate	1	0		
Coal	4	8	6	9

On Old House Branch, 1.75 miles northwest of Bald Knob P. O., an opening was made in the Coalburg seam, as follows:

Wharton Estate Prospect Opening—No. 113 on Map II.

Located on Old House Branch on the point between the two main forks, 1.75 miles northwest of Bald Knob P. O.; section taken by A. R. Montgomery; **Coalburg Coal**; elevation, 1720′ B.

			Ft.	In.
Slate roof				
Coal	0′	9″		
Slate	0	3½		
Coal	2	4¾		

			Ft.	In.
Slate	0′	3″		
Coal	2	3		
Slate	0	4		
Coal (slate floor)	1	3	[illegible]	6¼

Wharton Estate Opening—No. 114 on Map II.

Located on Bald Knob trail, 1.0 mile northeast of Bald Knob P. O.; section taken by Krebs; **Coalburg Coal**; elevation, 1819′ B.

			Ft.	In.
Sandy shale			10	0
Coal, cannel	0′	2″		
Shale, gray	1	6		
Coal, gas	2	0		
Slate	0	1		
Coal, block	0	6		
Fire clay	0	4		
Coal, splint	2	0		
Coal, splint, bony, (slate floor)	1	0	7	[illegible]

The coal has been mined here for local fuel use by **Wm. Price.**

Wharton Estate Prospect Opening—No. 114A on Map II.

Located on east side of Pond Fork. 0.8 mile north of Echart P. O.; section taken by A. R. Montgomery; **Coalburg Coal**; elevation, 2280′ B. (Teets).

			Ft.	In.
Sandstone roof				
Coal	2′	3½″		
Slate	0	2		
Coal	1	8		
Slate	0	2½		
Coal (slate bottom)	0	5½	4	9½

On West Fork several prospect openings have been made in the Coalburg seam, as follows:

Wharton Estate Opening—No. 115 on Map II.

Located on the south side of West Fork, 3.7 miles southeast of Gordon P. O.; section taken by Teets; **Coalburg Coal**; elevation, 1495′ B.

			Ft.	In.
1. Slate roof				
2. **Coal**, gas	1′	6″		
3. Slate, gray	0	0½		

			Ft.	In.
4. **Coal**, block	1′	3 ″		
5. Slate, dark	0	2		
6. **Coal**, splint	1	7		
7. Slate, gray	0	0½		
8. **Coal**, block	1	3		
9. **Coal**, gray splint (slate floor)	2	2	8	0

The analysis of a sample collected from Nos. 2, 4, 6, 8 and 9, as reported by Messrs. Hite and Krak, is published in the table of coal analyses under **No. 25.**

Wharton Estate Prospect Opening—No. 116 on Map II.

Located in the head of Jarrolds Branch of West Fork, 1.8 miles northeast of Chap P. O.; section taken by Teets; **Coalburg Coal**; elevation, 1600′ B.

			Ft.	In.
Slate roof				
Coal, block	2′	1 ″		
Slate, gray	0	0½		
Coal, block	0	10		
Slate, gray	0	4		
Coal, impure	0	6		
Slate, gray	0	6		
Coal, splint, visible	2	7½	6	11

James Creek flows into West Fork one-half mile southeast of Chap P. O. The Coalburg Coal has been opened on this creek:

Wharton Estate Prospect Opening—No. 117 on Map II.

Located on north side of Left Fork of James Creek, 1.0 mile east of Chap P. O.; section taken by Teets; **Coalburg Coal**; elevation, 1635′ B.

			Ft.	In.
Slate roof				
Coal, soft	1′	9 ″		
Slate, gray	0	0½		
Coal, splint	0	10		
Impure coal and niggerhead	1	6		
Coal, gray splint	1	10		
Slate, gray	0	1		
Coal, gray splint, visible	1	10½	7	11

PLATE XXIII(a)—Kanawha Series 0.5 mile north of Orange P. O.

PLATE XXIII(b)—View of River Terrace south of Madison.

PLATE XXIII(c)—Winifrede Coal opening, one mile east of Mud, at mouth of Ballard Fork of Mud River.

E. J. Berwind Prospect Opening—No. 118 on Map II.

Located on east side of West Fork, 2.5 miles southeast of Chap P. O.; section measured by Teets; **Coalburg Coal**; elevation, 1975' B.

			Ft.	In.
Sandy shale and sandstone roof				
Coal, impure	2'	1"		
Coal, gray splint	3	7		
Slate, gray	0	0¾		
Coal, gray splint	0	4¼		
Slate, gray	0	5		
Coal, gray splint, visible	2	7	9	1

E. J. Berwind Prospect Opening—No. 119 on Map II.

Located on west side of West Fork, 2.5 miles south of Chap P. O.; section taken by Krebs; **Coalburg Coal**; elevation, 1880' B.

				Ft.	In.
1.	Gray shale, full of fossil plants			6	0
2.	**Coal**, cannel	0'	1"		
3.	**Coal**, splint	0	1		
4.	Shale, dark	0	1		
5.	**Coal**, impure	0	2		
6.	Shale, gray	1	2		
7.	**Coal**, gas	2	3		
8.	Slate, gray	0	1		
9.	**Coal**, gas	0	4		
10.	**Coal**, block	0	1		
11.	Shale, gray	0	4		
12.	**Coal**, splint, harder	0	6		
13.	**Coal**, splint	1	6		
14.	Shale	0	2		
15.	**Coal**, splint, hard	1	0		
16.	**Coal**, block	0	6		
17.	**Coal**, splint (slate floor)	1	0	9	4

The analyses of samples Nos. 430K and 187T collected from Nos. 3, 7, 9, 10, 12, 13, 15, 16 and 17 by Krebs and Teets, respectively, as reported by Messrs. Hite and Krak, are published in the table of coal analyses under **Nos. 26 and 26A**, respectively.

The coal at this opening was sampled by J. S. Cunningham, agent for E. J. Berwind, in 1912, and analyzed by the Chief Chemist of the New River and Pocahontas Consolidated Coal Company at Berwind, W. Va.; also sampled by J. B. Dilworth for E. V. d'Invilliers, in 1903, and analyzed by McCreath, Harrisburg, Pa. The results of these two analyses are as follows:

	Moisture.	Volatile Matter.	Fixed Carbon.	Sulphur.	Ash.
J. S. Cunningham Sample..	1.47	33.71	60.08	0.75	3.99
J. B. Dilworth Sample.....	1.71	34.037	58.025	0.645	5.583

E. J. Berwind Opening—No. 120 on Map II.

Located on the west side of West Fork, opposite the mouth of James Creek; section by J. S. Cunningham; **Coalburg Coal**; elevation, 1750′ B.

			Ft.	In.
Coal	2′	4″		
Slate	1	3		
Coal, hard..............................	1	8		
Slate	0	5		
Coal, hard..............................	1	8	7	4

Another opening measured by Mr. Cunningham, located on Wm. Brown Hollow, on right bank of Sycamore Creek of West Fork, gives the following section 46 feet under cover:

E. J. Berwind Opening—No. 121 on Map II

			Ft.	In.
Coal, soft..............................	2′	2″		
Slate	1	3		
Coal, hard..............................	2	0		
Slate	0	4		
Coal, hard, 1810′ B....................	1	11	7	8

A sample collected by Mr. Cunningham and analyzed by the Chief Chemist of the New River and Pocahontas Consolidated Coal Company at Berwind, W. Va., gave the following results:

	Per cent.
Moisture	1.83
Volatile Matter..............................	32.60
Fixed Carbon..............................	60.86
Sulphur	0.784
Ash	3.926
Total	100.000

The above analysis shows the coal to be of excellent quality, being low in both ash and sulphur and high in fixed carbon.

E. J. Berwind Prospect Opening—No. 122 on Map II.

Located on the south side of James Creek; section taken by J. S. Cunningham; **Coalburg Coal**; elevation, 1720′ B.

			Ft.	In.
Coal, hard	2′	5″		
Slate	1	6		
Coal, soft	1	10		
Slate	0	10		
Coal, hard	1	5	8	0

E. J. Berwind Opening—No. 123 on Map II.

Located on south side of James Creek; section taken by J. S. Cunningham; **Coalburg Coal**; elevation, 1800′ B.

			Ft.	In.
Coal	1′	11″		
Slate	0	4		
Coal, hard	2	1		
Sandy shale	0	4		
Coal, hard	1	7	6	3

E. J. Berwind Opening—No. 124 on Map II.

Located on the west side of James Creek, 1.5 miles above section No. 123; measured by J. S. Cunningham; **Coalburg Coal**; elevation, 1690′ B.

			Ft.	In.
Coal, hard	1′	7″		
Slate	0	1		
Coal	0	10		
Slate	1	7		
Coal, hard	1	8		
Slate	0	3		
Coal, hard	2	0		
Slate	1	0		
Coal, hard	1	2	10	2

The above opening, while showing several slate partings, has nevertheless 7′ 3″ of coal.

E. J. Berwind Opening—No. 125 on Map II.

Located on left fork of Mats Creek, 1.5 miles above its mouth; section measured by J. S. Cunningham; **Coalburg Coal**; elevation, 1890′ B

				Ft.	In.
1.	**Coal**, hard	2′	0		
2.	Slate	0	5		

			Ft.	In.
3. Coal	2′	0″		
4. Coal, impure	0	6		
5. Slate	0	1½		
6. Coal	0	9		
7. Slate	0	1½		
8. Coal, hard	1	2	(	1

A sample collected by J. S. Cunningham from Nos. 1, 3, 6 and 8 and analyzed by the Chief Chemist of the New River and Pocahontas Consolidated Coal Company, at Berwind, in 1912, and also a sample collected from the same numbers by Mr. J. B. Dilworth for E. V. d'Invilliers and analyzed by A. S. McCreath, in 1903, gave the following results:

	Moisture.	Volatile Matter.	Fixed Carbon.	Sulphur.	Ash.
J. S. Cunningham Sample	3.72	33.09	58.62	0.615	3.955
J. B. Dilworth Sample	5.269	31.280	57.044	0.534	5.873

The above analyses show an excessive amount of moisture, which is due to the fact that the samples were from near the crop, and therefore badly weathered.

E. J. Berwind Opening—No. 126 on Map II.

Located on the head of Mats Creek; section No. 37 of J. S. Cunningham; **Coalburg Coal**; elevation, 2190′ B.

			Ft.	In.
Coal, hard	1′	7″		
Slate	0	6		
Coal, hard	2	0		
Slate	0	4		
Coal, hard	2	6	6	11

E. J. Berwind Opening—No. 127 on Map II.

Located on West Fork below the mouth of Mats Creek; section No. 39 of J. S. Cunningham; **Coalburg Coal**; elevation, 1800′ B.

			Ft.	In.
Coal, hard	2′	4″		
Slate	1	4		
Coal, hard	1	7		
Slate	0	4		
Coal, hard	1	2	6	9

E. J. Berwind Opening—No. 128 on Map II.

Located on West Fork, above mouth of Mats Creek; section No. 41 of J. S. Cunningham; **Coalburg Coal**; elevation, 1900' B.

			Ft.	In.
Coal	2'	2"		
Slate	1	7		
Coal	1	6		
Slate	0	4		
Coal	1	5	7	0

Near the head of Big Branch of Marsh Fork in Raleigh County, another section was measured by Mr. Cunningham as follows:

E. J. Berwind Opening—No. 129 on Map II.

			Ft.	In.
Coal, hard	1'	1"		
Slate	0	4		
Coal, hard	1	10		
Slate	0	4		
Coal, hard, 1900' B	1	10	5	5

A further discussion of the **Coalburg Coal**, with estimates of its probable tonnage, will be given in the Chapter on Coal.

THE LITTLE COALBURG COAL.

From 0 to 20 feet under the Coalburg Coal, and separated from the same by layers of impure fire clay and sandy shales, there often occurs a thin bed of coal that has been named the **Little Coalburg Coal**[2] from its proximity to the Coalburg seam, being a split from the latter.

This bed is thin in Boone County, and its structure and character are set forth in the general sections already given. It has little commercial value owing to the thinness of the seam and its impurity.

THE LOWER COALBURG SANDSTONE.

Underneath the Little Coalburg Coal, there occurs another sandstone that has been named the **Lower Coalburg**

[2]Hennen and Reger, Logan-Mingo Report, W. Va., Geol. Survey; 1914.

Sandstone. This bed is from 20 to 40 feet in thickness and forms rugged cliffs. It is medium coarse grained, buffish-gray and very hard. The character and general thickness of this sandstone, at different places in Boone County, are given in the general sections taken at different points, also the diamond core test holes that penetrate the same.

THE BUFFALO CREEK COAL.

From 5 to 9 feet under the Lower Coalburg Sandstone, there occurs a hard, splinty, multiple bedded coal from 0 to 2 feet in thickness that Ray V. Hennen has named the **Buffalo Creek Coal.**

In Boone County this bed is usually thin and of little commercial value, as already shown in the sections taken, and records of diamond core test holes, published on preceding pages of this Report.

THE BUFFALO CREEK LIMESTONE.

Underlying the Buffalo Creek Coal from 10 to 35 feet, there sometimes occurs a gray, hard, lenticular limestone from 1 to 2 feet thick that has been named the **Buffalo Creek Limestone**[3] from its proximity to the Buffalo Creek Coal. This limestone is impure in Boone County and quite often absent. It is often fossiliferous, containing marine fossils in Logan and Mingo Counties; and may contain fossils in a portion of Boone County, but the writer failed to find any.

THE UPPER WINIFREDE SANDSTONE.

Underlying the Buffalo Creek Limestone from 0 to 5 feet occurs a massive, yellowish-gray sandstone, fine grained below, but generally with a coarse grain in its upper half. This sandstone was named by Dr. I. C. White the **Upper Winifrede** from its occurrence over the Winifrede Coal bed. It generally has a smooth lower surface unlike most of the sandstones which form immediate roofs of coal beds, is very regular and

[3]Hennen and Reger, Logan-Mingo Report, W. Va. Geol. Survey; 1914.

does not cut into the underlying coal, but forms an even roof. The coal does not adhere or stick to the sandstone, but separates from it as freely as from a slate roof.

A short discussion of the sandstone bed will now be given by magisterial districts:

Peytona District.

In Peytona District the Upper Winifrede Sandstone comes out of Big Coal River just east of Brounland, and rising rapidly to the southeast, until, at the mouth of Fork Cree.., at the Kanawha-Boone County Line, the base of the sandstone is above the bed of Coal River.

This sandstone forms cliffs along Fork Creek to the south, as shown in the following section, measured at the mouth of Jimmy Fork:

	Feet.
Coal blossom, Coalburg..................................	4
Sandstone and concealed, Lower Coalburg and Upper Winifrede	104
Coal blossom, Winifrede, 900′ B......................	

In passing up Coal River from the mouth of Fork Creek in a southeastern direction, the measures rise more rapidly than the bed of the river, until at Dartmont the base of the Upper Winifrede Sandstone is 525 feet above the bed of Coal River, and 435 feet above the Campbell Creek (No. 2 Gas) Coal, as shown in the following section:

	Feet.
Sandstone, Upper Winifrede, massive, yellowish-gray, 1145′ B..	45
Concealed, Winifrede Coal horizon......................	15
Sandstone and concealed, Lower Winifrede Sandstone...	60

The following section at the head of Falling Rock Fork of Lick Creek exhibits its thickness there:

	Feet.
Sandstone, Coalburg..................................	
Sandy shale and concealed, Coalburg Coal horizon.......	15
Sandstone, Lower Coalburg and Upper Winifrede, massive, upper portion coarse grained.................	95
Sandy shale, Winifrede Coal horizon, 1255′ B...........	

This was taken about 0.5 mile west of the crest of the Warfield Anticline. The above section shows the thickness of the Upper Winifrede Sandstone greater than it is on Roundbottom or Whiteoak Creeks, where it ranges in thickness from 50 to 60 feet.

In passing to the southwest to the head of Brush Creek, the Upper Winifrede Sandstone is found near the tops of the hills, forming massive cliffs, as is shown in the following section:

	Feet.
Sandstone, Upper Winifrede, massive..................	85
Sandy shale, Winifrede Coal horizon....................	15
Sandstone, Lower Winifrede, massive, 1275′ B..........	75

This section was measured just north of the crest of the Warfield Anticline that crosses in a northeast and southwest direction on the divide between Brush and Drawdy Creeks.

Another section of the Upper Winifrede Sandstone, at the head of Drawdy Creek, on the crest of the Warfield Anticline, shows as follows:

	Feet.
Sandy shale and concealed, Coalburg Coal horizon......	
Sandstone, Upper Winifrede, massive..................	85
Sandy shale and concealed............................	30
Coal blossom, Winifrede................................	4
Sandstone, Lower Winifrede, massive, friable, 1325′ B..	2

Scott District.

The Upper Winifrede Sandstone comes out of Little Coal River just south of Pinnacle Tunnel of the Coal River Branch of the C. & O. Ry., and rises gradually above the bed of the river until at Altman the sandstone has an elevation of 735′ B. and is 50 feet thick.

The Upper Winifrede Sandstone rises faster than the bed of Little Coal River from Altman towards Madison, as is shown in the following section taken at Van Camp Station:

	Feet.
Coalburg Coal..	
Sandstone, sandy shale, Buffalo Creek and Upper Winifrede	100
Sandy shale, Winifrede Coal horizon, 1035′ B............	10

Another section on the head of Price Branch of Little Coal River, about 2.3 miles north of the Warfield Anticline, shows the following:

	Feet.
Coal blossom, Coalburg	
Sandstone, Upper Winifrede, massive, top portion coarse grained	85
Sandy shale	25
Bench, Winifrede Coal horizon	..
Sandstone, **Lower Winifrede,** 1370′ B	15

In crossing to the head of Workman Branch near the crest of the Warfield Anticline, the following section was obtained:

	Feet.
Coal blossom, Buffalo Creek	2
Sandstone, Upper Winifrede, massive, coarse grained	68
Sandy shale, **Winifrede Coal horizon**	10
Sandstone, **Lower Winifrede,** massive, 1535′ B	50

The Upper Winifrede Sandstone forms massive cliffs where it occurs on the spurs of ridges.

Just east of Danville, this sandstone occurs near the tops of the hills, 95 feet thick, its base coming 1320′ B., or 620 feet above the bed of the valley.

The following section is exposed on the road leading from Slippery Gut Branch to the head of Horse Creek:

	Feet.
Sandstone and sandy shale, **Buffalo Creek**	30
Coal blossom, Buffalo Coal horizon	3
Sandstone, Upper Winifrede, massive	35
Sandy shale	15
Sandstone and sandy shale, **Lower Winifrede,** 980′ B	57

The Upper Winifrede Sandstone forms cliffs along Horse Creek, south of Mistletoe P. O., ranging in thickness from 30 to 60 feet.

Washington District.

On Mud River the Upper Winifrede Sandstone rises out of that stream near the mouth of Ballard Fork, and is 40 to 60 feet thick.

The following section, taken in the head of Lukey Fork, illustrates its general character at this point:

	Feet.
Sandstone and concealed, **Coalburg**	55
Sandstone and sandy shale, Buffalo Creek and Upper Winifrede Sandstones	100
Bench, **Winifrede Coal horizon,** 1255′ B	10

Another section taken at the head of Mud River shows as follows:

	Feet.
Sandy shale, Coalburg Coal horizon	10
Sandstone, Upper Winifrede, massive, coarse grained	75
Sandy shale, Winifrede Coal horizon	10
Sandstone, Lower Winifrede	71
Coal blossom, Chilton "A," 1235' B	4

The Chilton "A" Coal is here 190 feet over the Dingess Limestone.

The Upper Winifrede Sandstone forms cliffs on Spruce Laurel Fork, ranging from 40 to 75 feet in thickness.

Sherman District.

In Sherman District the Upper Winifrede Sandstone crops in the hills from 400 to 600 feet above the bed of the valley.

The following section was measured just north of Racine:

	Feet.
Sandy shale	10
Sandstone, Upper Winifrede, massive, coarse grained	60
Coal blossom, Winifrede	2
Sandstone, massive, Lower Winifrede, 1135' B	80

The Winifrede Coal is 408 feet above the Campbell Cree' (No. 2 Gas) Coal at this point.

Near the head of Trace Fork of Joes Creek, the following section was measured:

	Feet.
Coal blossom, Coalburg	2
Sandy shale	2
Sandstone, Upper Winifrede, massive, medium coarse grained	73
Coal blossom, Winifrede	5
Sandy shale	2
Sandstone, Lower Winifrede, 1050' B	50

The Winifrede Sandstone forms abrupt bluffs and often cliffs in ascending Coal River from the mouth of Joes Creek. Near the head of Little Whiteoak Creek, the following section shows the character of the sandstone:

	Feet.
Coal blossom	...
Sandy shale and sandstone, Upper Winifrede	113
Coal blossom, Winifrede, 1165' B	5

Near the head of Cold Fork of Laurel Creek the following section is exposed:

	Feet.
Coal blossom, Buffalo Creek	2
Sandstone, Upper Winifrede, massive	47
Coal blossom, Winifrede, 1215′ B	2

Another section, just northeast of the mouth of Jarrolds Fork of Hopkins Fork, exhibits the following section:

	Feet.
Coal blossom, Coalburg	
Sandstone, massive, Buffalo Creek and Upper Winifrede	85
Coal blossom, Winifrede, 1280″ B	5

Near the head of Elk Run of Big Coal River, the following section was measured:

	Feet.
Coal blossom, Coalburg	...
Sandy shale and sandstone, Buffalo Creek and Upper Winifrede	79
Coal blossom, Winifrede	2
Sandy shale and concealed, Lower Winifrede	65
Coal blossom, Chilton "A," 1555′ B	1

The Chilton "A" Coal is 545 feet below the base of the East Lynn Sandstone at this point.

Crook District.

The Upper Winifrede Sandstone forms abrupt bluffs and often massive cliffs from 40 to 75 feet in thickness, as is shown in the sections taken in different parts of Crook District and already given on preceding pages of this Report.

On Pond Fork, in the northern part of the district, near the mouth of Workman Branch, this sandstone occurs about 850 feet above the bed of the stream, as is shown in the following section made near the crest of the Warfield Anticline:

	Feet.
Bench, Coalburg Coal horizon	...
Sandstone, massive, Upper Winifrede	100
Bench, Winifrede Coal horizon	...
Sandstone, Lower Winifrede, massive, 1505′ B	60

In passing up Coal River from this point, the strata dip southeastward into the Coalburg Syncline. The following section, taken 0.25 mile south of Lantie P. O., gives the following succession:

	Feet.
Coal blossom, Buffalo Creek	3
Sandstone and sandy shale, Upper Winifrede	91
Coal blossom, Winifrede, 1050' B	4

The following section, taken near the southern part of Crook District, at the mouth of Burnt Camp Branch, about one mile southeast of Echart P. O., gives the general structure of this sandstone in this part of the district:

	Feet.
Sandstone, massive, Upper and Lower Winifrede	200
Coal blossom, Chilton "A," 2115' B	5

THE WINIFREDE COAL.

Underlying the Upper Winifrede Sandstone, from 0 to 10 feet, is one of the most important coals in the Kanawha Series. It is a multiple bedded, hard, splinty coal with some layers of block coal which render it very desirable for domestic fuel and steam purposes.

This bed was named the **Winifrede Coal** from a mining village of that name on Fields Creek in Kanawha County, 15 miles south of Charleston, where the coal was first mined for commercial purposes as early as 1855.

In Boone County this seam is a very important bed, and will furnish a large proportion of the available coal tonnage. This bed will now be discussed by magisterial districts:

Peytona District.

The Winifrede **Coal** rises out of Big Coal River, between Brounland and the mouth of Fork Creek. It is mined on Brier Creek at Olcott and Dungriff in Kanawha County and is there named the "Black Band Coal." It is a hard, blocky, splint coal, and makes an excellent domestic fuel coal, and does not break or crush easily in transit from the mine to the consumer, and does not crumble when exposed to the weather.

The Forks Creek Coal Company has recently made several test openings on its property on Fork Creek. The tests thus far made prove the coal of excellent quality, but the bed is not very thick, as shown by the following openings:

Forks Creek Coal Company Opening—No. 130 on Map II.

Located on west side of Fork Creek, 0.25 mile above its mouth; section taken by E. B. Snider; **Winifrede Coal**; elevation, 630' B.

	Ft.	In.
Sandstone		
Coal, (slate floor)	1	8

Forks Creek Coal Company Opening—No. 131 on Map II.

Located on the west side of Fork Creek at the mouth of Locust Fork; section taken by Teets; **Winifrede Coal**; elevation, Lower Bench, 840' B.

				Ft.	In.
Sandstone roof					
Coal	0'	10"	Upper Bench	.4	3
Slate	0	3			
Coal	0	8			
Slate	1	4			
Coal	1	2			
Sandstone and concealed				16	9
Coal, splint	1'	3"	Lower Bench	2	7
Coal, soft	0	3			
Coal, splint, (slate floor)	1	1			

The coal appears in two benches at this point.

Forks Creek Coal Company Opening—No. 132 on Map II.

Located near head of Locust Fork of Fork Creek; section taken by E. B. Snider; **Winifrede Coal**; elevation, 910' B.

			Ft.	In.
Coal	0'	11"		
Slate	0	10		
Coal	1	5	3	2

Forks Creek Coal Company Opening—No. 133 on Map II.

Located on east side of River Fork of Fork Creek, about 1.0 mile from its mouth; section taken by E. B. Snider; **Winifrede Coal**; elevation, 675' B.

	Ft.	In.
Slate roof		
Coal, (slate floor)	2	4

Forks Creek Coal Company Opening—No. 134 on Map II.

Located on east side of Fork Creek, just north of Jimmy Fork; section measured by E. B. Snider; **Winifrede Coal**; elevation, 940' B.

	Ft.	In.
Slate roof		
Coal (slate floor)	2	8

In passing up Coal River from the mouth of Fork Creek to Dartmont, the Winifrede Coal rises more rapidly than the bed of the river.

Joseph Dart Opening—No. 135 on Map II.

Located on a branch of Left Fork of Bull Creek, 2.0 miles north of Dartmont; section taken by Teets; **Winifrede Coal**; elevation, 950' B.

			Ft.	In.
Slate roof				
Coal, gas	0'	6"		
Coal, splint, visible	3	6	4	0

Joseph Dart Opening—No. 136 on Map II.

Located on the north side of Bull Creek, 0.5 mile north of Dartmont; section by Dr. I. C. White, published in Volume II(A), W. Va Geological Survey, page 444; **Winifrede Coal**; elevation, 1060' B.

			Ft.	In.
Shale roof				
Coal, soft	0'	9"		
Coal, splint	3	0		
Shale	0	1		
Coal, soft	0	9	4	7

It is possible that the Winifrede **Coal** will be found of sufficient thickness and purity to be of commercial value throughout a considerable portion of Peytona District.

Scott District.

The Winifrede **bed** rises out of Little Coal River near Altman, in Scott District, but very little development has been made to determine its value as a commercial coal.

Horse Creek Coal and Land Company Opening. No. 137 on Map II.

Located on west side of Little Horse Creek, 1.0 mile south of Julian; section taken by Teets; Winifrede Coal; elevation, 820' B.

			Ft.	In.
Slate roof				
Coal, gas	0'	8"		
Fire clay	0	6		
Coal, semi-splint	1	3		
Slate, gray	0	4		
Coal, semi-splint (slate floor)	1	1	3	10

Mined for local fuel by Albert McMechen. In passing southwest to Big Horse Creek, the Winifrede Coal rises out of that stream about one mile south of Mistletoe P. O.

Horse Creek Coal and Land Company Opening. No. 138 on Map II.

Located on east side of Horse Creek, 1.0 mile south of Mistletoe P. O.; section taken by Krebs; Winifrede Coal; elevation, 750' B.

			Ft.	In.
Sandstone, massive			20	0
Coal, splint, visible	2'	0"		
Concealed by water	1	0	3	0

The coal has been mined for local fuel use along Horse Creek above this point.

Horse Creek Coal and Land Company Opening. No. 139 on Map II.

Located on Spruce Fork of Horse Creek, 2.25 miles southwest of Mistletoe P. O.; section taken by Krebs at local mine; Winifrede Coal; elevation 835' B.

			Ft.	In.
Slate roof				
Coal, splint	0'	6"		
Shale, gray	0	4		
Coal, splint	0	5		
Slate	0	3		
Coal, splint, visible	1	8	3	2

Horse Creek Coal and Land Company Opening. No. 140 on Map II.

Located on Jule Webb Fork of Horse Creek, 2.75 miles south of Mistletoe P. O.; section taken by Krebs; **Winifrede Coal**, elevation, 890′ B.

			Ft.	In.
Shale				
Coal	0′	6″		
Shale, gray	0	6		
Coal, splint	1	3	2	3

Near the head of Jule Webb Fork, the following opening was measured by Krebs:

Horse Creek Coal and Land Company Opening. No. 141 on Map II.

			Ft.	In.
Sandstone			1	0
Coal, splint	1′	8″		
Shale	0	4		
Coal, visible, 865′ B	0	10	2	10

From the sections given on the preceding pages it is evident that a portion of Scott District carries the Winifrede Coal of commercial thickness and purity.

Washington District.

The Winifrede Coal rises out of Mud River at the mouth of Ballard Fork, about one mile east of the Boone-Lincoln County Line.

Little Coal Land Company Opening—No. 142 on Map II.

Located on Ballard Fork of Mud River, about 500 feet above the mouth of Ballard Fork; section taken by Krebs; **Winifrede Coal**; elevation, 840′ B.

			Ft.	In.
1. Sandstone, massive				
2. **Coal**, splint	1′	4″		
3. Shale, gray	0	4		
4. **Coal**, splint (gray slate floor)	2	0	3	8

PLATE XXIV.—Winifrede Coal in mine of Anchor Coal Co., at High Coal.

The analysis of a sample collected from Nos. 2 and 4, as reported by Messrs. Hite and Krak, is published in the table of coal analyses under No. 27.

A sample of the same numbers analyzed for Clark and Krebs by Paul Demler, Charleston, W. Va., gave the following results:

	Per cent.
Moisture	2.51
Volatile Matter	36.70
Fixed Carbon	57.42
Ash	3.37
Total	100.00
Sulphur	0.514
Phosphorus	0.002
B. T. U.	13,551

The high moisture is due to the fact that the sample was taken from near the crop.

Little Coal Land Company Local Mine Opening. No. 143 on Map II.

Located on east side of Mud River, 1/8 mile south of the mouth of Ballard Fork; section taken by Krebs; Winifrede Coal; elevation, 850' B.

				Ft.	In.
1.	Sandstone, massive				
2.	Coal	1'	4"		
3.	Slate	0	2		
4.	Coal, splint	2	4	3	10

The analysis of a sample collected from Nos. 2 and 4 of the above section, as reported by Messrs. Hite and Krak, is given in the table of coal analyses under No. 27A.

In passing up Mud River, the Winifrede Coal rises faster than the bed of the stream.

Another opening, on Lukey Fork, 0.5 mile south of its mouth, shows the following section:

Little Coal Land Company Local Mine Opening. No. 144 on Map II.

			Ft.	In.
Sandstone roof				
Coal	1'	2"		
Slate	0	4		
Coal, 880' B	2	2	3	8

The Winifrede Coal is high in the hills along the Warfield Anticline, and very little development has been made on same, until south of the Coalburg Syncline, south of Clothier.

Chilton, MacCorkle and Chilton and Meany Opening. No. 145 on Map II.

Located on Right Fork of Whiteoak Branch of Spruce Laurel Fork, 1.0 mile from its mouth; section taken by Krebs; **Winifrede Coal;** elevation, 1328′ B.

			Ft.	In.
Sandstone			6	0
Shale, dark			0	2
Coal	0′	5″		
Shale, dark	0	10		
Coal, splint	0	10		
Coal, impure	0	4		
Coal, hard splint	1	0		
Shale, gray	0	1		
Coal, hard splint (slate floor)	0	8	4	2

The coal was mined here for fuel used in drilling core test hole No. 10 (113).

The following results were obtained in drilling core test hole No. 4 (136), at a depth of 481′ 3″, located on top of divide between Whiteoak Branch and Casey Creek:

Chilton, MacCorkle and Chilton and Meany Opening. No. 146 on Map II.

			Ft.	In.
Sandstone			8	6
Coal	0′	3″		
Slate	0	7		
Coal	2	0		
Bone	0	1		
Coal	0	8	3	[illegible]
Fire clay			2	0

On Spruce Laurel Fork, a number of openings have been made in the Winifrede seam by the Boone County Coal Corporation under the direction of A. R. Montgomery, who has kindly furnished the writer with a copy of the results of these openings:

Cassingham Prospect Opening—No. 147 on Map II.

Located on Spruce Laurel Fork, 0.5 mile below mouth of Dennison Fork; section taken by A. R. Montgomery; **Winifrede Coal.**

	Ft.	In.
Sandstone		
Coal	5	5

The coal shows considerable impurities.

Cassingham Prospect Opening—No. 148 on Map II.

Located on west side of Spruce Laurel Fork, just below mouth of Dennison Fork; section taken by A. R. Montgomery; **Winifrede Coal;** elevation, 1570′ B.

			Ft.	In.
Sandstone roof				
Coal, splint	0′	8″		
Slate	1	2		
Coal, splint	3	0		
Slate	0	2		
Coal, impure	0	8	5	8

Cassingham Prospect Opening—No. 149 on Map II.

Located on east side of Dennison Fork, 0.5 mile from its mouth; section by A. R. Montgomery; **Winifrede Coal.**

			Ft.	In.
Sandstone roof				
Coal, splint	0′	10″		
Slate	1	2		
Coal, splint	3	6		
Coal, impure	0	9	6	3

Calvin Pardee Prospect Opening—No. 150 on Map II.

Located on main Lefthand Fork of Dennison Fork, on north side of same; section measured by A. R. Montgomery; **Winifrede Coal.**

	Ft.	In.
Sandstone roof		
Coal, splint, (slate floor)	4	0

Calvin Pardee Prospect Opening—No. 151 on Map II.

Located at mouth of main fork of Dennison Fork; section by A. R. Montgomery; **Winifrede Coal.**

	Ft.	In.
Sandstone roof		
Coal, splint, (slate floor)	4	0

Cassingham Prospect Opening—No. 152 on Map II.

Located on south side of Spruce Laurel Fork, 0.5 mile above the mouth of Dennison Fork; section by A. R. Montgomery; **Winifrede Coal.**

			Ft.	In.
Sandstone roof				
Coal, splint			5	5

A small slate parting occurs in the coal.

Cassingham Prospect Opening—No. 153 on Map II.

Located on north side of Spruce Laurel Fork about 0.75 mile above mouth of Dennison Fork; section by A. R. Montgomery; **Winifrede Coal.**

			Ft.	In.
Sandstone roof				
Coal, splint, (slate floor)			5	5

There is a small parting of slate in the coal.

Cassingham Prospect Opening—No. 154 on Map II.

Located on south side of Spruce Laurel Fork, 0.8 mile above mouth of Dennison Fork; section taken by A. R. Montgomery; **Winifrede Coal.**

			Ft.	In.
Sandstone roof				
Coal, impure	0′	6″		
Slate	0	6		
Coal, splint	1	4		
Coal, impure	0	4		
Coal, splint (slate floor)	3	10	6	6

Cassingham Prospect Opening—No. 155 on Map II.

Located on north side of Spruce Laurel Fork, 1.0 mile above mouth of Dennison Fork; section by A. R. Montgomery; **Winifrede Coal.**

			Ft.	In.
Sandstone roof				
Coal, splint	0′	8″		
Slate	0	4		
Coal, hard, splint, (slate floor)	4	4	5	4

Cassingham Opening—No. 156 on Map II.

Located on north side of Big North Branch of Spruce Laurel Fork, 1.5 miles above mouth of Dennison Fork; section taken by Krebs; **Winifrede Coal**; elevation, 1670′ B.

			Ft.	In.
Sandstone, massive, visible			4	0
Coal, splint	0′	2″		
Slate	0	1		
Coal, splint	0	4		
Shale, gray	0	5		
Coal, splint	0	10		
Slate	0	1		
Coal, gray splint, hard	1	6		
Coal, gas	0	4		
Coal, splint, hard, (slate floor)	1	6	5	3

Coal mined for boiler fuel in sinking diamond core test hole, Cassingham No. 1 (131), and driven under cover about 20 feet.

Cassingham Prospect Opening—No. 157 on Map II.

Located on south side of Big North Branch of Spruce Laurel Fork, 1.5 miles above the mouth of Dennison Fork; section by Krebs; **Winifrede Coal**; elevation, 1675′ B.

				Ft.	In.
1.	Sandstone, massive				
2.	**Coal**, splint	0′	6″		
3.	Shale, gray	0	2		
4.	**Coal**, gray splint	1	2		
5.	**Coal**, impure, iron pyrites	0	1		
6.	**Coal**, splint, hard	2	6		
7.	Slate, gray	0	1		
8.	**Coal**, hard splint, (slate floor)	1	3	5	[illegible]

The analysis of a sample collected from Nos. 2, 4, 6 and 8, as reported by **Messrs.** Hite and Krak, is published in the table of coal analyses under **No.** 28.

Cassingham Prospect Opening—No. 158 on Map II.

Located on lefthand hollow of Big North Branch of Spruce Laurel Fork, 1.7 miles above mouth of Dennison Fork; section measured by Krebs; **Winifrede Coal**; elevation, 1685′ B.

			Ft.	In.
Sandstone roof				
Coal	0′	7″		
Slate	0	3		
Coal	1	2		

			Ft.	In.
Slate	0′	1″		
Coal	2	4		
Slate	0	1		
Coal	1	2	5	8

Cassingham Prospect Opening—No. 159 on Map II.

Located on north side of Big North Branch of Spruce Laurel Fork, 2.4 miles east of mouth of Dennison Fork; section by Krebs; **Winifrede Coal**; elevation, 1710′ B.

			Ft.	In.
Sandstone roof				
Coal, splint	0′	6″		
Slate	0	4		
Coal, splint	1	8		
Slate	0	1		
Coal, splint, hard	2	4	4	11

Cassingham Prospect Opening—No. 159A on Map II.

Located on the north side of North Branch of Spruce Laurel Fork, about 3.5 miles east of the mouth of Dennison Fork; section taken by A. R. Montgomery; **Winifrede Coal.**

			Ft.	In.
Slate roof				
Coal, splint	0′	8″		
Slate	0	7		
Coal, splint, (slate floor)	4	9	6	0

The foregoing sections exhibit the character, structure and thickness of the Winifrede Coal in Washington District, and from them it is evident that this coal, occurring as it does above drainage and only a few feet above the beds of the valleys, is a very important coal seam.

Sherman District.

The Winifrede Coal is an important seam in Sherman District, since it underlies in commercial thickness and purity a considerable area of the same. The Winifrede Coal Company has extended its mines through the hills from Fields Creek to Joes Creek, a tributary of Coal River, and has mined a considerable tonnage of coal from Boone County, which is credited to Kanawha County in the Report of the Department of Mines.

The Cabin Creek Consolidated Coal Company is also mining in the Boone County area from its Raccoon Mine No. 2, one mile south of Kayford.

Winifrede Coal Company Opening—No. 160 on Map II.

Located on Trace Fork of Joes Creek, 2.5 miles northeast of Comfort P. O., where a mine of the Winifrede Coal Company has pierced through the divide between Kanawha and Coal Rivers; **Winifrede Coal**; elevation, 1044′ B.

			Ft.	In.
Sandstone roof				
Coal	2′	5″		
Slate	0	2		
Coal, (shale floor)	2	6	5	1

Butts, N. 27° E.; faces, N. 63° W.; greatest rise, S. 35° E.

Winifrede Coal Company Opening—No. 161 on Map II

Located on Spicelick Fork of Joes Creek, where the mine of the Winifrede Coal Company tunnels through the divide from Fields Creek; section taken by Krebs; **Winifrede Coal**; elevation, 1040′ L.

			Ft.	In.
Sandstone roof				
Coal, splint	2′	4″		
Slate	0	8		
Coal, splint	2	5	5	5

Winifrede Coal Co. Prospect Opening—No. 162 on Map II.

Located on Left Fork of Joes Branch, 2.7 miles northeast of Racine P. O.; section taken by Teets; **Winifrede Coal**; elevation, 1090′ B.

			Ft.	In.
Sandstone, massive			10	0
Slate			0	4
Coal, splint	3′	1″		
Shale, gray	0	9		
Coal, splint, (slate floor)	2	2	6	0

Several more prospect openings have been made on Joes Creek by the Winifrede Coal Company, but when visited by the writer, these were fallen shut so that no good section could be measured.

Near the head of Joes Branch, where the mine from the

Winifrede Coal Company was driven through from Fields Creek, the following section was taken by Teets:

Winifrede Coal Company Opening—No. 163 on Map II.

			Ft.	In.
Sandstone roof				
Coal, splint	2'	6"		
Shale, gray	0	8		
Coal, splint (dark slate floor, 1095' B.)	1	6	4	8

On Sandlick Creek of Laurel Creek, the Winifrede Coal was once mined for local fuel use on the property of the Lackawanna Coal and Lumber Company.

Lackawanna Coal and Lumber Company Opening. No. 164 on Map II.

Located on the south side of Sandlick Creek, 2.0 miles west of Nelson P. O.; section taken by Krebs; **Winifrede Coal**; elevation, 1145' B.

			Ft.	In.
Sandstone, **Winifrede**, massive, visible			10	0
Coal, splint	2'	0"		
Slate	0	4		
Coal, splint, visible	1	2	3	6

South from the above opening across the divide on Griffiths Branch of Laurel Creek, the Winifrede Coal has been mined for local fuel use. The following opening, located on the south side of Griffiths Branch, 3.2 miles southwest of Nelson P. O., was measured by Krebs.

John Q. Dickinson Opening—No. 165 on Map II.

			Ft.	In.
Sandstone, massive, coarse grained			10	0
Shale, gray			0	10
Coal, splint	1'	0"		
Fire clay	0	3		
Coal, splint	0	6		
Slate	0	2		
Coal, splint, 1005' B	1	1	3	0

The coal is mined at this point by **Van Jarrell** for local fuel use.

PLATE XXV.—Hernshaw Coal opening on Little Ugly Branch of West Fork.

John Q. Dickinson Local Mine Opening—No. 166 on Map II.

Located on north side of Laurel Creek, 3.8 miles southwest of Nelson P. O.; section taken by Krebs; **Winifrede Coal**; elevation, 990' B.

				Ft.	In.
1.	Sandstone, visible			3	C
2.	Shale, gray			0	10
3.	**Coal**, impure	0'	3"		
4.	**Coal**, gas	0	3		
5.	Slate	0	1		
6.	**Coal**, splint	1	4		
7.	Shale, gray	0	8		
8.	**Coal**, splint	1	2		
9.	**Coal**, gas, (slate floor)	1	0	4	9

The analysis of a sample collected from Nos. 4, 6, 8 and 9, as reported by **Messrs.** Hite and Krak, is published in the table of coal analyses under **No. 29.**

John Q. Dickinson Opening—No. 167 on Map II.

Located on west side of Laurel Creek, 4.0 miles southwest of Nelson P. O.; section taken by Krebs; **Winifrede Coal**, elevation, 1015' B.

				Ft.	In.
1.	Sandstone, massive			5	0
2.	**Coal**	0'	2"		
3.	Shale, gray	1	0		
4.	Sandstone	1	1		
5.	Shale, gray	0	8		
6.	**Coal**, splint	0	7		
7.	**Coal**, block	0	8		
8.	Slate	0	1		
9.	**Coal**, block	1	2		
10.	Fire clay	0	1		
11.	Coal, block, (slate floor)	1	2	6	8

The analysis of a sample collected from Nos. 6, 7, 9 and 11, as reported by Messrs. Hite and Krak, is published in the table of coal analyses under **No. 29A.**

John Q. Dickinson Local Mine Opening—No. 168 on Map II.

Located on Laurel Creek, 6.2 miles southwest of Seth, near the mouth of Stolling Fork; section taken by Teets; **Winifrede Coal**; elevation, 1055' B.

			Ft.	In.
Sandstone, massive, visible			10	0
Coal, block	2'	10"		
Shale, gray	0	1		
Coal, hard splint, (slate floor)	1	8	4	[illegible]

The coal in this opening was sampled by J. B. Dilworth for E. V. d'Invilliers, and analyzed by A. S. McCreath, of Harrisburg, Pa., with the following results:

	Per cent.
Moisture	1.180
Volatile Matter	34.780
Fixed Carbon	53.996
Ash	8.740
Sulphur	1.304
Total	100.000

The ash in the above analysis is rather high, but this is very probably due to the sample being taken near the crop.

Near the head of Seng Creek of Coal River, the Winifrede Coal is being mined on a commercial scale by the Anchor Coal Company, where it exhibits the following structure:

Anchor Coal Company Mine—No. 169 on Map II.

Located at High Coal, on south side of Seng Creek; section taken by Krebs; Winifrede Coal; elevation, 1700' L.

				Ft.	In.
1.	Slate roof				
2.	**Coal**, splint	1'	6"		
3.	**Coal**, hard, splint	1	2		
4.	**Coal**, splint, glossy	2	0		
5.	**Coal**, splint, gray	1	4		
6.	Slate	0	1		
7.	**Coal**, splint, (slate floor)	1	3	7	4

Butts, N. 42° E.; faces, N. 48°W.

Coal shipped east and west for steam and domestic fuel; number of men employed, 100; daily capacity, 100 tons; authority, Lute Hornickel, President, and George Hornickel, Superintendent, of the Anchor Coal Company.

The analysis of a sample collected from Nos. 2, 3, 4, 5 and 7, as reported by Messrs. Hite and Krak, is published in the table of coal analyses under **No. 30.**

Samples of coal were taken from some railroad cars shipped from this mine by Mr. Lute Hornickel, the President of the Anchor Coal Company, and the analyses made by Messrs. Cornell and Murry, Chemists, Cleveland, Ohio, gave the following results, according to Mr. Hornickel:

Lab. No.	Coal.	Moisture.	Volatile Matter.	Fixed Carbon.	Ash.	Sulphur.	B. T. U.
10,913	R. O. M.	1.20	36.65	59.60	2.55	0.65	14,754
11,083	2″ Lump	0.95	35.40	58.77	4.82	0.56	14,360
11,215	2″ Lump	0.65	36.90	59.23	3.22	0.57	14,612
Average		0.93	36.32	59.20	3.55	0.59	14,575

Another analysis made at the Ewart Works at Indianapolis, Ind., by the Link-Belt Company, from a sample taken from a local car of "run-of-mine", gave the following, according to W. I. Balletine, Superintendent:

	Per cent.
Moisture	1.00
Volatile Matter	36.64
Fixed Carbon	57.59
Ash	4.24
Sulphur	0.53
Total	100.00

The foregoing analysis shows the coal to be of excellent quality for general fuel and steam purposes, being low in volatile matter, ash and sulphur, and high in fixed carbon and B. T. U.

Webb Fuel Company Mine—No. 170 on Map II.

Located on north side of Seng Creek, 2.0 miles west of High Coal; section taken by Teets; Winifrede Coal; elevation, 1648′ B.

				Ft.	In.
1.	Sandstone roof				
2.	Coal, block	1′	2″		
3.	Slate, gray	0	1		
4.	Coal, splint	4	1		
5.	Slate, gray	0	1		
6.	Coal, splint	0	8		
7.	Shale, dark	0	1		
8.	Coal, splint	0	10		
9.	Slate	0	1		
10.	Coal (to slate floor)	0	4	ı	5

Butts, N. 42° E.; faces, N. 48° W.; coal shipped east and west for steam and domestic fuel; number of men employed, 100; daily capacity, 800 tons; L. M. Webb, President; John Holmes, Manager.

The analysis of a sample taken from Nos. 2, 4, 6, 8 and 10, as reported by Messrs. Hite and Krak, is given in the table of coal analyses under **No. 30A.**

Seng Creek Coal Company Mine—No. 171 on Map II.

Located on east side of Coal River, 1.5 miles south of the mouth of Seng Creek at Whitesville P. O.; section taken by Teets; **Winifrede Coal**; elevation, 1590′ B.

				Ft.	In.
1.	Sandstone roof				
2.	**Coal**, block	0′	5″		
3.	**Coal**, gray splint	0	5		
4.	Bone	0	1		
5.	**Coal**, gray splint	1	1		
6.	**Coal**, block	2	1		
7.	Slate	0	2		
8.	**Coal**, (to slate floor)	0	2	4	5

Butts, N. 42° E.; faces, N. 48° W.; coal shipped east and west for steam and domestic fuel; number of men employed, 60; daily capacity, 500 tons; T. E. B. Siler, Charleston, W. Va., President.

The analysis of a sample collected from Nos. 2, 3, 5, 6 and 8, as reported by Messrs. Hite and Krak, is published in the table of coal analyses under **No. 31.**

Crook District.

The Winifrede Coal is one of the most important beds in Crook District, being of commercial thickness and purity in nearly the whole district. In the northern part of the district, the coal is high on the hills, but the measures dip rapidly through the Coalburg Syncline, so that quite a large area is underlain with this seam.

Squire Gibson Opening—No. 172 on Map II.

Located on the south side of Lick Branch of Pond Fork, about 1.1 miles from its mouth, and 0.9 mile north of Gordon P. O.; section taken by Teets; **Winifrede Coal**; elevation, 1055′ B.

				Ft.	In.
1.	Slate roof				
2.	**Coal**	0′	3 ″		
3.	Slate	0	8		
4.	**Coal**, block	1	8		
5.	**Coal**, gray splint	1	6		
6.	Slate, gray	0	0¾		
7.	**Coal**, gray splint, (slate floor)	1	7½	5	9⅛

The analysis of a sample collected from Nos. 2, 4, 5 and 7, as reported by **Messrs.** Hite and Krak, is published in the table of coal analyses under **No.** 32.

Lackawanna Coal and Lumber Company Local Mine Opening. No. 173 on Map II.

Located on head of Whites Branch of West Fork of Pond Fork, 3.0 miles northeast of Gordon P. O.; section by Krebs; **Winifrede Coal**; elevation, 1180′ B.

			Ft.	In.
Shale, gray				
Coal, impure	1′	0″		
Coal, splint	2	0		
Coal, block	1	0		
Shale, gray	0	4		
Coal, splint, visible	0	6	4	10

Lackawanna Coal and Lumber Company Local Mine Opening. No. 174 on Map II.

Located on east side of Whites Branch of West Fork, 2.8 miles northeast of Gordon P. O., and 0.2 mile southwest of opening No. 173; section taken by Krebs; **Winifrede Coal**; elevation, 1180′ B.

				Ft.	In.
1.	Sandstone, massive				
2.	Shale, gray			6	0
3.	**Coal**, impure	1′	0″		
4.	**Coal**, splint	2	6		
5.	Shale, gray	0	1		
6.	**Coal**	0	4		
7.	Shale, gray	0	4		
8.	**Coal**, splint	1	0		
9.	Shale, dark gray	2	0		
10.	**Coal**, splint, (slate floor)	1	3	8	6

Opening 45 feet above level of branch.

The analysis of a sample collected from Nos. 3, 4, 6, 8 and 10, as reported by **Messrs.** Hite and Krak, is published in the table of coal analyses under **No.** 33.

Lackawanna Coal and Lumber Company Prospect Opening. No. 175 on Map II.

Located on south side of Camp Branch of Whites Branch, 2.5 miles northeast of Gordon P. O.; section taken by Krebs; **Winifrede Coal**; elevation, 1240′ B.

			Ft.	In.
Slate roof				
Coal, splint	2′	2″		
Bone	0	1		
Coal, splint	0	7		
Slate	0	0½		
Coal	0	4		
Slate	0	4		
Coal, block	1	2		
Coal, impure, bone and slate	1	8		
Coal, hard splint, (slate floor)	1	4	[illegible]	8½

Lackawanna Coal and Lumber Company Opening. No. 176 on Map II.

Located on west side of Whites Branch, just north of mouth of Lick Branch, 1.5 miles northeast of Gordon P. O.; section taken by Krebs; **Winifrede Coal**; elevation, 1445′ B.

				Ft.	In.
1.	Shale, gray			4	0
2.	**Coal**, semi-cannel	0′	4″		
3.	Slate	0	0½		
4.	**Coal**	0	1½		
5.	Shale, dark	0	8		
6.	**Coal**, splint	1	0		
7.	**Coal**, gray splint, hard	3	0		
8.	Shale, gray, 1″ to	0	4		
9.	**Coal**, hard splint, (shale floor)	1	0	6	6

The analysis of a sample collected from Nos. 4, 6 and 7, as reported by Messrs. Hite and Krak, is published in the table of coal analyses under No. 34.

Near the head of Jacks Branch, the following section was measured in the Winifrede Coal bed:

Samuel Hunter Local Mine Opening—No. 177 on Map II.

Located near the head of Jacks Branch, 3.0 miles northeast of Clothier, and 0.5 mile west of the Washington-Crook District Line; section measured by Teets; **Winifrede Coal**; elevation, Lower Bench, 1080′ B.

				Ft.	In.
Sandstone, massive					
Coal, impure	0′	3″	**Upper Bench**	1	9
Coal, soft	1	6			

				Ft.	In.
Shale, gray				19	0
Slate				1	6
Coal	0'	6"	Lower Bench	5	8
Shale, gray	1	5			
Coal, hard splint	1	3			
Bone	0	2			
Coal, splint	1	1			
Slate	0	3			
Coal, splint, visible	1	0			

Marion Ferrell Prospect Opening—No. 178 on Map II.

Located on Right Fork of Cow Creek, 3.2 miles west of Bald Knob P. O.; section by J. S. Cunningham; **Winifrede Coal**; elevation, 1615' B.

			Ft.	In.
Sandstone				
Coal, splint	0'	5"		
Shale	0	4		
Coal, splint	3	0		
Shale	0	6		
Coal, splint	1	0		
Shale	0	2		
Coal	0	4	5	9

Wharton Estate Crop Opening—No. 179 on Map II.

Located on west side of Pond Fork, between the mouth of Grapevine Branch and Coal Branch; section by J. S. Cunningham; **Winifrede Coal**; elevation, 1555' B.

				Ft.	In.
Coal, splint	1'	7"	Upper Bench	5	6
Slate	0	2			
Coal, splint	0	4			
Slate	0	2			
Coal, splint	1	1			
Slate	1	1			
Coal, cannel	1	1			
Slate and coal				11	0
Coal, splint	0'	3"	Lower Bench	4	2
Slate	0	1			
Coal, hard splint	2	3			
Slate	0	1			
Coal, hard splint	1	6			

Here the Winifrede Coal appears in two benches, and farther up Grapevine Branch the same two benches occur, as is shown in the following section:

Cassingham Opening—No. 180 on Map II.

Located on Grapevine Branch, 2.0 miles west of Bald Knob P. O.; section taken by J. S. Cunningham; **Winifrede Coal**; elevation, Lower Bench, 1555' B.

					Ft.	In.
1.	Sandstone					
2.	**Coal**, hard splint	2'	0"	Upper Bench	4	7
3.	Slate	0	3			
4.	**Coal**, hard	2	4			
5.	Slate and concealed				10	0
6.	**Coal**	0'	5"	Lower Bench	5	4
7.	**Coal**, cannel	0	1			
8.	**Coal**, splint, hard	2	10			
9.	**Coal**, splint, hard	0	4			
10.	Slate	0	4			
11.	**Coal**, hard, splint	1	4			

The analysis of a sample collected from Nos. 6, 7, 8, 9 and 11 by J. S. Cunningham, made by the Chief Chemist of the New River and Pocahontas Consolidated Coal Company, gave the following results:

	Per cent.
Moisture	1.53
Volatile Matter	33.35
Fixed Carbon	59.61
Ash	5.51
Total	100.00
Sulphur	0.738

Wharton Estate Prospect Opening—No. 181 on Map II.

Located on south side of Old Camp Branch, 0.5 mile above its mouth; section taken by A. R. Montgomery; **Winifrede Coal**; elevation, 1510' B.

			Ft.	In.
Slate roof				
Coal	2'	6"		
Shale, gray	1	4		
Coal, cannel	2	0	5	10

Wharton Estate Prospect Opening—No. 182 on Map II.

Located on south side of Old Camp Branch in second hollow from the south above its mouth; section by A. R. Montgomery; **Winifrede Coal**; elevation, 1450' B.

			Ft.	In.
Sandstone				
Coal	0'	4 "		
Slate	0	0½		
Coal (fire clay bottom)	3	6½	3	11

Wharton Estate Prospect Opening—No. 183 on Map II.

Located on the south side of Old House Branch, 2.0 miles north of Bald Knob; section by A. R. Montgomery; **Winifrede Coal**; elevation, 1450′ B.

			Ft.	In.
Slate roof				
Coal, splint	2′	8″		
Slate	0	2		
Coal, splint, (hard blue clay floor)	1	7	4	5

Wharton Estate Prospect Opening—No. 184 on Map II.

Located on the south side of Old House Branch of Pond Fork, near head; section by A. R. Montgomery; **Winifrede Coal**; elevation, 1470′ B.

			Ft.	In.
Coal	1′	2″		
Slate	2	2		
Coal, splint	2	1		
Slate	0	0¼		
Coal	0	8¼		
Slate	0	2		
Coal, splint, (slate floor)	2	11	9	2½

Wharton Estate Opening—No. 185 on Map II.

Located on east side of Pond Fork, just east of Pond P. O.; section measured by A. R. Montgomery; **Winifrede Coal**; elevation, 1360′ B.

			Ft.	In.
Slate roof				
Coal, splint	2′	6″		
Slate	0	6		
Coal, cannel, (slate floor)	2	6	5	6

Near Bald Knob, the Winifrede Coal has been mined for local fuel use along the trail leading from Bald Knob P. O. to West Fork, on land of the Wharton Estate, and there it exhibits the following section:

Wharton Estate Opening—No. 186 on Map II.

Located on trail leading from Bald Knob P. O. to West Fork, 0.8 mile northeast of Bald Knob P. O.; section by Krebs; **Winifrede Coal**; elevation, 1689′ L.

	Ft.	In.
Sandstone, massive		
Sandy shale	4	0

			Ft.	In.
Coal, gas	1'	4"		
Slate	0	2		
Coal, splint	1	10		
Coal, cannel	0	10		
Coal, splint	0	10	5	0

Coal mined by **William Price** for local fuel use.

On up West Fork from the mouth of Pond Fork to James Creek, several prospect openings in the Winifrede Coal have been made, which show the coal to be of commercial thickness and purity.

E. J. Berwind Prospect Opening—No. 187 on Map II.

Located on the north side of Left Fork of James Creek, 0.8 mile up said creek; section taken by Teets; **Winifrede Coal**; elevation, 1525' B.

			Ft.	In.
Slate roof				
Coal, splint	1'	1"		
Slate, gray	0	1		
Coal, splint	1	9		
Slate, dark gray	0	7		
Coal	0	4		
Slate, gray	0	1		
Coal, block	0	7		
Slate, dark	0	4		
Coal, gray splint, visible	1	9	6	7

E. J. Berwind Opening—No. 188 on Map II.

Located on the east bank of James Creek, 1.5 miles southeast of Chap P. O.; section by J. S. Cunningham; **Winifrede Coal**; elevation, Lower Bench, 1630' B.

				Ft.	In.
Coal, soft	2'	6"	Upper Bench	4	5½
Slate	0	3½			
Coal, impure	0	5½			
Slate	0	3½			
Coal	0	4			
Slate	0	1			
Coal	0	6			
Concealed				6	0
Slate				0	10
Coal, splint, **Lower Bench**, (slate floor)				2	9

E. J. Berwind Prospect Opening—No. 189 on Map II.

Located on Mats Creek, 1.5 miles up same; section measured by J. S. Cunningham; **Winifrede Coal**; elevation, 1810′ B.

			Ft.	In.
Coal, soft	1′	9″		
Slate	0	2		
Coal, impure	0	3		
Slate	0	2		
Coal	0	3		
Slate	1	1		
Coal, splint	0	9		
Slate	0	4		
Coal, hard splint	3	4	8	1

Another opening on the head of Mats Creek shows the coal good in the upper bench, while the lower bench is thinner:

E. J. Berwind Prospect Opening—No. 190 on Map II.

Located on the head of Mats Creek; section by J. S. Cunningham; **Winifrede Coal**; elevation, Lower Bench, 2100′ B.

			Ft.	In.
Coal, soft	2′	10″		
Slate	0	3		
Coal, splint	1	6		
Slate	0	9		
Coal	0	7		
Slate	2	2		
Coal, splint, Lower Bench	1	7	9	8

E. J. Berwind Prospect Opening—No. 191 on Map II.

Located on the east side of West Fork, 0.7 mile south of Mats Creek, 2.6 miles northeast of Bald Knob P. O.; section taken by Teets; **Winifrede Coal**; elevation, Lower Bench, 1720′ B.

					Ft.	In.
1.	Sandstone, massive					
2.	**Coal**, splint	1′	0″	Upper Bench	2	11
3.	Slate, dark	1	7			
4.	**Coal**, splint	0	4			
5.	Shale, gray				5	0
6.	**Coal**, block	0′	11″	Lower Bench	4	[illegible]
7.	Slate, gray	0	1			
8.	**Coal**, splint	1	11			
9.	**Coal**, cannel	0	10			
10.	**Coal**, gas (slate floor)	0	10			

The analyses of two samples collected from Nos. 2, 4, 6, 8, 9 and 10, as reported by Messrs. Hite and Krak, are published in the table of coal analyses under **No. 35.**

White and Hopkins Opening—No. 191A on Map II.

Located on a branch on the north side of Pond Fork, 1.0 mile southwest of the mouth of West Fork; section taken by Teets; **Winifrede Coal**; elevation, Lower Bench, 1020' B.

					Ft.	In.
1.	Sandstone, massive					
2.	**Coal**, block	0'	8"	Upper Bench	3	4
3.	Shale, gray	0	5			
4.	**Coal**, gray splint	0	4			
5.	Shale, gray	0	2			
6.	**Coal**, block	0	6			
7.	Shale, gray	0	6			
8.	**Coal**, block	0	9			
9.	Shale, dark				5	0
10.	**Coal**	0'	1"	Lower Bench	5	0
11.	Slate, gray	1	3			
12.	**Coal**	0	4			
13.	Slate	0	2			
14.	**Coal**	0	2			
15.	Slate, gray	0	9			
16.	**Coal**, splint, visible	2	3			

The lower bench of this bed is mined by tenants for fuel use.

Mr. J. S. Cunningham collected a sample from Nos. 4, 6, 8 and 10 and the analysis made by A. S. McCreath gave the following results:

	Per cent.
Moisture	4.384
Volatile Matter	33.466
Fixed Carbon	57.214
Ash	4.331
Sulphur	0.605
Total	100.000

The excessive amount of moisture is due to the fact that the coal was sampled near the outcrop.

E. J. Berwind Opening—No. 192 on Map II.

Located on the west side of West Fork, 1.8 miles northeast of Bald Knob P. O.; section taken by Teets; **Winifrede Coal**; elevation, Lower Bench, 1710' B.

				Ft.	In.
Sandstone, massive					
Coal, soft	0'	9"	Upper Bench	2	10
Coal, splint, hard	2	1			
Slate				2	0
Sandstone and concealed				33	0
Coal, gray splint, **Lower Bench** (slate floor)				4	5

Driven in 60 feet; mined for local fuel use by **Wm. Miller**.

T. C. Jarrell Opening—No. 192A on Map II.

Located on the east side of small branch of Pond Fork, 0.5 mile above mouth of Old Camp Branch and 0.8 mile northwest of Pond P. O.; section by Teets; **Winifrede Coal**; elevation, 1360' B.

				Ft.	In.
1.	Sandstone, massive roof				
2.	**Coal**	0'	7"		
3.	Slate and shale	5	0		
4.	**Coal,** gas	2	6		
5.	Slate, gray	0	6		
6.	**Coal,** cannel (to slate floor)	3	2	11	9

The analysis of a sample collected from Nos. 2 and 4, as reported by Messrs. Hite and Krak, is published in the table of coal analyses under **No. 35A.**

The analysis of a sample collected from No. 6, as reported by Messrs. Hite and Krak, is published in the table of coal analyses under **No. 35B.**

E. J. Berwind Prospect Opening—No. 193 on Map II.

Located on Workman Branch of Pond Fork, 1.1 miles northeast of Bald Knob P. O.; section measured by Teets; **Winifrede Coal**; elevation, 1602' L.

			Ft.	In.
Slate roof				
Coal, gas	0'	11"		
Coal, splint, (slate floor)	3	0	3	11

E. J. Berwind Prospect Opening—No. 194 on Map II.

Located on the west side of West Fork at the mouth of James Creek, 0.5 mile south of Chap P. O.; section taken by J. S. Cunningham; **Winifrede Coal**; elevation, Lower Bench, 1640' B.

				Ft.	In.
Coal, Winifrede "Rider"				2	10
Shale and slate				12	0
Coal, splint	3'	2"	**Upper Bench**		
Slate	0	4			
Coal, soft	1	0		6	4
Coal, cannel	1	10			
Fire clay				3	10
Coal	0'	4"	**Lower Bench**		
Slate	0	5		4	3
Coal, splint	3	6			

E. J. Berwind Opening—No. 195 on Map II.

Located on the right side of Sycamore Fork, in William Brown Hollow; section taken by J. S. Cunningham; **Winifrede Coal**; elevation, Lower Bench, 1700′ B.

				Ft.	In.
1. **Coal**, hard splint...3′	9″		Upper Bench.....	5	4
2. **Coal**, cannel.......1	0				
3. **Coal**, soft.........0	7				
4. Shale and concealed				20	0
5. **Coal**, impure....0′	4″		Lower Bench.....	4	1½
6. **Coal** 0	10				
7. **Coal** and slate...0	3½				
8. **Coal**, splint.....2	8				

A sample collected by J. S. Cunningham from Nos. 1, 2 and 3 and analyzed by the Chief Chemist of the Pocahontas Consolidated Coal Company, Blume, W. Va., gave the following results:

	Per cent.
Moisture	3.11
Volatile Matter	33.51
Fixed Carbon	58.53
Sulphur	0.65
Ash	4.20
Total	100.00

E. J. Berwind Opening—No. 196 on Map II.

Located on west side of West Fork, 2.5 miles northeast of Bald Knob; section taken by J. S. Cunningham; **Winifrede Coal**; elevation, 1860′ B.

			Ft.	In.
Coal, splint	2′	1″		
Coal, cannel	0	3		
Coal, splint, hard	2	0		
Slate	2	10		
Coal, splint	1	7		
Slate	0	3		
Coal, splint	2	3	11	3

Horton Steel Company Prospect Opening—No. 197 on Map II.

Located on the west side of Pond Fork, opposite the mouth of Burnt Camp Branch, 1.0 mile southeast of Echart P. O.; section taken by Teets; **Winifrede Coal**; elevation, 2115′ B.

			Ft.	In.
Sandstone roof				
Coal, splint	1′	3″		
Shale, gray	0	1		
Coal, splint	0	9		
Shale, gray	0	1		
Coal, splint (slate floor)	2	7	4	9

The Rowland Land Company owns about 75,000 acres of land in Raleigh County on the east of Crook District. Their Engineers, W. P. Edwards and F. C. Colcord, have made a great many prospect openings on the property. A few of these sections in Raleigh County, adjoining Crook District, will now be given. These sections, being less than one mile from the Boone-Raleigh County Line, will give the structure, character and thickness of the Winifrede Coal along the county line:

Rowland Land Company Opening—No. 198 on Map II.

Located on the north side of Steer Hollow, 2.5 miles southwest of Jarrolds Valley, Marsh Fork District, Raleigh County, and 0.5 mile east of the Boone-Raleigh County Line; section by Edwards and Colcord; **Winifrede Coal**; elevation, 1725' B.

			Ft.	In.
Sandstone				
Shale			1	10
Slate			0	2
Coal, splint	0'	11"		
Coal	1	5½		
Slate	0	0½		
Coal, splint	3	2½	5	7½

Rowland Land Company Prospect Opening. No. 199 on Map II.

Located on the north side of Lower Big Branch, 3.1 miles southwest of Jarrolds Valley, and 0.6 mile east of the Boone-Raleigh County Line; section by Edwards and Colcord; **Winifrede Coal**; elevation, 1790' B.

			Ft.	In.
Sandstone, massive				
Slate			0	2
Coal	1'	0"		
Slate	0	1		
Coal, splint	4	1	5	2

Rowland Land Company Prospect Opening. No. 200 on Map II.

Located on head of Lower Big Branch of Marsh Fork, 3.5 miles southwest of Jarrolds Valley, and 0.25 mile east of Boone-Raleigh County Line; Marsh Fork District, Raleigh County; section taken by Edwards and Colcord; **Winifrede Coal**; elevation, 1785' B.

	Ft.	In.
Sandstone, massive		
Slate	0	6

			Ft.	In.
Coal, splint	0′	9½″		
Slate	0	2		
Coal, splint	1	9½		
Slate	0	1		
Coal, splint	2	0	4	10

Rowland Land Company Prospect Opening.
No. 201 on Map II.

Located on the west side of Upper Big Branch of Marsh Fork, 4.6 miles southwest of Jarrolds Valley, and 0.25 mile east of the Boone-Raleigh County Line; Marsh Fork District, Raleigh County; section taken by Messrs. Edwards and Colcord; **Winifrede Coal;** elevation, 1845′ B.

			Ft.	In
Shale roof				
Coal, splint	1′	1½″.		
Slate	0	2		
Coal, splint	1	10½		
Slate	0	3½		
Coal, splint	1	0½	4	6

Rowland Land Company Prospect Opening.
No. 202 on Map II.

Located on the east side of East Branch of Big Branch of Marsh Fork, 4.8 miles southwest of Jarrolds Valley, 5.6 miles east of the Boone-Raleigh County Line, Marsh Fork District, Raleigh County; section by Edwards and Colcord; **Winifrede Coal;** elevation, 1870′ B.

			Ft.	In.
Shale				
Coal, splint	1′	2 ″		
Slate	0	2		
Coal, splint	2	0½		
Slate	0	1		
Coal, splint	2	8	6	1½

Rowland Land Company Prospect Opening.
No. 203 on Map II.

Located on north side of Road Fork of Hazy Creek, 7.5 miles southwest of Jarrolds Valley, and 0.25 mile east of the Boone-Raleigh County Line, Marsh Fork District, Raleigh County; section by Edwards and Colcord; **Winifrede Coal;** elevation 1990′ B.

			Ft.	In.
Sandstone, massive				
Coal, splint	1′	4½″		
Slate	0	2½		
Coal, splint	2	4		
Slate	0	6		
Coal	2	8½	[illegible]	1½

PLATE XXVI.—Tipples of Peytona Block Coal Company, and topography of Kanawha Series.

A further discussion of the **Winifrede Coal** with estimate of the probable tonnage of available coal will be given in the Chapter on Coal.

THE LOWER WINIFREDE SANDSTONE.

Underneath the Winifrede Coal, and separated from same by impure fire clay and sandy shale from 2 to 5 feet, there occurs a bed of massive, gray sandstone that has been named the **Lower Winifrede.** This bed is often split up into several members of shale and sandstone, and it ranges in thickness from 20 to 40 feet. This sandstone bed appears to be different in texture and lithological aspect from the sandstone overlying the coal and marks the lower portion of the Upper Kanawha Series. This stratum rises above the beds of the streams in the northern part of Boone County, and is generally massive, forming cliffs along its outcrop.

THE CHILTON "A" COAL.

Underneath the Lower Winifrede Sandstone occurs a multiple-bedded seam of splint coal, ranging in thickness from 0 to 4 feet, that has been named the **Chilton "A" Coal.** This bed occurs in different parts of the county, but thus far it has furnished very little coal of commercial thickness and purity. It may be possible that when this coal has been thoroughly prospected, there will be considerable good coal found in this bed in the southern part of Boone County.

Washington District.

Yawkey and Freeman Prospect Opening—No. 204 on Map II.

Located on the north side of Burnt Cabin Branch of Spruce Laurel Fork, 2.4 miles northeast from Clothier; section taken by Teets; **Chilton "A" Coal**; elevation, 960' B.

	Ft.	In.
Sandstone, massive.............................		
Coal, splint (slate floor)......................	1	8

Another opening located on the north side of Spruce Laurel Fork, 2.5 miles southeast of Clothier, gives the following section as measured by Teets:

Yawkey and Freeman Opening—No. 205 on Map II.

			Ft.	In.
Slate roof				
Coal, splint	1′	8″		
Coal, gas, 1040′ B.	0	10	2	6

The coal was once mined here for local fuel purposes.

THE UPPER CHILTON SANDSTONE.

Separated by a stratum of impure fire clay and shale, 10 to 18 feet thick, from the Chilton "A" Coal, occurs the **Upper Chilton[4] Sandstone.**

This bed is a medium grained, hard, micaceous sandstone, and frequently forms massive cliffs from 20 to 40 feet thick.

Along Spruce Laurel Fork, southeast of Clothier, this sandstone makes prominent cliffs, and is possibly best exposed in Boone County as the Chesapeake and Ohio Railway Company has recently constructed its Spruce Laurel Branch a distance of 5 miles southeast from Clothier, and as a portion of the grade was cut through this sandstone, good exposures in this bed are shown there.

Peytona District.

In Peytona District, the **Upper Chilton Sandstone** comes above Coal River just east of Brounland, and rises rapidly above the same to the southeast as is shown in the following section, 0.5 mile southwest of Dartmont, taken by Teets:

	Ft.	In.
Sandstone, and concealed, **Winifrede Sandstone**	60	0
Sandstone, massive, Upper Chilton, 1035′ B., or 330′ above the Campbell Creek (No. 2 Gas) Coal	40	0

[4]Hennen and Reger, Logan and Mingo Report; 1914.

On the north side of Falling Rock Creek, the following section is exposed:

	Thickness Feet.	Total Feet.
Sandstone, massive, Upper Chilton........	60	60
Sandy shale, Chilton Coal horizon........	10	70
Sandstone and sandy shale, Lower Chilton and Hernshaw Sandstones............	80	150
Sandy shale..............................	12	162
Fire clay, Hernshaw Coal horizon..........	2	164

The Hernshaw Coal horizon occurs here 230 feet above the Campbell Creek (No. 2 Gas) Coal.

Another section in the head of Roundbottom Creek of Coal River, just northeast of the Peytona-Sherman District Line, was measured as follows:

	Thickness Feet.	Total Feet.
Coal blossom, Chilton "A".................	1	1
Sandy shale..............................	15	16
Sandstone, Upper Chilton.................	42	58
Coal, Chilton............................	3	61
Sandstone, Lower Chilton.................	37	98
Coal blossom, Little Chilton, 1150' B.......	2	100

Near the head of Drawdy Creek, the following section was measured, 1.75 miles west of Andrew P. O.:

	Thickness Feet.	Total Feet.
Coal blossom, Chilton "A" Coal............	2	2
Sandstone, Upper Chilton, massive, coarse grained	63	65
Sandy shale, Chilton Coal horizon, 1255' B., 315' above the Campbell Creek (No. 2 Gas) Coal..........................	3	68

Scott District.

The Upper Chilton Sandstone forms massive cliffs along the streams in Scott District, as is shown by the several sections given below:

On the south side of Camp Creek of Little Coal River, just southeast of the cross roads, the following section was measured:

	Feet.
Sandy shale, Chilton "A" Coal horizon..................	10
Sandstone and concealed..............................	100
Sandy shale, Hernshaw Coal horizon, 1015' B............	10

This horizon is 248 feet above the Campbell Creek (No. 2 Gas) Coal.

In the head of a branch of Rock Creek, in a road crossing the hill to Camp Creek, 2.6 miles east of Van Camp, the following section was measured:

	Feet.
Coal blossom, Winifrede	1
Sandy shale and sandstone, mostly sandstone	89
Coal blossom, Chilton "A"	3
Sandstone and sandy shale	67
Coal blossom, Hernshaw, 1040′ B	2
Sandstone, massive, coarse grained	58

In the hill just northeast of Van Camp, the following section was obtained:

	Feet.
Sandstone and concealed	110
Sandy shale, **Chilton Coal horizon**	10
Sandstone, Lower Chilton, massive and concealed to bench, **Little Chilton Coal horizon**	80
Sandstone and concealed	70
Coal, Hernshaw, 900′ B. (1′-7″)	1.6

Just east of Danville, the following section was measured:

	Feet.
Sandstone, **Lower Winifrede**	65
Sandy shale, **Chilton "A" Coal horizon**	5
Sandstone and concealed, **Upper Chilton**	40
Sandstone, Lower Chilton and Hernshaw, massive, cliff rock, medium grained	105
Sandy shale, **Williamson Coal horizon,** 1065′ B	15

The Williamson Coal horizon is 180 feet above the Campbell Creek (No. 2 Gas) Coal horizon.

The following section was measured near the crest of the Warfield Anticline at the head of Price Branch of Little Coal River, 1.75 miles northeast of Danville:

	Feet.
Sandstone, Upper Chilton	15
Sandy shale and concealed	35
Sandstone, Lower Chilton, massive, micaceous	40
Sandy shale and sandstone, **Hernshaw**	49
Coal blossom, Hernshaw, 1215′ B	1

The Hernshaw Coal is 235 feet above the Campbell Creek (No. 2 Gas) Coal.

On the north side of Workman Branch of Pond Fork, 2 miles east of Madison, the following section was measured:

	Feet.
Sandy shale and concealed, **Chilton "A" Coal horizon**	10
Sandstone, Upper Chilton, massive, medium grained	90
Sandstone, sandy shale and concealed	45
Coal blossom, Little Chilton, 1390′ B	1

Washington District.

The Upper Chilton Sandstone forms cliffs at several places in Washington District. A few sections will be given to illustrate the thickness and character of this sandstone.

On the Big Creek side, at the head of Lukey Fork of Mud River, about 2 miles northwest of Estep P. O., the following section was measured:

	Feet.
Sandstone, **Lower Winifrede,** massive	65
Sandy shale, **Upper Chilton**	42
Fire clay, **Chilton Coal horizon,** 1145′ B	3
Sandstone and concealed, **Lower Chilton**	45
Sandy shale	15

The Chilton Coal horizon or fire clay comes 226 feet above the fossiliferous **Dingess Limestone.**

The following section was measured on a slope just northwest of Anchor P. O.:

	Feet.
Concealed, **Upper Chilton Sandstone**	45
Sandy shale and sandstone, **Lower Chilton**	27
Coal blossom, Little Chilton, 1390′ B	3

In the head of the Right Fork of Dog Fork of Trace Fork of Big Creek of the Guyandot River, 1.5 miles southeast of Anchor P. O., on the crest of the Warfield Anticline, the following section was measured:

	Feet.
Sandstone, **Lower Winifrede,** massive	25
Sandy shale	60
Bench, **Chilton Coal horizon** at top	80
Coal blossom, Little Chilton, 1380′ B	2

The Chilton Coal is 305 feet above the Campbell Creek (No. 2 Gas) Coal.

The following section was measured at Mud Gap at the head of Mud River:

	Feet.
Sandstone, **Lower Winifrede**	71
Coal blossom, Chilton "A"	4
Sandstone, Upper Chilton	20
Coal blossom, Chilton, 1210′ B	3
Sandstone and sandy shale, **Lower Chilton**	32

The Chilton Coal is 165 feet above the Dingess Limestone fossil horizon.

At Clothier the Upper Chilton Sandstone is often broken up into shale and sandstone, but on Spruce Laurel Fork, this sandstone often forms massive cliffs.

One mile southeast of Clothier, the following section was measured:

	Feet.
Sandstone, massive, Upper Chilton	60
Coal blossom, Chilton	8

Another section measured at the mouth of Burnt Cabin Branch of Spruce Laurel Fork at the railroad grade is as follows:

		Feet.
Sandstone and concealed..30′	Upper Chilton	50
Sandstone, shelly.........20		
Coal, Chilton, 920′ B.		

At the mouth of Sandford Branch, the following section is exposed on the railroad grade:

		Feet.
Sandy shale and concealed 20′	Upper Chilton	50
Sandstone, massive.......30		
Coal, Chilton, 930′ B.		

Sherman District.

The Upper Chilton Sandstone usually forms massive cliffs in Sherman District.

On the north side of Coal River, just northwest of Racine, the following section was measured:

	Feet.
Sandstone, Lower Winifrede and Upper Chilton, massive, medium grained	88
Sandy shale, Chilton Coal horizon	15
Sandstone, Lower Chilton and Hernshaw	65
Coal, impure, Hernshaw, 1055′ B	2

The Hernshaw Coal is 240 feet above the Campbell Creek (No. 2 Gas) Coal.

In the head of Cold Fork of Laurel Creek of Coal River, 4.3 miles southwest of Nelson P. O., the following section was measured:

	Feet.
Coal, Winifrede, 1210′ B	2
Sandstone and concealed, Lower Winifrede	73
Sandstone, Upper and Lower Chilton	104

The Winifrede Coal is about 355 feet below the No. 5 Block Coal at this locality.

About one-fourth mile northwest of the mouth of Jarrolds Fork of Hopkins Fork, 1.5 miles northeast of Griffith P. O., the following section was measured:

	Feet.
Sandstone, Lower Winifrede, massive	67
Coal blossom, Chilton "A", 1235' B	3
Sandstone, Upper Chilton, massive	30
Concealed and sandy shale, Lower Chilton	70

The Chilton "A" Coal is 283 feet above the Upper Cedar Grove and 280 feet below the Stockton-Lewiston Coal.

Just east of Coal River Siding, the following section was measured:

	Feet.
Coal, Winifrede	4
Sandstone and concealed, Lower Winifrede	120
Concealed	20
Sandy shale	29
Coal blossom, Little Chilton, 1410' B	1
Sandstone	50

The Little Chilton Coal is 335 feet above the Campbell Creek (No. 2 Gas) Coal.

Crook District.

The Upper Chilton Sandstone usually forms massive cliffs in Crook District, as is shown in the different sections already given.

In the hill north of Coal River, just northwest of the mouth of Robinson Creek, 2.2 miles northwest of Lantie P. O., the following section was measured:

	Feet.
Sandstone, Lower Winifrede	60
Sandy shale	15
Sandstone, Upper Chilton, massive	69
Coal blossom, Chilton, 1085' B	1
Sandstone; Lower Chilton, massive	60

In the section at Bald Knob, this sandstone is 93′ 8″ thick, and at Workman Branch, it is nearly 100′ thick, and massive.

No quarries have been opened in this sandstone in Crook District, but it is a hard, micaceous rock, and would make a good building stone.

The Chilton Sandstones usually form massive cliffs in Crook District, as is shown in the different sections already given.

THE CHILTON COAL.

Underlying the Upper Chilton Sandstone, and separated from it by gray shale and impure fire clay, 0 to 8 feet thick, there occurs a multiple-bedded, splinty coal that has been designated the **Chilton Coal** by I. C. White from a small mining village of that name in Kanawha County, where it was first described.

This bed occurs from 70 to 150 feet under the Winifrede Coal, and has often been confused with the latter. It attains its greatest development in southern Boone and eastern Logan Counties, where it is mined on a commercial scale. This seam of coal is mined on Dingess Run, Rum and Buffalo Creeks in Logan County, and on Spruce Fork of Coal River in Boone and Logan Counties, and was formerly supposed to correlate with the Campbell Creek (No. 2 Gas) seam, but later detailed study has placed it in the **Chilton** horizon. This coal is an important bed in a portion of Boone County.

Peytona District.

The Chilton Coal in Peytona District appears to be thin and of little commercial value as far as developments have been made.

Peytona Coal Land Company Prospect Opening. No. 206 on Map II.

Located on a branch of Indian Creek, 1.0 mile northwest of Sterling P. O.; section taken by Krebs; **Chilton Coal**; elevation, 1120' B.

				Ft.	In.
Sandstone, massive					
Coal, splint	0'	9"	**Chilton**	5	4½
Shale, gray	1	7			
Coal, splint	0	6			
Shale, gray	0	0½			
Coal, splint, 1120' B.	2	6			
Fire clay and concealed				30	0
Coal, Little Chilton				2	5

PLATE XXVII.—Crop of Alma Coal in grade of Coal River Railroad at Ottawa.

E. T. Javins Opening—No. 207 on Map II.

Located 0.8 mile west of Cabell P. O., on land of E. T. Javins; section taken by Teets; **Chilton Coal**; elevation, 1180' B.

	Ft.	In.
Slate roof		
Coal, splint (slate floor)	2	1

The coal was formerly mined here by E. T. Javins for local fuel.

In prospecting the land of the Forks Creek Coal Company, Mr. E. B. Snider made several openings in the Chilton seam, and the measurements on some of these are given below:

Forks Creek Coal Company Prospect Opening. No. 208 on Map II.

Located on west side of Fork Creek, at the mouth of Locust Fork; section by E. B. Snider; **Chilton Coal**; elevation, 700' B.

	Ft.	In.
Slate roof		
Coal, splint	1	8

Forks Creek Coal Company Prospect Opening. No. 209 on Map II.

Located on the west side of Fork Creek, about 0.5 mile north of mouth of Jimmy Fork; section by E. B. Snider; **Chilton Coal**; elevation, 775' B.

			Ft.	In.
Slate roof				
Coal	0'	10"		
Slate	0	2		
Coal, splint, (slate floor)	2	0	3	0

Forks Creek Coal Company Prospect Opening. No. 210 on Map II.

Located on the east side of Fork Creek, at the mouth of Jimmy Fork; section taken by E. B. Snider; **Chilton Coal**; elevation, 805' B.

			Ft.	In.
Slate roof				
Coal, splint	0'	8"		
Slate	0	4		
Coal, splint	1	3		
Slate	0	2		
Coal, splint (slate floor)	1	1	3	6

Forks Creek Coal Company Prospect Opening. No. 211 on Map II.

Located on the west side of Wilderness Fork of Fork Creek; 0.25 mile south of Anderson Branch; section taken by E. B. Snider; **Chilton Coal**; elevation, 915' B.

			Ft.	In.
Slate roof				
Coal, splint	0'	5"		
Slate	0	2		
Coal, splint	1	8	2	3

Forks Creek Coal Company Prospect Opening. No. 212 on Map II.

Located on the west side of Jim Lick Branch of Wilderness Fork of Fork Creek, 2.0 miles south of mouth of Jimmy Fork; section by E. B. Snider; **Chilton Coal**; elevation, 1140' B.

			Ft.	In.
Slate roof				
Coal, splint	0'	6"		
Slate	1	8		
Coal	0	3		
Slate	0	3		
Coal, splint (slate floor)	1	4	4	0

Scott District.

Very little development has been made on the Chilton Coal in Scott District, and so far as developed the bed appears to be thin, but it is possible that there are places in the district where the coal will be found of commercial thickness and purity, when the entire area has been thoroughly prospected.

Samuel Cabell Opening—No. 213 on Map II.

Located on a small branch of Rock Creek, 1.0 mile east of Rock Creek Station; section taken by Teets; **Chilton Coal**; elevation, 1100' B.

			Ft.	In.
Sandstone, massive				
Coal, cannel	1'	4"		
Gray shale and fire clay	3	0		
Coal, splint (slate floor)	2	6	6	10

Washington District.

The Chilton Coal is an important bed in Washington District. It is mined by the Coal River Mining Company near Ottawa, at its upper mine, where the following section was measured:

Coal River Mining Company Mine—No. 214 on Map II.

Located on west side of Spruce Fork of Coal River at Ottawa; section measured by Krebs; Chilton Coal; elevation, 1130' B.

				Ft.	In.
1.	Slate roof				
2.	Coal	0'	1"		
3.	Shale, gray	0	9		
4.	Coal	0	1		
5.	Shale, gray	0	9		
6.	Coal, splint	0	1½		
7.	Shale, gray	1	4		
8.	Coal, hard splint	1	4		
9.	Coal, gray splint	1	4		
10.	Coal, gas	0	4		
11.	Coal, splint	0	3		
12.	Coal, gas (slate floor)	0	5	6	9½

Faces, N. 46° E.; butts N. 44° W.; capacity of the mine, about 100 tons daily; number of men employed, 15; coal shipped east and west for steam and domestic fuel coal; J. M. Moore, Superintendent, Ottawa, West Virginia.

The analysis of a sample collected from Nos. 8, 9, 10, 11 and 12, as reported by Messrs. Hite and Krak, is published in the table of coal analyses under No. 36.

The Chilton Coal is mined on the head of Mud River, near the divide between Mud River and Turtle Creek. The top portion of the bed has developed into a splendid cannel coal, which is being used for domestic fuel by the farmers in that vicinity.

Floyd Nelson Mine—No. 215 on Map II.

Located on head of Mud River, 2.0 miles southeast of Turtle Creek P. O.; section taken by Krebs; Chilton Coal; elevation, 1240' B.

				Ft.	In.
1.	Sandstone, massive				
2.	Coal, splint	0'	1"		
3.	Shale, gray	0	7		
4.	Coal, splint	1	9		
5.	Coal, cannel	3	8		
6.	Coal, splint (slate floor)	0	4	6	5

Butts, N. 50° W.; faces, N. 40° E.

The analysis of a sample collected from No. 5, as reported by **Messrs.** Hite and Krak, is published in the table of coal analyses under **No. 37.**

A sample was also collected by Clark and Krebs from No. 5 and analysis made of same by Mr. Paul Demler, Charleston, W. Va., with the following results:

	Per cent.
Moisture	1.88
Volatile Matter	40.69
Fixed Carbon	51.14
Ash	6.29
Total	100.00
Sulphur	0.657
Phosphorus	0.012
B. T U.	13,595

Near the head of Hunting Camp Branch of Mud River, the following prospect opening was measured:

Little Coal Land Company Prospect Opening. No. 216 on Map II.

Located on head of Hunting Camp Branch of Mud River, 3.0 miles southwest of Turtle Creek P. O.; section by Krebs; **Chilton Coal;** elevation 1250′ B.

				Ft.	In.
1.	Shale, gray				
2.	**Coal,** cannel	0′	4″		
3.	Shale, gray	1	4		
4.	**Coal,** splint	0	7		
5.	Shale, gray	1	6		
6.	**Coal,** splint	0	6		
7.	Slate	0	4		
8.	**Coal,** gas	0	10		
9.	**Coal,** cannel	1	0		
10.	Slate	0	1		
11.	**Coal,** splint (slate floor)	2	0	8	6

The analysis of a sample collected from Nos. 4, 6, 8, 9 and 11, as reported by **Messrs.** Hite and Krak, is published in table of coal analyses under **No. 38.**

A sample collected from the same numbers by **Messrs.** Clark and Krebs, and analysis made of same by Mr. Paul Demler gave the following results:

	Per cent.
Moisture	2.02
Volatile Matter	40.04
Fixed Carbon	53.49
Ash	4.45
Total	100.00
Sulphur	0.646
Phosphorus	0.004
B. T. U.	13,784

Near the head of Mud River the following section was measured:

Little Coal Land Company Opening—No. 217 on Map II.

Located on a small branch of Mud River, 0.9 mile southeast of mouth of Lukey Fork; section by Teets; **Chilton Coal**; elevation, 965′ B.

			Ft.	In.
Sandstone, massive			8	0
Coal, block	0′	7″		
Fire clay	1	6		
Slaty shale	7	6		
Coal	0	1		
Slate, black	0	0½		
Coal	0	1		
Slate, black	0	0½		
Coal, block	0	3		
Slate, gray	0	4		
Coal, block, visible	2	6	12	11

Total coal—3′ 6″; once mined for local fuel use.

Near the head of Sixmile Creek, the Chilton Coal is mined at several places for local fuel use:

Yawkey and Freeman Opening—No. 218 on Map II.

Located on the south side of Sixmile, near the divide between Sixmile and Big Creeks, 1.0 mile southwest of Anchor P. O.; section taken by Krebs; **Chilton Coal**, elevation, 1275′ B.

				Ft.	In.
1.	Sandstone, massive				
2.	Slate			0	2
3.	**Coal**, splint	0′	4″		
4.	Slate	0	1		
5.	**Coal**	0	1		
6.	Slate	0	0½		
7.	**Coal**	0	1		
8.	Slate	0	0½		
9.	**Coal**, splint	0	5		
10.	Slate	0	0½		
11.	**Coal**, splint, (slate floor)	2	6	3	7½

The analysis of a sample collected from Nos. 5, 7, 9 and 11, as reported by Messrs. Hite and Krak, is published in the table of coal analyses under **No. 39.**

Yawkey and Freeman Local Mine Opening. No. 219 on Map II.

Located on the head of Garrett Fork of Big Creek, 2.0 miles northeast of Curry P. O.; Chapmanville District, Logan County, and 1/8 mile west of the Boone-Logan County Line; section by Krebs; **Chilton Coal**; elevation, 1425' L.

			Ft.	In.
Sandstone, shelly				
Slate, gray			0	2
Sandstone			0	8
Coal, splint	0'	7"		
Coal, impure	0	6		
Coal, block, (slate floor)	2	2	3	3

Yawkey and Freeman Opening—No. 220 on Map II.

Located on the south side of Missouri Fork of Hewett Creek, 2.3 miles southwest of Hewett P. O.; section by Teets; **Chilton Coal**; elevation, 1380' B.

	Ft.	In.
Slate roof		
Coal, reported, (slate floor)	6	0

Opening fallen shut; unable to get the measurement. Blaine Ball once mined the coal here for local fuel use and reported the thickness given above. There is a probability that the section contains some impurities.

Ashford Ball Opening—No. 221 on Map II.

Located on the south side of Meadow Fork of Hewett Creek, 1.4 miles west of Hewett P. O.; section by Teets; **Chilton Coal**, elevation 1370' B.

				Ft.	In.
1.	Slate				
2.	**Coal**, block	1'	8"		
3.	Slaty shale	1	1		
4.	**Coal**	0	2		
5.	Slate, gray	0	6		
6.	**Coal**	0	5		
7.	Sandstone	5	0		
8.	Blue shale	2	0		
9.	**Coal**	0	1		
10.	Slaty shale	1	8		
11.	**Coal**, block	3	11		
12.	Shale, gray	0	4		
13.	**Coal**, block (slate floor)	1	7	18	5

The analysis of a sample collected from Nos. 11 and 13, as reported by **Messrs.** Hite and Krak, is published in the table of coal analyses under **No. 39A.**

From Ottawa the **Chilton Coal** falls very rapidly to the south, until at Clothier, where it crosses the Boone-Logan County Line, the coal is 50 to 75 feet above the bed of Spruce Fork. This coal is mined on a commercial scale on Spruce Fork in Logan County by several coal companies.

Up Spruce Laurel Fork from Clothier, the Chilton Coal rises about as fast as the bed of the stream, until near the mouth of Dennison, where the coal goes under water level. The seam has thinned out there and as far as it has been developed, it appears to be of little commercial value where it goes under Spruce Laurel Fork.

Yawkey and Freeman Opening—No. 222 on Map II.

Located on west side of Spruce Fork, 1.0 mile northeast of Clothier P. O.; section by Krebs; **Chilton Coal**; elevation, 860′ B.

			Ft.	In.
Sandstone, massive				
Coal, splint	0′	1″		
Shale, gray	0	6		
Coal, splint	0	1		
Shale, gray	0	3		
Coal, 1″ to	0	2		
Shale, gray	3	6		
Coal, splint	4	0		
Shale, gray	5	0		
Coal, splint, (gray shale floor)	0	6	14	1

The above measurement was taken from an exposure along the railroad grade.

Yawkey and Freeman Opening—No. 223 on Map II.

Located on the north side of Spruce Laurel Fork, on railroad grade, just below the mouth of Burnt Cabin Branch; section by Krebs; **Chilton Coal**; elevation, 890′ B.

			Ft.	In.
Sandstone, massive				
Coal, 0″ to	0′	4″		
Shale, dark gray	0	2		
Coal	0	1		
Shale, gray	0	6		
Sandstone	0	6		
Shale, gray	0	8		
Coal, visible	2	0	4	3

Yawkey and Freeman Opening—No. 224 on Map II.

Located on Right Fork of Burnt Cabin Branch; section by Teets; **Chilton Coal**; elevation, 935′ B.

		Ft.	In.
Shale, gray			
Coal, splint, visible		3	6

The opening had fallen in and it was impossible to get a complete section of same.

Yawkey and Freeman Local Mine Opening. No. 225 on Map II.

Located on south side of Spruce Laurel Fork, 2.5 miles southeast of Clothier; section by Krebs; **Chilton Coal**; elevation, 950′ B.

			Ft.	In.
Shale, gray			8	0
Coal, hard splint	0′	8″		
Coal, gray splint	0	8		
Coal, block (slate floor)	2	0	3	4

David Bias Local Mine Opening—No. 226 on Map II.

Located on north side of Spruce Laurel, just below the mouth of Whiteoak Branch; section taken by Krebs; **Chilton Coal**; elevation, 1015′ H. L.

			Ft.	In.
Sandstone, massive			10	0
Coal	0′	2 ″		
Shale, gray	0	4		
Coal, splint	0	3		
Shale, gray	0	1		
Coal, splint	0	2		
Shale, gray	0	7		
Coal	0	0½		
Shale, gray	0	1		
Coal, splint	0	8		
Shale, gray	0	1½		
Coal, hard splint (slate floor)	2	0	4	6

Chilton, MacCorkle, Chilton and Meany Opening. No. 226A on Map II.

Located on the north side of Jigly Branch of Skin Poplar Branch, 1.0 mile east of its mouth, 7.0 miles southeast of Clothier; section taken by Krebs; **Chilton Coal**; elevation, 1345′ B.

			Ft.	In.
1. Sandstone				
2. **Coal**, hard splint	0′	2″		
3. **Coal**, impure	0	1		

			Ft.	In.
4. **Coal**, splint	0′	4″		
5. Shale, dark	0	5		
6. **Coal**, hard splint	0	8		
7. **Coal**, gray splint	1	4		
8. Shale, dark	0	3		
9. **Coal**, splint (slate floor)	1	4	4	7

The analysis of a sample collected from Nos. 4, 6, 7 and 9, as reported by Messrs. Hite and Krak, is published in the table of coal analyses under **No. 39B.**

Chilton, MacCorkle, Chilton and Meany Opening. No. 227 on Map II.

Located on north side of Spruce Laurel Fork, at the mouth of Dennison; section taken by Krebs; **Chilton Coal**; elevation, 1460′ B.

	Ft.	In.
Sandstone, massive		
Coal, hard splint (shale, gray, floor)	2	0

The coal was mined for fuel in sinking the diamond core test hole, Cassingham No. 2 (135).

Sherman District.

The Chilton Coal has not been much explored in Sherman District, but so far as developed the bed shows no very thick coal.

LaFollette, Robson et al. Opening—No. 228 on Map II.

Located on north side of Whiteoak Creek, 1.3 miles southeast of Orange P. O.; section taken by Krebs; **Chilton Coal**; elevation, 955′ B.

			Ft.	In.
Sandstone, shelly			4	0
Coal, hard splint	1′	2″		
Shale, gray, 2″ to	0	4		
Coal, splint	0	6		
Shale	0	1		
Coal, impure (slate floor)	0	6	2	7

West of Coal River on Laurel Creek, the **Chilton Coal** crops at several places, and is mined for local fuel use:

J. W. Brinkley Local Mine Opening—No. 229 on Map II.

Located on north side of Laurel Creek of Coal River, 0.7 mile southwest of Nelson P. O.; section by Teets; **Chilton Coal**; elevation, 810' B.

			Ft.	In.
Gray shale roof				
Coal, hard splint	1'	0"		
Coal, gray splint (slate floor)	1	7	2	7

The foregoing openings give the approximate type section of the Chilton Coal in Sherman District.

Crook District.

The Chilton Coal appears in different parts of Crook District of thickness to be of commercial value, but thus far very little development has been made on it.

Lee Wills Local Mine Opening—No. 230 on Map II.

Located on west side of Jacks Branch, 3.0 miles southwest of Lantie P. O.; section taken by Krebs; **Chilton Coal**; elevation, 1110' B.

			Ft.	In.
Shale, gray			1	0
Coal, splint	1'	0"		
Coal, splint, harder	1	5		
Slate	0	2		
Coal, softer (slate floor)	1	0	3	7

Buck Dawes Local Mine Opening—No. 231 on Map II.

Located on south side of Jacks Branch, 1.0 mile from its mouth; section taken by Teets; **Chilton Coal**; elevation, 920' B.

	Ft.	In.
Sandstone, massive		
Coal, splint (slate floor)	2	2

Passing to the south, across the divide from Jacks Branch to Bull Creek, the following opening was measured:

Sidney White Local Mine Opening—No. 232 on Map II.

Located on east side of Bull Creek, 1.7 miles due south of Lantie P. O.; section taken by Teets; **Chilton Coal**; elevation, 1015' B.

	Ft.	In.
Sandstone		
Coal, hard splint (slate floor)	1	7

Near the head of Whites Branch of West Fork several prospect openings have been made in the **Chilton seam** by the Lackawanna Coal and Lumber Company, as follows:

Lackawanna Coal and Lumber Company Prospect Opening. No. 233 on Map II.

Located on north side of Whites Branch of West Fork of Pond Fork, 1.8 miles northeast of Gordon P. O.; section taken by Krebs; **Chilton Coal**; elevation, 1105′ B.

			Ft.	In.
Gray shale				
Coal, splint	0′	1″		
Shale, gray	0	2		
Coal, splint	0	10		
Slate	0	2		
Coal	0	1		
Slate, dark	0	3		
Fire clay shale	0	6		
Coal, splint (slate floor)	1	6	3	7

Lackawanna Coal and Lumber Company Local Mine Opening—No. 234 on Map II.

Located on Lick Branch of Whites Branch, 1.5 miles northeast of Gordon P. O.; section taken by Krebs; **Chilton Coal**; elevation, 1095′ B.

			Ft.	In.
Shale, gray				
Coal, splint	1′	0″		
Shale, gray, full of plant fossils	3	0		
Coal, splint (gray shale floor)	2	4	6	4

Butts, N. 50° W.; faces, N. 40° E.

Lackawanna Coal and Lumber Company Opening. No. 235 on Map II.

Located on north side of Whites Branch, 0.5 mile northeast of Gordon P. O.; section taken by Krebs; **Chilton Coal**; elevation, 1040′ B.

			Ft.	In.
Sandstone, massive, visible			10	0
Shale, gray, full of plant fossils			1	0
Coal, splint	1′	0″		
Slate	0	2		
Coal, splint (slate floor)	2	4	3	6

Mined for local fuel use by **J. H. Hendrickson.**

Wharton Estate Prospect Opening—No. 236 on Map II.

Located on south side of Greens Branch, just below mouth of Still Hollow, 2.3 miles northwest of Chap P. O.; section taken by Krebs; **Chilton Coal**; elevation, 1175' B.

			Ft.	In.
Sandstone, massive				
Shale, gray, full of plant fossils			5	0
Slate			0	2
Coal, splint	1'	8"		
Coal, splint, harder, (slate floor)	1	10	3	6

Lackawanna Coal and Lumber Company Opening. No. 237 on Map II.

Located on the south side of Jarrolds Branch of West Fork, 1.1 miles due north of Chap P. O.; section by Teets; **Chilton Coal**; elevation, 1235' B.

			Ft.	In.
1. Sandstone				
2. **Coal**, splint	2'	3"		
3. Slate, gray	2	0		
4. **Coal**, splint, softer (slate floor)	0	10	5	1

The analysis of a sample taken from Nos. 2 and 4, as reported by **Messrs. Hite and Krak**, is published in the table of coal analyses under **No. 40.**

The coal is mined for local fuel use by **Geo. Jarrold.**

Lackawanna Coal and Lumber Company Local Mine Opening—No. 238 on Map II.

Located on the north side of Jarrolds Branch, about 3.0 miles from mouth of same; section by Teets; **Chilton Coal**; elevation, 1190' B.

			Ft.	In.
1. Sandstone roof				
2. **Coal**, splint	2'	2"		
3. Shale, gray	0	2		
4. **Coal**, splint, (slate floor)	0	10	3	2

The analysis of a sample collected from Nos. 2 and 4, as reported by **Messrs. Hite and Krak**, is published in the table of coal analyses under **No. 41.**

Polly Miller Local Mine Opening—No. 239 on Map II.

Located on the west side of West Fork, just above mouth of Jarrolds Branch, 1.3 miles northwest of Chap P. O.; section taken by Teets; **Chilton Coal**; elevation, 1190' B.

			Ft.	In.
Sandstone roof				
Slate			0	2
Coal, splint	2'	3"		
Coal, impure	0	1		
Coal, gray splint, (slate floor)	0	10	3	2

Brenny Miller Local Mine Opening—No. 240 on Map II.

Located on Spruce Lick Fork of West Fork, about 0.6 mile west of Chap P. O.; section by Teets; **Chilton Coal**; elevation, 1285' B.

			Ft.	In.
Sandstone				
Coal, splint	2'	3"		
Shale, dark	0	1		
Coal, gray splint (slate floor)	1	1	3	5

Along James Creek, on the lands of E. J. Berwind, some prospecting has been done for the different coal seams:

E. J. Berwind Prospect Opening—No. 241 on Map II.

Located on the Lefthand Fork of James Creek of West Fork, 1.5 miles southeast of Chap P. O.; section by Teets; **Chilton Coal**; elevation, 1300' B.

			Ft.	In.
Sandstone				
Coal, splint	2'	4"		
Bone	0	2		
Coal, gray splint (slate floor)	1	3	3	9

E. J. Berwind Opening—No. 242 on Map II.

Located on the west side of James Creek, about 0.7 mile up same, and 1.1 miles southeast of Chap P. O.; section measured by Teets; **Chilton Coal**; elevation, 1425' B.

			Ft.	In.
Sandstone				
Coal, splint	0'	8"		
Shale, gray	0	1		
Coal, splint, hard (slate floor)	2	3	3	0

The coal is mined for local fuel use by Leander Dickens.

E. J. Berwind Prospect Opening—No. 243 on Map II.

Located on north side of James Creek, 1.2 miles southeast of Chap P. O.; **Chilton Coal**; elevation, 1420' B.

	Ft.	In.
Sandstone roof		
Coal, splint (slate floor)	3	9

Near the head of West Fork, the following opening was measured on Bowen Branch, where the coal is mined for local fuel use:

John Q. Dickinson Opening—No. 244 on Map II.

Located on Bowen Branch of West Fork, 4.0 miles southeast of Bald Knob P. O.; section taken by Krebs; **Chilton Coal**; elevation, 1825' B.

				Ft.	In.
1.	Sandstone, massive				
2.	**Coal**, splint, soft	0'	10"		
3.	Shale, gray, full of plant fossils	3	0		
4.	**Coal**, splint	0	5		
5.	Slate	0	4		
6.	**Coal**, splint	1	0		
7.	**Coal**, gray splint	0	8		
8.	**Coal**, block (slate floor)	1	4	7	[illegible]

Butts, N. 70° W.; faces, N. 20° E.

The analysis of a sample collected from Nos. 2, 4, 6, 7 and 8, as reported by Messrs. Hite and Krak, is published in the table of coal analyses under **No. 42**.

Wharton Estate Local Mine Opening—No. 245 on Map II.

Located on Dry Branch of Pond Fork, 1.8 miles north of Pond P. O.; section by Krebs; **Chilton Coal**; elevation, 1145' B.

			Ft.	In.
Sandstone, massive				
Coal, soft splint	1'	0"		
Shale, gray	0	6		
Coal, splint, visible	1	6		
Concealed	1	6		
Coal, splint	1	6		
Coal, splint, hard (slate floor)	1	2	7	2

Everett Workman Local Mine Opening—No. 246 on Map II.

Located on south side of Old Camp Branch, 1.5 miles northeast of Pond P. O.; section taken by Krebs; **Chilton Coal**; elevation, 1195' B.

			Ft.	In.
Sandstone, massive				
Shale, gray, full of plant fossils			5	0
Coal, soft splint	1'	4"		
Coal, gray splint	0	2		
Coal, splint, hard (slate floor)	2	0	3	6

Butts, N. 45° W.; faces, N. 45° E.

John Q. Dickinson Prospect Opening—No. 247 on Map II

Located on the head of Coal Branch of Pond Fork, 1.3 miles northeast of Bald Knob P. O.; section by Teets; **Chilton Coal**; elevation, 1525' B.

			Ft.	In.
Sandstone, massive				
Shale, gray			2	0
Coal, splint	0'	3"		
Shale, gray	0	0½		
Coal, gray splint	2	6½		
Slate, gray	0	1		
Coal, gray splint (slate floor)	1	4	4	3

This coal is 160 feet above the Hernshaw Coal seam at this point and 200 feet above the Dingess Limestone.

Wharton Estate Prospect Opening—No. 248 on Map II.

Located on the south side of Workman Branch of Pond Fork, 1.2 miles due north of Bald Knob P. O.; section by Teets; **Chilton Coal**; elevation, 1548' L.

			Ft.	In.
Sandstone roof				
Coal, gas	0'	6½"		
Slate, gray	0	0½		
Coal, gray splint	2	0		
Slate, gray	0	2		
Coal, gray splint (slate floor)	1	3	4	0

Near the head of West Fork, a few prospect openings have been made in the Chilton seam. The following section shows the general structure and thickness of the coal in that region:

Wharton Estate Opening—No. 249 on Map II.

Located just east of Jarrolds Flats, and 1.0 mile west of the Boone-Raleigh County Line, 3.5 miles southeast of Bald Knob P. O.; section by Teets; **Chilton Coal**; elevation, 2595' B.

			Ft.	In.
Slate roof				
Coal, splint	1'	0"		
Slate	0	2		
Coal, splint, visible	2	3	3	5

The coal was formerly mined here for local fuel use, but the opening is now abandoned.

The composition and calorific value of the coal in the commercial mines and some of the local country banks will be discussed fully in a subsequent chapter, together with an estimate of the probable available area and tonnage of the bed by magisterial districts.

THE LOWER CHILTON SANDSTONE.

From 0 to 5 feet under the Chilton Coal, there often occurs a massive sandstone, from 5 to 40 feet in thickness, that has been designated the **Lower Chilton[5] Sandstone.** This is a hard micaceous sandstone, and is often current bedded. It usually quarries very well and will make a good building stone.

Its thickness and structure are shown in the various sections already published on preceding pages of this Report.

THE LITTLE CHILTON COAL.

From 5 to 40 feet under the Chilton Coal, there often occurs another bed of coal that is possibly a split from the Chilton, and this bed has been designated as the **Little Chilton[6] Coal.**

It is multiple bedded, splinty, and often contains layers of softer coal between the splint layers. It was not observed by the writer in **Peytona or Scott Districts.** If present it is represented by thin bands of slate and coal. However, in the

[5]Hennen and Reger, Logan and Mingo Report; 1914.
[6]Hennen and Reger, Logan and Mingo Report; 1914.

PLATE XXVIII.—Campbell Creek (No. 2 Gas) Coal opening at Uneeda.

remaining districts of Boone County, this bed appears to be present.

The sections of the coal will now be given by magisterial districts:

Washington District.

In passing up Spruce Laurel Fork from Clothier, for about two miles, the **Chilton Coal** appears above the C. & O. Railway grade, after which, for nearly one mile, the coal is in the railroad cuts.

The lower member of the bed, ranging in thickness from one to two feet, is seen to be diverging and getting farther under the main Chilton bed, being separated from same by shale and sandstone, from two feet in a section 2.0 miles east of Clothier, to eight feet in a section 2.5 miles east of Clothier, until at the mouth of Whiteoak Branch, 5.0 miles southeast of Clothier, the **Little Chilton Coal** appears 30 feet under the Chilton bed, being represented by slate and coal about two feet thick.

Sherman District.

In Sherman District, the Little **Chilton Coal** was observed at several places, where it is usually from one to two feet thick and usually divided by a bed of slate. Its horizon is given in the various sections already published on preceding pages of this Report.

Crook District.

The **Little Chilton Coal** was observed at several points in Crook District, varying in thickness from one to three feet.

Bedford et al. Prospect Opening—No. 250 on Map II.

Located 1.0 mile west of Workman Knob, and 2.0 miles southeast of Madison; section by Teets; **Little Chilton Coal**; elevation, 1310′ B.

			Ft.	In.
Sandstone, massive				
Coal, hard splint	1′	0″		
Slate, gray	0	2		
Coal, splint, visible	0	6	1	8

Near the mouth of Cow Creek the following opening was measured in the Little Chilton bed:

Wharton Estate Prospect Opening—No. 251 on Map II.

Located on south side of Cow Creek, 2.9 miles southwest of Pond P. O.; section by Teets; elevation, 1425' B.

				Ft.	In.
1.	Slate roof				
2.	Coal, splint	1'	1 "		
3.	Coal, gray splint	1	0		
4.	Shale, gray	0	0½		
5.	Coal, splint	0	3½		
6.	Shale, gray	0	2		
7.	Coal, gray splint	1	3	3	10

The analysis of a sample collected from Nos. 2, 3, 5 and 7, as reported by Messrs. Hite and Krak, is published in the table of coal analyses under No. 43.

Wharton Estate Prospect Opening—No. 252 on Map II.

Located on the south side of Workman Branch, 1.1 miles due north of Bald Knob P. O.; section by Teets; Little Chilton Coal; elevation, 1451' L.

	Ft.	In.
Sandstone		
Coal, splint, (sandstone floor)	2	4

At the mouth of Greens Branch the following section was measured:

Wharton Estate Local Mine Opening—No. 253 on Map II.

Located on north bank of Greens Branch of West Fork, 2.2 miles northwest of Chap P. O.; section by Krebs; Little Chilton Coal; elevation, 1120' B.

	Ft.	In.
Sandstone, massive, visible	0	10
Coal, splint, (slate floor)	3	0

From the foregoing data, it is readily seen that the Little Chilton Coal has a considerable area of commercial thickness in Crook District. A further discussion of this bed will follow in a subsequent Chapter on Coal.

THE HERNSHAW SANDSTONE.

The next member of the Kanawha Series in descending order is a massive sandstone, varying in thickness from 20 to 40 feet. It has been designated the **Hernshaw Sandstone**[7] from a coal of that name which it overlies.

The base of the sandstone varies from 250 to 320 feet over the Campbell Creek (No. 2 Gas) Coal. Detailed sections of the sandstone are given in the sections published on preceding pages of this Report.

THE HERNSHAW COAL.

The next coal of importance in descending order is the **Hernshaw Coal.** This coal was so named from **Hernshaw**[8], a small mining village on Lens Creek, Kanawha County, where it is mined on a commercial scale, by the Marmet Coal Company, and has been variously designated the "Black Band Seam" and the "Block Seam". It is a hard, splinty, multiple bedded coal, the lower portion of which usually breaks into large lumps or blocks which do not break easily, and make an excellent domestic coal.

The coal usually has a gray shale roof, containing plenty of fossil plants, but at times this is displaced with a sandstone and thin coal, varying from 0 to 1 foot, separated by gray shale, 2 to 10 inches thick, from the main bed of coal, the latter being 2 feet to 3 feet 6 inches thick, with a gray shale floor.

The sections of the coal will now be taken up by magisterial districts:

Peytona District.

From the prospecting that has been done in Peytona District, it is evident that the **Hernshaw Coal** contains considerable fuel of commercial value in this district.

---

[7]Hennen and Reger, Logan and Mingo Report; 1914.
[8]Hennen and Reger, Logan and Mingo Report; 1914.

L. E. Epling Local Mine Opening—No. 254 on Map II.

Located on a south branch of Lick Creek, 2.5 miles northeast of Peytona; section by Teets; **Hernshaw Coal**; elevation, 1105' B.

			Ft.	In.
Sandstone, massive				
Coal	0'	3"		
Slate	0	0½		
Coal, mining in large blocks, splinty	1	6		
Coal, hard, (slate floor)	0	5	2	2½

Butts, N. 48° W.; faces, N. 42° E.

The Boone and Kanawha Land and Mining Company owns a tract of coal on the waters of Roundbottom, Lick and Short Creeks, north of Peytona P. O. The owners have recently made some prospect openings on this land. Several of these coal sections have recently been measured by J. M. Clark, of the firm of Clark and Krebs, Engineers, Charleston, and the results are given below:

J. W. Wade Prospect Opening—No. 255 on Map II.

Located on Left Fork of Roundbottom Creek, 0.5 mile above Wade's house; section taken by J. M. Clark; **Hernshaw Coal**; elevation, 1115' B.

	Ft.	In.
Sandstone		
Coal, cannel, (slate floor)	2	6

Charles Kirk Prospect Opening—No. 256 on Map II.

Located on right side of Tug Fork of Lick Creek, above Charles Kirk's house; section taken by J. M. Clark; **Hernshaw Coal**; elevation, 1105' B.

			Ft.	In.
Slate roof				
Coal, splint	0'	8½"		
Slate, gray	0	4		
Coal, block, hard, (slate floor)	2	10	3	10½

Near the head of Fork Creek, the following section was measured:

Forks Creek Coal Company Opening—No. 257 on Map II.

Located on Fork Creek, 0.5 mile north of mouth of Jimmy Branch; section taken by Teets; Hernshaw Coal; elevation, 770' B.

				Ft.	In.
1.	Sandstone, massive				
2.	Coal, splint	0'	3"		
3.	Shale, gray	0	8		
4.	Coal, gray splint	1	2		
5.	Coal, semi-splint, (fire clay floor)	0	8	2	9

Butts, N. 70° W.; faces, N. 20° E.

The coal is mined for domestic fuel by Rome Barker.

The analysis of a sample collected from Nos. 2, 4 and 5, as reported by Messrs. Hite and Krak, is published in the table of coal analyses under No. 44.

Scott District.

Very few openings in the Hernshaw Coal were observed in Scott District. It can not be determined whether or not this bed of coal extends into the hills in Scott District, of commercial thickness and purity, until considerable more prospect work has been done to test the seam.

In the southern part of Scott District, at the head of Workman Branch, there is a cannel coal that seems to correlate with the Hernshaw bed, and there the following section was measured:

Bedford et al. Opening—No. 258 on Map II.

Located at the head of Workman Branch of Pond Fork, 3.0 miles east of Madison; section by Krebs; Hernshaw Coal; elevation, 1250' B.

			Ft.	In.
Slate roof				
Coal, cannel	4'	2"		
Coal, bituminous, (slate floor)	0	2	4	4

The analysis of a sample collected from No. 2 of section, as reported by Messrs. Hite and Krak, is published in the table of coal analyses under No. 44A.

The cannel coal is mined for local fuel at present. Several

years ago the coal was mined and transported in wagons to Madison and loaded into railroad cars. No tests have been made to determine the area covered at this point by the cannel bed given above.

No more openings were observed in the Hernshaw Coal in Scott District, except those given in the general sections on preceding pages.

Washington District.

In Washington District, the **Hernshaw Coal** appears to be one of the important coals as far as developments have been made. Sections will now be given showing its thickness, character and structure within that area.

John Robertson Local **Mine Opening—No. 259 on Map II.**

Located on east side of Twin Branch of Mud River, 0.5 mile from its mouth, and 3.0 miles west of Mud Gap; section by Teets; **Hernshaw Coal**; elevation, 955' B.

				Ft.	In.
1.	Sandstone roof				
2.	Coal, splint	0'	7"		
3.	Coal, gray splint	0	5		
4.	Coal, hard, block, (slate floor)	1	2	2	2

A sample collected from Nos. 2, 3 and 4 by Clark and Krebs and analyzed by Mr. Paul Demler gave the following results:

	Per cent.
Moisture	2.54
Volatile Matter	39.68
Fixed Carbon	55.01
Ash	2.77
Total	100.00
Sulphur	0.876
Phosphorus	0.004
B. T. U.	13,674

Near the head of Cox Fork of Turtle Creek and the main Turtle Creek, several openings have been made in the Hernshaw bed, a few sections of which will now be given.

Little Coal Land Company Opening—No. 260 on Map II.

Located on the head of Cox Fork of Turtle Creek on the south side of same, 1.2 miles southwest of Turtletown; section taken by Krebs; **Hernshaw Coal**; elevation, 1250' B.

			Ft.	In.
Shale, gray			8	0
Coal, splint	0'	5"		
Shale, gray	0	5		
Coal, splint	0	2		
Shale, dark	0	1		
Coal, splint, hard	1	5		
Shale, dark	0	1		
Coal, gray splint, (slate floor)	1	8	4	3

The coal is mined for local fuel use at this point by Green Polly.

Columbus Miller Opening—No. 261 on Map II.

Located near the head of Right Fork of Turtle Creek, 1.8 miles southwest of Turtletown; section by Teets; **Hernshaw Coal**; elevation, 1275' B.

	Ft.	In.
Sandstone roof		
Coal, splint, blocky, (slate floor)	3	0

The coal was once mined here for local fuel use.

Noah Bias Opening—No. 262 on Map II.

Located on the east side of Turtle Creek, 2.0 miles southwest of Turtletown; section measured by Teets; **Hernshaw Coal**; elevation, 1265' B.

	Ft.	In.
Sandstone roof		
Coal, splint, blocky, (slate floor)	2	6

Butts, N. 48° W.; faces, N. 42° E.

Burrell Bias Local Mine Opening—No. 263 on Map II.

Located on the head of Low Gap Creek, 1.7 miles southwest of Low Gap Station; section by Teets; **Hernshaw Coal**; elevation, 1265' B.

	Ft.	In.
Sandstone roof		
Coal, splint, blocky, (slate floor)	2	6

Susan Miller Local Mine Opening—No. 264 on Map II.

Located on the north side of Low Gap Creek, 1.7 miles southwest of Low Gap Station; section by Teets; **Hernshaw Coal**; elevation, 1260′ B.

			Ft.	In.
Sandstone roof				
Coal, splint, blocky, (slate floor)			2	7

Lena Webb Local Mine Opening—No. 265 on Map II.

Located on the north side of Trace Fork of Big Creek, 2.0 miles southwest of Anchor P. O.; section by Teets; **Hernshaw Coal**; elevation, 1235′ B.

			Ft.	In.
Sandstone, massive				
Coal, splint	0′	4″		
Shale, gray	0	2		
Coal, splint	0	6		
Shale, gray	0	1		
Coal, block	1	2		
Shale, gray	0	0½		
Coal, gray splint, (slate floor)	1	11	4	2½

Albert Ellis Local Mine Opening—No. 266 on Map II.

Located on the north side of Trace Fork of Big Creek, 1.0 mile southwest of Anchor P. O.; section by Teets; **Hernshaw Coal**; elevation, 1012′ B.

			Ft.	In.
Sandstone roof				
Coal, block	1′	9″		
Coal, gray splint	0	4		
Slate, gray	0	1		
Coal, gray splint, visible	0	9	2	11

Butts, N. 53° W.; faces, N. 37° E.

James Workman Opening—No. 267 on Map II.

Located on the Left Fork of Dog Fork of Trace Fork of Big Creek, 0.7 mile due south of Anchor P. O.; section by Teets; **Hernshaw Coal**; elevation, 1280′ B.

			Ft.	In.
1. Sandstone roof				
2. **Coal**, splint	2′	3″		
3. Shale, gray	0	1		
4. **Coal**, block	0	6		
5. Shale, gray	0	1		
6. **Coal**, block, (slate floor)	0	5	3	4

PLATE XXIX.—Dingess Fossiliferous Limestone and Shale below Lantie on Pond Fork.

The analysis of a sample collected from Nos. 2, 4 and 6, as reported by Messrs. Hite and Krak, is given in the table of coal analyses under **No. 45.**

Kirby Hill Local Mine Opening—No. 268 on Map II.

Located on the Left Fork of Trace Fork of Big Creek; 0.6 mile due south of Anchor P. O.; section by Teets; **Hernshaw Coal**; elevation, 1280′ B.

			Ft.	In.
Sandstone roof				
Coal, splint, soft	0′	8″		
Coal, block	2	0		
Slate, gray	0	1		
Coal, gray splint, (slate floor)	0	5	3	2

Near the head of Sixmile Creek, several openings have been made in the Hernshaw Coal, as follows:

U. S. Hoge Prospect Opening—No. 269 on Map II.

Located on north side of Sixmile Creek, 1.5 miles west of Ramage; section by Teets; **Hernshaw Coal**; elevation, 1210′ B.

			Ft.	In.
Sandy shale roof				
Coal, impure	1′	4″		
Fire clay	0	2		
Coal, impure	0	1		
Shale, gray	0	5		
Coal, impure	0	1		
Shale, gray	0	4		
Coal, impure	0	2		
Fire clay	1	1		
Coal, splint, block, (fire clay and slate floor)	2	5	6	1

H. Hager Local Mine Opening—No. 270 on Map II.

Located on the south side of Sixmile Creek, 0.7 mile southwest of Havana P. O.; section by Teets; **Hernshaw Coal**; elevation, 1215′ B.

	Ft.	In.
Slate roof		
Coal, hard, visible	1	8

H. F. Cook Local Mine Opening—No. 271 on Map II.

Located on the south side of Missouri Fork of Hewett Creek, 2.8 miles southwest of Hewett P. O.; section by Teets; **Hernshaw Coal**; elevation, 1275' B.

	Ft.	In.
Sandstone roof		
Coal, block, visible	2	3

Henry Keadle Opening—No. 272 on Map II.

Located on the south side of Hewett Creek, near the head of same: section taken by Teets; **Hernshaw Coal**; elevation, 1275' B.

			Ft.	In.
Shale roof				
Coal, impure	0'	1"		
Shale, dark	0	1		
Coal, impure	0	1		
Shale, gray	0	6		
Coal, impure, block, (slate floor)	3	6	4	3

Emma Hager Opening—No. 273 on Map II.

Located on Right Fork of Meadow Fork of Hewett Creek, 1.5 miles southeast of Anchor P. O.; section by Krebs; **Hernshaw Coal**; elevation, 1245' B.

			Ft.	In.
1. Sandy shale				
2. Slate				
3. **Coal**, splint	0'	8"		
4. Slate	0	2		
5. **Coal**, splint, (slate floor)	3	0	3	10

The analysis of a sample collected from Nos. 3 and 5, as reported by Messrs. Hite and Krak, is given in the table of coal analyses under **No. 46.**

Between Clothier and the mouth of Sycamore Fork, no openings were observed in the Hernshaw bed.

Chilton, MacCorkle, Chilton and Meany Opening. No. 274 on Map II.

Located on North Fork of Sycamore Fork of Spruce Laurel Fork, about 0.5 mile from its mouth; section by Krebs; **Hernshaw Coal**; elevation, 1145' B.

	Ft.	In.
1. **Sandstone**, massive	15	0
2. Shale, dark	0	2

				Ft.	In.
3.	Coal, splint	0′	4″		
4.	Shale, gray	0	4		
5.	Shale, dark	0	3		
6.	Coal, splint	1	0		
7.	Shale dark	0	1		
8.	Coal, hard splint	0	2		
9.	Shale	0	1		
10.	Coal, hard splint	1	8		
11.	Shale, gray	0	1		
12.	Coal, splint, (slate floor)	0	4	4	4

The coal is mined for local use by **Reed Jarrell.**

The analysis of a sample collected from Nos. 3, 6, 8, 10 and 12, as reported by **Messrs.** Hite and Krak, is given in the table of coal analyses under **No. 47.**

Chilton, MacCorkle, Chilton and Meany Opening. No. 275 on Map II.

Located on Bear Hollow of Spruce Laurel Fork, about 0.25 mile from its mouth; section taken by Krebs; **Hernshaw Coal**; elevation, 1170′ B.

			Ft.	In.
Sandstone, massive				
Shale, dark			0	2
Coal, cannel	0′	1″		
Shale, dark	0	2		
Coal, splint	0	6		
Shale, dark	0	8		
Coal, splint	1	0		
Shale, dark	0	1		
Coal, impure	0	3		
Coal, hard splint, (slate floor)	1	8	4	5

The coal was mined at this point by **David Bias** for fuel in sinking diamond core test hole.

Sherman District.

The **Hernshaw Coal**, so far as it has been developed, appears to be of commercial thickness and purity in a portion of Sherman District. A few sections will now be given, showing the character, structure and thickness of its bed in that region:

Boone and Kanawha Land and Mining Company Prospect Opening—No. 276 on Map II.

Located on the head of Short Creek, 2.5 miles northeast of Racine; section taken by Krebs; **Hernshaw Coal**; elevation, 1060′ B.

			Ft.	In.
Sandstone, flaggy			10	0
Shale, sandy			7	0
Coal, gray splint	0′	6″		
Shale, dark	0	1½		
Coal, splint, (slate floor)	2	0	2	7½

Morris Bradshaw Local Mine Opening—No. 277 on Map II.

Located on head of Short Creek, 0.8 mile southeast of the Peytona-Sherman District Line, on the dividing ridge between Boone and Kanawha Counties; section taken by Teets; **Hernshaw Coal**; elevation, 1035′ B.

			Ft.	In.
Shale roof				
Coal, soft	0′	2″		
Shale, gray	0	0½		
Coal, splint	0	4		
Shale, gray	0	1½		
Coal, splint, visible	2	4	3	0

Butts, N. 45° W.; faces, N. 45° E.

C. P. Payne Local Mine Opening—No. 278 on Map II.

Located on head of Short Creek, on south side of same, and 0.3 mile south of Peytona-Sherman District Line, corner of the Boone-Kanawha County Line; section by Teets; **Hernshaw Coal**; elevation, 1040′ B.

			Ft.	In.
Slate roof				
Coal, splint	0′	6″		
Slate	0	1		
Coal, splint, visible	2	1	2	8

Butts, N. 44° W.; faces, N. 46° E.

Alice Snodgrass Local Mine Opening—No. 279 on Map II.

Located in the head of Left Branch of Short Creek, 2.0 miles northeast of Racine; section by Teets; **Hernshaw Coal**; elevation, 1035′ B.

			Ft.	In.
Slate roof				
Coal, splint, softer	0′	8″		
Slate, gray	0	2		
Coal, splint, visible	2	2	3	0

About five miles southeastward on Joes Creek, the Hernshaw Coal has been mined for local fuel use, as follows:

LaFollette, Robson et al. Opening—No. 280 on Map II.

Located on the west side of Joes Creek, 2.5 miles southeast of Seth; section by Krebs; **Hernshaw Coal**; elevation, 885' B.

			Ft.	In.
Shale, gray			2	0
Coal, hard splint	2'	6"		
Shale, dark	0	1		
Coal, splint, (slate floor)	0	10	3	5

LaFollette, Robson et al. Opening—No. 281 on Map II.

Located on east side of Joes Creek, 3.6 miles southeast of Seth; section taken by Krebs; **Hernshaw Coal**; elevation, 895' B.

			Ft.	In.
Shale, gray			1	0
Coal, hard splint	3'	0"		
Shale, gray	0	2		
Coal, splint, (slate floor)	0	10	4	0

Coal mined for local fuel use by Alexander Smoot.

Broun Coal Prospect Opening—No. 282 on Map II.

Located on a small branch flowing into Coal River from the north, 0.25 mile north of Seth; opening 0.5 mile northeast of Seth; section by Krebs; **Hernshaw Coal**; elevation, 870' B.

			Ft.	In.
Sandstone				
Coal, splint	2'	8"		
Slate, dark	0	2		
Coal, visible	0	6	3	4

Sutphin Local Mine Opening—No. 283 on Map II.

Located on east side of Coal River, 1.0 mile southeast of Seth; section by Krebs; **Hernshaw Coal**; elevation, 860' B.

			Ft.	In.
Sandstone, massive				
Coal, splint	2'	10"		
Shale, dark	0	2		
Coal, splint, (slate floor)	0	8	3	8

Lackawanna Coal and Lumber Company Opening. No. 284 on Map II.

Located on west side of Coal River at Seth; section taken by Teets; **Hernshaw Coal**; elevation, 860' B.

			Ft.	In.
Sandstone				
Coal, gray splint	0'	8"		
Shale, gray	0	7		
Coal, splint, visible	1	6	2	9

The coal was formerly mined here for local fuel use.

Southeast from Seth, up Coal River, the Hernshaw Coal is opened along the river to the mouth of Whiteoak Creek, and coal is being mined at different points from this seam for local fuel, by the tenants living on the lands of the large holding companies.

LaFollette, Robson et al. Local Mine Opening. No. 285 on Map II.

Located on east side of Coal River, 1.5 miles southeast of Seth; section taken by Teets; **Hernshaw Coal**; elevation, 865' B.

			Ft.	In.
Sandstone, massive, visible			30	0
Coal, splint	1'	1"		
Coal, gray splint	1	4		
Bone **coal**	0	2		
Coal, gray splint, (slate floor)	0	8	3	3

LaFollette, Robson et al. Opening—No. 286 on Map II.

Located on east side of Coal River, 1.9 miles southeast of Seth; section taken by Teets; **Hernshaw Coal**; elevation, 865' B.

				Ft.	In.
1.	Sandstone, massive				
2.	**Coal**, splint	1'	8"		
3.	**Coal**, gray splint	1	1		
4.	Shale, gray	0	3		
5.	**Coal**, gray splint	0	3		
6.	**Coal**, gray splint, (slate floor)	0	3	3	6

The analysis of a sample collected from Nos. 2, 3, 5 and 6, as reported by **Messrs.** Hite and Krak, is given in the table of coal analyses under **No. 48.**

The coal is mined for local fuel use at this point by **James Farrell.**

LaFollette, Robson et al. Opening—No. 287 on Map II.

Located on east side of Coal River, 1.9 miles southeast of Seth; section by Teets; **Hernshaw Coal**; elevation, 875' B.

				Ft.	In.
1.	Sandstone				
2.	Coal, splint	2'	0"		
3.	Coal, gray splint	1	3		
4.	Slate	0	1		
5.	Coal, splint, (slate floor)	0	6	3	10

The coal is mined here for local fuel use by **Sweet Dotson**. The analysis of a sample collected from Nos. 2, 3 and 5, as reported by **Messrs**. Hite and Krak, is given in the table of coal analyses under **No. 49**.

LaFollette, Robson et al. Opening—No. 288 on Map II.

Located on east side of Coal River, on a small branch flowing from the east, about 0.3 mile from its mouth, and 2.0 miles southeast from Seth; section taken by Teets; **Hernshaw Coal**; elevation, 870' B.

			Ft.	In.
Sandstone roof				
Coal, splint	1'	6"		
Coal, gray splint, (slate floor)	1	8	3	2

The coal is mined by **Blair Kuhn** for local fuel.

H. A. Robson Local Mine Opening—No. 289 on Map II.

Located on Horse Branch of Coal River, about 0.75 mile from its mouth, and 1.5 miles northwest of Orange P. O.; section measured by Krebs; **Hernshaw Coal**; elevation, 840' B.

			Ft.	In.
Sandstone, massive				
Coal, splint	3'	0"		
Bone, from 2" to	0	3		
Coal	0	10	4	1

Near the mouth of Haggle Branch, the Hernshaw Coal is being mined for local fuel use by Blackburn Cooper:

Blackburn Cooper Opening—No. 290 on Map II.

Located just south of the mouth of Haggle Branch, 2.0 miles northwest of Orange P. O.; section by Krebs; **Hernshaw Coal**; elevation, 865′ B.

			Ft.	In.
Sandstone, massive				
Coal, hard splint	3′	0″		
Shale, dark	0	6		
Coal, splint	1	0	4	6

H. A. Robson Local Mine Opening—No. 291 on Map II.

Located on the east side of Coal River, 1.2 miles northwest of Orange P. O.; section by Krebs; **Hernshaw Coal**; elevation, 835′ B.

			Ft.	In.
1. Sandstone, massive				
2. **Coal**, splint, softer	1′	0″		
3. **Coal**, gray splint	1	8		
4. Shale, dark	0	4		
5. **Coal**, semi-cannel, (slate floor)	0	10	3	10

The analyses of two samples collected from Nos. 2, 3 and 5 of the above section, as reported by Messrs. Hite and Krak, are given in the table of coal analyses under **Nos. 49B and 49D.**

F. W. Mason Opening—No. 291A on Map II.

Located on the south side of Big Coal River on a branch flowing northeast, 2.2 miles southwest from Orange P. O.; section by Teets; **Hernshaw Coal**; elevation, 895′ B.

			Ft.	In.
1. Sandstone roof				
2. **Coal**, block	3′	10″		
3. Shale	0	2		
4. **Coal** (to slate floor)	0	2	4	2

The analysis of a sample collected from Nos. 2 and 4, as reported by Messrs. Hite and Krak, is published in the table of coal analyses under **No. 49C.**

Lackawanna Coal and Lumber Company Local Mine Opening. No. 292 on Map II.

Located on the east side of Laurel Creek, about 0.5 mile south of the mouth of Sandlick Creek, and 2.0 miles southwest from Seth; section measured by Teets; **Hernshaw Coal**; elevation, 840′ B.

	Ft.	In.
Sandstone, massive		
Coal, splint, visible	2	2

On up Laurel Creek to the mouth of Cold Fork, the Hernshaw Coal remains above the bed of the stream and is often exposed to view on either bank of the stream.

Lackawanna Coal and Lumber Company Opening. No. 293 on Map II.

Located on the first branch flowing from the north, above the mouth of Cold Fork of Laurel; section by Teets; **Hernshaw Coal**; elevation, 855' B.

	Ft.	In.
Sandstone, massive		
Coal, splint, visible	1	8

The opening had fallen shut and it was not possible to get a total section. The coal was once mined here for local fuel.

Lackawanna Coal and Lumber Company Opening. No. 294 on Map II.

Located at the south side of Hopkins Fork, 0.8 mile southwest of Nelson P. O.; section taken by Teets; **Hernshaw Coal**; elevation, 895' B.

	Ft.	In.
Sandstone, massive		
Coal, splint (slate floor)	2	4

The coal is mined for local fuel use by Lorenza Michael.

Lackawanna Coal and Lumber Company Local Mine Opening. No. 295 on Map II.

Located on John Branch of Hopkins Fork of Laurel Creek, 0.5 mile east of Nelson P. O.; section by Teets; **Hernshaw Coal**; elevation, 855' B.

			Ft.	In.
Sandstone				
Coal, splint	0'	8"		
Shale, dark	0	0¼		
Coal, splint	1	6		
Coal, gray splint	0	8		
Shale, gray	0	2		
Coal, gray splint, (slate floor)	0	6	3	6¼

Butts, N. 62° W.; faces, N. 28° E.

Lackawanna Coal and Lumber Company Prospect Opening. No. 296 on Map II.

Located on the east side of Hopkins Fork of Laurel Creek, 0.5 mile south of the mouth of Jarrolds Branch; section by Teets; **Hernshaw Coal**; elevation, 995' B.

			Ft.	In.
Sandstone				
Coal, splint	0'	8"		
Slate	0	1		
Coal, splint	2	1	2	10

Lackawanna Coal and Lumber Company Opening. No. 297 on Map II.

Located on the west side of Hopkins Fork of Laurel Creek, at the mouth of Jarrolds Branch; section by Teets; **Hernshaw Coal**; elevation, 950' B.

			Ft.	In.
1. Sandstone roof				
2. **Coal**, splint	1'	10 "		
3. Slate, gray	0	1½		
4. **Coal**, gray splint	0	8		
5. **Coal**, splint (slate floor)	1	7	4	2½

The analysis of a sample collected from Nos. 2, 4 and 5, as reported by Messrs. Hite and Krak, is given in the table of coal analyses under **No. 49A.**

From the foregoing sections of the Hernshaw Coal, it will be observed that it covers a large area where it is of commercial thickness and purity in Sherman District.

Crook District.

Crook District is rich in coal, having possibly a greater number of seams of coal of commercial thickness and purity than any of the other magisterial districts in Boone County.

The Hernshaw **Coal** appears to be a very important one in this district. This bed has usually been classified in this region as the **Cedar Grove seam,** by mining experts who have made reports on the coals of this district; while sometimes it has been correlated as the Winifrede **bed.**

From sections given on preceding pages of this Report, from the study of the stratigraphy, and from the occurrence of the **Dingess Limestone,** it appears to be the Hernshaw **Coal.**

Sections in different parts of the district will now be given, showing the location, character and structure of the coal in this area:

D. M. Arbogast Local Mine Opening—No. 298 on Map II.

Located on the south side of Robinson Creek, 0.7 mile from its mouth, and 1.5 miles north of Lantie P. O.; section taken by Krebs; **Hernshaw Coal**; elevation, 850′ B.

	Ft.	In.
Sandstone with **coal** streaks	20	0
Coal, splint, visible	2	0

Yawkey and Freeman Opening—No. 299 on Map II.

Located on the west side of Pond Fork, just south of the mouth of Jacks Branch, at Lantie P. O.; section by Krebs; **Hernshaw Coal**; elevation, 790′ B.

			Ft.	In.
Sandstone, massive				
Coal, splint	0′	4″		
Slate, gray	1	2		
Coal, splint	1	3		
Slate, gray	0	6		
Coal, impure (slate floor)	0	7	3	10

Coal mined for local fuel use by Van Jarrell.

Ira Sutphin Opening—No. 300 on Map II.

Located on the west side of Pond Fork, 0.7 mile north of the mouth of West Fork; section by Teets; **Hernshaw Coal**; elevation, 800′ L.

			Ft.	In.
1. Sandstone, massive				
2. **Coal**, gas	0′	3″		
3. **Coal**, splint	2	1		
4. **Coal**, gray splint (slate floor)	0	8	3	0

The analysis of a sample collected from Nos. 2, 3 and 4, as reported by Messrs. Hite and Krak, is given in the table of coal analyses under No. 50.

Lackawanna Coal and Lumber Company Prospect Opening. No. 301 on Map II.

Located on west side of West Fork, 0.25 mile east of the mouth of West Fork; section by Krebs; **Hernshaw Coal**; elevation, 845' B.

			Ft.	In.
Sandstone, massive				
Coal, splint	0'	5"		
Shale, gray	0	5		
Coal, gray splint (slate floor)	1	11	2	9

On up Pond Fork from the mouth of West Fork, the Hernshaw Coal occurs above the bed of the creek, until at the mouth of Casey Creek it is from 60 to 70 feet above water level.

Chilton, MacCorkle, Chilton and Meany Local Mine Opening. No. 302 on Map II.

Located on the east side of Casey Creek, 0.5 mile from its mouth and 3.0 miles northwest of Pond P. O.; section by Krebs; **Hernshaw Coal**; elevation, 970' B.

			Ft.	In.
Shale, gray, with plant fossils			4	0
Coal, hard splint	2'	3"		
Shale, dark gray	0	4		
Coal, hard splint (slate floor)	0	7	3	2

Wharton Estate Local Mine Opening—No. 303 on Map II.

Located on the east side of Coal Branch of Pond Fork, near its head, 1.3 miles northwest of Bald Knob P. O.; section taken by Teets; **Hernshaw Coal**; elevation, 1360' B.

			Ft.	In.
1. Sandstone				
2. **Coal**, gray splint	1'	3"		
3. **Coal**, splint	0	6		
4. **Coal**, gas	0	3		
5. **Coal**, gray splint	2	3		
6. **Coal**, gas (slate floor)	1	3	5	6

The analysis of a sample collected from Nos. 2, 3, 4, 5 and 6, as reported by Messrs. Hite and Krak, is given in the table of coal analyses under **No. 51.**

Wharton Estate Prospect Opening—No. 304 on Map II.

Located near the head of Right Fork of Workman Branch, 1.1 miles northwest of Bald Knob P. O.; section by Teets; **Hernshaw Coal**; elevation, 1388′ B.

			Ft.	In.
Gray shale roof				
Coal, splint, blocky	2′	0″		
Coal, gray splint (slate floor)	2	1	4	1

Wharton Estate Prospect Opening—No. 305 on Map II.

Located on the south side of Workman Branch, 1.1 miles due north of Bald Knob P. O.; section taken by Teets; **Hernshaw Coal**; elevation, 1390′ L.

			Ft.	In.
Sandstone				
Coal, gas	0′	9″		
Coal, gray splint (sandstone floor)	3	1	3	10

The above opening is 60 feet above the Dingess Limestone and 388 feet above the Campbell Creek (No. 2 Gas) Coal at this point.

Cassingham Prospect Opening—No. 306 on Map II.

Located on the Left Fork of Jasper Workman Branch, 1.0 mile southwest of Bald Knob P. O.; section by Krebs; **Hernshaw Coal**; elevation, 1345′ B.

			Ft.	In.
Sandstone, massive, coarse grained			4	0
Shale, gray, full of plant fossils			1	8
Coal, splint, hard	2′	4″		
Coal, gray splint	0	6		
Coal, splint, block (slate floor)	1	6	4	4

Lackawanna Coal and Lumber Company Opening. No. 307 on Map II.

Located at the mouth of Whites Branch of West Fork; section measured by Krebs in bed of creek; **Hernshaw Coal**; elevation, 840′ L.

			Ft.	In.
Sandstone, massive				
Coal, splint	0′	6″		
Slate	0	4		
Coal, gray splint, visible	1	6	2	4

Up West Fork from the mouth of Whites Branch, the Hernshaw Coal rises rapidly above the bed of the stream, and has been mined for local fuel by the residents at several points along West Fork:

Pond Fork Coal and Land Company Local Mine Opening. No. 308 on Map II.

Located on the north side of Duck Brown Hollow of West Fork, 1.8 miles southeast from Gordon P. O.; section measured by Krebs; **Hernshaw Coal**; elevation, 960' B.

			Ft.	In.
Sandstone, massive				
Coal, splint	0'	2"		
Shale, gray	0	2		
Coal, splint (slate floor)	2	8	3	0

Pond Fork Coal and Land Company Opening. No. 309 on Map II.

Located on the south side of Duck Brown Hollow, about 1.0 mile from its mouth; section taken by Teets; **Hernshaw Coal**; elevation, 970' B.

			Ft.	In.
1. Sandstone				
2. **Coal**, splint	0'	3"		
3. Shale, gray	0	2		
4. **Coal**, splint, blocky	2	1		
5. **Coal**, impure (slate floor)	0	3	2	9

Coal mined for local fuel use by **Joel Miller.**

Analysis of sample collected from Nos. 2, 4 and 5, as reported by **Messrs.** Hite and Krak, is given in the table of coal analyses under **No. 52.**

Pond Fork Coal and Land Company Local Mine Opening. No. 310 on Map II.

Located on the east side of West Fork, 2.1 miles southeast of Gordon P. O; section by Teets; **Hernshaw Coal**; elevation, 970' B.

			Ft.	In.
Sandstone roof				
Coal, splint	2'	4"		
Bone	0	2		
Coal, hard splint (slate floor)	0	9	3	3

Pond Fork Coal and Land Company Opening. No. 311 on Map II.

Located on south side of Browns Branch, about 1.0 mile above its mouth, and 3.5 miles southeast of Gordon P. O.; section measured by Teets; **Hernshaw Coal**; elevation, 1085' B.

			Ft.	In.
Sandstone roof				
Coal, splint, block	2'	6"		
Coal, impure	0	4		
Coal, gas, (slate floor)	0	2	3	0

The coal is mined here by **Garland Brown** for local fuel.

E. J. Berwind Local Mine Opening—No. 312 on Map II.

Located on trail leading from West Fork of Little Ugly Branch to Bald Knob P. O., 1.5 miles south of Chap; section by Krebs; **Hernshaw Coal**; elevation, 1370' B.

	Ft.	In.
Gray shale roof		
Coal, splint, mining in large blocks	3	0

Wharton Estate Opening—No. 313 on Map. II.

Located on the south side of Old Camp Branch, 1.5 miles from its mouth and 2.0 miles east of Pond P. O.; section by Krebs; **Hernshaw Coal**; elevation, 1325' B.

			Ft.	In.
Slate roof				
Coal, gas	1'	2"		
Coal, splint, blocky, visible	2	0	3	2

Coal mined for local fuel use by **Tom Price.**

The composition and calorific value of the coal will be given in a subsequent chapter on coal resources; also an estimate will be given on the probable available area of the coal bed by magisterial districts.

THE NAUGATUCK SANDSTONE.

Underlying the Hernshaw Coal, and separated from same by from 2 to 5 feet of shale and fire clay, there occurs a medium coarse, micaceous sandstone, that has been named the **Naugatuck Sandstone.**[9] This bed varies in thickness from 10 to 30

[9]Hennen and Reger, Logan and Mingo Report; 1914.

feet in Boone County, and often makes massive cliffs. Measurements showing its thickness at different points in Boone have been given in the sections on preceding pages of this Report.

THE DINGESS COAL.

Underneath the Naugatuck Sandstone occurs a thin multiple bedded coal that was once mined at Dingess in Mingo County, from whence it takes its name.[10] This bed of coal is thin in Boone County so far as it has been developed, and occurs from 20 to 35 feet under the Hernshaw Coal.

The following section gives the general character of the coal:

Yawkey and Freeman Local Mine Opening.
No. 314 on Map II.

Located at the mouth of Jacks Branch of Pond Fork, 0.5 mile northwest of Lantie P. O.; section by Krebs; **Dingess Coal.**

	Ft.	In.
Sandstone, massive, Naugatuck..................		
Coal, gas..	1	6
Slate	1	0
Sandstone, massive, Williamson................		

THE WILLIAMSON SANDSTONE.

Separated from the Dingess Coal by gray and dark shale, from one to 5 feet, there occurs another sandstone in descending order, from 5 to 20 feet thick, that has been designated the **Williamson Sandstone,** from the town of that name in Mingo County. This sandstone is often massive, and, together with the overlying Naugatuck Sandstone, forms massive cliffs along the streams in Boone County.

THE DINGESS LIMESTONE.

The next important horizon in descending order is the Dingess Limestone.[11] This is possibly one of the most im-

[10]Hennen and Reger, Logan and Mingo Report; 1914.

[11]Hennen and Reger, Logan and Mingo Report; 1914.

PLATE XXX.—View where Dingess Limestone Fossil Horizon goes under Pond Fork, 1.5 miles south of mouth of Robinson Creek.

portant horizons in Boone County; not for its intrinsic value, for the limestone is impure and of little value, but from its position as a "marker" or "key" rock in the Kanawha Series, by means of which great assistance is given in the correlation of the different coal seams, both above and below this horizon. This bed is a dark gray, impure, lenticular limestone, and nearly always contains marine fossils. It derives its name from Dingess, a small village in Mingo County. The limestone occurs in two levels, separated by dark gray shale with nodular limestones ("Turtle Backs"), in same, from 0 to 25 feet, each horizon containing marine fossils.

A few sections will now be given showing the occurrence of this bed in the different magisterial districts of Boone County:

Peytona District.

The Dingess Limestone rises out of the bed of Coal River, about one mile north of Emmons Station, and 1.5 miles southeast of the mouth of Fork Creek. There the following section was measured by Teets:

	Thickness Ft.	In.	Total Ft.	In.
Sandstone	5	0	5	0
Shale, gray	19	0	24	0
Slaty shale	8	0	32	0
Limestone, impure, full of marine fossils, Dingess	1	0	33	0
Shale, dark	1	4	34	4
Sandy shale	1	8	36	0
Coal ....0' 7" / Fire clay....0 2 / **Coal, (gray shale floor, 647' L.)**....0 11½ } **Williamson**	1	8½	37	8½

Specimens of this limestone were collected by Teets and sent to the office of the Survey at Morgantown, to be examined by Dr. W. Armstrong Price, Paleontologist of the Survey.

Up Coal River from Emmons, the Dingess Limestone soon rises above the bed of the river, and on the crest of the Warfield Anticline, its horizon occurs from 420 to 450 feet above the river, at an elevation of 1060 to 1100 feet above tide. From this axis, it dips southeastward until at Seth near the mouth of Laurel Creek, the limestone is only 30 to 40 feet above water level.

Scott District.

The Dingess Limestone comes out of Little Coal River between Harless and Lowrey, and gradually rises above the bed of the river until at Low Gap, on the crest of the Warfield Anticline, its horizon is more than 500 feet above the bed of Spruce Fork.

The following section was measured by Krebs along the road descending into Rock Creek, 2.5 miles northeast from Danville:

	Ft.	In.
Sandstone, massive, **Williamson**	6	0
Shale, dark gray	5	0
Limestone, impure, full of marine fossils, Dingess	1	0
Slate and fire clay	2	0
Coal 1' 6" / Slate 0 2 / **Coal** 0 6 } **Williamson** 1156' B.	2	2

The Dingess Limestone occurs about 400 feet over the Eagle Limestone just north of Danville.

Washington District.

The Dingess Limestone rises out of Mud River, just west of the mouth of Bearcamp Branch, about one mile west of Mud Gap. The stratum is a hard, bluish-gray, impure limestone containing many **marine fossils.** It occurs 190 feet under the Chilton Coal. The following section was measured by Teets:

	Thickness		Total	
	Ft.	In.	Ft.	In.
Sandstone	10	0	10	0
Dark gray shale	15	0	25	0
Limestone, fossiliferous, impure, Dingess	0	6	25	6
Gray shale, 1040' B.	1	6	27	0

Samples of the Dingess Limestone were collected by W. Armstrong Price and Teets for the purpose of studying the marine fossils.

Southwest about 2.0 miles, near the head of Big Ugly Creek, in Harts Creek District, Lincoln County, 2.0 miles northwest of Estep P. O., and 1.0 mile east of Trace Branch, the **Dingess Limestone** appears on the north side of the creek. There the following section was measured by Krebs:

	Ft.	In.
Sandy shale and concealed		
Shale, dark, with iron ore nodules	3	0
Limestone, impure, with iron ore nodules and marine fossils, Dingess	1	0

Local Mine Opening No. 315 on Map II.

	Ft.	In.		Ft.	In.
Coal, semi-cannel	0′	4″	**Williamson** 940′ B.	6	11
Shale, gray	0	3			
Coal, cannel	0	4			
Slate	0	3			
Coal, cannel, impure	0	6			
Slate	0	3			
Coal, semi-cannel	0	4			
Slate	0	2			
Coal	0	6			
Slate, dark, full of plant fossils	2	6			
Coal, splint, visible	1	6			

The **Dingess Limestone** is impure, sandy, and contains iron nodules and marine fossils.

Another section measured about 100 feet east of the foregoing section is as follows:

	Ft.	In.
Sandstone, massive	20	0
Shale, dark	2	0
Shale, dark, with iron ore nodules	4	0
Sandstone, limy, with iron ore nodules and marine fossils, Dingess	1	6

Opening No. 316 on Map II.

	Ft.	In.		Ft.	In.
Coal, gas, hard	0′	6″	**Williamson** 940′ B.	[illegible]	11½
Shale, dark gray	0	3			
Coal, gas, hard	0	4			
Shale, dark gray	0	3			
Coal, splint	0	8			
Shale, dark gray	0	3			
Coal, splint	0	5			
Shale, dark gray	0	1½			
Coal, splint	0	6			
Fire clay shale, dark gray	2	6			
Coal, splint	1	2			
Dark shale	0	1			
Coal, splint (shale, gray, floor	0	11			

Another hand level section of the Dingess Limestone measures was made by Hennen, Reger and Krebs, showing its relation to the coal mined at Ramage, as follows:

Section at Ramage.

			Thickness Ft.	In.	Total Ft.	In.
Shale, sandy			3	0	3	0
Sandstone, iron ore nodules			1	0	4	0
Shale, gray			1	6	5	6
Sandstone, ferruginous, marine fossils, Dingess Limestone horizon			1	6	7	0
Opening No. 317 on Map II.						
Coal, gas	0′	2″				
Slate, gray	0	0½				
Coal, gas, hard	1	5				
Shale, dark gray, 1″ to	0	2				
Coal, gas, soft	0	10	**Williamson**	7	11½	14 11½
Fire clay shale	3	4				
Coal, semi-splint	0	6				
Bone, ½″ to	0	1				
Coal, semi-splint	1	5				
Slate			1	0½	16	0
Sandstone, massive			5	0	21	0
Concealed			30	0	51	0
Coal blossom, Cedar Grove, Upper Bench			2	0	53	0
Sandstone and concealed			28	0	81	0
Coal, splint	0′	4″	**Cedar Grove, Lower Bench**			
Coal, gas, hard	1	2		1 6	82	6
Fire clay, soft			2	0	84	6
Shale, gray			28	0	112	6
Opening No. 318 on Map II.						
Coal, semi-splint	1′	0″				
Slate, bony	0	2				
Coal, gas	0	5½	**Alma**	4	4½	116 10½
Coal, splint, hard	0	5				
Coal, gas	0	2				
Coal, splint	2	2				
Slate, 900′ B			1	1½	118	0

The Alma Coal was once mined at this place by the Spruce River Coal Company.

Passing up Spruce Fork, the **Dingess Limestone** goes under the bed of the stream just below the mouth of Spruce Laurel Fork.

This limestone was encountered in four different core test holes recently sunk by owners of the land of the Cassingham Coal and Land Company, near the head of Spruce Laurel Fork.

Through the courtesy of Mr. O. R. Ross, the driller in charge, the writer was enabled to get correct locations and samples from each of these diamond core test holes.

The **Dingess Limestone** as found in the diamond core test holes is as follows:

	Diamond Core Test Holes.		
	No. 1 (131).	No. 2 (130).	No. 3 (123).
Elevation of Surface, Ft. A. T.	1570	1400	1410
Dingess Limestone, Depth, Ft.	203.7	57.7	213.5
Dingess Limestone, Thickness, Inches	12	3	7

A sample (No. 406-K) of the Dingess Limestone from No. 3 (123) was collected by Krebs and it yielded the following results on analysis by Messrs. Hite and Krak:

	Per cent.
Silica (SiO_2)	49.95
Ferric Iron (Fe_2O_3)	3.20
Alumina (Al_2O_3)	7.63
Calcium Carbonate ($CaCO_3$)	35.19
Magnesium Carbonate ($MgCO_3$)	0.08
Phosphoric Acid (P_2O_5)	0.25
Loss on ignition	1.47
Total	97.77

The cores of these test holes were examined by the writer, and samples from Nos. 1, 2 and 3 were sent to W. Armstrong Price, Paleontologist of the Survey, for study.

The distance from the section measured at Big Ugly to diamond core test hole No. 1 on the head of Spruce Laurel Fork is 16.5 miles, thus showing the continuity of the **Dingess Fossiliferous Limestone horizon** throughout Washington District.

Sherman District.

The following section, measured by Krebs on the west side of Laurel Creek, one mile southwest of Seth, exhibits the general structure of the Dingess Limestone at that point:

	Thickness Ft. In.	Total Ft. In.
Shale, dark gray, full of limestone nodules, "turtle-backs"	40 0	40 0
Iron ore nodules with marine fossils, Dingess	1 0	41 0
Coal, ¼" to	0 1	41 1
Green shale	1 11	43 0
Sandstone and concealed to creek, 715′ L.	10 0	53 0

The following section was measured by Teets on the east side of Coal River at Seth:

	Thickness Ft. In.	Total Ft. In.
Sandstone, massive	10 0	10 0
Limestone, impure, ferriferous, marine fossils, Dingess	0 6	10 6
Shale and concealed, 750′ B.	9 6	20 0

Near the head of Seng Creek, the **Dingess Limestone** is exposed on the east side of the tunnel of the C. & O. Railway.

The following hand level section was measured by Teets one mile west of Kayford and 0.25 mile east of the Boone-Kanawha County Line, descending eastward to an opening in the Campbell Creek (No. 2 Gas) Coal, 0.25 mile west of Kayford:

Section 1.0 Mile West of Kayford and 0.25 Mile East of the Boone-Kanawha County Line.

	Thickness Ft. In.	Total Ft. In.
Concealed	100 0	100 0
Sandy shale	10 0	110 0
Dark gray, slaty limestone, Dingess, marine fossils	0 6	110 6
Shale, dark	1 6	112 0
Coal, gas, Williamson	2 10	114 10
Shale, gray	1 2	116 0
Sandstone, massive, **Upper Cedar Grove**	24 0	140 0
Coal, splint, visible, Cedar Grove	2 0	142 0
Shale and concealed	1 0	143 0
Sandstone and sandy shale	4 0	147 0
Coal, splint, Cedar Grove, Lower Bench	1 0	148 0
Slate	1 0	149 0
Sandstone, massive	45 0	194 0
Limestone, dark, impure, iron ore nodules, and marine fossils, Campbell Creek	1 0	195 0
Sandy shale and concealed	91 0	286 0
Coal, Campbell Creek (No. 2 Gas), mined at Kayford, 1335′ L.	4 0	290 0

The foregoing section puts the **Dingess Limestone** 180 feet above the Campbell Creek (No. 2 Gas) bed. However, the section is taken to the rise, for about one-half mile, which will make the interval between the Dingess Limestone and the Campbell Creek (No. 2 Gas) Coal about 210 feet, allowing for the usual dip in the seam in this vicinity.

The **Campbell Creek Limestone** becomes fossiliferous here at 91 feet above the No. 2 Gas Coal, the interval between it and the underlying coal having thickened up greatly over that (35 feet) at the type locality near Malden, Kanawha County.

Crook District.

The **Dingess Limestone** enters Crook District from the south on the Warfield Anticline at 500 to 600 feet above the bed of Pond Fork and gradually dips from this point southeast into the Coalburg Syncline, until it descends to the level of Pond Fork, about two miles northwest of Lantie P. O., or 1.5 miles southeast of the mouth of Robinson Creek.

The section at the mouth of Robinson Creek, published on page 171 of this Report, gives the occurrence of the **Dingess Limestone** there, together with its horizon above the Campbell Creek (No. 2 Gas) Coal encountered in the Arbogast well.

The following section was measured by Krebs on the north side of Pond Fork, 1.5 miles south of the mouth of Robinson Creek:

			Thickness Ft.	Thickness In.	Total Ft.	Total In.
Sandstone, massive			20	0	20	0
Shale, gray, iron ore nodules and plant fossils			10	0	30	0
Limestone, impure, ferriferous, marine fossils, **Dingess**			1	0	31	0
Shale, gray, plant fossils			0	6	31	6
Opening No. 320 on Map II.						
Coal, gas	0'	2"				
Shale, gray	1	0				
Coal, gas	1	0	**William-son**			
Shale, gray	0	5		2	11	34 5
Coal	0	4				
Shale, gray				2	7	37 0
Concealed to Pond Fork, 778' B.				15	0	52 0

Passing up Pond Fork from this point, the **Dingess Limestone** dips under the bed of the creek for about four miles, and then rises out of the bed of the creek about 1.5 miles south of the mouth of West Fork, where the following section was measured by Teets:

	Thickness Ft.	In.	Total Ft.	In.
Sandstone, massive	30	0	30	0
Shale, gray	17	0	47	0
Limestone, impure, Dingess, marine fossils	0	4	47	4
Shale, gray	0	6	47	10
Limestone	1	0	48	10
Shale, dark gray	9	2	58	0
Limestone, hard, ferriferous, full of marine fossils, Dingess, 843′ L.	1	0	59	0

Here there are three layers of the Dingess Limestone.

The **Dingess Limestone** is again exposed on Pond Fork near Casey Creek, as shown in the following section by Teets:

				Thickness Ft.	In.	Total Ft.	In.
Sandstone, massive				20	0	20	0
Shale, gray				17	0	37	0
Limestone, impure, marine fossils	0′	6″	**Dingess**	1	6	38	6
Limestone, ferriferous, 880′ B.	1	0					
Shale, gray				0	10	39	4
Sandy shale				10	8	50	0
Shale, gray				1	5	51	5
Opening No. 321 on Map II.							
Coal	0′	3″	**Williamson**	9	6	60	11
Shale	0	2					
Coal	0	5					
Shale, gray	1	5					
Sandstone	2	0					
Gray shale	4	0					
Coal	0	3					
Slate	0	2					
Coal	0	2					
Slate	0	7					
Coal	0	1					

Another section measured by Teets on Beaver Pond Branch of Pond Fork, 0.5 mile from its mouth, shows the following:

PLATE XXXI.—View where the Dingess Fossiliferous Limestone Horizon comes out of Pond Fork above Lantie, 1.5 miles south of mouth of West Fork.

	Thickness Ft.	In.	Total Ft.	In.
Sandstone, massive	20	0	20	0
Shale, gray	12	0	32	0
Limy sandstone	1	0	33	0
Shale, gray	13	0	46	0
Limestone, impure, marine fossils, Dingess, 1060' B	2	0	48	0
Shale, gray	2	0	50	0
Sandstone and concealed	9	0	59	0
Opening No. 322 on Map II.				
Coal, visible, **Williamson**	1	3	60	3

In passing to the east of Pond Fork, 1.5 miles up Old Camp Branch, near the water mill of **N. M.** Price, the following section was measured by Krebs and Teets:

	Thickness Feet.	Total Feet.
Coal, Hernshaw	3	3
Sandy shale and concealed	10	13
Limestone, dark, marine fossils, Dingess	1	14
Gray limy shale	8	22
Limestone, dark gray, marine fossils, Dingess	0.5	22.5
Gray shale	27	49.5
Coal 0' 3"; Shale 0 4; **Coal** 0 1 — **Williamson.**	0.7	50.2
Shale, gray	4.8	55
Sandstone and concealed	27	82
Limestone, dark gray, marine fossils, Seth.	0.4	82.4
Coal 0' 1"; Shale 0 2; **Coal** 0 3 — **Cedar Grove**	0.5	82.9
Gray shale to creek, 1072' L	18.1	101

Sample No. 405-K of the Dingess Limestone in the above section was collected by Krebs and was analyzed by **Messrs.** Hite and Krak who report the following results:

	Per cent.
Silica (SiO_2)	48.46
Ferric Iron (Fe_2O_3)	15.29
Alumina (Al_2O_3)	9.65
Calcium Carbonate ($CaCO_3$)	13.42
Magnesium Carbonate ($MgCO_3$)	3.01
Phosphoric Acid (P_2O_5)	0.20
Loss on ignition	7.98
Total	98.01

The **Dingess Limestones** were also found on Workman Branch along the east side of Pond Fork. For a detailed section of same, see Workman Branch Section on preceding pages on this Report.

On West Fork, the **Dingess Limestone** rises out of that stream between the mouths of Browns and Jarrolds Branches. It gradually rises a few feet above the bed of the stream and remains above for more than 7 miles, until the limestone goes under again near the head of same, about one-fourth mile above Bowen Branch, and five miles southeast of Chap P. O.

The **Dingess Limestone** occurs on Petry Fork of Mats Creek at an elevation of 1445′ B., where the following section was measured by Krebs and Teets:

	Feet.
Sandy shale	2
Limestone, dark gray, marine fossils, Dingess	2
Sandy shale	4
Sandstone and concealed to top of Diamond Core Test Hole, E. J. Berwind No. 3 (135)	85

The Dingess Limestone contains many marine fossils at this point.

THE WILLIAMSON COAL.

There usually occurs at 0 to 20 feet under the Dingess Limestone a multiple bedded coal from 1 to 8 feet in thickness, which has been designated the **Williamson Coal** by Hennen and Reger,[12] from its occurrence near the town of that name in **Mingo** County. This coal is usually splinty, and often contains some layers of cannel coal. In Boone County, the bed is generally impure, having slate and coal mixed in such manner as to render it of little commercial value. A few more sections of this bed will be given, in addition to those already shown in connection with the Dingess Limestone:

Peytona District.

The **Williamson Coal** occurs from 200 to 450 feet above the bed of Coal River in Peytona District, and very little exposure or few openings were found in this seam in this

[12]Logan and Mingo Report; 1914.

district. On the head of Drawdy Creek, along the road leading to Rock Creek, the following measurement was made by Teets:

Terry Dawson Prospect Opening—No. 324 on Map II.

			Ft.	In.
Sandy shale				
Coal, splint	0′	4″		
Shale, gray, plant fossils	0	4		
Coal, splint	0	5		
Shale, dark	0	0½		
Coal, splint	0	2		
Shale, gray	0	4		
Coal, splint	0	3		
Shale, dark	0	1½		
Coal, splint	0	3		
Shale, dark	0	1		
Coal, splint, (slate floor, 1116′ L.)	1	0		
			3	4

Scott District.

The Williamson Coal rises out of Little Coal River, between Harless and Lowrey, and rises rapidly above the bed of the stream to the south, as it approaches the Warfield Anticline.

The following section was measured by Teets, about one-third mile up Browns Branch, and 0.5 mile east of Van Camp Station:

Pryor and Allen Prospect Opening—No. 325 on Map II.

			Ft.	In.
Shale, gray				
Coal, splint	0′	1″		
Slate, dark	0	4		
Coal	0	1		
Slate, dark	0	0½		
Coal, splint	0	10½		
Slate, dark (900′ B)	0	1	1	6

Washington District.

The horizon of the Williamson Coal gets above water level in the head of Mud River, near the mouth of Bearcamp Branch, but no coal was observed at that point. The Williamson Coal comes out of the bed of Ugly Creek, about 2.5 miles west of Estep P. O., and rises rapidly eastward above the bed of the stream.

Colway Bias Opening—No. 326 on Map II.

Located on the east side of Hewett Creek, 2.0 miles northwest of Clothier; section taken by Teets; **Williamson Coal**; elevation, 1075' B.

	Ft.	In.
Slate roof		
Coal, splint	4	0

The coal was once mined at this place for local fuel by **Colway Bias**, and comes here 110 feet above the Alma Coal.

Isaac Gore Local Mine Opening—No. 327 on Map II.

Located on the head of Missouri Branch of Hewett Creek, 4.4 miles west of Ottawa P. O.; section by Teets; **Williamson Coal**; elevation, 1175' B.

			Ft.	In.
1. Sandstone				
2. **Coal**, splint	0'	10"		
3. Shale, gray	0	1		
4. **Coal**, splint	1	8		
5. Shale	0	6		
6. **Coal**, visible	1	6	4	7

The analysis of a sample collected from Nos. 2, 4 and 6, as reported by **Messrs.** Hite and Krak, is given in the table of coal analyses under **No. 53.**

Ashford Ball Local Mine Opening—No. 328 on Map II.

Located on north side of Meadow Fork of Hewett Creek, 1.3 miles due west of Hewett P. O.; section by Teets; **Williamson Coal**; elevation, 1140' B.

			Ft.	In.
1. Gray shale roof				
2. **Coal**, gas	0'	8"		
3. **Coal**, splint	1	2		
4. **Coal**, gas	0	4		
5. Shale, gray	0	10		
6. **Coal**, splint (slate floor)	0	8	3	8

The analysis of a sample collected from Nos. 2, 3, 4 and 6, as reported by Messrs. Hite and Krak, is given in the table of coal analyses under **No. 54.**

Ashford Ball Local **Mine Opening—No. 329 on Map II.**

Located at the south side of Meaks Fork of Meadow Fork of Hewett Creek, 2.8 miles west of Jeffery P. O.; section by Teets; **Williamson Coal**; elevation, 1120' B.

			Ft.	In.
Shale, gray, roof				
Coal, splint	0'	9"		
Shale, gray	0	1		
Coal, hard splint (slate floor)	2	2	3	0

In passing up Spruce Fork from Ramage, the **Williamson Coal** goes under the bed of the stream between Ottawa and the mouth of Spruce Laurel, and does not rise above the bed of the latter stream any more. The core drill holes sunk on lands of Chilton, MacCorkle, Chilton and Meany, on Spruce Laurel, and those on lands of the Cassingham Coal and Land Company, published on preceding pages of this Report, give the thickness and depth of this coal, to which the reader is referred.

Crook District.

The **Williamson Coal** enters Crook District from the south, near the crest of the Warfield Anticline, from 500 to 600 feet above the bed of Pond Fork, and dips gradually southeastward, going under Pond Fork about 1.5 miles southeast of the mouth of Robinson Creek, near which point it is a cannel bed, and is mined about 0.5 mile below the mouth of Robinson Creek, on the land of W. P. Croft, where the following measurement was made by Krebs:

W. P. **Croft Opening—No. 330 on Map II.**

			Ft.	In.
1. Dark shale, plant fossils			3	0
2. **Coal,** splint	0'	9"		
3. Shale, gray	0	3		
4. **Coal,** splint	0	1		
5. **Coal,** semi-cannel	0	7		
6. Shale and fire clay	0	10		

			Ft.	In.
7. **Coal,** splint	0′	3″		
8. Shale, gray, full of plant fossils	1	0		
9. **Coal,** cannel, 3′-8″ to	3	0	6	9
10. Fire clay floor, 845′ B.				

This cannel seam has long been mined for domestic fuel by W. P. Croft. A sample of No. 9, taken by Mr. Croft, was analyzed by Messrs. Hite and Patton, and the results given on page 399 of Volume II (A), West Virginia Geological Survey Reports, being republished in the table of coal analyses under **No. 55** in this volume.

In passing up Pond Fork about one mile above the Croft mine, the **Williamson Coal** goes under the bed of the stream, as is shown in a previous section giving the Dingess Limestone. It comes up out of Pond Fork again about 1.5 miles above the mouth of West Fork.

Wharton Estate Prospect Opening—No. 331 on Map II.

Located on the West Fork of Pond Fork, opposite the mouth of Burnt Camp Branch, 0.8 mile southeast of Echart P. O.; section by Teets; **Williamson Coal**; elevation, 1890′ B.

			Ft.	In.
Sandy shale				
Coal, splint	0′	2″		
Shale, gray	0	5		
Coal, splint	0	3		
Slate	0	4		
Coal, splint	1	11		
Coal, gray splint	1	8		
Coal, block	0	1		
Coal, gray splint	0	8	5	6

The **Williamson Coal** comes out of West Fork of Pond Fork, near the mouth of Jarrolds Branch, about 1.5 miles northwest of Chap P. O., and rises a little faster than the bed of the stream to the southeast, for more than seven miles, where it goes under the bed of the stream again.

The following section was measured by Teets on the west side of West Fork, 1.5 miles above the mouth of Mats Creek, and 3.0 miles south of Chap P. O.:

E. J. Berwind Opening—No. 332 on Map II.

				Thickness Ft.	In.	Total Ft.	In.
Shale, gray				20	0	20	0
Limestone, impure, marine fossils, Dingess				0	6	20	6
Shale, gray, plant fossils				0	3	20	9
Coal, splint	0'	4"	Williamson				
Shale, gray	5	11					
Coal, splint	0	6		6	9	27	6
Concealed to creek, 1400' B.				0	6	28	0

Another measurement by Teets on the east side of West Fork, 2.8 miles above the mouth of Mats Creek, on land of John Q. Dickinson, gives the following:

John Q. Dickinson Opening—No. 333 on Map II.

				Thickness Ft.	In.	Total Ft.	In.
Gray shale, limestone concretions and iron ore nodules				35	0	35	0
Limestone, ferriferous, marine fossils, Dingess				0	3	35	3
Coal	0'	2"	Williamson				
Shale, gray	11	7					
Coal	0	3		12	10	38	1
Slate	0	6					
Coal	0	4					
Shale, gray, to creek, 1510' B.				3	11	42	0

The Williamson Coal was encountered in the three core drill holes sunk on Casey Creek, on the land of Chilton, MacCorkle, Chilton and Meany, the records of which are published on preceding pages of this volume, to which the reader is referred for detailed sections of the coal.

THE UPPER CEDAR GROVE SANDSTONE.

At 5 to 20 feet under the Williamson Coal occurs a grayish, medium grained sandstone, from 20 to 40 feet thick, known as the **Upper Cedar Grove Sandstone.** This sandstone is usually of a soft nature, and does not always form massive cliffs. Its thickness at different points is given in the general sections published on preceding pages of this Report.

THE CEDAR GROVE COAL.

At 20 to 50 feet under the Upper Cedar Grove Sandstone, and separated from it by thick gray shale containing iron ore nodules and plant fossils, occurs one of the **Cedar Grove** seams of coal. This bed is one of importance in the Kanawha Series, and was named by I. C. White from a small mining village on the north side of the Kanawha River, at the mouth of Kellys Creek, where the coal was first mined and shipped down the Kanawha and Ohio Rivers in barges.

While this seam is only from 2 feet 6 inches to 3 feet 6 inches thick in the Kanawha Valley, yet it is quite an important coal, on account of its purity and excellence as a steam and general fuel coal.

The following is a section taken by Krebs at Cedar Grove, where it was first mined, and represents the type section of the bed:

		Thickness Ft. In.	Total Ft. In.
1.	Sandstone, massive, **Upper Cedar Grove**	30 0	30 0
2.	Shale, gray, iron ore nodules and plant fossils	12 0	42 0
3.	**Coal**, splint, 4" to 0' 9"		
4.	Shale, gray, plant fossils, 2' to 4 0		
5.	**Coal**, splint, 1' to 1 6		
6.	**Coal, splint, soft,** 0" to 1 0	**Cedar Grove.** 7 11	49 11
7.	**Coal**, splint, breaking in blocks 0 8		
8.	Soft fire clay floor	2 1	52 0

The floor of the bed is usually of soft fire clay for about 2 to 3 feet.

Peytona District.

The Cedar Grove Coal comes out of Coal River near Emmons Station, and rises rapidly above the bed of the stream in passing to the southeast until at the crest of the Warfield Anticline, it is about 350 feet above the river.

From the prospect openings and the surface exposures in this seam, it is evident that the **Cedar Grove** is an important coal in Peytona District.

Sections will now be given taken at different points in the district to show its general structure and thickness:

The Boone and Kanawha Land and Mining Company owns a large area of land on the north side of Coal River, between Lick and Short Creeks. It has recently done considerable prospecting in opening the coal seams, and J. S. Cunningham of Charleston has made a report on the property for the Company, a copy of which it has kindly placed at the disposal of the Survey, giving sections of the different openings, etc., from which the following results are taken:

Boone and Kanawha Land and Mining Company Prospect Opening—No. 334 on Map II.

Located on Abshire Branch of Roundbottom Creek, 1.5 miles northeast of Peytona; section by H. A. Henderson for J. S. Cunningham; **Cedar Grove Coal**; elevation, 950' B.

			Ft.	In.
Slate roof				
Coal, splint	1'	8"		
Slate	0	3		
Coal, splint	1	0		
Slate	0	2		
Coal, splint, blocky	0	9		
Slate	0	3		
Coal, splint	0	8	4	9

Boone and Kanawha Land and Mining Company Prospect Opening—No. 335 on Map II.

Located on Abshire Branch of Roundbottom Creek, about 1,000 feet northwest from Opening No. 334; section by H. A. Henderson for J. S. Cunningham; **Cedar Grove Coal**; elevation, 950' B.

			Ft.	In.
Slate roof				
Coal, splint	2'	10"		
Slate	0	1½		
Coal, splint	0	11½		
Slate	0	1		
Coal, splint	0	10	4	10

Boone and Kanawha Land and Mining Company Prospect Opening—No. 336 on Map II.

Located on head of Roundbottom Creek, 1.5 miles north of Peytona; section by H. A. Henderson for J. S. Cunningham; **Cedar Grove Coal**; elevation, 950′ B.

			Ft.	In.
Slate roof				
Coal, splint	2′	11″		
Slate	0	1		
Coal, splint	0	7		
Slate	0	3		
Coal, splint	0	11	4	9

The coal is mined for local fuel and opening driven 50 feet under cover.

Boone and Kanawha Land and Mining Company Opening. No. 337 on Map II.

Located near the head of Roundbottom Creek, 1.4 miles northeast of Peytona; section by Teets; **Cedar Grove Coal**; elevation, 970′ B.

			Ft.	In.
Dark shale roof				
Coal, splint, soft	0′	7″		
Coal, impure	0	2		
Coal, block	0	10		
Slate, gray	0	1		
Coal, gray splint	1	3		
Slate, dark	0	1		
Coal, gray splint	0	3		
Bone	0	2		
Coal, gray splint	0	4		
Slate, gray	0	1		
Coal, impure, (fire clay floor)	0	6	4	4

Butts, N. 48° W.; faces, N. 42° E.

Boone and Kanawha Land and Mining Company Prospect Opening—No. 338 on Map II.

Located on the north side of Roundbottom Creek, 1.1 miles north of Peytona; section by H. A. Henderson for J. S. Cunningham; **Cedar Grove Coal**; elevation, 960′ B.

			Ft.	In.
Slate roof				
Coal, splint	3′	9″		
Slate	0	2		
Coal	0	9	4	8

Northwest about 0.75 mile, on the north side of Brier Branch of Coal River, another section gives the following:

Boone and Kanawha Land and Mining Company Opening. No. 339 on Map II.

Located on the north side of Brier Branch of Coal River, 2.0 miles northwest of Peytona; section by H. A. Henderson for J. S. Cunningham; **Cedar Grove Coal**; elevation, 980′ B.

			Ft.	In.
Slate roof				
Coal, splint	3′	3″		
Slate	0	1		
Coal, (slate floor)	0	10	4	2

Coal mined for local fuel use and opening driven in 60 feet.

Boone and Kanawha Land and Mining Company Prospect Opening—No. 340 on Map II.

Located on the south side of Curtis Branch, 0.75 mile northeast of Brushton Station; section by H. A. Henderson for J. S. Cunningham; **Cedar Grove Coal**; elevation, 950′ B.

			Ft.	In.
Slate roof				
Coal, splint	1′	6″		
Slate	0	1		
Coal, splint, (slate floor)	1	7½	3	2½

J. T. Price Local Mine Opening—No. 341 on Map II.

Located on a small branch of Lick Creek, 2.5 miles north of Peytona; section by Teets; **Cedar Grove Coal**; elevation, 970′ B.

			Ft.	In.
Slate roof				
Coal, splint	2′	2″		
Slate	0	0¼		
Coal, gray splint, (slate floor)	1	1	3	3¼

Butts, N. 50° W.; faces N. 40° E.

Boone and Kanawha Land and Mining Company Prospect Opening—No. 342 on Map II.

Located on the south side of Peytona Branch of Lick Creek, 2.5 miles northeast of Peytona; section by H. A. Henderson for J. S. Cunningham; **Cedar Grove Coal**; elevation, 950′ B.

			Ft.	In.
Slate roof				
Coal, splint	2′	7″		
Slate	0	9½		
Coal, splint	2	2½	5	'

William Barker Opening—No. 343 on Map II.

Located on the head of Falling Rock Fork of Lick Creek, 3.5 miles northeast of Peytona; section by Teets; **Cedar Grove Coal**; elevation, 950″ B.

	Ft.	In.
Slate roof		
Coal, splint, (slate floor)	3	0

F. G. Kirk Local Mine Opening—No. 344 on Map II.

Located in the head of a ravine, on the south branch of Lick Creek, 1.2 miles southwest of the common corner of the Sherman-Peytona District Line in the Boone-Kanawha County Line; section by Teets; **Cedar Grove Coal**; elevation, 860′ B.

				Ft.	In.
1.	Slate roof				
2.	**Coal**, splint	2′	4″		
3.	Fire clay, slaty	1	4		
4.	**Coal**, gray splint	0	1		
5.	Slate, gray	0	1		
6.	**Coal**, gray splint, (slate floor)	1	0	4	10

Butts, N. 43° W.; faces, N. 47° E.

The analysis of a sample collected from Nos. 2, 4 and 6, as reported by Messrs. Hite and Krak, is given in the table of coal analyses under **No. 56**.

Boone and Kanawha Land and Mining Company Prospect Opening—No. 345 on Map II.

Located on the south side of Lick Creek, 0.5 mile east of mouth of same; section by H. A. Henderson for J. S. Cunningham; **Cedar Grove Coal**; elevation 910′ B.

	Ft.	In.
Slate roof		
Coal, splint	3	5

Boone and Kanawha Land and Mining Company Opening. No. 346 on Map II.

Located on the south side of Lick Branch, about 0.25 mile southeast of Ashford Station; section by H. A. Henderson for J. S. Cunningham; **Cedar Grove Coal**; elevation, 915′ B.

			Ft.	In.
Slate roof				
Coal, splint	1′	5″		
Slate	0	0½		
Coal, (slate floor)	1	1½	2	7

Coal mined for local fuel use, and opening driven in 100 feet.

Coal River Mining Company Prospect Opening. No. 347 on Map II.

Located on the west side of Coal River, opposite the mouth of Brier Branch, 0.5 mile southeast of Brushton Station; section measured by Teets; **Cedar Grove Coal**; elevation, 950′ B.

			Ft.	In.
Slate roof with plant fossils				
Coal, gas	0′	5″		
Slate, gray	0	1		
Coal, splint	0	9		
Slate, dark	0	0½		
Coal, splint	0	7		
Slate, gray	0	1		
Coal, gray splint, (fire clay floor)	0	8	2	7½

Butts, N. 50° W.; faces, N. 40° E.

Minerva Hill Opening—No. 348 on Map II.

Located on the head of Brier Branch of Coal River, 1.25 miles north of Peytona P. O.; section by Teets; **Cedar Grove Coal**; elevation, 980′ B.

				Ft.	In.
1.	Slate roof				
2.	**Coal**, gray splint	0′	6″		
3.	**Coal**, hard, gnarly	0	5		
4.	**Coal**, gas	0	11		
5.	**Coal**, block	1	4		
6.	**Coal**, bony	0	4		
7.	Slate, gray	0	5		
8.	**Coal** gas, (slate floor)	0	3	4	2

Butts, N. 43° W.; faces, N. 47° E.

The analysis of a sample collected from Nos. 2, 3, 4, 5, 6 and 8, as reported by Messrs. Hite and Krak, is given in the table of coal analyses under **No. 57**.

Near the head of Whiteoak Creek, several openings have been made in this bed, as follows:

Coal River Mining Company Opening—No. 349 on Map II.

Located at the head of Left Fork of Whiteoak Branch, 1.5 miles west of Brushton; section by Teets; **Cedar Grove Coal**; elevation, 840′ B.

			Ft.	In.
Slate roof				
Coal, block	0′	7″		
Slate, gray	0	1		
Coal, hard	0	5		
Slate, gray	0	1		
Coal, splint	1	1		
Slate	0	0½		
Coal, visible	0	8½	3	0

Butts, N. 43° W.; faces, N. 47° E.

Coal River Mining Company Opening—No. 350 on Map II.

Located at the head of Right Fork of Whiteoak Creek, 2.5 miles southwest of Peytona; section by Teets; **Cedar Grove Coal**; elevation 990′ B.

	Ft.	In.
Coal (reported) (opening fallen shut and abandoned)	3	0

Edgar Williams Local Mine Opening—No. 351 on Map II.

Located on the north side of Whiteoak Creek, 1.0 mile west of Peytona; section by Teets; **Cedar Grove Coal**; elevation, 990′ B.

	Ft.	In.
Slate roof		
Coal, splint, (slate floor)	4	0

S. M. Javins Local Mine Opening—No. 352 on Map II.

Located near the head of Left Fork of Whiteoak Creek; section by Teets; **Cedar Grove Coal**; elevation, 1015′ B.

			Ft.	In.
1. Slate roof				
2. **Coal**, gray splint	0′	7″		
3. **Coal**, splint, blocky	1	10		
4. Slate	0	0¾		
5. **Coal**, gray splint, (slate floor)	0	9	3	2¾

Butts, N. 43° W.; faces, N. 47° E.

Analysis of sample collected from Nos. 2, 3 and 5, as reported by Messrs. Hite and Krak, is given in the table of coal analyses under **No. 58.**

Marion Turley Opening—No. 353 on Map II.

Located on the south side of Brush Creek, 0.5 mile west of Cabell P. O.; section measured by Teets; **Cedar Grove Coal.**

	Ft.	In.
Slate roof......................................		
Coal, splint, visible; once mined for fuel locally..	2	6

Coal River Mining Company Opening—No. 354 on Map II.

Located on the south side of Brush Creek, 1.5 miles southwest of Brushton; section taken by Krebs; **Cedar Grove Coal**; elevation, 948′ B.

	Ft.	In.
Slate roof......................................		
Coal, splint (slate floor)......................	3	0

The Coal River Mining Company owns a tract of land on the waters of Brush Creek, and it has recently made some prospect developments in the different coal seams, the sections of which were kindly placed at the disposal of the Survey, as follows:

Coal River Mining Company Prospect Opening. No. 355 on Map II.

Located on the south side of Brush Creek, 0.5 mile northeast of Cabell P. O.; section by H. C. Pierson for the Coal River Mining Company; **Cedar Grove Coal**; elevation, 915′ B.

	Ft.	In.
Slate roof......................................		
Coal, splint (slate floor)......................	2	10

Coal River Mining Company Prospect Opening. No. 356 on Map II.

Located on the south side of Brush Creek, 0.5 mile northeast of Cabell P. O.; section by H. C. Pierson; **Cedar Grove Coal**; elevation, 915′ B.

	Ft.	In.
Slate roof......................................		
Coal, splint, (slate floor)......................	3	3

Coal River Mining Company Prospect Opening. No. 357 on Map II.

Located on the west side of Honeycamp Fork of Brush Creek, 1.5 miles southwest of Brushton Station; section by H. C. Pierson; **Cedar Grove Coal**; elevation, 949' B.

	Ft.	In.
Slate roof		
Coal (slate floor)	3	2

Coal River Mining Company Opening—No. 358 on Map II.

Located on the west side, in the head of Honeycamp Fork of Brush Creek, 1.5 miles southeast of Cabell P. O.; section by H. C. Pierson; **Cedar Grove Coal**; elevation, 1018' B.

	Ft.	In.
Slate roof		
Coal, splint, (slate floor)	3	3

Coal River Mining Company Opening—No. 359 on Map II.

Located on the Left Fork of Honeycamp Fork of Brush Creek, 2.0 miles east of Cabell P. O.; section by H. C. Pierson; **Cedar Grove Coal**; elevation, 1018' B.

	Ft.	In.
Slate roof		
Coal splint (slate floor)	3	3

Coal River Mining Company Prospect Opening. No. 360 on Map II.

Located on a branch of Whiteoak Branch, 1.8 miles south of Dartmont; section by Teets; **Cedar Grove Coal**; elevation, 840' B.

			Ft.	In.
Slate roof				
Coal, splint	0'	11"		
Slate	0	1		
Coal, medium hard splint (slate floor)	2	1	3	1

Butts, N. 41° W.; faces, N. 49° E.

Peytona Land Company Opening—No. 361 on Map II.

Located on the head of Morgan Branch of Drawdy Creek, 1.5 miles southeast of Andrew P. O.; section by Teets; **Cedar Grove Coal**; elevation, 1015' B.

			Ft.	In.
Slate roof				
Coal, splint	1'	3"		
Slate, gray	0	2		
Coal, splint, softer	0	4		
Coal, splint, visible	3	4	5	1

PLATE XXXII.—Alma Coal where it comes out of Pond Fork at mouth of Beaver Pond Branch.

The coal is mined for local fuel by **Milton Meadows.**

Elmer Harless Local **Mine Opening—No. 362 on Map II.**

Located on Tony Branch of Drawdy Creek, 1.0 mile northwest of Andrew P. O.; section by Krebs; **Cedar Grove Coal**; elevation, 1050' B.

				Ft.	In.
1.	Slate roof, gray, plant fossils			2	0
2.	**Coal**, splint	1'	2 "		
3.	Slate	0	0½		
4.	**Coal**, gray splint, (slate floor)	1	11	3	1½

The analysis of a sample collected by Krebs from Nos. 2 and 4 of the above section, as reported by **M**essrs. Hite and Krak, is given in the table of coal analyses under **No. 58A.**

Eliza Abshire **Opening—No. 363 on Map II.**

Located on the west side of Left Fork of Tony Fork, 1.2 miles northwest of Andrew P. O.; section by Krebs; **Cedar Grove Coal**; elevation, 1050' B.

				Ft.	In.
1.	Slate roof, plant fossils			2	0
2.	**Coal**	1'	0"		
3.	Slate	0	1		
4.	**Coal**, splint	1	6		
5.	**Coal**, gas, (slate floor)	0	11	3	6

Analysis of sample collected from Nos. 2, 4 and 5, as reported by **M**essrs. Hite and Krak, is given in the table of coal analyses under **No. 59.**

Peytona Land **Company Opening—No. 364 on Map II.**

Located on the south side of Coal River, 0.5 mile southeast of Peytona P. O.; section taken by Krebs; **Cedar Grove Coal**; elevation, 950' B.

			Ft.	In.
Slate, gray			3	0
Coal, cannel	1'	4 "		
Slate, gray	0	0⅛		
Coal, semi-cannel, (slate floor)	0	8	2	0⅛

Capacity of mine, about 25 to 50 tons daily; number of men employed, 20; W. H. Warner, Cleveland, Ohio, President; C. H. Meade, Superintendent.

The above section was taken in the old mine where cannel coal was first mined on Coal River, more than fifty years ago, and shipped down that stream to the Kanawha and Ohio in boats. The Peytona Block Coal Company is now mining the coal at this point and ships it by rail.

Scott District.

Scott District lies west of Peytona District and the **Cedar Grove Coal** bed extends into this district and forms one of its important commercial coals.

The **Cedar Grove** seam comes out of Little Coal River just south of Lowrey Station, and rising rapidly southward, gets from 450 to 500 feet above the bed of the river just south from Madison on the crest of the Warfield Anticline.

The following section was measured by Teets just southwest of Lowrey Station, where the coal appears in the grade of the Coal River R. R.:

Opening No. 365 on Map II.

				Ft.	In.
Shale, greenish-gray				25	0
Coal, semi-splint	1'	1"	Cedar Grove	3	1
Shale, gray	0	1			
Coal, splint	1	3			
Bone **coal**	0	1½			
Coal, gas	0	6½			
Fire clay, soft				3	11
Sandy shale to level of railroad, 662' L				8	0

The coal rises very rapidly at this point, there being a local rise of 38 feet in 400.

Joseph Hill Local Mine Opening—No. 366 on Map II.

Located on the east side of Little Coal River, just above the mouth of Camp Creek, and 0.25 mile southeast of Lowrey Station; section by Teets; **Cedar Grove Coal**; elevation, 690' B.

			Ft.	In.
Slate roof				
Coal, splint	1'	1"		
Shale, gray	0	1		
Coal, semi-splint	1	3		
Bone	0	1		
Coal, gas, (fire clay floor)	0	7	3	1

Pryor and Allen Prospect Opening—No. 367 on Map II.

Located on the south side of Camp Creek, about one-third mile southeast of Lowrey Station; section by Teets; **Cedar Grove Coal**; elevation, 720′ B.

			Ft.	In.
Slate roof				
Coal, visible, ("reported" 3 feet)			2	0

W. H. Dolan Local Mine Opening—No. 368 on Map II.

Located on the south side of Camp Creek, about 0.75 mile southeast of Lowrey Station; section by Krebs; **Cedar Grove Coal**; elevation, 780′ B.

			Ft.	In.
Shale, gray, plant fossils				
Coal, hard splint	1′	0″		
Shale, gray	0	1		
Coal, semi-splint	1	5		
Bone	0	1		
Coal, semi-splint, (fire clay floor)	0	5	3	0

Julia Woodrum Local Mine Opening—No. 369 on Map II.

Located on the south side of Camp Creek, about 0.8 mile southeast of Lowrey Station; section by Teets; **Cedar Grove Coal**; elevation, 795′ B.

			Ft.	In.
1. Slate roof				
2. **Coal**, hard splint	0′	11″		
3. Shale, gray	0	1		
4. **Coal**, splint	1	3		
5. Bone	0	1		
6. **Coal**, splint, (soft fire clay floor)	1	10	4	2

Analysis of a sample collected from Nos. 2, 4 and 6, as reported by Messrs. Hite and Krak, is given in the table of coal analyses under **No. 60.**

Patrick Kerse Opening—No. 370 on Map II.

Located on the north side of Camp Creek, 1.9 miles east of Lowrey Station, on a branch flowing from the north, and about 0.5 mile up same; section by Teets; **Cedar Grove Coal**; elevation, 835′ B.

			Ft.	In.
Slate roof				
Coal, impure, splint	1′	0″		
Fire clay	0	4		
Coal, splint, (slate floor)	1	7	2	11

Butts, N. 50° W.; faces, N. 40° E.

Joseph Eustes Local Mine Opening—No. 371 on Map II.

Located in the head of Camp Creek, 1.2 miles southwest of Cabell P. O.; section by Teets; **Cedar Grove Coal**; elevation 900′ B.

	Ft.	In.
Slate roof..		
Coal, splint, (slate floor)........................	2	2

A. J. Kerse Local Mine Opening—No. 372 on Map II.

Located in the head of Camp Creek, 1.6 miles due south of Cabell P. O.; section by Teets; **Cedar Grove Coal**; elevation, 1045′ B.

			Ft.	In.
1. Slate roof..				
2. **Coal**, gas.........................0′	3″			
3. Bone 0	2			
4. **Coal**, splint.........................0	9			
5. Shale, gray.........................0	4			
6. **Coal**, hard, splint, visible..........2	2		3	8

Butts, N. 40° W.; faces, N. 50° E.

Analysis of sample collected from Nos. 2, 4 and 6, as reported by **Messrs**. Hite and Krak, is given in the table of coal analyses under **No. 61.**

Westward down Camp Creek, about 0.2 mile from the Kerse Opening, Dr. A. J. Dolan is mining the Cedar Grove seam of coal for local fuel where the following section was measured by Teets:

Dr. A. J. Dolan Opening—No. 373 on Map II.

		Ft.	In.
Slate roof..			
Coal, splint.........................0′	2″		
Bone 0	2		
Coal, splint.........................0	10		
Shale, gray.........................0	4		
Coal, splint, (slate floor, 1050′ B.).....2	1	3	[illegible]

Southward about a mile, near the head of Hubbard Fork of Rock Creek, several openings have been made in the Cedar Grove bed and the coal mined for local fuel, as follows:

Frank Javin Opening—No. 374 on Map II.

Located on the head of Hubbard Fork of Rock Creek, 2.0 miles northeast of Foster P. O.; section by Krebs; **Cedar Grove Coal**; elevation, 1050' B.

			Ft.	In.
Slate roof				
Coal, splint	0'	9"		
Shale, dark	0	0½		
Coal, gray splint	0	11		
Coal, hard block, glossy	0	10		
Coal, gray splint, (slate floor)	0	6	3	0½

Butts, N. 42° W.; faces, N. 48° E.

The Rock Creek Colliery Company of Lynchburg, Virginia, owns a tract of several thousand acres of land lying on the waters of Rock Creek. Mr. M. W. Venable, of Charleston, W. Va., has made a report of the coals on this area and the report has been kindly furnished the Survey by Mr. M. B. Handy, President of the Rock Creek Colliery Company, from which the following data are taken:

Rock Creek Colliery Company Opening—No. 375 on Map II.

Located on the south side of Rock Creek, just opposite A. B. Elkin's house, about 1.7 miles northwest of Foster P. O.; section by M. W. Venable; **Cedar Grove Coal**; elevation, 1035' B. (Krebs).

	Ft.	In.
Slate roof		
Coal, clean, soft, (slate floor)	3	6

A sample collected by Mr. Venable and analyzed by A. S. McCreath & Son, of Harrisburg, Pa., gave the following results:

	Per cent.
Moisture	3.692
Volatile Matter	36.248
Fixed Carbon	56.980
Sulphur	0.660
Ash	2.420
Total	100.000

The excessive amount of moisture is due to the fact that the sample was taken near the outcrop of the coal. The above analysis shows a very small percentage of ash.

William Eckels Local Mine Opening—No. 376 on Map II.

Located at the head of a northern tributary of Rock Creek, 2.5 miles northeast of Rock Creek Station; section by Teets; **Cedar Grove Coal**; elevation, 895′ B.

			Ft.	In.
Slate roof				
Coal, splint	1′	1″		
Coal, impure	0	6		
Coal, splint	1	2		
Bone	0	2		
Coal, splint, (fire clay floor)	0	3	3	2

Butts, N. 38° W.; faces, N. 52° E.

Samuel Cabell Local Mine Opening—No. 377 on Map II.

Located on a branch of Rock Creek flowing from the north, 1.0 mile east of Rock Creek Station; section by Teets; **Cedar Grove Coal**; elevation, 870′ B.

			Ft.	In.
Shale, gray, plant fossils				
Coal, splint	1′	2″		
Shale, gray	0	2		
Coal, splint, (slate floor)	1	5	2	9

Vernon Vandelinde Local Mine Opening—No. 378 on Map II.

Located on the south side of Rock Creek, 1.2 miles southeast from Rock Creek Station; section by Teets; **Cedar Grove Coal**; elevation, 890′ B.

			Ft.	In.
1. Slate roof				
2. **Coal**, splint	1′	3″		
3. Slate	0	1		
4. **Coal**, gray splint, (fire clay floor)	1	2	2	6

Butts, N. 48° W.; faces, N. 42° E.

Analysis of sample collected from Nos. 2 and 4, as reported by Messrs. Hite and Krak, is given in the table of coal analyses under **No. 62.**

J. B. Price Local Mine Opening—No. 379 on Map II.

Located on the head of Low Gap Branch of Rock Creek, 1.2 miles north of Danville; section by Teets; **Cedar Grove Coal**; elevation, 920′ B.

			Ft.	In.
Slate roof				
Coal, splint	1′	8″		
Slate, gray	0	2		
Coal, splint, (slate floor)	0	10	2	8

Rock Creek Colliery Company Opening—No. 380 on Map II.

Located on the head of Adkins Hollow of Rock Creek; section by M. W. Venable; **Cedar Grove Coal**; elevation, 1040′ B.

			Ft.	In.
1. Sandstone				
2. Draw slate			1	0
3. **Coal**, soft	3′	10½″		
4. Slate, soft	0	3½		
5. **Coal**, splint, (slate floor)	1	4½	5	6½

A sample collected from Nos. 3 and 5 by Mr. Venable and analyzed by A. S. McCreath & Son, of Harrisburg, Pa., gave the following results:

	Per cent.
Moisture	2.214
Volatile Matter	36.136
Fixed Carbon	56.394
Sulphur	0.686
Ash	4.570
Total	100.000

The above analysis shows the coal to be an excellent steam and fuel coal.

A. H. Williams Opening—No. 381 on Map II.

Located on the head of Left Fork of Rock Creek, 2.0 miles southeast of Foster P. O; section by Krebs.

				Ft.	In.
Dark shale roof					
Coal, splint	1′	1″	**Cedar Grove, Upper Bench**	3	0
Shale, gray	0	2			
Coal, splint	1	1			
Shale, gray	0	4			
Coal, splint	1	0			
Fire clay, soft				2	0
Sandstone				7	0

				Ft.	In.
Coal, splint, Cedar Grove, Lower Bench				2	0
Concealed				2	0
Sandstone, massive				4	0
Slate, gray				1	0
Coal, splint	1′	4″			
Shale, gray	0	6	**Alma**, 1067′ L.	4	4
Coal, splint, visible	2	6			

The above section shows both the Cedar Grove and Lower Cedar Grove Coals and also the Alma, the partings between the coals having thinned away greatly at this point.

Price Branch flows into Little Coal River from the northeast at Danville. Several openings have been made in the **Cedar Grove Coal** on this branch.

Jas. P. **Mullens** Local **Mine Opening—No. 382 on Map II.**

Located in the head of Price Branch about 1.2 miles northeast of Danville; section by Teets; **Cedar Grove Coal**; elevation, 1050′ B.

			Ft.	In.
Slate roof				
Coal, splint, hard	0′	8 ″		
Slate, gray	0	0½		
Coal, block	0	6		
Slate, gray	0	1		
Coal, splint	1	1		
Coal, semi-cannel, (slate floor)	0	8	3	0½

Butts, N. 40° W.; faces, N. 50° E.

W. T. **Smoot** Heirs Local **Mine Opening—No. 383 on Map II.**

Located on the north side of Price Branch, about 0.7 mile northeast of Danville; section by Teets; **Cedar Grove Coal**; elevation, 1030′ B.

			Ft.	In.
Slate roof				
Coal, splint	1′	2″		
Slate	0	1		
Coal, splint	1	0		
Slate, gray	0	1		
Coal, semi-cannel (slate floor)	0	2	2	6

Butts, N. 40° W.; faces, N. 50° E.

On the west side of Coal River, about 3.0 miles northwest of the above opening, the following section was measured by Teets:

Lewis Hill Local Mine Opening—No. 384 on Map II.

Located 1.0 mile southwest of Lowrey Station; **Cedar Grove Coal**; elevation, 705′ B.

			Ft.	In.
Slate roof				
Coal, gas	0′	3″		
Slate, gray	0	0½		
Coal, splint	0	9		
Slate, dark	0	1		
Coal, splint	1	0		
Slate, gray	0	0½		
Coal, splint, (slate floor)	0	9	2	11

Samuel Griffith Local Mine Opening—No. 385 on Map II.

Located on the west side of Coal River, 0.5 mile southwest of Lowrey Station; section by Teets; **Cedar Grove Coal**; elevation, 679′ L.

			Ft.	In.
Slate roof				
Coal, hard splint	0′	11″		
Slate, gray	0	1		
Coal, hard splint, (slate floor)	1	11	2	11

H. F. Stollings Local Mine Opening—No. 386 on Map II.

Located on the east side of Little Coal River, 0.5 mile south of Rock Creek Station; section by Teets; **Cedar Grove Coal**; elevation, 835′ B.

			Ft.	In.
Slate roof				
Coal, hard splint	1′	0″		
Shale, gray	0	1		
Coal, gray splint	1	7		
Coal, gas	0	4		
Coal, gray splint	0	10		
Coal, splint, hard, (slate floor)	0	8	4	6

Geo. Hill Local Mine Opening—No. 387 on Map II.

Located on the east side of Little Coal River, just below the mouth of Rock Creek; section by Teets; **Cedar Grove Coal**; elevation, 800′ B.

			Ft.	In.
Slate roof				
Coal, splint	0′	9″		
Shale, gray	0	1		
Coal, gray splint	0	6		
Slate, dark	0	0¼		

			Ft.	In.
Coal, gas	0′	5″		
Coal, gray splint	1	1		
Slate, gray	0	1		
Coal, splint, (slate floor)	0	7	3	6¼

Riley Hill Local Mine Opening—No. 388 on Map II.

Located on the east side of Little Coal River, 0.2 mile south of Van Camp Station; section by Teets; **Cedar Grove Coal**; elevation, 730′ B.

			Ft.	In.
Slate roof				
Coal, splint	1′	0″		
Slate, gray	0	1		
Coal, gray splint, (slate floor)	2	1	3	2

Daniel Smoot Local Mine Opening—No. 389 on Map II.

Located on the south side of Browns Branch of Little Coal River, just east of Van Camp Station; section by Teets; **Cedar Grove Coal**; elevation, 760′ B.

			Ft.	In.
Slate roof				
Coal, splint	1′	0″		
Slate, gray	0	1		
Coal, hard splint, (slate floor)	2	2	3	3

Griffith Brothers Local Mine Opening—No. 390 on Map II.

Located on the north side of the first branch from the west, flowing into Little Coal River, north of the mouth of Rock Creek; section by Teets; **Cedar Grove Coal**; elevation, 780′ B.

			Ft.	In.
Slate roof				
Coal, splint	1′	1″		
Shale, gray	0	1		
Coal, splint, (slate floor)	1	11	3	1

Dora Hopkins Opening—No. 391 on Map II.

Located on the west side of Little Coal River, just below the mouth of Lick Creek; section by Teets; **Cedar Grove Coal**; elevation, 845′ B.

			Ft.	In.
1. Slate roof				
2. Coal, hard splint	1′	1″		
3. Shale, gray	0	1		

			Ft.	In.
4. **Coal**, gray splint	0′	5″		
5. **Coal**, gas	0	6		
6. **Coal**, gray splint	0	11		
7. **Coal**, splint, hard, (slate floor)	0	6	3	6

Analysis of sample collected from Nos. 2, 4, 5, 6 and 7, as reported by Messrs. Hite and Krak, is given in the table of coal analyses under **No. 63.**

George Weseron Prospect Opening—No. 392 on Map II.

Located at the forks of Slippery Gut Branch of Little Coal River 1.0 mile west of Rock Creek Station; section by Teets; **Cedar Grove Coal**; elevation, 750′ B.

			Ft.	In.
Slate roof				
Coal, splint	0′	10″		
Slate, gray	0	2		
Coal, hard splint	1	5		
Coal, gray splint	0	2		
Coal, gas, (slate floor)	0	4	2	11

Fred Chambers Prospect Opening—No. 393 on Map II.

Located on the south side of Lick Creek, 1.7 miles southwest of Hopkins Station; section by Teets; **Cedar Grove Coal**; elevation, 855′ B.

			Ft.	In.
Gray shale full of fossil plants				
Coal, semi-cannel	0′	1″		
Slate	0	0½		
Coal, block	0	2		
Slate	0	3		
Coal, hard splint	0	3		
Coal and slate, interlaminated	0	5		
Coal, splint	0	3		
Gray shale	0	3		
Coal, hard splint	0	10		
Slate	0	0½		
Coal, splint, hard (slate floor)	0	8	3	3

John Tawney Local Mine Opening—No. 394 on Map II.

Located on the south side of Lick Creek, 3.0 miles southwest of Hopkins Station; section by Krebs; **Cedar Grove Coal**; elevation, 820′ B.

	Ft.	In.
Sandstone, massive, visible	8	0
Shale, greenish-gray	0	6

			Ft.	In.
Coal, splint	1'	6"		
Shale, gray	0	10		
Coal, splint, (slate floor)	1	4	3	[illegible]

Walker McKinney Local Mine Opening—No. 395 on Map II.

Located on the Left Fork of Lick Creek, 2.0 miles southwest of Hopkins Station; section by Krebs; **Cedar Grove Coal**; elevation, 960' B.

			Ft.	In.
Gray shale			15	0
Dark slate, plant fossils			1	0
Coal, splint	1'	0"		
Dark shale	0	4		
Coal, splint, (slate floor)	1	0	2	4

Burrell Miller Local Mine Opening—No. 396 on Map II.

Located on Left Fork of Lick Creek, 1.7 miles southwest of Hopkins Station; section by Krebs; **Cedar Grove Coal**; elevation, 940' B.

			Ft.	In.
Sandy shale			10	0
Coal, splint	1'	0"		
Dark shale	0	4		
Coal, semi-splint, (slate floor)	1	0	2	4

From the sections given in the foregoing openings, it will be seen that the Cedar Grove Coal is an important seam in Scott District.

Washington District.

Washington District lies west and south of Scott, and the **Cedar Grove** bed extends into this district. In the northern part of the district, the Cedar Grove Coal has attained sufficient thickness and purity to be of commercial value, while in the southern part of the district, so far as the coal has been prospected, it is unusually thin and impure and of very little economic value.

Sections will now be given showing the structure, character and thickness of this bed:

W. C. Miller Local Mine Opening—No. 397 on Map II.

Located on the north side of Long Branch of Turtle Creek, 2.0 miles southwest of Danville; section by Krebs; **Cedar Grove Coal**; elevation, 960′ B.

			Ft.	In.
Gray slate roof				
Coal, splint	1′	4″		
Slate, dark	1	6		
Coal, splint, (slate floor)	1	6	4	4

Ballard Prospect Opening—No. 398 on Map II.

Located on the west side of Cox Fork of Turtle Creek, 1.2 miles northwest of Turtletown; section taken by Krebs; **Cedar Grove Coal**; elevation, 1125′ B.

			Ft.	In.
Sandstone, brown, **marine fossils, Dingess**			1	0
Concealed and sandstone			58	0
Shale, gray			6	0
Coal, semi-splint	0′	9½″		
Slate, black	0	1		
Coal, gas	1	2½		
Coal, splint, (fire clay floor)	1	0	3	1

Perry A. Miller Local Mine Opening—No. 399 on Map II.

Located on west side of Indian Grove Branch of Turtle Creek, 1.0 mile northeast of Turtletown; section by Teets; **Cedar Grove Coal**; elevation, 1080′ B.

			Ft.	In.
Sandstone roof				
Coal, soft	0′	5″		
Slate, gray	0	4		
Coal, gas	1	1		
Slate, gray	0	1		
Coal, hard splint, (slate floor)	1	2	3	1

Butts, N. 55° W.; faces, N. 35° E.

Charles Ball Opening—No. 400 on Map II.

Located on the east side of Turtle Creek, 0.6 mile north of Olive Branch Church; section taken by Teets; **Cedar Grove Coal**; elevation, 1060′ B.

			Ft.	In.
1. Sandstone roof				
2. **Coal**, hard splint	0′	5 ″		
3. Shale, gray	0	3		
4. **Coal**, gas	1	0		
5. Shale, gray	0	0½		
6. **Coal**, hard splint, (slate floor)	1	4	3	0½

Butts, N. 50° W.; faces, N. 40° E.

Analysis of sample taken from Nos. 2, 4 and 6, as reported by **Messrs.** Hite and Krak, is given in the table of coal analyses under **No. 64.**

Silas Billetts Local **Mine Opening—No. 401** on **Map II.**

Located on the north side of Turtle Creek, 2.5 miles southwest of Danville; section taken by Teets; **Cedar Grove Coal**; elevation, 995' B.

				Ft.	In.
1.	Sandstone, massive				
2.	**Coal**, splint	0'	11"		
3.	Shale, gray	0	0½		
4.	**Coal**	0	0½		
5.	Niggerhead	0	2		
6.	**Coal**, splint	1	2		
7.	Shale, gray	0	0½		
8.	**Coal**, hard splint (slate floor)	1	1	3	5½

Butts, N. 50° W.; faces, N. 40° E.

Analysis of sample collected from Nos. 2, 6 and 8, as reported by **Messrs.** Hite and Krak, is given in the table of coal analyses under **No. 65.**

Hayes Hager **Prospect Opening—No. 402** on **Map II.**

Located on the south side of Sixmile Creek, 0.6 mile southwest of Havana P. O.; section measured by Teets; **Cedar Grove Coal**; elevation, 1155' B.

			Ft.	In.
Slate roof				
Coal, splint	0'	10"		
Slate, gray	0	2		
Coal, gray splint (slate floor)	1	10	2	10

Ballard Hager **Opening—No. 403** on **Map II.**

Located in the head of a hollow, on the south side of Dog Fork of Trace Fork of Big Creek, 1.1 miles southwest of Anchor P. O.; section taken by Teets; **Cedar Grove Coal**; elevation, 1180' B.

				Ft.	In.
1.	Slate roof				
2.	**Coal**, splint	0'	8"		
3.	Shale, gray	0	1		
4.	**Coal**, splint, visible	1	8	2	5

Analysis of sample collected from Nos. 2 and 4, as reported by Messrs. Hite and Krak, is given in the table of coal analyses under No. 66.

Lee Harless Local Mine Opening—No. 404 on Map II.

Located on Left Fork of Dog Fork of Trace Fork of Big Creek, 0.7 mile south of Anchor P. O.; section by Teets; **Cedar Grove Coal**; elevation, 1200′ B.

			Ft.	In.
Slate roof				
Coal, splint	0′	7″		
Shale, gray	0	1		
Coal, splint	1	0		
Coal, gray splint	1	6	3	2

James Workman Local Mine Opening—No. 405 on Map II.

Located on the south side of the Left Fork of Dog Fork of Trace Fork of Big Creek, about 0.6 mile due south of Anchor P. O.; section measured by Teets; **Cedar Grove Coal**; elevation, 1200′ B.

			Ft.	In.
Slate roof				
Coal, splint	0′	8″		
Shale, gray	0	1		
Coal, gas (slate floor)	2	6	3	3

Butts, N. 36° W.; faces, N. 54° E.

Edward Bell Local Mine Opening—No. 406 on Map II.

Located on the north side of the Right Fork of Dog Fork of Trace Fork of Big Creek; section by Teets; **Cedar Grove Coal**; elevation, 1200′ B.

			Ft.	In.
Slate roof				
Coal, splint	0′	10″		
Slate	0	1		
Coal, blocky, glossy	1	2		
Coal, splint, (slate floor)	0	10	2	11

Near Ramage, the section given on page 364 of this Report shows the Cedar Grove Coal in two benches, the Upper Bench being two feet thick and the Lower Bench one foot six inches. Continuing to the south from this point, the core test holes drilled by Messrs. Chilton, MacCorkle, Chilton and Meany, on Spruce Laurel Fork, demonstrate that the Cedar Grove Coal thins away and almost entirely disappears in the southern part of Washington District.

Sherman District.

Sherman District lies south and east of Peytona, and in the northern part of this district, the **Cedar Grove Coal** appears to be of commercial thickness, while in the southern part of the same, it is quite thin so far as has been prospected, or sections of it have been exposed.

George Evan Local Mine Opening—No. 407 on Map II.

Located on the north side of Short Creek, 2.0 miles northeast of Racine P. O.; section taken by Krebs; **Cedar Grove Coal**; elevation, 980′ B.

			Ft.	In.
Sandstone, massive				
Coal, gas	0′	5″		
Shale, dark	0	0½		
Coal, splint	0	3		
Slate, dark	0	2		
Coal, splint	0	7		
Concealed by water	0	11	2	4½

Boone and Kanawha Land and Mining Company Prospect Opening—No. 408 on Map II.

Located on the Left Fork of Short Creek, about 1.0 mile north of Racine P. O.; section by H. A. Henderson for J. S. Cunningham; **Cedar Grove Coal**; elevation, 915′ B.

			Ft.	In.
Slate roof				
Coal, splint	2′	0″		
Slate	0	1		
Coal, splint	1	8		
Slate	0	1		
Coal, splint (slate floor)	1	4	5	2

Boone and Kanawha Land and Mining Company Prospect Opening—No. 409 on Map II.

Located on the Left Fork of Short Creek, near the head of same, 1.0 mile north of Racine P. O.; section by H. A. Henderson for J. S. Cunningham; **Cedar Grove Coal**; elevation, 920′ B.

			Ft.	In.
1. Slate roof				
2. **Coal**, splint	1′	10″		
3. Slate	0	1		
4. **Coal**	0	11		
5. Slate	0	1		
6. **Coal** (slate floor)	0	11	3	10

PLATE XXXIII.—Sandstone over Upper Campbell Creek (No. 2 Gas) Coal on Pond Fork, below Grapevine Branch.

A sample collected from Nos. 2, 4 and 6 by Mr. Henderson and analyzed by the Chief Chemist of the New River and Pocahontas Consolidated Coal Company, at Berwind, gave the following results:

	Per cent.
Moisture	2.16
Volatile Matter	36.02
Fixed Carbon	52.68
Ash	9.14
Total	100.00
Sulphur	0.985

Boone and Kanawha Land and Mining Company Prospect Opening—No. 410 on Map II.

Located on the west side of Coal River, 0.25 mile west of Racine P. O.; section by H. A. Henderson for J. S. Cunningham; **Cedar Gróve Coal**; elevation, 915′ B.

			Ft.	In.
Slate roof				
Coal, splint	1′	6″		
Fire clay	0	2		
Coal, splint	1	10		
Slate	0	3		
Coal, splint	0	8		
Slate	0	4		
Coal, splint	0	8	5	5

Peytona Land Company Opening—No. 411 on Map II.

Located on Indian Creek, 0.25 mile west of Sterling P. O.; section by Krebs; **Cedar Grove Coal**; elevation, 900′ B.

			Ft.	In.
Slate, gray				
Coal, splint	0′	6″		
Slate, gray	0	0½		
Coal, splint (slate floor)	2	6	3	0½

Lackawanna Coal and Lumber Company Prospect Opening. No. 412 on Map II.

Located on Sandlick Creek of Laurel Creek, 2.0 miles southwest of Seth P. O.; section taken by Krebs; **Cedar Grove Coal**; elevation, 820′ B.

			Ft.	In.
Slate roof				
Coal, splint	0′	4″		
Shale, gray	0	8		
Coal, splint	1	4	2	4

LaFollette, Robson et al. Opening—No. 413 on Map II.

Located on the east side of Coal River, 1.6 miles southeast of Seth P. O.; section by Krebs; **Cedar Grove Coal**; elevation, 1715' B.

			Ft.	In.
Sandstone				
Coal, splint	0'	6"		
Slate	0	2		
Coal, splint	0	4		
Slate	0	0½		
Coal, gray splint	0	5		
Coal, splint (fire clay floor)	0	3½	1	9

Crook District.

The Cedar Grove Coal enters Crook District on the crest of the Warfield Anticline, 500 to 550 feet above the bed of Pond Fork, and dips southward into the Coalburg Syncline, the coal going under Pond Fork near the mouth of Robinson Creek, where it is mined for local fuel by **Patrick Hamilton**, on the land of D. **M.** Arbogast, where the following section was measured by Krebs:

D. M. Arbogast Local Mine Opening—No. 414 on Map II.

				Ft.	In.
1.	Shale, gray, full of plant fossils				
2.	**Coal**, gray splint	0'	2"		
3.	**Coal**, block	1	10		
4.	**Coal**, gray splint	0	10	2	10
5.	Slate floor, 755' L				

Analysis of sample collected from Nos. 2, 3 and 4, as reported by **Messrs.** Hite and Krak, is given in the table of coal analyses under **No. 67.**

Daniel Gore Local Mine Opening—No. 415 on Map II.

Located on the east side of Pond Fork, 0.8 mile southeast of Uneeda; section by Krebs; **Cedar Grove Coal**; elevation, 810' B.

			Ft.	In.
Slate roof				
Coal, splint	1'	0"		
Slate	0	1		
Coal, splint	1	6		
Slate	0	6		
Coal, splint (slate floor)	0	6	3	7

The **Cedar Grove Coal** is thin in the southern part of Crook District so far as the coal has been developed or openings explored. Its horizon and thickness are given in the generalized sections of Crook District. A further description of the character, quality and quantity of this coal available in Boone County will be given in a subsequent Chapter on Coal.

THE MIDDLE CEDAR GROVE SANDSTONE.[13]

Underneath the Upper Cedar Grove Coal, from 2 to 10 feet, appears a sandstone sometimes containing thin layers of coal. It ranges in thickness from 10 to 40 feet, and often forms rugged cliffs along the slopes of the hills where it is exposed. Its thickness at different points is given in the sections already published on preceding pages of this Report.

THE LOWER CEDAR GROVE COAL.

Underneath the Middle Cedar Grove Sandstone, there often occurs a multiple bedded, splinty coal that has been named the **Lower Cedar Grove Coal.**[14] In Boone County, this seam is thin and of little economic value so far as the coal has been prospected and the bed is exposed. Sections of this coal are given in the general sections, and in the records of the diamond core test holes already shown on preceding pages of this Report.

THE LOWER CEDAR GROVE SANDSTONE.

Underneath the Lower Cedar Grove Coal, and separated from same by 1 to 10 feet of fire clay and impure shale, there occurs a medium grained, gray sandstone, from 20 to 30 feet in thickness, known as the **Lower Cedar Grove Sandstone.**[15]

In Boone County this sandstone usually contains some layers of shale and slate and is very seldom massive. Measurements of this sandstone, showing its location and thickness.

---

[13]Hennen and Reger, Logan and Mingo Report; 1914.
[14]Hennen and Reger, Logan and Mingo Report; 1914.
[15]Hennen and Reger, Logan and Mingo Report; 1914.

have already been given in the general sections published on preceding pages of this Report.

THE ALMA COAL.

From 30 to 60 feet underneath the Upper Cedar Grove Coal, there occurs a multiple bedded, splinty coal in Boone County that has been named the Alma Coal.[16] This seam has been called by various names in different parts of the southwestern portion of West Virginia, as the "Draper" at Logan, the "Eagle" at Clothier and Ramage, and the "Campbell Creek," "No. 2 Gas" at Big Creek in Logan County, and possibly the "Peerless" of Kanawha County. The sections showing thickness, location and character of this coal will be taken up by magisterial districts:

Peytona District.

In Peytona District the Alma Coal is usually from 1 to 2 feet thick, and is a hard splint, multiple bedded coal. The coal comes out of Coal River south of Emmons Station, and continues to rise to the southeast until, on the Warfield Anticline, the coal is about 300 feet above the bed of the river.

In Diamond Core Test Hole (100) on Fork Creek, pages 61-2 of this volume, the Alma Coal is 1′ 6″ thick, and comes 50 feet under the Cedar Grove Coal, and 85 feet over the Campbell Creek (No. 2 Gas) Coal.

A section 2 miles south of Cabell P. O., published on page 64 of this Report, shows the Alma one foot thick and 68 feet over the Campbell Creek (No. 2 Gas) Coal.

Scott District.

The Alma Coal comes out of Little Coal River near Van Camp Station, and rises rapidly above the same to the southeast, until just south of Madison, the coal is 475 to 500 feet above the bed of Coal River.

[16] I. C. White, Volume II(A), p. 403; 1908.

J. A. Cantley Local Mine Opening—No. 416 on Map II.

Located on the east side of Hubbard Fork of Rock Creek, 0.4 mile northeast of Foster P. O.; section by Teets; **Alma Coal**; elevation, 1220' B.

			Ft.	In.
Sandstone roof				
Coal, dark splint	0'	7"		
Coal, cannel (slate floor)	2	2	2	9

Butts, N. 36° W.; faces, N. 54° E.

Jackson Darlow Local Mine Opening—No. 417 on Map II.

Located on the north side of Rock Creek, just below the mouth of Hubbard Fork, 0.2 mile west of Rock Creek; section by Teets; **Alma Coal**; elevation, 1065' B.

			Ft.	In.
Sandstone roof				
Coal, splint	0'	9"		
Coal, cannel (slate floor)	2	3	3	0

Butts, N. 50° W.; faces, N. 40° E.

William Holstein Local Mine Opening—No. 418 on Map II.

Located on the north side of Rock Creek, just north of Foster P. O.; section measured by Krebs; **Alma Coal**; elevation, 1055' B.

			Ft.	In.
Sandstone, brownish			1	0
Slate, grayish-green, roof			8	0
Coal, splint	1'	0"		
Coal, cannel (slate floor)	1	10	2	10

Samuel Carpenter Local Mine Opening—No. 419 on Map II.

Located on the north side of Rock Creek, a little northwest of Foster P. O.; section taken by Krebs; **Alma Coal**; elevation, 1055' B.

			Ft.	In.
Shale, gray, roof			10	0
Coal, splint	1'	5"		
Coal, cannel (slate floor)	2	6	3	11

D. J. Smoot Opening—No. 420 on Map II.

Located on the west side of Little Coal River, 1.0 mile west of Danville, on a little branch flowing from the west into Little Coal River; section measured by Teets; **Alma Coal**; elevation, 920' B.

			Ft.	In.
Slate roof				
Coal, splint	0'	11"		
Shale, gray	0	3		
Coal, splint (slate floor)	2	1	3	3

Julian Hill Local Mine Opening—No. 421 on Map II.

Located in the head of Rucker Branch of Little Coal River, 1.4 miles northeast of Madison; section by Teets; **Alma Coal**; elevation, 1175' B.

			Ft.	In.
1. Gray shale roof				
2. **Coal**, hard splint	1'	8"		
3. Slate, gray	0	1		
4. **Coal**, cannel (slate floor)	1	6	3	3

Analysis of sample collected from No. 2, as reported by Messrs. Hite and Krak, is given in the table of coal analyses under **No. 68**, and the analysis of a sample collected from No. 4, as reported by Hite and Krak, is given under **No. 69**.

R. P. Chew Opening—No. 422 on Map II.

Located in the head of Workman Branch, 0.5 mile north of Workman Knob; section by Teets; **Alma Coal**; elevation, 1150' B.

			Ft.	In.
Slate roof				
Coal, gas	0'	9 "		
Slate, black	0	0½		
Slate, greenish	1	6		
Shale	0	1		
Coal, block	1	8	4	0½

Dr. H. Lon Carter Opening—No. 423 on Map II.

Located on the west side of Coal River, 0.5 mile northwest of Madison; section by Teets; **Alma Coal**; elevation, 1090' B.

			Ft.	In.
Slate roof				
Coal, splint	1'	0"		
Slate	0	2		
Coal, splint	2	6	3	[illegible]

Coal formerly mined for local fuel use, but now abandoned.

J. C. Ballard Opening—No. 424 on Map II.

Located on the north side of Turtle Creek, 1.5 miles southwest of Danville; section by Krebs; **Alma Coal**; elevation, 1030' B.

			Ft.	In.
Sandstone, massive			8	0
Shale, grayish-green			5	0
Coal, hard splint	0'	7"		
Shale, gray	0	5		
Coal, hard splint	1	0		
Coal, cannel, visible	1	3	3	3

J. C. Ballard Opening—No. 425 on Map II.

Located on the south side of Turtle Creek, 0.8 mile southwest of Danville; section by Krebs; **Alma Coal**; elevation, 1005' B.

			Ft.	In.
Sandy shale, greenish-gray			6	0
Coal, splint	1'	6"		
Coal, cannel, visible	2	0	3	6

Washington District.

The **Alma Coal** is one of the most important coals in Washington District. In the northern part, it is the lowest seam of coal of sufficient thickness and purity to be of commercial value. It occurs about 460 feet above Coal River at Madison and rises to the crest of the Warfield Anticline at Low Gap, where it occurs at an elevation of 1215' B. From this point, it dips to the southeast until it is mined at Ottawa at tipple height and then dips very rapidly into the Coalburg Syncline at Clothier.

B. R. Vandeline Local Mine Opening—No. 426 on Map II.

Located on the point of the ridge, just south of the mouth of Pond and Spruce Forks, 0.5 mile south of Madison; section taken by Krebs; **Alma Coal**; elevation, 1175' B.

			Ft.	In.
Shale, grayish-green, large concretions			20	0
Coal, splint, impure	0'	8"		
Shale, gray	0	1		
Coal, blocky	0	2		
Shale, gray	0	1		
Coal, hard splint, glossy	1	9		
Shale, gray	0	6		
Coal, splint (slate floor)	1	2	4	5

Butts, N. 60° W.; faces, N. 30° E.

R. B. Allen Opening—No. 427 on Map II.

Located on the east side of Spruce Fork, 0.5 mile east of Low Gap Station and near crest of Warfield Anticline; section taken by Krebs; **Alma Coal**; elevation, 1215′ B.

No.	Layer			Ft.	In.
1	Shale, gray, with concretions			20	0
2.	Slate			1	0
3.	**Coal**, impure	0′	3″		
4.	**Coal**, splint	0	7		
5.	Slate	0	1		
6.	**Coal**, splint	0	2		
7.	**Coal**, impure	0	3		
8.	**Coal**, splint, glossy	2	3		
9.	**Coal**, impure	0	3		
10.	Slate	0	3		
11.	**Coal**, gas (slate floor)	0	4	4	5

Analysis of sample collected from Nos. 4, 6, 8 and 11, as reported by Messrs. Hite and Krak, is given in the table of coal analyses under **No. 70.**

Albert Meyers Local Mine Opening—No. 428 on Map II.

Located on south side of Sparrow Creek, 1.5 miles southwest of Madison; section by Teets; **Alma Coal**; elevation, 1150′ B.

No.	Layer			Ft.	In.
1.	Slate roof				
2.	**Coal**, hard splint	0′	9″		
3.	Slate, gray	0	2		
4.	**Coal**, splint	2	2		
5.	Bone	0	2		
6.	Slate	0	8		
7.	**Coal**, hard splint (slate floor)	1	7	5	6

Butts, N. 48° W.; faces, N. 42° E.

Analysis of sample collected from Nos. 2, 4 and 7, as reported by Messrs. Hite and Krak, is given in the table of coal analyses under **No. 71.**

John Kessinger Local Mine Opening—No. 429 on Map II.

Located on the north side of Laurel Branch of Spruce Fork, 2.2 miles southwest of Madison; section by Teets; **Alma Coal**; elevation, 1170′ B.

Layer			Ft.	In.
Slate roof				
Coal, hard splint	0′	7″		
Slate, dark	0	2		
Coal, hard splint, blocky	2	3		
Slate	0	10		
Coal, hard splint (slate floor)	1	5	5	[illegible]

John Kessinger Local Mine Opening—No. 430 on Map II.

Located on the west side of Spruce Fork, about 1.5 miles southwest of Madison; section taken by Teets; **Alma Coal**; elevation, 1150' B.

			Ft.	In.
Slate roof				
Coal, splint	0'	9"		
Coal, block, greenish	2	0		
Slate and fire clay	1	9		
Coal, gray splint	1	8	6	2

P. K. Miller Opening—No. 431 on Map II.

Located on the north side of Low Gap Creek, 0.5 mile west of Low Gap Station; section by Teets; **Alma Coal**; elevation, 1180' B.

			Ft.	In.
Slate roof				
Coal, splint	2'	0"		
Slate	1	6		
Coal, splint (slate floor)	2	0	5	6

John Stollings Opening—No. 432 on Map II.

Located on the east side of Low Gap Creek, 1.0 mile southwest of Low Gap Station; section by Teets; **Alma Coal**; elevation, 1175' B.

			Ft.	In.
1. Slate roof				
2. **Coal**, hard splint	0'	9"		
3. Slate, dark	0	1		
4. **Coal**	1	10		
5. Fire clay	2	0		
6. **Coal**, gray splint (slate floor)	2	1	6	9

Butts, N. 48° W.; faces, N. 42° E.

Analysis of a sample taken from Nos. 2, 4 and 6, as reported by **Messrs.** Hite and Krak, is given in the table of coal analyses under **No. 72.**

Jefferson Powell Local Mine Opening—No. 433 on Map II.

Located on the west side of Spruce Fork, 0.5 mile southeast of Powell Branch, and about 3.8 miles south of Madison; section taken by Teets; **Alma Coal**; elevation, 1080' B.

			Ft.	In:
Slate roof				
Coal, splint	1'	0"		
Slate, gray	0	2		
Coal, gray splint (slate floor)	3	8	4	10

Kelley Allen Opening—No. 434 on Map II.

Located on the north side of Big Branch of Spruce Fork, 0.5 mile north of Greenview; section measured by Teets; **Alma Coal**; elevation, 945' B.

			Ft.	In.
1. Slate roof				
2. **Coal**, hard splint	0'	11"		
3. Slate, gray	0	1		
4. **Coal**, splint, hard	1	0		
5. **Coal**, splint (slate floor)	2	7	4	7

Analysis of sample collected from Nos. 2, 4 and 5, as reported by Messrs. Hite and Krak, is given in the table of coal analyses under **No. 73.**

Rufus Ball Opening—No. 435 on Map II.

Located on the north side of Big Branch of Spruce Fork, 1.9 miles northwest of Greenview; section by Teets; **Alma Coal**; elevation, 1140' B.

			Ft.	In.
Slate roof				
Coal, splint	0'	10"		
Slate	0	1		
Coal, block, greenish (slate floor)	1	10	2	9

Reuben Connelly Local Mine Opening—No. 436 on Map II.

Located near the head of Big Branch of Spruce Fork, 2.6 miles northwest of Greenview; section by Teets; **Alma Coal**; elevation, 1160' B.

			Ft.	In.
1. Slate roof				
2. **Coal**, splint	0'	11"		
3. Shale, gray	0	1		
4. **Coal**, hard splint (slate floor)	1	9	2	9

Analysis of sample collected from Nos. 2 and 4, as reported by Messrs. Hite and Krak, is given in the table of coal analyses under **No. 74.**

M. P. Pauley Opening—No. 437 on Map II.

Located on the north side of Big Branch, 2.0 miles south of Low Gap Station; section measured by Krebs; **Alma Coal**; elevation, 1200' B.

			Ft.	In.
Light sandy shale			10	0
Slate			0	2
Coal, splint	0'	11"		
Slate	0	1		
Coal, splint (slate floor)	1	7	2	7

On the head of Ugly Creek, several openings have been made in the Alma Coal and mined for local fuel by the farmers:

Isaac Bell Local Mine Opening—No. 438 on Map II.

Located on the head of Ugly Creek on the north side of same, 1.5 miles southeast of Estep P. O.; section by Krebs; **Alma Coal**; elevation, 1150' B.

			Ft.	In.
Gray shale, plant fossils			5	0
Slate roof			0	5
Coal, splint	0'	11"		
Slate, dark	0	2		
Coal, splint (slate floor)	1	8	2	9

Isaac Bell Opening—No. 439 on Map II.

Located in the head of Ugly Creek, on the south side of same, 1.5 miles southeast of Estep P. O.; section by Krebs; **Alma Coal**; elevation, 1140' B.

			Ft.	In.
1. Sandy shale				
2. Slate roof			0	10
3. **Coal**, splint	0'	11"		
4. Slate	0	1		
5. **Coal**, splint (slate floor)	1	11	2	11

Analysis of sample collected from Nos. 3 and 5, as reported by Messrs. Hite and Krak, is given in the table of coal analyses under **No. 75.**

Elisha Miller Local Mine Opening—No. 440 on Map II.

Located on the north side of Ugly Creek, 0.5 mile southeast of Estep P. O.; section by Krebs; **Alma Coal**; elevation, 1055′ B.

			Ft.	In.
Sandstone, massive			10	0
Slate			0	2
Coal, splint	1′	1″		
Slate	0	2		
Coal, splint	0	8		
Coal, gray splint (slate floor)	1	1	3	0

Barker Miller Opening—No. 441 on Map II.

Located on the south side of Ugly Creek, 0.8 mile southeast of Estep P. O.; section by Teets; **Alma Coal**; elevation, 1080′ B.

			Ft.	In.
Slate roof				
Coal, splint	0′	10″		
Dark slate	0	1		
Coal, gas	0	5		
Coal, gray splint	1	4	2	8

Butts, N. 45° W.; faces, N. 45° E.

J. E. Chapman Local Mine Opening—No. 442 on Map II.

Located on the south side of Ugly Creek at Estep P. O.; section measured by Krebs; **Alma Coal**; elevation, 1020′ L.

			Ft.	In.
Gray shale			8	0
Slate			0	2
Coal, hard splint	1′	1″		
Slate, dark	0	1		
Coal, soft	0	4		
Coal, hard splint (slate floor)	1	3	2	9

J. E. Chapman Opening—No. 443 on Map II.

Located on the north side of Ugly Creek, 0.5 mile west of Estep P. O.; section by Krebs; **Alma Coal**; elevation, 985′ B.

				Ft.	In.
1.	Sandstone roof				
2.	Slate			1	0
3.	**Coal**, cannel	0′	1″		
4.	**Coal**, splint	0	11		
5.	Slate	0	0½		
6.	**Coal**, splint, hard	0	8		
7.	**Coal**, gray splint (slate floor)	1	0	2	8½

Analysis of sample collected from Nos. 4, 6 and 7, as reported by Messrs. Hite and Krak, is given in the table of coal analysis under **No. 76.**

J. E. Chapman Local Mine Opening—No. 444 on Map II.

Located on the north side of Ugly Creek, 0.6 mile west of Estep P. O.; section by Krebs; Alma Coal; elevation, 985' B.

			Ft.	In.
Slate roof				
Coal, splint	0'	11"		
Dark slate	0	1		
Coal, hard splint (slate floor)	1	9	2	9

On Big Creek, several openings have also been made in the Alma seam by the farmers, and the coal mined for local fuel, as follows:

John Thomas Local Mine Opening—No. 445 on Map II.

Located on the north side of North Fork of Big Creek, Chapmanville District, Logan County, 1.3 miles west of Logan-Boone County Line; section measured by Krebs; Alma Coal; elevation, 810' B.

			Ft.	In.
Shale, gray			9	0
Slate			1	0
Coal, splint	0'	7¾"		
Slate, dark	0	0¼		
Coal, gas	1	1		
Coal, splint	1	0		
Slate, gray	0	5		
Coal, splint (slate floor)	1	0	4	2

J. Vickers Local Mine Opening—No. 446 on Map II.

Located on the north side of Big Creek, 3.0 miles southwest of Estep P. O.; Chapmanville District, Logan County, 1.0 mile west of the Logan-Boone County Line; section by Krebs; Alma Coal; elevation, 845' B.

			Ft.	In.
Slate roof				
Coal, gas	0'	9¾"		
Slate	0	0¼		
Coal, splint, hard	0	7		
Coal, gas	0	5		
Coal, splint	1	8		
Shale, gray	0	11		
Coal, splint, hard	0	4	4	9

Local Mine Opening—No. 447 on Map II.

Located on North Fork of Big Creek, 3.0 miles southwest of Estep P. O., and 0.8 mile west of the Logan-Boone County Line; section by Krebs; **Alma Coal**; elevation, 865' B.

			Ft.	In.
Sandstone				
Coal, gas, hard	0'	9"		
Slate, dark	0	1		
Coal, splint	2	0		
Concealed by water	1	0	3	10

Ed. Fowler Prospect Opening—No. 448 on Map II.

Located on the north side of North Fork of Big Creek, 1.2 miles southwest of Estep P. O.; section by Krebs; **Alma Coal**; elevation, 905' B.

				Ft.	In.
1.	Shale, gray			4	0
2.	**Coal, gas**	0'	3"		
3.	Slate	0	1		
4.	**Coal, gas**	0	2		
5.	Shale, with **coal** streaks	0	5½		
6.	**Coal, splint**	0	3½		
7.	Shale, with **coal** streaks	0	7		
8.	**Coal, splint**	0	10		
9.	Shale, gray	0	5		
10.	**Coal, splint (slate floor)**	1	2	4	3

The analysis of a sample collected from Nos. 2, 4, 6, 8 and 10, as reported by Messrs. Hite and Krak, is given in the table of coal analyses under **No. 76A.**

Prospect Opening—No. 449 on Map II.

Located on the south side of North Fork of Big Creek, 1.0 mile southwest of Estep P. O.; section taken by Krebs; **Alma Coal**; elevation, 920' B.

			Ft.	In.
Shale, gray			1	0
Coal, splint	0'	5"		
Slate, dark	0	1		
Coal, splint	0	7		
Slate	0	2		
Coal, splint	0	2		
Concealed by water	0	10	2	3

On Trace Fork of Big Creek, several openings have been made in the Alma seam, and the coal mined for local fuel by farmers:

Barn Hager Local Mine Opening—No. 450 on Map II.

Located on the south side of Trace Fork, 1.5 miles east of Manila P. O.; section by Krebs; **Alma Coal**; elevation, 1145' B.

				Ft.	In.
1.	Sandstone			6	0
2.	Shale			0	2
3.	**Coal**, splint	0'	6"		
4.	Slate	0	1		
5.	**Coal**, splint	2	9	3	4

Analysis of sample taken from Nos. 3 and 5, as reported by Messrs. Hite and Krak, is given in the table of coal analyses under No. 77.

Roland Bias Opening—No. 451 on Map II.

Located on the north side of Trace Fork of Big Creek, near the mouth of Hurricane Branch; section by Krebs; **Alma Coal**; elevation, 915' B.

				Ft.	In.
1.	Gray shale			6	0
2.	**Coal**, impure	0'	4"		
3.	Gray shale, full of plant fossils	3	0		
4.	**Coal**	0	1		
5.	Shale	0	1		
6.	**Coal**	0	4		
7.	Slate	0	2		
8.	**Coal**	0	1		
9.	Slate	0	2		
10.	**Coal**, splint	0	6		
11.	**Coal**, hard, block (slate floor)	1	6	6	3

Analysis of sample collected from Nos. 6, 8, 10 and 11, as reported by Messrs. Hite and Krak, is given in the table of coal analyses under No. 78.

Elbert Haner Opening—No. 452 on Map II.

Located on the head of Hurricane Branch of Trace Fork of Big Creek, 2.0 miles northeast of Manila P. O.; section by Krebs; **Alma Coal**; elevation, 940' B.

				Ft.	In.
1.	Shale, gray, full of plant fossils			6	0
2.	Cannel slate			0	4
3.	**Coal**, hard splint	1'	0"		
4.	Slate	0	1		
5.	**Coal**, gas	0	6		
6.	**Coal**, hard splint (slate floor)	1	6	3	1

Analysis of sample taken from Nos. 3, 5 and 6, as reported by **Messrs.** Hite and Krak, is given in the table of coal analyses under **No. 71A.**

U. S. Hatfield Local **Mine Opening—No. 453 on Map II.**

Located at the mouth of Trace Fork of Big Creek, 0.25 mile west of Manila P. O.; section taken by Krebs; **Alma Coal**; elevation, 950′ B

			Ft.	In.
Shale, greenish-gray			0	8
Coal, splint	0′	9″		
Slate	0	2		
Coal, gas	2	0		
Coal, splint	1	2		
Fire clay	0	8		
Coal, (slate), visible	1	0	5	9

Elsie Garrett Local **Mine Opening—No. 454 on Map II.**

Located on the south side of Big Creek, just below the mouth of Cloverlick Branch, 2.0 miles northwest of Curry P. O., Chapmanville District, Logan County, and one mile south of the Logan-Boone County Line; section measured by Krebs; **Alma Coal**; elevation, 990′ B.

			Ft.	In.
Gray sandy shale			10	0
Soft shale			0	8
Coal, splint	0′	4″		
Coal, impure	0	6		
Coal, splint	2	2		
Gray shale	0	6		
Coal, splint, visible	1	0	4	6

M. D. **Garrett** Local **Mine Opening—No. 455 on Map II.**

Located on the south side of Big Creek, 1.0 mile northwest of Curry P. O., Chapmanville District, Logan County, 0.75 mile south of the Logan-Boone County Line; section by Krebs; **Alma Coal**; elevation, 1020′ B.

			Ft.	In.
Slate roof				
Coal, impure	0′	8″		
Coal, splint	2	2		
Slate	0	2		
Coal, splint (slate floor)	1	0	4	0

On Sixmile Creek, several openings have been made in the Alma Coal for local fuel use by the farmers, as follows:

John Ballard Opening—No. 456 on Map II.

Located on the south side of Sixmile Creek, 1.5 miles southwest of Havana P. O.; section by Krebs; **Alma Coal**; elevation, 1230′ B.

			Ft.	In.
Shale, greenish-gray			3	0
Slate roof			0	2
Coal, hard splint	1′	0″		
Shale, gray	0	3		
Coal, splint (slate floor)	1	2	2	5

Yawkey and Freeman Opening—No. 457 on Map II.

Located on the south side of Hewett Creek, 1.0 mile southwest of Jeffery P. O.; section by Krebs; **Alma Coal**; elevation, 1010′ B.

			Ft.	In.
Greenish gray shale			5	0
Dark slate roof			1	0
Coal, splint	0′	10″		
Slate	0	1		
Coal, splint (dark slate floor)	1	10	2	9

James Allen Local Mine Opening—No. 458 on Map II.

Located in the first deep branch from the south into Sixmile Creek, 0.7 mile southwest of Ramage; section by Teets; **Alma Coal**; elevation, 930′ B.

			Ft.	In.
1. Slate roof				
2. **Coal**, splint	0′	9″		
3. Slate	0	1		
4. **Coal**, gnarly, dark (slate floor)	2	3	3	1

Analysis of sample taken from Nos. 2 and 4, as reported by Messrs. Hite and Krak, is given in the table of coal analyses under No. 79.

G. T. Hill Local Mine Opening—No. 459 on Map II.

Located on the head of Sixmile Creek, 0.8 mile east of Anchor P. O., and 2.0 miles southwest of Turtletown; section by Teets; **Alma Coal**; elevation, 1200′ B.

			Ft.	In.
Slate roof				
Coal, splint	1′	2″		
Slate, gray	0	1		
Coal, splint	0	10		
Coal, gray splint (slate floor)	0	11	3	0

Prospect Opening—No. 460 on Map II.

Located on head of Turtle Creek, 1.5 miles southwest of Turtletown P. O.; section by Krebs; **Alma Coal**; elevation, 1050′ B.

			Ft.	In.
Slate, gray, greenish, marine fossils			1	10
Coal, semi-splint	1′	0″		
Slate, black	0	1		
Coal, gas	0	7		
Coal, splint (slate floor)	1	4	3	0

Hewett Creek flows from the west and empties into Spruce Fork at Jeffery, 2 miles north of the Boone-Logan County Line. Considerable prospecting work has been done in the Alma seam on this branch. Also the coal is mined for local fuel at different points by the farmers.

Emma Hager Local Mine Opening—No. 461 on Map II.

Located on Right Fork of Meadow Fork of Hewett Creek, 1.5 miles southeast of Anchor P. O.; section by Krebs; **Alma Coal**; elevation, 1150′ B.

			Ft.	In.
Slate roof				
Coal, splint	0′	8″		
Slate	0	6		
Coal, splint	1	2		
Slate	0	6		
Coal, splint	1	2		
Slate	0	4		
Coal, splint (slate floor)	0	3	4	[illegible]

Yawkey and Freeman Local Mine Opening. No. 462 on Map II.

Located on the south side of Hewett Creek, about 1.0 mile south west of Jeffery P. O.; section by Krebs; **Alma Coal**; elevation, 925′ B.

			Ft.	In.
Slate roof				
Coal, splint	1′	2″		
Coal, impure	0	1		
Coal, splint (slate floor)	2	6	3	9

Edward Dent Opening—No. 463 on Map II.

Located in the head of the Left Fork of Meadow Fork of Hewett Creek, 3.2 miles west of Jeffery P. O.; section by Teets; **Alma Coal**; elevation, 1200′ B.

				Ft.	In.
1.	Sandy shale roof				
2.	**Coal**, gas	0′	11″		
3.	Slate, dark	0	1		
4.	**Coal**, splint	1	2		
5.	**Coal**, block, hard, greenish	1	3		
6.	**Coal**, gray splint (slate floor)	0	11	4	4

Analysis of sample collected from Nos. 2, 4, 5 and 6, as reported by Messrs. Hite and Krak, is given in the table of coal analyses under **No. 80.**

J. D. Dall Opening—No. 464 on Map II.

Located on the north side of the Left Fork of Meadow Fork of Hewett Creek, 4.2 miles due west of Jeffery P. O.; section by Teets; **Alma Coal**; elevation, 1100′ B.

				Ft.	In.
1.	Slate roof, plant fossils				
2.	**Coal**, block	0′	3 ″		
3.	Slate, dark	0	4		
4.	**Coal**	0	0½		
5.	Shale, gray	0	4½		
6.	**Coal**, splint	1	2		
7.	**Coal**, gray splint	1	1	3	3

Analysis of sample taken from Nos. 2, 4, 6 and 7, as reported by Messrs. Hite and Krak, is given in the table of coal analyses under **No. 81.**

Wilburn Estep Opening—No. 465 on Map II.

Located on the east side of Hewett Creek, 0.3 mile south of Hewett P. O., and just northeast of mouth of Missouri Fork; section measured by Teets; **Alma Coal**; elevation, 940′ B.

				Ft.	In.
1.	Slate roof				
2.	**Coal**, splint	1′	0″		
3.	Slate, gray	0	1		
4.	**Coal**, hard splint	3	4		
5.	Slate, gray	0	1		
6.	**Coal**, gray splint	0	4		
7.	**Coal**, hard, block (slate floor)	0	6	5	4

Butts, N. 78° W.; faces N. 12° E.

Analysis of sample collected from Nos. 2, 4, 6 and 7, as reported by **Messrs.** Hite and Krak, is given in the table of coal analyses under **No. 82.**

Calvary Bias Local Mine Opening—No. 466 on Map II.

Located on the east side of Hewett Creek, about 0.5 mile northeast of the Boone-Lincoln County Line, and 1.0 mile south of Hewett P. O.; section taken by Teets; **Alma Coal**; elevation, 965' B.

				Ft.	In.
1.	Slate roof				
2.	**Coal**, splint	1'	0"		
3.	Slate, dark	0	1		
4.	**Coal**, semi-splint	2	11		
5.	Slate, gray	0	9		
6.	**Coal**, visible	1	1	5	10

Butts, N. 79° W.; faces, N. 11° E.

Analysis of sample collected from Nos. 2, 4 and 6, as reported by **Messrs.** Hite and Krak, is given in the table of coal analyses under **No. 83.**

William Bias Local Mine Opening—No. 467 on Map II.

Located on the west side of Hewett Creek, 2.0 miles west of Clothier P. O., and about 0.2 mile north of the Boone-Logan County Line; section measured by Teets; **Alma Coal**; elevation, 980' B.

				Ft.	In.
1.	Slate roof				
2.	**Coal**, splint	1'	1 "		
3.	Slate	0	0½		
4.	**Coal**, gray splint	2	9		
5.	Slate	0	10		
6.	**Coal**, gray splint (slate floor)	1	2	5	10½

Analysis of sample collected from Nos. 2, 4 and 6, as reported by **Messrs.** Hite and Krak, is given in the table of coal analyses under **No. 84.**

O. B. Bias Opening—No. 468 on Map II.

Located on the west side of Hewett Creek, at the mouth of Morrison Fork, about 3.0 miles southwest of Hewett P. O.; section by Teets; **Alma Coal**; elevation, 950' B.

				Ft.	In.
1.	Slate roof				
2.	**Coal**, gas	0'	10"		
3.	Slate, gray	0	1		

			Ft.	In.
4. **Coal**, splint	2'	11"		
5. Gray shale	0	4		
6. **Coal**, splint (slate floor)	2	4	6	6

Analysis of sample collected from Nos. 2, 4 and 6, as reported by Messrs. Hite and Krak, is given in the table of coal analyses under **No. 85.**

J. H. Ellis Prospect Opening—No. 469 on Map II.

Located on the south side of Missouri Fork of Hewett Creek, 1.1 miles southwest of Hewett P. O.; section by Teets; **Alma Coal**; elevation, 1000' B.

			Ft.	In.
Slate roof				
Coal, impure	0'	11"		
Slate, gray	0	1		
Coal, splint (slate floor)	2	5	3	5

Augustus Ball Local Mine Opening—No. 470 on Map II.

Located on the north side of Missouri Fork of Hewett Creek, 1.2 miles southwest of Hewett P. O.; section by Teets; **Alma Coal**; elevation, 1000' B.

			Ft.	In.
Slate roof				
Coal, gas	0'	10"		
Slate	0	1		
Coal, hard, block	2	3		
Slate, gray	0	10		
Coal, splint	1	6	5	6

Ab. Adkins Local Mine Opening—No. 471 on Map II.

Located on the north side of Missouri Fork of Hewett Creek, about 1.3 miles southwest of Hewett P. O., and about 700 feet west of Augustus Ball's opening; section by Teets; **Alma Coal**; elevation, 1010' B.

			Ft.	In.
Slate roof				
Coal, gas	0'	10"		
Slate, gray	0	1		
Coal, block	2	4		
Slate	0	9		
Coal, splint (slate floor)	1	6	5	6

Elmer Ellishire Local Mine Opening—No. 472 on Map II.

Located on the south side of Missouri Fork of Hewett Creek, 1.5 miles southwest of Hewett P. O.; section by Teets; **Alma Coal**; elevation, 1030' B.

			Ft.	In.
Slate roof				
Coal, gas	0'	11"		
Slate, dark	0	1		
Coal, block	2	2		
Slate	0	10		
Coal, reported	1	6	5	6

The lower member, 1' 6" of coal, is not mined at this point, but from the amount of exposure seen and "reported," it is more than probable that it is present.

Joe Ellis Opening—No. 473 on Map II.

Located on the south side of Missouri Fork of Hewett Creek, 2.0 miles southwest of Hewett P. O.; section measured by Teets; **Alma Coal**; elevation, 1060' B.

			Ft.	In.
1. Slate floor				
2. **Coal**, splint	0'	7"		
3. Slate	0	1		
4. **Coal**, splint (slate floor)	2	6	3	2

Analysis of a sample collected from Nos. 2 and 4, as reported by Messrs. Hite and Krak, is given in the table of coal analyses under **No. 86.**

Charles Spencer Opening—No. 474 on Map II.

Located on the south side of Meadow Fork of Hewett Creek, 0.5 mile west of Hewett P. O.; section taken by Teets; **Alma Coal**; elevation, 945' B.

			Ft.	In.
1. Slate roof				
2. **Coal**, gas	0'	11"		
3. **Coal**, splint	3	1		
4. Slate	0	10		
5. **Coal**, splint, "reported"	1	6	6	4

Analysis of sample collected from Nos. 2 and 3, as reported by Messrs. Hite and Krak, is given in the table of coal analyses under **No. 87.**

Quinn Jeffery Heirs Local Mine Opening. No. 475 on Map II.

Located on the south side of Hewett Creek at the mouth of Obe Branch, 0.2 mile west of Jeffery P. O.; section by Teets; **Alma Coal**; elevation, 845' B.

				Ft.	In.
1.	Slate roof				
2.	**Coal**, gas	1'	0"		
3.	**Coal**, splint	1	6		
4.	Slate, gray	0	0½		
5.	**Coal**, splint, visible	0	9	3	3½

Analysis of sample collected from Nos. 2, 3 and 5, as reported by Messrs. Hite and Krak, is given in the table of coal analyses under **No. 88.**

Henry Keadle Prospect Opening—No. 476 on Map II.

Located in the head of Obe Branch of Hewett Creek, 2.0 miles northwest of Jeffery P. O.; section by Teets; **Alma Coal**; elevation, 1040' B.

			Ft.	In.
Slate roof				
Coal, cannel	0'	6"		
Coal, impure (slate floor)	3	5	3	11

B. W. Bias Local Mine Opening—No. 477 on Map II.

Located on Obe Branch of Hewett Creek, 1.2 miles due west of Jeffery P. O.; section by Teets; **Alma Coal**; elevation, 930' B.

				Ft.	In.
1.	Slate roof				
2.	**Coal**, gas	0'	11½"		
3.	Slate, gray	0	1		
4.	**Coal**, splint	2	11		
5.	Slate, gray	0	7		
6.	**Coal**, gray splint (slate floor)	0	9	5	3½

The analysis of a sample collected from Nos. 2, 4 and 6, as reported by Messrs. Hite and Krak, is given in the table of coal analyses under **No. 88A.**

Napoleon Sutphin Local Mine Opening—No. 478 on Map II.

Located on the head of Hunters Branch, 1.5 miles northeast of Greenview P. O.; section by Krebs; **Alma Coal**; elevation, 955′ B.

			Ft.	In.
Shale, gray			0	10
Coal, gas	1′	2″		
Slate	0	2		
Coal, gray splint	1	3		
Coal, cannel (slate floor)	0	6	3	1

Park Ball Local Mine Opening—No. 479 on Map II.

Located on the east side of Spruce Fork, 0.5 mile northeast of Greenview P. O.; section by Krebs; **Alma Coal**; elevation, 940′ B.

			Ft.	In.
1. Gray shale				
2. **Coal**, gas	0′	9″		
3. **Coal**, impure	0	3		
4. **Coal**, splint (slate floor)	3	5	4	5

Butts, N. 42° W.; faces, N. 48° E.

Analysis of sample collected from Nos. 2 and 4, as reported by Messrs. Hite and Krak, is given in the table of coal analyses under **No. 89.**

White and Hopkins Local Mine Opening—No. 480 on Map II.

Located on the north side of Hunters Branch, 1.4 miles north of Greenview P. O.; section by Krebs; **Alma Coal**; elevation, 970′ B.

			Ft.	In.
Dark shale			6	0
Coal, hard splint	1′	3″		
Dark shale	0	2		
Coal, gas	0	1		
Coal, splint, hard	1	0		
Shale, dark	0	0½		
Coal, splint	1	1		
Dark slate	0	1		
Coal, splint (slate floor)	0	10	4	6½

J. H. Hatfield Local Mine Opening—No. 481 on Map II.

Located on the east side of Spruce Fork, 1.0 mile north of Greenview P. O.; section by Krebs; **Alma Coal**; elevation, 1065' B.

			Ft.	In.
Shale, gray				
Coal, splint	0'	11"		
Shale, gray	0	2		
Coal, splint	2	8		
Shale, gray	0	6		
Coal, splint (slate floor)	0	8	4	11

Park Ball Prospect Opening—No. 482 on Map II.

Located on the east side of Spruce Fork at Greenview P. O.; section by Krebs; **Alma Coal**; elevation, 925' B.

			Ft.	In.
1. Slate roof				
2. **Coal**, splint	0'	10"		
3. Slate	0	4		
4. **Coal**, splint	3	10	5	0

Analysis of sample collected from Nos. 2 and 4, as reported by **Messrs.** Hite and Krak, is given in the table of coal analyses under **No. 90.**

A sample collected by Clark and Krebs from Nos. 2 and 4, and analyzed by Mr. Paul Demler, Charleston, W. Va., gave the following results:

	Per cent.
Moisture	1.76
Volatile Matter	40.10
Fixed Carbon	54.09
Ash	4.05
Total	100.00
Sulphur	1.54
Phosphorus	0.006

M. Bennett Local Mine Opening—No. 483 on Map II.

Located on the east side of Spruce Fork, 0.25 mile southeast of Greenview P. O.; section taken by Krebs; **Alma Coal**; elevation, 900' B.

			Ft.	In.
1. Slate roof				
2. **Coal**, splint	1'	2"		
3. Slate	0	2		
4. **Coal**, splint (slate floor)	3	8	5	0

The analysis of a sample collected from Nos. 2 and 4 of above section, as reported by Messrs. Hite and Krak, is given in the table of coal analyses under **No. 90A.**

Spruce River Coal Company Mine—No. 484 on Map II.

Located on the east side of Spruce Fork, 0.5 mile southeast of Ramage; section by Krebs; **Alma Coal**; elevation, 900′ B.

				Ft.	In.
1.	Slate roof				
2.	**Coal**, splint	1′	0″		
3.	Slate	0	1		
4.	**Coal**, splint (slate floor)	3	4	4	5

Coal mined for steam and domestic use; number of men employed, 40; capacity, 300 tons per day; President, E. J. Corbett, Detroit, Michigan; General Manager, F. F. Tagart, Massillon, Ohio; Superintendent, J. B. Ramage, Ramage, W. Va.

Analysis of sample collected from Nos. 2 and 4, as reported by Messrs. Hite and Krak, is given in the table of coal analyses under **No. 91.**

Coal River Mining Company Mine—No. 485 on Map II.

Located at Ottawa; section measured by Krebs; **Alma Coal**; elevation, 800′ B.

				Ft.	In.
1.	Shale, gray, roof				
2.	**Coal**, hard splint	0′	9″		
3.	**Coal**, softer	0	7		
4.	**Coal**, splint	2	6		
5.	Bone	0	2		
6.	**Coal**, hard splint, (slate floor)	0	7	4	7

Coal mined for steam, domestic and by-product purposes, and shipped east and west; capacity of mine, 500 tons daily; number of men employed, 80; President, Thomas Nichol, Glen Jean, W. Va.; Superintendent, J. M. Moore, Ottawa, W. Va.

Analysis of sample collected from Nos. 2, 3, 4 and 6, as reported by Messrs. Hite and Krak, is given in the table of coal analyses under **No. 91A.**

Yawkey and Freeman Opening—No. 486 on Map II.

Located on Bias Branch, 1.0 mile northeast of Jeffery P. O.; section taken by Teets; Alma Coal; elevation, 870' B.

				Ft.	In.
1.	Slate roof				
2.	Coal, splint	0'	9"		
3.	Slate, gray	0	2		
4.	Coal, splint	1	5		
5.	Slate, gray	0	1		
6.	Coal, splint, (slate floor)	0	11	3	4

Analysis of sample collected from Nos. 2, 4 and 6, as reported by Messrs. Hite and Krak, is given in the table of coal analyses under **No. 92.**

Passing to the south of Ottawa, the **Alma Coal** dips very rapidly to the southeast into the Coalburg Syncline near Clothier, then gradually rises to the southeast in going up Spruce Laurel Fork, but does not come above the bed of the creek any more in Washington District.

Several core holes drilled at Clothier and on Spruce Laurel Fork give the sections and depths of the coal at these points, as follows:

Core Drill Hole No. 2 (109)—No. 487 on Map II.

Located 0.3 mile northwest of Clothier; Alma Coal, 143' 4" deep.

			Ft.	In.
Sandstone, with shale streaks, roof				
Coal	1'	10"		
Fire clay	0	3		
Bone	0	1		
Coal, (fire clay floor)	1	4	3	6

Core Drill Hole No. 4 (110)—No. 488 on Map II.

Located 0.2 mile northwest of Clothier; Alma Coal, 221' 5" deep.

			Ft.	In
Shale			0	7
Coal	3'	4 "		
Fire clay	0	2½		
Coal and bone	0	1½		
Coal, (fire clay floor)	1	11	5	7

A shaft for mining the Alma seam has recently been sunk by James Clark et al., on the west side of Spruce Fork, 0.75 mile northwest of Clothier, where the following section was measured by Teets:

James Clark et al. Shaft—No. 489 on Map II.

	Thickness Ft.	Thickness In.	Total Ft.	Total In.
Surface	8	0	8	0
Shale, blue	9	6	17	6
Coal, Cedar Grove	2	1	19	7
Sandstone, very hard	7	5	27	0
Shale, dark	0	8	27	8
Sandstone, bluish-gray, very hard and close grained	19	8	47	4
Sandstone, with slate parting	3	2	50	6
Sandstone, very hard, bluish-gray	2	0	52	6
Slate, dark	4	6	57	0
Coal, impure..0′ 1″ } Lower Coal splint....2 0 } Cedar Grove	2	1	59	1
Fire clay	3	0	62	1
Shale, sandstone streaks	26	10	88	11
Coal, gas...........3′ 10″ Shale, gray........0 1 **Coal, gas..........1 6 } Alma**	5	5	94	4
Fire clay floor, elevation, 818′ B.				

The coal is being mined by Mr. Clark and at about 100 feet from the shaft the total section has increased to a thickness of 5′ 10″.

Boone County Coal Corporation Core Test Hole No. 5B (111). No. 490 on Map II.

Alma Coal, 255′ 8″ deep.

			Ft.	In.
Sandstone roof				
Coal	1′	5″		
Shale	0	3		
Coal, (slate floor)	0	9	2	5

Spruce Laurel Core Test Hole No. 9 (112). No. 491 on Map II.

Located on Jerry Fork of Spruce Laurel Fork on the D. Bias farm, 4.0 miles S. 60° E. from Clothier; **Alma Coal, 209′ 8″ deep.**

			Ft.	In.
Sandy shale roof				
Coal	0′	1″		
Fire clay	0	5		
Shale	0	9		
Coal	0	1		
Fire clay	1	10		
Sandy shale	2	6		
Slate	0	4		
Coal and bone	0	4		
Coal, (fire clay floor)	2	9	9	1

Diamond Core Test Hole No. 1 (115)—No. 492 on Map II.

Located on Spruce Laurel Fork, 4.6 miles southeast of Clothier; **Alma Coal, 139′ 9″ deep.**

			Ft.	In.
Shale roof				
Coal	1′	3 ″		
Slate	0	3½		
Coal	0	3		
Shale	0	8½		
Coal	0	4		
Slate	3	7		
Coal, (sandy shale floor)	0	1	6	6

Diamond Core Test Hole No. 16 (116)—No. 493 on Map II.

Located on Spruce Laurel Fork, 5.0 miles southeast of Clothier; **Alma Coal, 142′ 2″ deep.**

			Ft.	In.
Sandy shale				
Coal	1′	0″		
Coal and bone, interlaminated	0	4		
Coal	0	4		
Shale	1	2		
Coal	0	4		
Sandy shale	6	7		
Coal and bone, interlaminated	0	5	10	2
Sandy shale floor				

Diamond Core Test Hole No. 4 (117)—No. 494 on Map II.

Located on Spruce Laurel Fork, 5.1 miles southeast of Clothier; **Alma Coal, 183′ 5″ deep.**

			Ft.	In.
Slate roof				
Coal	0′	1″		
Shale	0	9		
Coal	1	2		
Shale	0	3		
Coal	0	4		
Shale	0	6		
Coal, (sandy shale floor)	0	3	3	4

Diamond Core Test Hole No. 5 (118)—No. 495 on Map II.

Located on Spruce Laurel Fork, 5.2 miles S. 49° 30′ E. from Clothier; **Alma Coal, 136′ 4″ deep.**

			Ft.	In.
Sandy shale roof				
Coal	0′	6″		
Shale	0	1		
Coal	0	7		
Shale	0	3		
Coal	0	3		
Shale	0	1		
Coal	0	2		
Coal and slate, interlaminated	0	3		
Shale	0	11		
Coal	0	5		
Shale	2	0		
Coal	0	2		
Shale	1	8		
Coal and bone, interlaminated, (slate floor)	0	2	7	6

Diamond Core Test Hole No. 11 (119)—No. 496 on Map II.

Located on Bear Hollow of Spruce Laurel Fork, 5.0 miles southeast from Clothier; **Alma Coal, 227′ deep.**

			Ft.	In.
Sandy shale				
Coal	1′	0″		
Shale	0	3		
Coal	0	5		
Coal and bone, interlaminated	0	2		
Coal	0	7		
Shale	2	2		
Coal	0	3		
Shale	0	6		
Coal	0	2		
Shale	1	3		
Coal	0	2		
Shale	0	2		
Coal and bone, (shale floor)	0	3	7	4

Diamond Core Test Hole No. 6 (120)—No. 497 on Map II.

Located on Spruce Laurel Fork, 5.5 miles S. 49° E. from Clothier; **Alma Coal, 126' 8" deep.**

			Ft.	In.
Sandy shale				
Coal	0'	10"		
Shale	0	9		
Coal	0	10		
Shale	2	0		
Coal, (sandy shale floor)	0	3	4	8

Diamond Core Test Hole No. 7 (121)—No. 498 on Map II.

Located at the mouth of Skin Poplar Branch of Spruce Laurel Fork, 5.5 miles S. 42° 30' E. from Clothier; **Alma Coal, 130' 6" deep.**

			Ft.	In.
Sandy shale				
Coal	1'	1"		
Shale	0	6		
Coal	0	9		
Coal and bone, interlaminated	0	1		
Coal	0	3		
Shale	1	6		
Coal	0	5		
Shale	1	1		
Coal and bone, interlaminated	0	2		
Shale	0	11		
Coal and bone, interlaminated, (sandstone floor)	0	1	6	10

Diamond Core Test Hole No. 15 (122)—No. 499 on Map II.

Located on Skin Poplar Branch of Spruce Laurel Fork, 6.3 miles S. 40° 30' E. from Clothier; **Alma Coal, 151' 9" deep.**

			Ft.	In.
Shale				
Coal	1'	0"		
Shale	1	6		
Coal	0	4		
Shale	0	2		
Coal	1	0		
Shale	0	10		
Slate	1	8		
Coal, (shale floor)	0	4	6	10

Diamond Core Test Hole, Cassingham No. 3 (123). No. 500 on Map II.

Located on the East Fork of Jigly Branch of Skin Poplar Branch of Spruce Laurel Fork, 7.0 miles S. 44° E. from Clothier; **Alma Coal, 348′ 3″ deep.**

			Ft.	In.
Sandy shale and sandstone streaks				
Coal, cannel	0′	3″		
Coal	1	3		
Coal and bone, interlaminated	0	4		
Shale	3	4		
Coal (fire clay bottom)	0	2	5	4

Diamond Core Test Hole No. 18 (124)—No. 501 on Map II.

Located on Spruce Laurel Fork, at mouth of Trough Fork, 5.9 miles S. 37° 30′ E. from Clothier; **Alma Coal, 161′ 9″ deep.**

			Ft.	In.
Sandy shale				
Coal and bone, interlaminated	0′	4″		
Shale	0	3		
Fire clay	0	4		
Coal and bone, interlaminated	0	4		
Coal	0	7		
Coal and bone, interlaminated	0	3		
Fire clay	2	3		
Coal (fire clay bottom)	0	4	4	8

Diamond Core Test Hole No. 13 (125)—No. 502 on Map II.

Located on Trough Fork of Spruce Laurel Fork, about 0.5 mile from its mouth, and 5.8 miles S. 32° E. from Clothier; **Alma Coal, 278′ 5″ deep.**

			Ft.	In.
Sandy shale				
Coal	1′	0″		
Shale	0	6		
Coal	1	0		
Coal and bone, interlaminated	0	4		
Shale	2	7		
Coal and bone, interlaminated	0	8		
Shale	1	8		
Coal, (shale bottom)	0	2	7	11

PLATE XXXV.—Outcrop of Eagle Fossiliferous Limestone and Shale Horizon, just north of Madison.

Diamond Core Test Hole No. 12 (126)—No. 503 on Map II.

Located on Spruce Laurel Fork, 6.3 miles S. 34° E. from Clothier; **Alma Coal, 141' 4" deep.**

		Ft.	In.
Sandy shale			
Coal	0' 10"		
Shale	1 1		
Coal	0 8	2	'
Shale		2	2

Diamond Core Test Hole No. 14 (127)—No. 504 on Map II.

Located on Spruce Laurel Fork, 6.9 miles S. 31° 30' E. from Clothier; **Alma Coal, 154' 10" deep.**

		Ft.	In.
Shale			
Coal	0' 10"		
Shale	2 2		
Coal	1 0	4	0
Fire clay		1	6

Diamond Core Test Hole No. 15 (128)—No. 505 on Map II.

Located on the south side of Spruce Laurel Fork, 7.7 miles E. 39° E. from Clothier; **Alma Coal, 120' 4" deep.**

		Ft.	In.
Sandstone and shale			
Coal	0' 11"		
Shale	0 5		
Fire clay	1 2		
Coal	0 10		
Shale	0 6		
Coal	0 3		
Shale	0 3		
Coal (fire clay floor)	0 2	4	6

Diamond Core Test Hole, Cassingham No. 2 (130). No. 506 on Map II.

Located on Spruce Laurel Fork, about one-third mile below the mouth of Dennison Fork, 9.0 miles S. 33° E. from Clothier; **Alma Coal, 188' 8" deep.**

		Ft.	In.
Sandstone			
Coal	1' 5"		
Coal and bone, interlaminated	0 2		
Dark shale	3 4		
Coal and bone, interlaminated	0 3	5	2

Diamond Core Test Hole, Cassingham No. 1 (131). No. 507 on Map II.

Located on Spruce Laurel Fork, 10.0 miles S. 39° E. from Clothier; **Alma Coal, 352′ 8″ deep.**

			Ft.	In.
Sandy shale				
Coal	3′	1″		
Coal, interlaminated	0	3	3	4
Sandy shale				

In the sections given in the core test holes south of Whiteoak Creek, it will be noted that the Alma **Coal** contains considerable impurities and is therefore of very little commercial or economic value as it will be difficult to separate the slate and other impurities from the coal in mining same, so as to make a marketable fuel.

Sherman District.

The Alma Coal in Sherman District is usually thin and of little economic importance, so far as the coal has been prospected or the sections exposed, except along Coal River, south of Orange P. O., where the coal is of sufficient thickness and purity to be of commercial value, as follows:

George Evan Local Mine Opening—No. 508 on Map II.

Located on south side of Short Creek, 2.5 miles northeast from Racine P. O.; section by Krebs; **Alma Coal**; elevation, 896′ B.

			Ft.	In.
Slate roof				
Coal, splint	1′	8″		
Slate	0	2		
Coal, cannel (slate floor)	0	10	2	8

George Evan Local Mine Opening—No. 509 on Map II.

Located on north side of Short Creek, 2.0 miles northeast of Racine P. O.; section by Krebs; **Alma Coal**; elevation, 890′ B.

			Ft.	In.
Sandstone, massive			40	0
Coal, gas	0′	5 ″		
Slate	0	0½		

			Ft.	In.
Coal, splint	0'	3"		
Slate, dark	0	2		
Coal, splint	0	7		
Concealed by water	0	11	2	4½

Southward about 4 miles, near Sterling on Indian Creek, the following prospect opening was made by the Hickory Ash Coal Company, on its lease from the Peytona Land Company:

Hickory Ash Coal Company Prospect Opening. No. 510 on Map II.

Section by Krebs; Alma Coal; elevation, 945' B.

			Ft.	In.
Slate roof				
Coal, impure	0'	1"		
Shale, gray	0	3		
Coal, cannel (fire clay and shale floor)	2	0	2	4

On up Coal River above Racine P. O., the Alma Coal is thin and of little importance from an economic standpoint.

The coal appears in the Coal River Branch of the C. & O. Ry. on Seng Creek, and varies from 2 to 3 feet in thickness.

On the north side of Stover Branch of Coal River, 2 miles southeast of Orange P. O., the coal is mined for local fuel on the land of LaFollette, Robson, et al., where the following section was measured by Teets:

LaFollette, Robson et al. Local Mine Opening. No. 511 on Map II.

			Ft.	In.
Sandstone, massive				
Sandy shale			1	2
Coal, splint	1'	10"		
Slate	0	3		
Coal, gas, visible, 945' B.	2	3	4	4

The Alma Coal comes out of Coal River at Orange P. O. and rising above the water, the coal is more than 65 feet above the bed of the stream at the mouth of Seng Creek.

Shonk and Garrison Prospect Opening—No. 511A on Map II.

Located on the north side of Pack Branch of Coal River, 1.5 miles northwest of the mouth of Seng Creek; section by Teets; **Alma Coal**; elevation, 935' B.

			Ft.	In.
Shale and sandstone roof				
Coal, gas	0'	6"		
Shale, gray	0	6		
Coal, block, glossy	2	4		
Shale, gray	0	4		
Coal, gas	1	2	4	10

LaFollette, Robson et al. Prospect Opening. No. 511B on Map II.

Located on the east side of Coal River, just below the mouth of Stover Branch; section by Krebs; **Alma Coal**; elevation, 755' B.

			Ft.	In.
Sandy shale roof				
Coal, gas	0'	4"		
Slate	0	1		
Coal, gas	0	4		
Slate	0	2		
Coal, gas	1	3		
Slate	0	7		
Coal, splint	0	10		
Slate	0	1		
Shale, gray	0	5		
Coal, gas, (slate floor)	0	3	4	4

LaFollette, Robson et al. Local Mine Opening. No. 511C on Map II.

Located near the head of Mill Branch of Coal River, 2.5 miles northwest from the mouth of Seng Creek; section by Teets; **Alma Coal**; elevation, 855' B.

			Ft.	In.
Shale roof				
Coal	0'	4"		
Slate and bone, interlaminated	0	6		
Coal, gas	2	0		
Slate, gray	0	4		
Coal, splint, visible	1	2	4	4

The coal in this opening was sampled by E. V. d'Invilliers and analyzed by A. S. McCreath & Son, of Harrisburg, Pa., with the following results:

	Per cent.
Moisture	0.724
Volatile Matter	39.526
Fixed Carbon	51.533
Sulphur	1.757
Ash	6.460
Total	100.000

Pond Creek Coal and Land Company Opening. No. 511D on Map II.

Located on the west side of Elk Run of Coal River, 1.9 miles southwest of Jarrolds Valley; section by Teets; **Alma Coal**; elevation, 1125' B.

				Ft.	In.
1.	Shale roof				
2.	**Coal**	0'	1"		
3.	Sandstone	0	1		
4.	**Coal, block**	0	5		
5.	Slate	0	2		
6.	**Coal, block, (slate floor)**	2	5	3	2

Analysis of sample collected from Nos. 2, 4 and 6, as reported by Messrs. Hite and Krak, is given in the table of coal analyses under **No. 92A.**

Crook District.

The Alma **Coal** occurs from 450 to 500 feet above the bed of Coal River in the northern part of Crook District, on the crest of the Warfield Anticline south of **M**adison, and from this point dips to the southeast into the Coalburg Syncline near Van P. O., going under Pond Fork, near the mouth of Robinson Creek. The coal rises out of Pond Fork, east from the Coalburg Syncline, near the mouth of Beaver Pond Branch, about 3 miles northwest of Pond P. O.

Some sections will now be given showing its thickness, structure and elevation, in Crook District:

R. L. Curry Local Mine Opening—No. 512 on Map II.

Located on the south side of Pond Fork, 0.8 mile southeast of Madison P. O.; section by Teets; **Alma Coal**; elevation, 1180' B.

				Ft.	In.
1.	Slate roof				
2.	**Coal**, splint	1'	2"		
3.	Shale, gray	0	2		
4.	**Coal**, splint	0	1		
5.	**Coal**, impure	0	2		
6.	**Coal**, gas	1	8		
7.	**Coal**, gray splint (slate floor)	1	10	5	1

Analysis of sample collected from Nos. 2, 4, 6 and 7, as reported by **Messrs.** Hite and Krak, is given in the table of coal analyses under **No. 93.**

Laura Thompson Local Mine Opening—No. 513 on Map II.

Located on the south side of Pond Fork, 0.7 mile southeast of Madison P. O.; section taken by Teets; **Alma Coal**; elevation, 1180' B.

				Ft.	In.
1.	Slate roof				
2.	**Coal**, splint	1'	2"		
3.	Slate, gray	0	2		
4.	**Coal**, impure	0	2		
5.	**Coal**, splint	1	9		
6.	**Coal**, gray splint (slate floor)	1	10	5	1

Analysis of sample collected from Nos. 2, 5 and 6, as reported by **Messrs.** Hite and Krak, is given in the table of coal analyses under **No. 94.**

Milton Smoot Local Mine Opening—No. 514 on Map II.

Located on the south side of Pond Fork, 1.6 miles southeast of Madison P. O.; section by Teets; **Alma Coal**; elevation, 1180' B.

			Ft.	In.
Slate roof				
Coal, hard splint	0'	10"		
Slate, gray	0	2		
Coal, block	1	1		
Coal, impure	0	2		
Coal, block	1	10		
Coal, gray splint (slate floor)	1	2	5	3

Butts, N. 40° W.; faces, N. 50° E.

R. Price Local Mine Opening—No. 515 on Map II.

Located on the south side of Pond Fork, 2.0 miles southeast of Madison P. O.; section by Krebs; **Alma Coal**; elevation, 1180' B.

			Ft.	In.
Shale roof				
Coal, impure	0'	9"		
Slate, gray	0	2		
Coal, impure	0	3		
Coal, splint, visible	3	0	4	2

In passing up Pond Fork to the mouth of Beaver Pond Branch, where the Alma Coal rises out of that branch, several openings have been made in this seam and the coal mined for local fuel, as follows:

Grover Green Local Mine Opening—No. 516 on Map II.

Located on the west side of Pond Fork, just north of the mouth of Beaver Pond Branch, 2.5 miles northwest from Pond P. O.; section taken by Teets; **Alma Coal**; elevation, 922' L.

				Ft.	In.
1.	Slate roof				
2.	**Coal**, splint	1'	6"		
3.	Niggerhead	0	4		
4.	**Coal**, splint	1	10		
5.	Slate, gray	0	2		
6.	**Coal**	0	2		
7.	Slate, gray	0	2		
8.	**Coal**, impure, visible	1	6	5	8

Analysis of sample collected from Nos. 2, 4, 6 and 8, as reported by **M**essrs. Hite and Krak, is given in the table of coal analyses under **No. 95.**

Manley Jarrold Local Mine Opening—No. 517 on Map II.

Located on the north side of Pond Fork, just north of the mouth of Dry Branch, 1.8 miles northwest of Pond P. O.; section measured by Krebs; **Alma Coal**; elevation, 920' L.

			Ft.	In.
Shale, dark			4	0
Coal, splint	1'	4"		
Shale, dark	2	0		
Shale, sandy	6	0		
Shale, gray, plant fossils	2	0		
Coal, splint	1	6		
Coal, impure	0	4		
Coal, splint	1	10		

			Ft.	In.
Fire clay	0'	3"		
Coal, hard splint	0	2		
Shale, gray	0	3		
Coal, splint (fire clay floor)	0	1	15	9

South from the mouth of Dry Branch, the **Alma Coal** has about the same rise as the bed of Pond Fork, as far as the mouth of Old Camp Branch, and from that point the coal bed rises faster and soon appears in the hills above the bed of the stream.

The following section was measured by Krebs at the mouth of Old Camp Branch, at an exposure on the west side of Pond Fork, 1.0 mile north of Pond P. O.:

Wharton Estate Opening—No. 518A on Map II.

				Ft.	In.
Sandstone, massive					
Coal, splint, Alma "Rider"				1	0
Shale, dark gray				2	0
Sandstone, massive				10	0
Coal, splint	1'	6"	**Alma**	4	0
Shale, dark gray, 0" to	0	6			
Coal, splint, visible	2	0			
Concealed by water, 930' B.					

Butts, N. 40° W.; faces, N. 50° E.

Cassingham Core Test Hole No. 5 (156)—No. 522 on Map II.

Located on Cow Creek, just above the mouth of Big Abe Fork, 3.5 miles southwest of Pond P. O.; **Alma Coal, 188' 11" deep.**

	Ft.	In.
Sandstone		
Shale	1	7
Coal, (slate floor)	2	7

Wharton Estate Opening—No. 523 on Map II.

Located on the north side of Grapevine Branch of Pond Fork, 0.25 mile from its mouth and 3.5 miles northwest from Bald Knob P. O.; section by Krebs; **Alma Coal**; elevation, 1188' B.

	Ft.	In.
Sandstone, massive		
Coal, splint, visible, (slate floor)	1	2

Three diamond core test holes were drilled along Casey Creek on the lands of Chilton, MacCorkle, Chilton and Meany.

and the Alma Coal was encountered in them as given by depth and thickness in the following sections:

Diamond Core Test Hole No. 2 (137)—No. 524 on Map II.

Located on Casey Creek, 1.3 miles from its mouth; Alma Coal, 178′ 1½″ deep.

			Ft.	In.
Shale				
Coal and bone, interlaminated	2′	2½″		
Shale	0	4		
Coal	2	0		
Shale	0	5		
Coal (sandstone floor)	2	1	[illegible]	0½

John Q. Dickinson Opening—No. 524A on Map II.

Located on Rocklick Branch of Pond Fork, about 3 miles southeast from Bald Knob P. O., and 1 mile from mouth of Rocklick; section by Krebs; Alma Coal; elevation, 1325′ B.

	Ft.	In.
1. Sandstone roof		
2. Shale, gray	0	2
3. Coal, hard splint, (to slate floor)	2	6

The analysis of a sample collected from No. 3 of above section, as reported by Messrs. Hite and Krak, is published in the table of coal analyses under No. 95B.

Diamond Core Test Hole No. 1 (138)—No. 525 on Map II.

Located on Casey Creek, 1.0 mile south of Core Test Hole No. 2 (137); Alma Coal, 211′ 1″ deep.

			Ft.	In.
Shale				
Coal and bone, interlaminated	0′	8½″		
Slate	0	1½		
Coal and bone	0	9		
Shale	0	2		
Coal and bone, interlaminated	0	5½		
Coal, good	1	10		
Shale	0	5		
Coal	0	2½		
Shale	0	1½		
Coal	1	2½		
Shale	0	0½		
Coal (fire clay floor)	0	4½	6	5

Diamond Core Test Hole No. 3 (139)—No. 526 on Map II.

Located on Casey Creek, 0.6 mile south of Core Test Hole No. 2 (138), and 4.8 miles N. 59° W. from Bald Knob P. O.; **Alma Coal, 265′ 7″ deep.**

			Ft.	In.
Slate				
Coal	0′	6 ″		
Slate	0	0½		
Coal	3	9		
Shale	2	3		
Coal	0	1		
Shale	0	9		
Coal (fire clay floor)	1	9½	9	2

In passing to the southeast from Pond P. O. to the head of Pond Fork, the **Alma Coal** appears to be split up by shale and slate and does not carry its full thickness that it developed in the northern part of Crook District. This bed of coal rises out of West Fork about 0.5 mile north of the mouth of Jarrolds Branch, 1.75 miles north of Chap P. O. Here the following section was measured on the creek bank by Teets:

Wharton Estate Opening—No. 527 on Map II.

			Ft.	In.
Slate roof				
Coal	0′	9″		
Slate	0	1		
Coal, splint	0	9		
Slate	0	2		
Coal, splint, 1078′ L.	1	3	3	0

This coal rises above the bed of West Fork and remains above the surface until two miles south of Chap P. O.

The following section was measured in the creek bank by Krebs at the mouth of Little Ugly Branch, 3.0 miles northeast from Bald Knob P. O.:

Wharton Estate Opening—No. 528 on Map II.

			Ft.	In.
Gray slate				
Coal, gas	0′	7″		
Slate	0	1		
Coal, gas (slate floor) 1200′ B.	1	4	2	0

Butts, N. 50° W.; faces, N. 40° E.

E. J. Berwind Diamond Core Test Hole No. 4 (133). No. 529 on Map II.

			Ft.	In.
Shale roof				
Coal	0′	11″		
Shale	8	4		
Coal	0	11	10	2

The top of the Alma **Coal** was encountered at a depth of 28′ 3″ from the surface.

Grant Perry **Opening—No. 529A on Map II.**

Located on the west side of James Creek of West Fork, 1.2 miles southeast from Chap P. O.; section by Teets; **Alma Coal**; elevation, 1245′ B.

				Ft.	In.
1.	Sandstone roof				
2.	**Coal**	0′	9″		
3.	Sandstone and shale	6	0		
4.	Shale	1	0		
5.	**Coal**, gas	1	1		
6.	**Coal**, gray splint	1	9		
7.	**Coal**, gas (to slate floor)	0	3	10	10

The analysis of a sample collected from Nos. 5, 6 and 7, as reported by **Messrs.** Hite and Krak, is published in the table of coal analyses under **No. 95A.**

The character, composition and calorific value of the Alma **Coal** at the commercial mines and the local openings will be discussed in a subsequent chapter on the coal resources of Boone County. An estimate will also be given on the probable available area of this coal by magisterial districts.

THE LOGAN SANDSTONE.

Underneath the Alma Coal from 0 to 5 feet, there occurs a massive, bluish-gray, medium grained, micaceous sandstone from 20 to 40 feet thick, which has been designated the *Logan **Sandstone** by Hennen and Reger[17], from a town of that name in Logan County, where it is quarried for building purposes. In Boone County, this sandstone usually forms abrupt bluffs

[17]Logan and Mingo Report; 1914.

*—Now changed to **Monitor** because the former was preoccupied.—I. C. W.

and massive cliffs. It is hard and does not erode easily. It furnishes a good building stone and has been quarried at several points in the county for building bridge piers, etc.

This sandstone was quarried near the mouth of Rock Creek for the abutments of the Coal River Railroad bridge across Coal River at that point. Here it forms a cliff about 30 feet high.

The thickness and occurrence of this sandstone are given in the sections taken at different points in the county, and published on preceding pages of this Report.

THE LITTLE ALMA COAL.

From 20 to 50 feet underneath the Alma Coal, there often occurs a multiple bedded, impure coal, varying in thickness from 1 to 3 feet, which has been named the **Little Alma Coal**[18].

In Boone County, this coal does not attain sufficient thickness and purity to make it a coal of commercial value, being usually thin and impure.

Sections of this coal are shown in the general sections taken at different points and given on preceding pages of this Report.

THE MALDEN SANDSTONE.

Underneath the Little Alma Coal, there occurs a massive sandstone that has been named the **Malden Sandstone** by I. C. White from the village of that name in Kanawha County. This sandstone varies in thickness from 5 to 50 feet, is grayish-blue, micaceous, and often contains layers of slate bands.

THE CAMPBELL CREEK LIMESTONE.

Underlying the Malden Sandstone is a dark, grayish, earthy, impure limestone, which has been named the **Campbell Creek Limestone** from a creek of that name in Kanawha County, where the bed was first studied by Dr. I. C. White It frequently takes the form of turtle-shaped concretions, and often exhibits the "cone-in-cone" structure, and is also occasionally fossiliferous, as near Kayford.

---

[18]Hennen and Reger, Logan and Mingo Report; 1914.

The following section was measured at Dana on Campbell Creek:

	Thickness Feet.	Total Feet.
Sandstone, Malden	10	10
Shale, gray	20	30
Limestone, bluish-gray, Campbell Creek	2	32
Gray shale	14	46
Coal, Campbell Creek (No. 2 Gas), 610′ B.	4	50

This bed is fairly persistent in Boone County, and the stratum is seldom more than 2 feet thick. It is often found in large concretions like "Turtle-backs", being from 2 to 6 feet in length, and 2 to 4 feet in width, and 1 to 2 feet thick.

The limestone comes out of Coal River near Bull Creek at Dartmont, and rises rapidly to the southeast toward the Warfield Anticline.

The following section was measured by Krebs on the head of Drawdy Creek, 1.0 mile northwest from Andrew P. O.:

		Thickness Feet.	Total Feet.
1.	Sandstone, Malden		
2.	Shale, gray	10	10
3.	Limestone, dark gray, Campbell Creek	1.5	11.5
4.	Dark gray shale	7	18.5
5.	Sandstone, gray, fine grained	5	23.5
6.	Shale, gray, plant fossils	5	28.5
7.	Coal, Campbell Creek (No. 2 Gas), 940′ B.		

A sample (No. 384-K) was collected from No. 3 of the above section and analysis by Messrs. Hite and Krak shows as follows:

	Per cent.
Silica (SiO_2)	18.69
Ferric Iron (Fe_2O_3)	4.32
Alumina (Al_2O_3)	6.63
Calcium Carbonate ($CaCO_3$)	64.22
Magnesium Carbonate ($MgCO_3$)	1.89
Phosphoric Acid (P_2O_5)	0.66
Loss on ignition	2.63
Total	99.04

The Campbell Creek Limestone goes under the Coal River Railroad grade, just opposite the mouth of Tony Creek, where the following section was measured:

	Thickness Feet.	Total Feet.
Shale, gray	10	10
Limestone, Campbell Creek	2	12
Shale and concealed to railroad	18	30
Concealed	5	35
Coal, Campbell Creek (No. 2 Gas), 685' B...		

On Coal River near the head of Tenmile Creek, the following section was measured by Teets along the Cabin Creek Branch of the C. & O. Railway:

	Thickness Ft.	In.	Total Ft.	In.
Concealed	100	0	100	0
Sandy shale	10	0	110	0
Dark gray, slaty limestone (Seth), marine fossils	0	6	110	6
Shale, dark	1	6	112	0
Coal, gas (**Cedar Grove**)	2	10	114	10
Shale, gray	1	2	116	0
Sandstone, massive, **Cedar Grove**	24	0	140	0
Coal, splint, visible........2' / Shale and concealed.......1 / Sandstone and sandy shale 4 / **Coal**, splint...............1 } **Alma Coal**	8	0	148	0
Slate	1	0	149	0
Sandstone, massive, **Malden**	45	0	194	0
Limestone, dark, impure, iron ore nodules, and marine fossils (Campbell Creek)	1	0	195	0
Sandy shale and concealed	91	0	286	0
Coal, Campbell Creek (No. 2 Gas), mined at Kayford at an elevation of 1335' L	4	0	290	0

Samples of the limestone were collected here for study by Dr. Wm. Armstrong Price.

THE CAMPBELL CREEK (No. 2 GAS) COAL.

The next coal of economic value in descending order is the **Campbell Creek (No. 2 Gas) Coal.** It is a multiple-bedded seam at its type locality on Campbell Creek, 6 miles above Charleston, and it often splits up and becomes more complex to the southeast, along the Kanawha River and its tributaries, where it is often separated into two or three merchantable beds of coal.

This coal bed is different in type from the hard splint coals, like the Coalburg or Winifrede, and is generally known in the market as **No. 2 Gas Coal.** The coal from this bed has long been used for gas making and by-product coal. The coal is soft in texture to the southeast, and cokes well, so that its type is entirely different from the coals above its horizon.

The following section was measured by Krebs on the land of **C. C. Dickinson** at **Malden**, 1.0 mile south of Dana, near the type locality of the coal:

			Thickness Ft.	In.	Total Ft.	In.
1.	**Limestone, Campbell Creek**		1	0	1	0
2.	Shale and sandstone		18	0	19	0
3.	**Coal, gas (Peerless)**		1	9	20	9
4.	Shale, dark, with coal streaks	0' 8"				
5.	**Coal**, semi-splint	0 5				
6.	Slate, dark, with coal streaks	0 4	4	3	25	0
7.	**Coal**, gas	1 5				
8.	Slate, black	0 2				
9.	**Coal**, splint (slate floor) 640' B.	1 3				

The foregoing represents its **type section.** No. 4 often thickens to 50 feet, and is then represented by a heavy sandstone.

This is one of the most important coals in Boone County, being near the water level and frequently under the beds of the streams, nearly the entire county carrying its horizon, very little of its bed being eroded.

Sections showing its thickness, structure and elevation will now be given by magisterial districts:

Peytona District.

The **Campbell Creek (No. 2 Gas) Coal** comes out of Coal River about 0.5 mile west of Dartmont Station, and rises rapidly above the bed of the stream to the southeast to the crest of the Warfield Anticline where it is nearly 200 feet above the river.

Joseph Dart Local **Mine Opening—No. 531 on Map II.**

Located on the north side of Bull Creek, 0.5 mile northeast of Dartmont Station; section by Krebs; **Campbell Creek (No. 2 Gas) Coal**; elevation, 685' B.

			Ft.	In.
Sandstone, massive				
Coal, splint	0'	10"		
Slate	0	4		
Coal, splint, visible	2	0	3	2

Hamilton Morris Local Mine Opening—No. 532 on Map II.

Located on the north side of Coal River, about 0.5 mile below Dartmont Station; section by Teets; **Campbell Creek (No. 2 Gas) Coal**; elevation, 645' B.

			Ft.	In.
Sandstone roof				
Coal, gas	0'	5"		
Fire clay	0	4		
Coal, splint	0	11		
Slate	0	2		
Coal, splint, visible	2	10	4	8

Joseph Dart Local Mine Opening—No. 533 on Map II.

Located on the north side of Coal River, opposite the mouth of Whiteoak Branch, 1.25 miles southeast from Dartmont Station; section by Krebs; **Campbell Creek (No. 2 Gas) Coal**; elevation, 730' B.

			Ft.	In.
Sandstone roof				
Coal, impure	0'	2"		
Coal, gas	1	0		
Slate	0	2		
Coal, splint, visible	2	6	3	10

Noah Booker Local Mine Opening—No. 534 on Map II.

Located on the south side of Coal River, 1.0 mile south of Dartmont Station; section by Teets; **Campbell Creek (No. 2 Gas) Coal**; elevation, 750' B.

			Ft.	In.
Gray shale roof				
Coal, splint	0'	7"		
Shale, gray	0	2		
Coal, splint	1	6		
Coal, impure (fire clay floor)	0	2	2	5

PLATE XXXVI.—View of sandstone about 30 feet above the Eagle Fossils in roadside just below Madison.

Broun Heirs Local Mine Opening—No. 535 on Map II.

Located on Whiteoak Branch, about 0.5 mile from its mouth, and 1.0 mile south of Dartmont Station; section by Teets; **Campbell Creek (No. 2 Gas) Coal**; elevation, 740' B.

			Ft.	In.
Sandstone roof				
Coal, gas	1'	2"		
Slate	0	0¾		
Coal (slate floor)	0	11¼	2	2

Boone and Kanawha Land and Mining Company Local Mine Opening—No. 536 on Map II.

Located on the south side of Lick Creek, 0.25 mile east of Ashford P. O.; section by Krebs; **Campbell Creek (No. 2 Gas) Coal**; elevation, 830' B.

			Ft.	In.
Slate roof				
Coal	0'	2"		
Shale, gray	0	2		
Coal, splint	0	9		
Slate, gray	0	4		
Coal, splint	1	0		
Coal, block (slate floor)	1	0	3	5

Boone and Kanawha Land and Mining Company Prospect Opening—No. 537 on Map II.

Located on the south side of Lick Creek, 0.25 mile east of Ashford P. O.; section by Krebs; **Campbell Creek (No. 2 Gas) Coal**; elevation, 840' B.

			Ft.	In.
Coal	0'	2"		
Slate	0	0½		
Coal, splint	0	8		
Slate	0	2		
Coal, splint	1	9		
Coal, block (slate floor)	1	0	3	9½

Boone and Kanawha Land and Mining Company Prospect Opening—No. 538 on Map II.

Located on south side of Lick Creek, 0.5 mile east of Ashford P. O.; section by H. A. Henderson for J. S. Cunningham; **Campbell Creek (No. 2 Gas) Coal**; elevation, 825' B.

			Ft.	In.
Slate roof				
Coal	0'	3"		
Slate	0	1		
Coal	0	11		
Slate	0	2		
Coal, splint, (slate floor)	2	9	4	2

Boone and Kanawha Land and Mining Company Prospect Opening—No. 539 on Map II.

Located on the south side of Lick Creek, 1.5 miles east from Ashford P. O.; section by H. A. Henderson for J. S. Cunningham; **Campbell Creek (No. 2 Gas) Coal**; elevation, 840' B.

			Ft.	In.
Slate roof				
Coal	0'	2½"		
Slate	0	2½		
Coal, splint	3	7	4	0

Walter Kirk Opening—No. 540 on Map II.

Located on a branch of Lick Creek, 0.8 mile northeast of Ashford P. O. and 1.3 miles southwest from Dartmont Station; section measured by Teets; **Campbell Creek (No. 2 Gas) Coal**; elevation, 815' B.

				Ft.	In.
1.	Sandstone roof				
2.	**Coal**, gas	1'	1"		
3.	Slate	0	2		
4.	**Coal**, hard splint	1	6		
5.	**Coal**, gas, (slate floor)	0	9	3	6

Butts, N. 40° W.; faces, N. 50° E.

Analysis of sample collected from Nos. 2, 4 and 5, as reported by Messrs. Hite and Krak, is given in the table of coal analyses under **No. 99**.

George M. McDermott Opening—No. 541 on Map II.

Located on a branch of Lick Creek, 1.8 miles southeast of Dartmont Station; section by Teets; **Campbell Creek (No. 2 Gas) Coal**; elevation, Lower Bench, 818' B.

				Ft.	In.
Slate roof					
Coal, splint	0'	11"	Upper Bench	3	2
Slate, gray	0	3			
Coal, hard	1	1			
Coal, gray splint	0	4			
Coal, gas	0	7			
Slate and concealed				20	0
Coal	0'	6"	Lower Bench	2	0
Slate, gray	0	0¼			
Coal, gas	0	10			
Slate, gray	0	0½			
Coal, gas	0	5			
Slate, dark gray	0	0¼			
Coal, splint, (fire clay floor)	0	2			

Hughey Snodgrass Local Mine Opening—No. 542 on Map II.

Located on a branch of Lick Creek, 2.5 miles southeast of Dartmont Station; section by Teets; **Campbell Creek (No. 2 Gas) Coal**; elevation, 840′ B.

				Ft.	In.
Sandstone roof					
Coal	0′	5″	Lower Bench	2	0
Slate	0	1			
Coal, splint, (slate floor)	1	6			

Jacob Barker Local Mine Opening—No. 543 on Map II.

Located on a tributary of Lick Creek, flowing from the south, 2.25 miles southeast from Dartmont Station; section measured by Teets; **Campbell Creek (No. 2 Gas) Coal**; elevation, 880′ B.

			Ft.	In.
Sandstone roof				
Coal, gas	1′	11″		
Slate, gray	0	2½		
Coal, block	0	3½		
Coal, gas, (slate floor)	0	8	3	1

Butts, N. 40° W.; faces, N. 50° E.

Boone and Kanawha Land and Mining Company Local Mine Opening—No. 544 on Map II.

Located on the east side of Roundbottom Creek, 1.3 miles northeast from Peytona; section by Teets; **Campbell Creek (No. 2 Gas) Coal**; elevation, 875′ B.

			Ft.	In.
Slate roof				
Coal, block	0′	9″		
Slate	0	2		
Coal, gray splint	0	5		
Coal, gas	1	0		
Bone	0	1		
Coal, gray splint, (slate floor)	1	4	3	9

Boone and Kanawha Land and Mining Company Opening. No. 545 on Map II.

Located on the east side of Roundbottom Creek, 1.25 miles north of Peytona; section by H. A. Henderson for J. S. Cunningham; **Campbell Creek (No. 2 Gas) Coal**; elevation, 870′ B.

			Ft.	In.
1. Slate roof				
2. **Coal**	0′	4″		
3. Slate	0	9		
4. **Coal**, (slate floor)	3	9	4	10

Analysis of sample collected by Mr. Henderson from Nos. 2 and 4 by the Chief Chemist of the New River and Pocahontas Consolidated Coal Company gave the following results:

	Per cent.
Moisture	2.30
Volatile Matter	33.57
Fixed Carbon	57.25
Ash	6.88
Total	100.00
Sulphur	0.98
B. T. U.	13,732

Boone and Kanawha Land and Mining Company Prospect Opening—No. 546 on Map II.

Located on Curtis Branch, about 1.0 mile southeast from Ashford P. O.; section by H. A. Henderson for J. S. Cunningham; **Campbell Creek (No. 2 Gas) Coal**; elevation, 875′ B.

			Ft.	In.
Slate roof				
Coal	0′	3″		
Slate	0	2		
Coal	1	1		
Slate	0	2		
Coal	2	1	3	9

E. G. Harless Local Mine Opening—No. 547 on Map II.

Located on Falling Rock Fork of Lick Creek, about 0.2 mile from its mouth; section by Teets; **Campbell Creek (No. 2 Gas) Coal**; elevation, 860′ B.

			Ft.	In.
Slate roof				
Coal, gas	1′	0″		
Slate, gray	0	4		
Coal, gas	0	11		
Slate, gray	0	1		
Coal, splint	0	6		
Bone	1	1		
Coal, gray splint, (slate floor)	0	1	4	0

Butts, N. 40° W.; faces, N. 50° E.

Daniel Harless Opening—No. 548 on Map II.

Located on the east side of Falling Rock Fork of Lick Creek, 3.2 miles north of Peytona; section by Teets; **Campbell Creek (No. 2 Gas) Coal**; elevation, 860′ B.

			Ft.	In.
Sandstone roof				
Coal, block	0′	9″		
Slate	0	3		
Coal, gas	1	0		
Coal, bony	0	1		
Coal, gas, splint, (slate floor)	0	11	3	0

Boone and Kanawha Land and Mining Company Prospect Opening—No. 549 on Map II.

Located on Brier Branch of Coal River, 1.5 miles north of Peytona P. O.; section by Teets; **Campbell Creek (No. 2 Gas) Coal**; elevation, 870′ B.

			Ft.	In.
Dark shale roof				
Coal, gas	0′	8″		
Slate, gray	0	3		
Coal, gas	0	2		
Coal, impure	0	2		
Coal, gas, splint, visible	1	6	2	9

Boone and Kanawha Land and Mining Company Opening. No. 550 on Map II.

Located on the south side of Brier Branch, about 0.7 mile from its mouth; section by Teets; **Campbell Creek (No. 2 Gas) Coal**; elevation, 860′ B.

			Ft.	In.
1. Slate roof				
2. **Coal**, splint	1′	8″		
3. Slate, gray	0	1		
4. **Coal**	0	1		
5. Slate, gray	0	2		
6. **Coal**, (slate floor)	0	7	2	[illegible]

Butts, N. 50° W.; faces, N. 40° E.

Analysis of sample collected from Nos. 2, 4 and 6, as reported by Messrs. Hite and Krak, is given in the table of coal analyses under **No. 100.**

West of Coal River considerable prospect work has been done in the Campbell Creek (No. 2 Gas) Coal on Whiteoak and Brush Creeks, since this seam is located from 100 to 200 feet above water level, much of the district is underlain with this coal, and cropping near the base of the hills, the coal can be easily and cheaply mined.

E. A. Holstein Local Mine Opening—No. 551 on Map II.

Located on the south side of Whiteoak Branch, 1.5 miles southwest from Dartmont Station; section by Teets; **Campbell Creek (No. 2 Gas) Coal**; elevation, 750′ B.

				Ft.	In.
1.	Slate roof				
2.	**Coal**, block	0′	8″		
3.	**Coal**, hard splint	1	10		
4.	**Coal**, impure, (fire clay floor)	0	5	2	11

Butts, N. 41° W.; faces, N. 49° E.

Analysis of sample collected from Nos. 2 and 3, as reported by Messrs. Hite and Krak, is given in the table of coal analyses under **No. 101.**

Crawford and Ashby Local Mine Opening. No. 552 on Map II.

Located about 1,000 feet northeast from Opening No. 551 on Whiteoak Branch; section by Teets; **Campbell Creek (No. 2 Gas) Coal**; elevation, 735′ B.

			Ft.	In.
Slate roof				
Coal, block	0′	9″		
Slate	0	3		
Coal, splint	1	8		
Coal, bony	0	6	3	2

James Stewart Opening—No. 553 on Map II.

Located on the west side of Coal River, about 0.3 mile southwest from Ashford Station; section by Teets; **Campbell Creek (No. 2 Gas) Coal**; elevation, 810′ B.

			Ft.	In.
Sandstone roof				
Coal, gas	1′	2 ″		
Slate, gray	0	2½		
Coal, gas	1	5½		
Coal, impure, (fire clay floor)	0	5	3	3

Brush Creek flows into Coal River from the southwest at Brushton. The Coal River Mining Company has made several openings in the Campbell Creek (No. 2 Gas) Coal recently, along the waters of Brush Creek, and has placed the results at the disposal of the Survey. A portion of these sections had already been measured in the field work in that portion of the district.

Coal River Mining Company Opening—No. 554 on Map II.

Located on the south side of Brush Creek, 0.8 mile southwest of Brushton Station; section by Teets; **Campbell Creek (No. 2 Gas) Coal**; elevation, 810′ B.

			Ft.	In.
Slate roof				
Coal, impure	0′	7″		
Slate	0	1		
Coal, gas, (fire clay floor)	1	5½	2	1½

Coal River Mining Company Prospect Opening. No. 555 on Map II.

Located in Loggy Hollow of Honeycamp Fork of Brush Creek, 3.0 miles southwest of Brushton Station; section by Teets; **Campbell Creek (No. 2 Gas) Coal; Lower Bench**; elevation, 865′ B.

			Ft.	In.
1. Sandstone roof				
2. **Coal**, gas	0′	8″		
3. Slate	0	0½		
4. **Coal**, (slate floor)	0	8½	1	5

The analysis of a sample collected by Krebs from Nos. 2 and 4 of above section, as reported by Messrs. Hite and Krak, is given in the table of coal analyses under **No. 101A.**

E. M. Kinder Local Mine Opening—No. 556 on Map II.

Located on the east side of Honeycamp Fork of Brush Creek, 3.0 miles southwest from Brushton Station; section by Teets; **Campbell Creek (No. 2 Gas) Coal**; elevation, 875′ B.

			Ft.	In.
1. Shale roof				
2. **Coal**, gray splint	0′	6″		
3. **Coal**, hard, block, (slate floor)	1	8	2	2

Analysis of sample collected from Nos. 2 and 3, as reported by Messrs. Hite and Krak, is given in the table of coal analyses under **No. 102.**

Coal River Mining Company Local Mine Opening. No. 557 on Map II.

Located on the east side of Brush Creek, 2.0 miles south of Cabell P. O.; section measured by Teets; **Campbell Creek (No. 2 Gas) Coal**; elevation, 860′ B.

				Ft.	In.
1.	Slate roof				
2.	**Coal**, hard block	0′	9″		
3.	Slate, gray	0	3		
4.	**Coal**, hard block, (slate floor)	2	7	3	7

Butts, N. 35° W.; faces, N. 55° E.

The analysis of a sample collected from Nos. 2 and 4 of above section, as reported by Messrs. Hite and Krak, is given in the table of coal analyses under **No. 102A.**

George Estep Opening—No. 558 on Map II.

Located on the west side of Brush Creek, 1.0 mile south of Cabell P. O.; section taken by Teets; **Campbell Creek (No. 2 Gas) Coal**; elevation, 890′ B.

				Ft.	In.
1.	Slate roof				
2.	**Coal**, splint	0′	8 ″		
3.	Slate, gray	0	0½		
4.	**Coal**, splint	0	2		
5.	Slate, gray	0	3		
6.	**Coal**, hard splint, visible	2	4	3	5½

Analysis of sample collected from Nos. 2, 4 and 6, as reported by Messrs. Hite and Krak, is given in the table of coal analyses under **No. 103.**

Charles Walker Local Mine Opening—No. 559 on Map II.

Located on the west side of Brush Creek, 1.5 miles south of Cabell P. O.; section taken by Teets; **Campbell Creek (No. 2 Gas) Coal**; elevation, 935′ B.

				Ft.	In.
1.	Slate roof				
2.	**Coal**, block	0′	9 ″		
3.	Slate, gray	0	0½		
4.	**Coal**, splint	0	1		
5.	**Coal**, gray splint	0	4		
6.	**Coal**, hard splint, (slate floor)	0	7	1	9½

Analysis of sample collected from Nos. 2, 4 and 6 by G. R. Krebs and analysis made for Clark and Krebs by Mr. Paul Demler, Charleston, W. Va., gave the following results:

	Per cent.
Moisture	1.05
Volatile Matter	34.80
Fixed Carbon	62.87
Ash	1.28
Total	100.00
Sulphur	0.69
Phosphorus	0.001
B. T. U.	15,240

The above analysis shows unusually good results for this seam, the coal being low in moisture, ash and sulphur, and high in fixed carbon and B. T. U., and it is unfortunate that the bed is not of present minable thickness.

Turley Opening—No. 560 on Map II.

Located on the east side of Brush Creek, 0.5 mile northeast from Cabell P. O.; section by Krebs; **Campbell Creek (No. 2 Gas) Coal**; elevation, 920' B.

				Ft.	In.
1.	Shale roof				
2.	**Coal**, hard, block	1'	10 "		
3.	Slate	0	0½		
4.	**Coal**, splint	0	4		
5.	Slate, gray	2	4		
6.	**Coal**, block	0	4		
7.	Niggerhead	0	5		
8.	**Coal**, gas	0	10		
9.	**Coal**, splint	0	1		
10.	Slate, gray	0	3		
11.	**Coal**	0	1		
12.	Slate	0	2		
13.	**Coal**, gas, (slate floor)	0	4	6	0½

Butts, N. 50° W.; faces, N. 40° E.

Analyses of samples collected from Nos. 2, 4, 6, 8, 9, 11 and 13, as reported by Messrs. Hite and Krak, are given in the table of coal analyses under **Nos. 104 and 104A.**

The analysis of a sample collected by Krebs from No. 7 of above section, as reported by Messrs. Hite and Krak, is given in the table of coal analyses under **No. 104B.**

A sample collected by G. R. Krebs, and analysis made for Clark and Krebs, by Mr. Paul Demler, Charleston, W. Va., gave the following results:

	Per cent.
Moisture	0.90
Volatile Matter	35.30
Fixed Carbon	59.62
Ash	4.18
Total	100.00
Sulphur	0.73
Phosphorus	0.002
B. T. U.	14,780

The above analysis indicates that the coal is excellent for domestic fuel and steam purposes, being low in moisture, sulphur and ash, and high in fixed carbon and B. T. U.

Curtis Blake Opening—No. 561 on Map II.

Located on the head of Drawdy Creek, 1.0 mile west of Andrew P. O.; section measured by Krebs; **Campbell Creek (No. 2 Gas) Coal**; elevation, 945' B.

				Ft.	In.
1.	Sandstone roof				
2.	**Coal**, block	0'	8"		
3.	Shale, gray	0	4		
4.	**Coal**, splint	0	11		
5.	**Coal**, gas, (slate floor)	1	1	3	0

Analysis of sample collected from Nos. 2, 4 and 5, as reported by Messrs. Hite and Krak, is given in the table of coal analyses under **No. 105.**

Irvin Green Opening—No. 562 on Map II.

Located on the north side of Drawdy Creek, at the mouth of Tony Branch, 0.75 mile northwest of Andrew P. O.; section measured by Krebs; **Campbell Creek (No. 2 Gas) Coal**; elevation, Lower Bench, 935' B.

					Ft.	In.
1.	Gray shale, full of plant fossils					
2.	**Coal**, gray splint	0'	9"	Upper Bench	4	4
3.	Shale, gray	0	3			
4.	**Coal**, hard splint	2	0			
5.	**Coal**, impure	0	3			
6.	**Coal**, gas	1	1			
7.	Shale, gray				10	0
8.	**Coal**, gas, (gray shale floor), **Lower Bench**				2	0

Analysis of sample collected from Nos. 2, 4 and 6, as reported by Messrs. Hite and Krak, is given in the table of coal analyses under **No. 106.**

J. A. Ball Local Mine Opening—No. 563 on Map II.

Located on the west side of Drawdy Creek, 0.5 mile northwest of Andrew P. O.; section by Krebs; **Campbell Creek (No. 2 Gas) Coal;** elevation, 910′ B.

			Ft.	In.
Shale, gray				
Coal, block	0′	9″		
Shale, gray	0	3		
Coal, splint, (slate floor)	3	2	4	2

Terry Dawson Local Mine Opening—No. 564 on Map II.

Located on the south side of the county road, near the head of Drawdy Creek, 1.5 miles west of Andrew P. O.; section by Teets; **Campbell Creek (No. 2 Gas) Coal;** elevation, 940′ B.

			Ft.	In.
Slate roof				
Coal, block	0′	8″		
Slate	0	4		
Coal, hard splint	2	3		
Bone	0	2		
Coal, splint, (slate floor)	0	8	4	1

Wesley Harless Opening—No. 565 on Map II.

Located on the south side of Drawdy Creek, 0.25 mile west of Andrew P. O.; section by Teets; **Campbell Creek (No. 2 Gas) Coal;** elevation, 935′ B.

			Ft.	In.
Slate roof				
Coal, splint	0′	10″		
Slate, gray	0	2		
Coal, splint, (fire clay floor)	2	2	3	2

Olsey Perdue Local Mine Opening—No. 566 on Map II.

Located on the south side of Drawdy Creek at Andrew P. O.; section measured by Teets; **Campbell Creek (No. 2 Gas) Coal;** elevation, 920′ B.

			Ft.	In
1. Slate floor				
2. **Coal,** splint	0′	8″		
3. Slate, gray	0	4		
4. **Coal,** splint, (slate floor)	2	3	3	3

Analysis of sample collected from Nos. 2 and 4, as reported by Messrs. Hite and Krak, is given in the table of coal analyses under **No. 107**.

T. L. Lewis Local Mine Opening—No. 567 on Map II.

Located on the north side of Whiteoak Creek, near its mouth, 0.7 mile southwest of Peytona Station; section by Teets; **Campbell Creek (No. 2 Gas) Coal**; elevation, 905' B.

			Ft.	In.
Slate roof				
Coal, block	0'	8"		
Slate	0	2		
Coal, splint, (slate floor)	2	3	3	1

John Kinder Local Mine Opening—No. 568 on Map II.

Located on the west side of Coal River, 1.1 miles northwest of Peytona P. O.; section by Teets; **Campbell Creek (No. 2 Gas) Coal**; elevation, 900' B.

			Ft.	In
Slate roof				
Coal, block	0'	9"		
Shale, gray	0	2		
Coal, hard splint	1	7		
Coal, impure	0	1		
Coal, gas	0	5		
Coal, impure	0	1		
Coal, gas, (slate floor)	0	5	3	6

The above section was taken near the crest of the Warfield Anticline.

Thomas Epling Local Mine Opening—No. 569 on Map II.

Located on the south side of Whiteoak Creek, about 1.0 mile above its mouth, and 1.6 miles southwest of Peytona; section measured by Teets; **Campbell Creek (No. 2 Gas) Coal**; elevation, 910' B.

			Ft.	In.
1. Sandstone, massive				
2. **Coal**, gray	0'	6 "		
3. Slate, dark	0	0½		
4. **Coal**, splint, (slate floor)	1	7	2	1½

Butts, N. 40° W.; faces, N. 50° E.

Analysis of sample collected from Nos. 2 and 4, as reported by Messrs. Hite and Krak, is given in the table of coal analyses under **No. 108**.

Frank **Epling** Local **Mine Opening—No.** 570 **on Map II.**

Located on Whiteoak Creek, 2.0 miles southwest of Peytona P. O.; section taken by Teets; **Campbell Creek (No. 2 Gas) Coal;** elevation, 910′ B.

			Ft.	In.
Sandstone roof				
Coal, block	0′	7″		
Slate, gray	0	3		
Coal, splint, (slate floor)	1	5	2	3

Daniel **Keefer** Local **Mine Opening—No.** 571 **on Map II.**

Located on the north side of Whiteoak Creek, 2.3 miles southwest of Peytona P. O.; section by Teets; **Campbell Creek (No. 2 Gas) Coal; Upper Bench;** elevation, 940′ B.

			Ft.	In.
Slate				
Coal, block	0′	9″		
Slate, gray	0	3		
Coal, hard splint, visible	1	9	2	9

Several core drill holes have been put down on Fork Creek in the northern part of Peytona District, and **Core** Drill Hole **No. 100,** located 1.2 miles west of Emmons Station, gives the following section of the **Campbell Creek (No. 2 Gas) Coal,** at a depth of 230′ from the surface:

			Ft.	In
Shale roof				
Coal	0′	4″		
Fire clay	1	8		
Coal, (sandstone floor)	1	0	3	0

Another **Core Test Hole (No. 102),** located at the mouth of Jimmy Fork of Fork Creek, gives the following section of the **Campbell Creek (No. 2 Gas) Coal,** at a depth of 192′ 9″ from the surface:

			Ft.	In.
Shale roof				
Coal	1′	10″		
Shale	0	1		
Coal	3	9	5	8

It is noted that this core test hole was drilled with a shot drill and as it is difficult to get a perfect core with this drill, no total section of the coal is reported. If a test were made with diamond drill, it might be possible that the core would show somewhat different from the one reported in the section above.

Peytona Block Coal Company Mine—No. 571A on Map II.

Located on the south side of Coal River, 1.0 mile east of Peytona P. O.; section by Teets, in main heading; **Campbell Creek (No. 2 Gas) Coal**; elevation, 820′ B.

			Ft.	In.
1. Shale, gray, roof				
2. **Coal**, block	0′	10″		
3. Shale, gray	0	2		
4. **Coal**, hard splint	1	9		
5. **Coal**, gray splint, (slate floor)	1	0	3	9

Coal mined for steam and domestic fuel; shipped by rail, East and West; capacity, 400 tons daily; men employed, 75; W. H. Warner, President, Cleveland, Ohio; C. H. Meade, Superintendent, Peytona, W. Va.

Analysis of sample collected from Nos. 2, 4 and 5, as reported by **Messrs.** Hite and Krak, is given in the table of coal analyses under **No. 108A.**

Another section of the same opening was taken by Krebs at the drift mouth, as follows:

			Ft.	In.
Gray shale, with many plant fossils			2	0
Coal, gas, bright	0′	11″		
Shale, gray	0	2		
Coal, semi-splint	1	6		
Coal, splint	1	6		
Shale, soft, gray, 3″ to	0	6		
Coal, gray splint, reported	0	3	4	10

From the sections of the **Campbell Creek (No. 2 Gas) Coal** given on preceding pages, it will be noted that this bed is the most important coal in Peytona District.

Scott District.

Scott District lies west of Peytona, and the **Campbell Creek (No. 2 Gas) Coal** extends through a portion of the eastern part of the district, of sufficient thickness and purity to be of commercial value, while in the western and southwestern part, the bed appears to be thin and impure, so far as it has been developed by prospect openings and surface exposures.

This coal comes out of Little Coal River just north of the mouth of Rock Creek, and rises rapidly southeastward to the crest of the Warfield Anticline.

The following section was measured by Krebs on the west side of Little Coal River at the mouth of Rock Creek:

Griffith Heirs Opening—No. 572 on Map II.

			Ft.	In.
Sandstone			5	5
Shale, gray			1	6
Coal, splint	1'	4"		
Shale, gray	1	4		
Coal, gas, (gray shale floor), 690' B.	1	4	4	0

The **Campbell Creek (No. 2 Gas) Coal** rises out of Camp Creek about 1.2 miles east of its mouth, and keeps above the bed of the same for about 3 miles. Several openings have been made in this bed along Camp Creek, where the coal is mined for local fuel, as follows:

Jacob Conners Local Mine Opening—No. 573 on Map II.

Located on Camp Creek, 1.2 miles east from the mouth of the creek; section by Teets; **Campbell Creek (No. 2 Gas) Coal**; elevation, 688' L.

			Ft.	In.
Dark shale roof				
Coal, splint	0'	8"		
Shale, gray	0	6		
Coal, splint, visible	2	2	3	4

Mandaville Dolan Local Mine Opening—No. 574 on Map II.

Located on the south side of Camp Creek, 1.8 miles east from its mouth; section by Teets; **Campbell Creek (No. 2 Gas) Coal**; elevation, 735' B.

			Ft.	In.
1. Dark shale				
2. **Coal**, block	0'	8"		
3. Shale, gray	0	8		
4. **Coal**, splint, (fire clay floor)	2	2	3	6

Analysis of sample collected from Nos. 2 and 4, as reported by Messrs. Hite and Krak, is given in the table of coal analyses under **No. 109.**

Peter Dolan Local Mine Opening—No. 575 on Map II.

Located on the north side of Camp Creek, 1.8 miles east of the mouth of same; section by Teets; **Campbell Creek (No. 2 Gas) Coal;** elevation, 760′ B.

			Ft.	In.
Sandstone				
Shale			1	0
Coal, splint	0′	7″		
Slate	0	4		
Coal, splint, (slate floor)	2	2	3	1

"Doc" Smith Local Mine Opening—No. 576 on Map II.

Located on the north side of Camp Creek, near the school house, about 2.0 miles east from its mouth; section by Teets; **Campbell Creek (No. 2 Gas) Coal;** elevation, 780′ B.

			Ft.	In.
Slate roof				
Coal, hard splint	0′	6″		
Shale, gray	0	6		
Coal, splint, (fire clay floor)	2	1	3	1

Butts, N. 40° W.; faces, N. 50° E.

David Austin Opening—No. 577 on Map II.

Located on the north side of Camp Creek, and about 1.0 mile west of the Scott-Peytona District Line; section by Teets; **Campbell Creek (No. 2 Gas) Coal;** elevation, 830′ B.

			Ft.	In.
Slate roof				
Coal, hard splint	0′	9″		
Shale, gray	0	4		
Coal, splint	0	11		
Shale, dark	0	3		
Coal, hard, gnarly, (slate floor)	1	6	3	9

George W. Dolan Local Mine Opening—No. 578 on Map II.

Located on the south side of Camp Creek, 1.0 mile west of the Scott-Peytona District Line; section by Teets; **Campbell Creek (No. 2 Gas) Coal;** elevation, 960′ B.

			Ft.	In.
Slate roof				
Coal, block	0′	1″		
Slate, gray	0	2		

PLATE XXXVII.—Malden Sandstone at Peytona.

Coal, splint	0′	11″		
Slate, gray	0	3		
Coal, splint, (slate floor)	1	5	2	10

Southeast from Camp Creek about one mile, near the head of Hubbard Fork of Rock Creek, the **Campbell Creek (No. 2 Gas) Coal** rises out of the bed of the stream, 1.2 miles east of Foster P. O., where the following section was measured by Krebs on the north side of the stream:

Rock Creek Collieries Company Local Mine Opening. No. 579 on Map II.

				Ft.	In.
1.	Sandstone, massive				
2.	Shale, gray			0	3
3.	**Coal,** hard splint	0′	8″		
4.	Slate	0	0½		
5.	**Coal**	0	1		
6.	Slate, dark	0	3		
7.	**Coal,** hard splint	2	4		
8.	**Coal,** gray splint	0	2		
9.	**Coal,** hard splint, 945′ B., (slate floor)	1	3	4	9½

Analysis of sample taken from Nos. 3, 5, 7, 8 and 9, as reported by Messrs. Hite and Krak, is given in the table of coal analyses under **No. 110.**

Mr. M. W. Venable, of Charleston, W. Va., collected a sample from the above numbers and the analysis made by A. S. McCreath & Son, of Harrisburg, Pa., gave the following results:

	Per cent.
Moisture	1.89
Volatile Matter	39.04
Fixed Carbon	50.97
Sulphur	1.82
Ash	6.28
Total	100.00

The above results show the coal to be excellent for both steam and domestic purposes.

A. B. Elkins Local Mine Opening—No. 580 on Map II.

Located on the south side of Hubbard Fork of Rock Creek, 1.2 miles east of Foster P. O; section by Krebs; **Campbell Creek (No. 2 Gas) Coal**; elevation, 945' B.

				Ft.	In.
1.	Sandstone, massive				
2.	**Coal**, hard, block	0'	8"		
3.	Slate	0	1		
4.	**Coal**	0	1		
5.	Slate	0	3		
6.	**Coal**, splint, (slate floor)	3	9	4	10

Analysis of sample collected from Nos. 2, 4 and 6 by M. W. Venable, made by A. S. McCreath & Son, of Harrisburg, Pa., gave the following results:

	Per cent.
Moisture	1.72
Volatile Matter	38.65
Fixed Carbon	49.50
Sulphur	2.94
Ash	7.19
Total	100.00

The excessive moisture, ash and sulphur in the foregoing analysis are due to the fact that the sample was taken near the outcrop of the coal seam.

Mandaville Hopkins Local Mine Opening—No. 581 on Map II.

Located on the west side of Little Coal River, 0.5 mile north of Hopkins Station; section by Krebs; **Campbell Creek (No. 2 Gas) Coal**; elevation, 825' B.

			Ft.	In.
Slate roof				
Coal, splint	0'	6"		
Coal, impure	0	1		
Coal, splint	0	7		
Slate	0	1		
Coal, hard splint	1	4		
Slate	0	1		
Coal, splint, (slate floor)	0	11	3	7

C. C. Chambers Local Mine Opening—No. 582 on Map II.

Located on Joes Branch of Lick Creek, 0.75 mile northwest from Hopkins Station; section by Krebs; **Campbell Creek (No. 2 Gas) Coal**; elevation, 795' B.

			Ft.	In.
Shale, gray, roof				
Coal, splint	1'	2"		
Shale	0	1		
Coal, gas	1	3		
Coal, gray splint	0	2		
Slate	0	1		
Coal, splint, (slate floor)	0	5	3	2

The foregoing sections give the structure, thickness and character of the **Campbell Creek (No. 2 Gas) Coal** in Scott District. It will be noted that no sections were obtained on this coal in the vicinity of Danville or Madison, and the tests made thus far in prospecting for this seam indicate that the coal is thin and impure in a portion of the district.

Washington District.

Washington District lies south and west of Scott, and so far as tests have been made by prospect openings, surface exposures and diamond core test holes, it has been proved that in the northern part of the district the **Campbell Creek (No. 2 Gas) Coal** is thin, impure and of very little commercial value, but in the southern part of the district, the diamond core test holes thus far drilled give this coal its usual thickness and purity and indicate it as a valuable bed in this part of the district.

Sections will now be given showing its structure, character and thickness as observed in different parts of the district:

Chambers Prospect Opening—No. 583 on Map II.

Located on the north side of Cox Fork of Turtle Creek, 0.25 mile north of the mouth of Mud Fork; section by Krebs; **Campbell Creek (No. 2 Gas) Coal**; elevation, 900' B.

			Ft.	In.
Sandstone roof				
Coal, splint	0'	5 "		
Slate and bone	0	4½		
Coal, splint	1	2½		
Shale, gray	0	3		
Coal, splint, (fire clay floor)	0	10	3	1

Opening—No. 584 on Map II.

Located on the east side of Cox Fork of Turtle Creek, 1.0 mile northwest of Turtletown; section by Krebs; **Campbell Creek (No. 2 Gas) Coal**; elevation, 975' B.

			Ft.	In.
Sandstone, massive				
Shale, gray			6	0
Coal, splint, (dark shale floor)			1	4

Opening—No. 585 on Map II.

Located on Chapman Branch of North Fork of Big Creek, 3.0 miles southwest of Estep P. O.; Chapmanville District, Logan County, and 0.8 mile west of the Boone-Logan County Line; section by Krebs; **Campbell Creek (No. 2 Gas) Coal**; elevation, 760' B.

			Ft.	In.
Shale, dark gray, full of iron ore nodules			8	0
Coal	0'	1"		
Shale, gray	0	6		
Coal, gas	0	6		
Coal, impure	0	7		
Coal, gas	0	5		
Coal, hard splint, (gray shale floor)	1	9	3	10

Prospect Opening—No. 586 on Map II.

Located on the north side of Turtle Creek, 1.5 miles southwest of Turtletown; section by Krebs; **Campbell Creek (No. 2 Gas) Coal**; elevation, 1013' L.

			Ft.	In.
Gray shale				
Limestone, dark gray, **Campbell Creek**			2	0
Sandstone			8	0
Coal, splint	1'	0"		
Slate	0	1		
Coal, gas, visible	0	6	1	7

Yawkey and Freeman Opening—No. 587 on Map II.

Located on the west side of Spruce Fork, 0.5 mile south of Jeffery P. O.; section by Krebs; **Campbell Creek (No. 2 Gas) Coal**; elevation, 790' B.

			Ft.	In.
Sandstone, massive			40	0
Coal, gas	0'	4"		
Shale, gray	0	2		
Coal, gas, (sandstone floor)	0	2	0	8

Measurement was taken in the side of the railroad grade where the section is exposed.

Yawkey and Freeman Opening—No. 588 on Map II.

Located on the west side of Spruce Fork, 0.6 mile south of Jeffery P. O.; section by Krebs; **Campbell Creek (No. 2 Gas) Coal**; elevation, 780′ L.

			Ft.	In.
Sandy shale				
Coal, gas	0′	2″		
Sandstone, 2″ to	0	10		
Coal, gas	0	4		
Coal, splint, (fire clay floor)	0	6	1	10

The **Campbell Creek (No. 2 Gas) Coal** dips under Spruce Fork about 0.8 mile south of Jeffery P. O., where the following section was measured by Krebs:

			Ft.	In.
Sandstone, massive			30	0
Shale, gray			5	0
Coal, gas	0′	10″		
Shale, gray	0	2		
Coal, gas, visible, to creek, 775′ B	0	8	1	8

Sections will now be given of the **Campbell Creek (No. 2 Gas) Coal** showing its thickness and depth below the surface in the diamond core test holes that have been drilled by Messrs. Chilton, MacCorkle, Chilton and Meany, on Spruce Laurel Fork, southeast from Clothier, as follows:

Spruce Laurel Diamond Core Test (112)—No. 589 on Map II.

Located on Jerry Fork, 4.0 miles southeast of Clothier; **Campbell Creek (No. 2 Gas) Coal, 287′ 4¼″ deep.**

			Ft.	In.
Slate roof				
Coal	2′	6″		
Bone	0	0⅛		
Coal	0	10½		
Bone	0	0⅛		
Coal	1	0½		
Coal and bone, interlaminated	0	2		
Coal, (shale floor)	1	10	6	5¼

Spruce Laurel Diamond Core Test (113)—No. 590 on Map II.

Located 5.5 miles almost due east of Clothier; **Campbell Creek (No. 2 Gas) Coal, 454′ 7″ deep.**

			Ft.	In.
Slate roof				
Coal	2′	5¼″		
Shale and **coal**	0	2¾		
Shale	14	1		
Coal, (fire clay floor)	3	10	20	7

Spruce Laurel Diamond Core Test (114)—No. 591 on Map II.

Located on Spruce Laurel Fork, 4.2 miles southeast of Clothier; **Campbell Creek (No. 2 Gas) Coal, 217′ 4″ deep.**

	Ft.	In.
Slate roof		
Coal, (fire clay floor)	6	5

Spruce Laurel Diamond Core Test (115)—No. 592 on Map II.

Located on the east side of Spruce Laurel Fork, 4.7 miles southeast of Clothier; **Campbell Creek (No. 2 Gas) Coal, 236′ 0″ deep.**

	Ft.	In.
Shale roof		
Coal, (shale floor)	7	3

Spruce Laurel Diamond Core Test (116)—No. 593 on Map II.

Located on Spruce Laurel Fork, 5.0 miles southeast from Clothier; **Campbell Creek (No. 2 Gas) Coal, 234′ 3″ deep.**

			Ft.	In.
Slate roof				
Coal	0′	3 ″		
Slate	0	1½		
Coal	2	6½		
Shale	0	0¼		
Coal	1	8		
Bone	0	0¾		
Coal	0	6		
Coal and slate, interlaminated	0	4½		
Shale	0	1½		
Coal, (fire clay floor)	2	3	7	11

Spruce Laurel Diamond Core Test (117)—No. 594 on Map II.

Located on Sycamore Fork, 5.1 miles southeast from Clothier; **Campbell Creek (No. 2 Gas) Coal, 274′ 7″ deep.**

				Ft.	In.
1.	Slate roof				
2.	**Coal**	0′	5″		
3.	Slate	0	1		
4.	**Coal**	5	4		
5.	Slate	0	2		
6.	**Coal**, (fire clay floor)	2	2	8	2

Analysis of sample taken from Nos. 2, 4 and 6 made by A. S. McCreath & Son, of Harrisburg, Pa.; also sample taken from the same numbers and analyzed by **Messrs.** Hite and Krak, as reported by them, is given in the table of coal analyses under **No. 111.**

McCreath's analysis is as follows:

	Per cent.
Moisture	1.195
Volatile Matter	36.123
Fixed Carbon	56.335
Ash	6.347
Total	100.000
Sulphur	0.616

Spruce Laurel Diamond **Core** Test (118)—**No.** 595 on Map II.

Located on Spruce Laurel Fork, 5.2 miles southeast from Clothier; **Campbell Creek (No. 2 Gas) Coal, 227′ 3 15/16″ deep.**

			Ft.	In.
Slate roof				
Coal	3′	5¾″		
Shale	0	0⅛		
Coal	0	0⅞		
Slate	0	0 1/16		
Coal	0	1½		
Coal and shale, interlaminated	0	4½		
Slate	0	0⅜		
Coal	0	7½		
Slate and coal streaks	0	2		
Coal and bone, interlaminated	0	2¾		
Slate	0	3		
Coal	0	0⅜		
Slate	0	0⅛		
Coal, (fire clay floor)	2	3	[illegible]	7 15/16

Spruce Laurel Diamond **Core** Test (119)—**No.** 596 on Map II.

Located in Bear Hollow, 5.0 miles southeast from Clothier; **Campbell Creek (No. 2 Gas) Coal, 315′ 3″ deep.**

			Ft.	In.
Slate roof				
Coal	0′	3½″		
Slate	0	1		
Coal	0	1		
Coal and bone, interlaminated	0	1½		
Coal	4	9		
Coal and bone, interlaminated	0	5		
Slate	0	2½		
Coal, (shale floor)	2	0½	[illegible]	0

Spruce Laurel Diamond Core Test (120)—No. 597 on Map II.

Located on Spruce Laurel Fork, 5.5 miles southeast from Clothier; **Campbell Creek (No. 2 Gas) Coal, Lower Bench, 224′ 2″ deep.**

				Ft.	In.
Slate roof					
Coal	0′	6″	**Upper Bench**	5	11¾
Bone and **coal**, interlaminated	0	3			
Coal	2	5			
Shale	0	0¼			
Coal	2	2½			
Coal and bone, interlaminated	0	2½			
Slate	0	2½			
Coal	0	2			
Fire clay				0	3¼
Shale				1	2
Sandstone				8	6
Shale				0	8
Coal and bone, interlaminated, **Lower Bench**, (shale floor)				2	0

Spruce Laurel Diamond Core Test (121)—No. 598 on Map II.

Located at the mouth of Skin Poplar Branch of Spruce Laurel Fork, 5.5 miles southeast from Clothier; **Campbell Creek (No. 2 Gas) Coal, Lower Bench, 228′ deep.**

				Ft.	In.
Slate roof					
Coal	0′	8¼″	**Upper Bench**	[illegible]	0
Bone and **coal**, interlaminated	0	2			
Coal	2	11¼			
Bone	0	0¾			
Coal	1	4			
Bone	0	0¼			
Coal	1	3			
Bone	0	3			
Coal and bone, interlaminated	0	3½			
Sandy shale				9	0½
Coal and bone, interlaminated	0′	3″	**Lower Bench**	2	2½
Coal	0	5			
Bone	0	0½			
Coal	0	8			
Coal and bone, interlaminated	0	1			
Shale	0	7			
Coal and bone, interlaminated. (shale floor)	0	2			

Spruce Laurel Diamond Core Test (122)—No. 599 on Map II.

Located on Skin Poplar Branch, 6.3 miles southeast from Clothier; **Campbell Creek (No. 2 Gas) Coal, 229′ 5″ deep.**

			Ft.	In.
Slate roof				
Coal	1′	1″		
Slate	0	1½		
Coal and bone, interlaminated	0	2½		
Coal	0	8½		
Coal streaks with sulphur	0	0½		
Coal	3	6		
Shale with **coal** streaks, (shale floor)	1	0	6	8

Spruce Laurel Diamond Core Test (124)—No. 601 on Map II.

Located on Spruce Laurel Fork, at mouth of Trough Fork, 5.9 miles southeast of Clothier; **Campbell Creek (No. 2 Gas) Coal, 245′ 5″ deep.**

			Ft.	In.
Slate roof				
Coal and bone, interlaminated	0′	6″		
Coal	0	7½		
Shale	0	3		
Coal	1	2		
Coal and bone, interlaminated	0	4		
Coal	0	6		
Bone	0	0½		
Coal	0	3½		
Bone	0	0½		
Coal	0	7		
Sulphur	0	1		
Coal	1	0		
Shale	0	1½		
Coal and bone, interlaminated	0	2		
Coal	0	0¾		
Shale	0	3		
Coal and bone, interlaminated, (sandstone floor)	0	5¾	6	6

Spruce Laurel Diamond Core Test (125)—No. 602 on Map II.

Located on Trough Fork of Spruce Laurel Fork, 5.8 miles southeast of Clothier; **Campbell Creek (No. 2 Gas) Coal, 352′ 1¼″ deep.**

			Ft.	In.
Slate roof				
Coal	1′	0″		
Fire clay	0	1		
Coal	2	6½		
Bone	0	0¼		
Coal, (fire clay floor)	2	1½	5	9¼

Spruce Laurel Diamond Core Test (126)—No. 603 on Map II.

Located on Spruce Laurel Fork, 6.3 miles southeast of Clothier; **Campbell Creek (No. 2 Gas) Coal, 222' 2" deep.**

			Ft.	In.
Slate roof				
Coal	0'	4 "		
Bone	0	0½		
Coal	1	0		
Shale	0	0¼		
Coal	2	5½		
Bone	0	0¼		
Coal, (shale floor)	1	9½	5	8

Spruce Laurel Diamond Core Test (127)—No. 604 on Map II.

Located 6.8 miles southeast from Clothier; **Campbell Creek (No. 2 Gas) Coal, 242' 2" deep.**

			Ft.	In.
Slate roof				
Coal	0'	1 "		
Sulphur	0	1½		
Bone	0	1		
Coal	0	5½		
Sulphur	0	0½		
Coal	0	3½		
Shale	0	2		
Coal	2	4½		
Shale	0	0½		
Coal, (shale floor)	2	2	5	10

Spruce Laurel Diamond Core Test (128)—No. 605 on Map II.

Located on Spruce Laurel Fork, 7.8 miles southeast from Clothier; **Campbell Creek (No. 2 Gas) Coal, 199' 11" deep.**

	Ft.	In.
Shale, black, roof		
Coal, (shale floor)	6	9

Cassingham Diamond Core Test (130)—No. 606 on Map II.

Located on Spruce Laurel Fork, below mouth of Dennison Fork of Spruce Laurel, 9.0 miles southeast from Clothier; **Campbell Creek (No. 2 Gas) Coal, 329' 3" deep.**

			Ft.	In.
Dark shale roof				
Coal and bone, interlaminated	0'	3"		
Coal, (shale floor)	1	2	1	5

Cassingham Core Test Hole (131)—No. 607 on Map II.

Located on Spruce Laurel Fork, 3.25 miles southwest of Bald Knob P. O., and 10.0 miles southeast of Clothier; **Campbell Creek (No. 2 Gas) Coal, 454' 3" deep.**

			Ft.	In.
Slate roof				
Bone	0'	1"		
Coal	1	0		
Fire clay	1	0		
Shale, dark and sandy	2	2		
Coal, and bone, interlaminated, (sandstone floor)	1	0	5	3

Cassingham Core Test Hole (155)—No. 608 on Map II.

Located on Skin Poplar Branch of Spruce Laurel Fork, 1.3 miles southeast from its mouth; **Campbell Creek (No. 2 Gas) Coal, 247' 2½" deep.**

			Ft.	In.
Slate roof				
Coal	0'	5½"		
Coal and bone, interlaminated	0	0½		
Coal	0	5½		
Coal and bone, interlaminated	0	1		
Coal	0	7		
Coal and shale, interlaminated	0	4½		
Coal	0	1		
Sulphur	0	0½		
Coal, (shale floor)	4	4	6	5½

From the foregoing sections obtained in the diamond core test holes, it is evident that there is quite a large area of thick **Campbell Creek (No. 2 Gas) Coal** in the southern part of Washington District.

Sherman District.

Sherman District lies south and east of Peytona and w— of Cabin Creek District, Kanawha County. Its location and the sections of coal obtained indicate that the **Campbell Cree-(No. 2 Gas) Coal** is an important bed in this district.

George C. Evans Local Mine Opening—No. 609 on Map II.

Located on the south side of Short Creek, about 2.0 miles northeast from Racine P. O.; section by Krebs; **Campbell Creek (No. 2 Gas) Coal**; elevation, 820' B.

			Ft.	In.
Shale roof, plant fossils			4	0
Coal, gas	0'	3"		
Shale, gray, 1" to	0	2		
Coal, gas	1	0		
Shale, gray	0	3		
Coal, semi-splint	1	3		
Coal, splint, hard, (gray shale floor)	1	3	4	2

William Gadd Local Mine Opening—No. 610 on Map II.

Located on the north side of Short Creek, 1.6 miles northeast from Racine P. O.; section by Krebs; **Campbell Creek (No. 2 Gas) Coal**; elevation, Lower Bench, 795' B.

				Ft.	In.
Shale, gray, roof					
Coal, splint	0'	4"	Upper Bench	4	3
Shale, gray	0	3			
Coal, splint	0	11			
Shale, dark	0	2			
Coal, splint	2	7			
Slate and concealed				5	0
Sandstone, massive				21	0
Shale				1	0
Coal, semi-splint	0'	11½"	Lower Bench	2	8
Bone	0	0½			
Coal, splint, (slate floor)	1	8			

Boone and Kanawha Land and Mining Company Prospect Opening—No. 611 on Map II.

Located on the north side of Short Creek, just below the mouth of Snodgrass Branch, 1.5 miles northeast from Racine P. O.; section by H. A. Henderson for J. S. Cunningham; **Campbell Creek (No. 2 Gas) Coal**; elevation, 800' B.

			Ft.	In.
Slate roof				
Coal, block	0'	9"		
Shale, gray	0.	4		
Coal, splint, (slate floor)	2	6	3	7

Boone and Kanawha Land and Mining Company Opening. No. 612 on Map II.

Located on the north side of Short Creek, 0.8 mile northeast of Racine P. O.; section by H. A. Henderson for J. S. Cunningham; **Campbell Creek (No. 2 Gas) Coal**; elevation, 820′ B.

				Ft.	In.
1.	Slate roof				
2.	**Coal**, block	0′	10½″		
3.	Shale, gray	0	3		
4.	**Coal**, splint, (slate floor)	2	6½	3	8

A sample collected from Nos. 2 and 4 by E. V. d'Invilliers, and analyzed by A. S. McCreath & Son, Harrisburg, Pa., gave the following results:

	Per cent.
Moisture	1.286
Volatile Matter	35.794
Fixed Carbon	54.994
Sulphur	1.411
Ash	6.515
Total	100.000

Boone and Kanawha Land and Mining Company Opening. No. 613 on Map II.

Located on the east side of Left Fork of Short Creek, 1.0 mile north of Racine P. O.; section by H. A. Henderson for J. S. Cunningham; **Campbell Creek (No. 2 Gas) Coal**; elevation, 820′ B.

			Ft.	In.
Slate roof				
Coal	0′	4 ″		
Shale, grayish-green	0	3		
Coal, splint	2	8½		
Slate	0	1		
Coal, gas, (slate floor)	0	9	4	1½

Boone and Kanawha Land and Mining Company Local Mine Opening—No. 614 on Map II.

Located on the north side of Mill Branch of Coal River, 0.25 mile northwest from Racine P. O.; section by H. A. Henderson for J. S. Cunningham; **Campbell Creek (No. 2 Gas) Coal**; elevation, 820′ B.

			Ft.	In.
Slate roof				
Coal, splint	0′	10½″		
Shale, gray	0	2½		
Coal, splint, (slate floor)	2	5½	3	6½

Boone and Kanawha Land and Mining Company Prospect Opening—No. 615 on Map II.

Located on the north side of Coal River, 0.3 mile west from Racine P. O.; section by H. A. Henderson for J. S. Cunningham; **Campbell Creek (No. 2 Gas) Coal**; elevation, 830′ B.

			Ft.	In.
Slate roof				
Coal, splint	0′	9″		
Shale, gray	0	3		
Coal, splint, (slate floor)	2	8½	3	8½

Boone and Kanawha Land and Mining Company Opening. No. 616 on Map II.

Located on the east side of Whetstone Branch of Coal River, 0.6 mile southwest from Racine P. O.; section by H. A. Henderson for J. S. Cunningham; **Campbell Creek (No. 2 Gas) Coal**; elevation, 835′ B.

			Ft.	In.
Slate roof				
Coal, splint	0′	10″		
Shale, gray	0	2		
Coal, splint, (slate floor)	2	9	3	9

Samuel Gillispie Opening—No. 617 on Map II.

Located on the south side of Short Creek, 1.2 miles northeast from Racine P. O.; section by Teets; **Campbell Creek (No. 2 Gas) Coal**; elevation, 805′ B.

				Ft.	In.
1.	Slate roof				
2.	**Coal**, block	0′	4″		
3.	Shale, gray	1	4		
4.	**Coal**, gas	0	5		
5.	Shale, gray	0	4		
6.	**Coal**, gas	1	8		
7.	**Coal**, gray splint, visible	0	10	4	11

Butts, N. 50° W.; faces, N. 40° E.

Analysis of sample collected from Nos. 2, 4, 6 and 7, as reported by Messrs. Hite and Krak, is given in the table of coal analyses under **No. 112**.

Witcher Snodgrass Local Mine Opening—No. 618 on Map II.

Located on the south side of Short Creek, 1.6 miles northeast from Racine P. O.; section by Teets; **Campbell Creek (No. 2 Gas) Coal**; elevation, 810' B.

			Ft.	In.
Slate roof				
Coal, block	0'	3"		
Shale, gray	0	9		
Coal, gnarly	0	6		
Shale, gray	0	3		
Coal, gas	1	5		
Coal, gray splint, (slate floor)	1	3	4	4

Local Mine Opening—No. 619 on Map II.

Located on the south side of Coal River opposite Racine P. O.; section by Krebs; **Campbell Creek (No. 2 Gas) Coal**; elevation, 820' L.

			Ft.	In.
Shale, gray, roof				
Coal, splint	0'	10"		
Shale, dark gray	0	3		
Coal, gas	1	9		
Coal, gray splint, (slate floor)	0	11	3	9

Hickory Ash Coal Company Opening—No. 620 on Map II.

Located on Indian Creek of Coal River at Sterling P. O.; section by Teets taken in Mine No. 2 main west heading; **Campbell Creek (No. 2 Gas) Coal**; elevation, 775' B.

			Ft.	In.
1. Slate roof				
2. **Coal**, block	0'	10 "		
3. Shale, gray	0	2½		
4. **Coal**, gas	1	10		
5. **Coal**, gray splint	0	8		
6. **Coal**, gas, (slate floor)	0	1	3	7½

Butts, N. 60° W.; faces, N. 30° E.; coal mined and shipped east and west for fuel and domestic purposes; capacity of mine, 500 tons daily; number of men employed, 80; H. A. Goddard, General Manager, Wellston, Ohio; C. L. Voglesang, Superintendent, Sterling, W. Va.

Analysis of a sample collected from Nos. 2, 4, 5 and 6, as reported by Messrs. Hite and Krak, is given in the table of coal analyses under **No. 113**.

Hickory Ash Coal Company Mine No. 1—No. 621 on Map II.

Located on Indian Creek of Coal River at Sterling P. O.; section by Teets taken in the main heading; **Campbell Creek (No. 2 Gas) Coal;** elevation, 775' B.

				Ft.	In.
1.	Slate roof				
2.	**Coal,** block	0'	10"		
3.	Shale, gray	0	3		
4.	**Coal,** gas	1	10		
5.	**Coal,** gray splint	0	9		
6.	**Coal,** gas, (slate floor)	0	1	3	9

The analyses of two samples collected from Nos. 2, 4, 5 and 6, as reported by Messrs. Hite and Krak, are given in the table of coal analyses under No. 114.

Hickory Ash Coal Company Mine No. 1—No. 622 on Map II.

Located on Indian Creek of Coal River at Sterling, P. O.; section by Krebs, taken at the drift mouth of Mine No. 1; **Campbell Creek (No. 2 Gas) Coal;** elevation, 775' B.

			Ft.	In.
Shale, dark gray			10	0
Sandstone, iron ore nodules			1	0
Shale, dark gray			0	4
Coal, gas, hard	0'	9"		
Shale, dark	0	0½		
Coal, gas	0	0½		
Shale, dark gray	0	2½		
Coal, gray splint, (fire clay floor)	3	6	4	6½

Mines Nos. 1 and 2 are connected and shipment is made over the same railroad tipple for both mines.

Perdue Heirs Local Mine Opening—No. 623 on Map II.

Located on the north side of Coal River, 1.0 mile east of Racine P. O.; section by Krebs; **Campbell Creek (No. 2 Gas) Coal;** elevation, 720' B.

			Ft.	In.
Gray shale roof				
Coal, hard block	0'	10"		
Shale, gray	0	2		
Coal, gray splint, (slate floor)	2	6	3	6

PLATE XXXVIII.—Campbell Creek (No. 2 Gas) Coal opening below mouth of Joes Creek.

George Midkiff Local Mine Opening—No. 624 on Map II.

Located on the north side of Coal River, about 0.25 mile below the mouth of Joes Branch; section by Krebs and Teets; **Campbell Creek (No. 2 Gas) Coal**; elevation, 715' B.

				Ft.	In.
1.	Sandstone, massive				
2.	Shale, gray, full of plant fossils			4	0
3.	**Coal**, hard, block	0'	10"		
4.	Shale, gray	0	1		
5.	**Coal**, impure	0	1		
6.	**Coal**, splint, (slate floor)	2	1	3	1

Analysis of sample collected from Nos. 3, 5 and 6, as reported by **Messrs.** Hite and Krak, is given in the table of coal analyses under **No. 115.**

Robert Tony Opening—No. 625 on Map II.

Located on the north side of Coal River, 1.9 miles southeast of Racine P. O.; section by Teets; **Campbell Creek (No. 2 Gas) Coal**; elevation, 715' B.

				Ft.	In.
1.	Slate roof				
2.	**Coal**, block	0'	10"		
3.	Shale, gray	0	3		
4.	**Coal**, gas	1	7		
5.	**Coal**, gray splint	0	4		
6.	**Coal**, gas, (slate floor)	0	2	3	2

Analysis of sample collected from Nos. 2, 4, 5 and 6, as reported by **Messrs.** Hite and Krak, is given in the table of coal analyses under **No. 116.**

Lackawanna Coal and Lumber Company Local Mine Opening. No. 626 on Map II.

Located on the south side of Coal River, 1.4 miles southeast from Racine P. O.; section by Krebs; **Campbell Creek (No. 2 Gas) Coal**; elevation, Lower Bench, 750' B.

				Ft.	In.
Shale, gray roof					
Coal, block	0"	10"	Upper Bench		
Shale, gray	0	2	Upper Bench	4	4
Coal, splint	3	4	Upper Bench		
Slate				1	0
Concealed, mostly sandstone				12	0
Sandstone roof				2	0
Coal, splint, (slate floor), **Lower Bench**				2	0

A sample collected from the coal in the Upper Bench by Mr. E. V. d'Invilliers and analyzed by A. S. McCreath & Son, of Harrisburg, Pa., gave the following results:

	Per cent.
Moisture	1.129
Volatile Matter	35.774
Fixed Carbon	57.584
Sulphur	0.743
Ash	4.770
Total	100.000

William Forber Opening—No. 627 on Map II.

Located on the east side of Tony Creek, 1.6 miles northeast from Racine P. O.; section by Teets; **Campbell Creek (No. 2 Gas) Coal**; elevation, 780′ B.

				Ft.	In.
1.	Sandstone roof				
2.	**Coal**, block	0′	3″		
3.	Shale, dark	0	8		
4.	**Coal**, block	0	10		
5.	Shale, gray	0	4		
6.	**Coal**, gas	1	5		
7.	**Coal**, gray splint	0	4		
8.	**Coal**, gas, (slate floor)	0	2	4	0

Analysis of sample collected from Nos. 2, 4, 6, 7 and 8, as reported by Messrs. Hite and Krak, is given in the table of coal analyses under **No. 117.**

Harry Belcher Opening—No. 628 on Map II.

Located on the east side of Tony Creek, 2.2 miles northeast of Racine P. O.; section by Teets; **Campbell Creek (No. 2 Gas) Coal**; elevation, 785′ B.

				Ft.	In.
1.	Slate roof				
2.	**Coal**, gas	0′	4″		
3.	Shale, gray	0	1		
4.	**Coal**, splint	1	1		
5.	Shale, gray	0	4		
6.	**Coal**, hard splint, (slate floor)	2	3	4	1

Analysis of sample collected from Nos. 2, 4 and 6, as reported by Messrs. Hite and Krak, is given in the table of coal analyses under **No. 118.**

The **Campbell Creek (No. 2 Gas) Coal** dips under Coal River between the mouth of Joes Branch and the mouth of

Joes Creek, and remains under said river through the Coalburg Syncline. The coal was encountered in the Winifrede Well **No. 1 (56)** located on Trace Fork of Joes Creek, at a depth of 255 feet, and shows a thickness of 5 feet as reported by the driller.

The coal was encountered in the Lackawanna Well **No. 1 (54)** located at Seth, as follows:

	Ft.	In.
Slate		
Coal, Upper Bench	3	6
Slate	32	6
Coal, Lower Bench	3	6

The depth of the lower coal is 163′ 6″ from the surface.

The coal was encountered in the Lackawanna Well **No. 2 (57),** located on Sandlick Creek, at a depth of 199 feet, with a thickness of 4 feet.

LaFollette, Robson et al. Core Test Hole No. 3 (141). No. 629 on Map II.

Located on Haggle Branch, 0.25 mile from its mouth and 2.2 miles northwest from Orange P. O.; **Campbell Creek (No. 2 Gas) Coal,** 97′ 2″ deep.

			Ft.	In.
Dark shale roof				
Coal	2′	11″		
Fire clay	0	9		
Coal	2	9		
Slate	0	1		
Coal, (fire clay floor)	0	8	6	2

The **Campbell Creek (No. 2 Gas) Coal** rises out of Coal River about 2 miles south of Orange P. O., and about 0.5 mile north of Gidd Branch, soon appearing in the hills above the bed of the creek and at the mouth of Seng Creek, and the coal is more than 200 feet above the bed of the river.

Lackawanna Coal and Lumber Company Prospect Opening. No. 630 on Map II.

Located on Mill Branch of Coal River, about 0.5 mile from its mouth, and 2.5 miles northwest from the mouth of Seng Creek; section taken by Krebs; **Campbell Creek (No. 2 Gas) Coal;** elevation, 875′ B.

			Ft.	In.
Slate roof				
Coal, gas	2′	6″		
Coal, impure	0	6		
Coal, gas, visible	0	6	3	6

Nancy Kuhn Opening—No. 631 on Map II.

Located at the mouth of Roundbottom Branch of Coal River, 3.0 miles northwest of Jarrolds Valley; section by Teets; **Campbell Creek (No. 2 Gas) Coal**; elevation, 910′ B.

			Ft.	In.
1. Slate roof				
2. **Coal**, gas	2′	6″		
3. **Coal**, gray splint, (slate floor)	0	6	3	0

Analysis of sample taken from Nos. 2 and 3, as reported by Messrs. Hite and Krak, is given in the table of coal analyses under No. 119.

Coal River Coal and Mining Company Opening. No. 632 on Map II.

Located on the east side of Coal River, 1.75 miles northwest of Jarrolds Valley; section by Teets; **Campbell Creek (No. 2 Gas) Coal**; elevation, 1010′ B.

			Ft.	In.
Slate roof				
Coal, block	2′	11″		
Slate, gray	0	4		
Coal	0	2		
Slate, dark	0	2		
Coal, (slate floor)	0	1	3	8

J. Q. Dickinson Prospect Opening—No. 633 on Map II.

Located on a small branch of Coal River, 0.5 mile northwest from the mouth of Seng Creek; section by Teets; **Campbell Creek (No. 2 Gas) Coal**; elevation, 945′ B.

			Ft.	In.
Slate roof				
Coal, gas	1′	1″		
Coal, block, (slate floor)	1	6	2	[illegible]

J. Q. Dickinson Prospect Opening—No. 634 on Map II.

Located on Petes Branch of Coal River, 0.7 mile northwest from mouth of Seng Creek; section by Krebs; **Campbell Creek (No. 2 Gas) Coal**; elevation, Lower Bench, 950′ B.

				Ft.	In.
Slate roof					
Coal, gas	3′	2″	**Upper Bench**	4	6
Coal, reported	1	4			
Concealed				30	0
Slate roof					
Coal, gas	2′	6″	**Lower Bench**	4	6
Slate	1	0	**Powellton? (I. C. W.)**		
Coal	1	0			

Pond Creek Coal and Land Company Opening. No. 635 on Map II.

Located on the west side of Elk Run of Coal River, 1.8 miles due west of Jarrolds Valley; section by Teets; **Campbell Creek (No. 2 Gas) Coal**; elevation, 1080' B.

				Ft.	In.
1.	Slate roof				
2.	**Coal**, gas	0'	8"		
3.	Slate, black	0	3		
4.	**Coal**, gray splint	1	2		
5.	**Coal**, block, visible	1	6	3	

Analysis of sample taken from Nos. 2, 4 and 5, as reported by Messrs. Hite and Krak, is given in the table of coal analyses under **No. 120.**

Shonk and Garrison Local Mine Opening—No. 636 on Map II.

Located on the north side of Seng Creek, 2.0 miles north of Jarrolds Valley; section by Teets; **Campbell Creek(No. 2 Gas) Coal**; elevation, 950' B.

	Ft.	In.
Slate roof		
Coal, gas, (slate floor)	2	3

Coal River Land and Mining Company Opening. No. 637 on Map II.

Located on the south side of Seng Creek, just south of its mouth; section by Krebs; **Campbell Creek (No. 2 Gas) Coal**; elevation, Lower Bench, 963' B.

				Ft.	In.
Sandstone, massive					
Shale, gray				4	0
Coal, impure	0'	4"	Upper Bench	3	4
Coal, gas	3	0			
Shale and concealed				41	0
Coal, gas	1'	6"	Lower Bench Powellton? (I. C. W.)	11	8
Coal, splint	1	0			
Coal, gas	1	0			
Shale, gray	1	0			
Coal	0	4			
Shale, gray	4	0			
Concealed	1	0			
Shale, dark	1	0			
Coal, gas	0	10			

Upper bench, prospect opening; lower bench, exposed in railroad grade.

Crook District.

The **Campbell Creek (No. 2 Gas) Coal** horizon in the northern part of Crook District on the Warfield Anticline occurs about 400 feet above the bed of Pond Fork, and dips rapidly to the southeast into the Coalburg Syncline. The coal goes under Pond Fork just south of Uneeda P. O., and remains under the stream for about eleven miles to the southeast to a point near Pond P. O., about three and one-fourth miles northwest of Bald Knob P. O. The coal seam does not rise above the bed of West Fork at all.

Reynolds Local Mine Opening—No. 638 on Map II.

Located on Jarrell Branch of Pond Fork, 3.0 miles southeast of Madison; section by Teets; **Campbell Creek (No. 2 Gas) Coal**; elevation, 850' B.

			Ft.	In.
Slate roof				
Coal	0'	2"		
Coal, block	0	4		
Slate, gray	0	1		
Coal	0	5		
Slate, gray	0	1		
Coal, block, visible	1	6	2	7

R. P. Chew Local Mine Opening—No. 639 on Map II.

Located on the east side of Pond Fork, 0.5 mile south of Uneeda P. O.; section by Krebs; **Campbell Creek (No. 2 Gas) Coal**; elevation, 740' B.

			Ft.	In.
1. Slate roof				
2. **Coal**, block	0'	2"		
3. Shale, gray	0	3		
4. **Coal**, gas	1	0		
5. **Coal**, splint	0	10		
6. **Coal**, impure	0	2		
7. **Coal**, gray splint, (slate floor)	0	9	3	2

The analysis of a sample collected from Nos. 2, 4, 5 and 7, as reported by Messrs. Hite and Krak, is given in the table of coal analyses under **No. 121.**

The **Arbogast Well (63)**, located at the mouth of Robinson Creek, encountered the Campbell Creek (No. 2 Gas) Coal at a depth of 113 feet from the surface with a thickness of 3 feet as reported by the driller.

Three diamond core test holes were drilled on the land of Chilton, MacCorkle, Chilton and Meany, on Casey Creek, and the Campbell Creek (No. 2 Gas) Coal was encountered in each of these core test holes, as follows:

Casey Diamond Core Test (137)—No. 640 on Map II.

Located on Casey Creek, 1.3 miles above its mouth; **Campbell Creek (No. 2 Gas) Coal, 349′ 11″ deep.**

			Ft.	In.
Shale roof				
Coal, (shale floor)			4	9

Casey Diamond Core Test (138)—No. 641 on Map II.

Located on Casey Creek, 5.2 miles northwest of Bald Knob P. O.; **Campbell Creek (No. 2 Gas) Coal, 289′ 9½″ deep.**

			Ft.	In.
Shale roof				
Coal	0′	1″		
Shale	0	0¼		
Coal	0	7		
Shale	0	1¼		
Coal	0	3½		
Shale	0	2½		
Coal, (shale floor)	2	10	4	1½

Casey Diamond Core Test (139)—No. 642 on Map II.

Located on Casey Creek, 4.8 miles northwest of Bald Knob P. O.; **Campbell Creek (No. 2 Gas) Coal, 345′ 4½″ deep.**

			Ft.	In.
Shale roof				
Coal	1′	10″		
Shale	0	1		
Coal	0	4		
Shale	0	8		
Coal, (fire clay floor)	3	9½	6	8½

Wharton Estate Local Mine Opening—No. 518 on Map II.

Located on the west side of Pond Fork, just above Pond P. O.; section taken by Teets; **Campbell Creek (No. 2 Gas) Coal**; elevation, 980′ L.

			Ft.	In.
Sandstone roof				
Coal, splint	1′	8″		
Shale	1	2		
Coal, splint	1	7		
Shale	0	6		
Coal, splint, visible	2	8	[illegible]	[illegible]

Butts, N. 40° W.; faces, N. 50° E.

J. T. Hatfield Opening—No. 519 on Map II.

Located on the south side of Cow Creek of Pond Fork, 0.8 mile southwest of Pond P. O.; section by Teets; **Campbell Creek (No. 2 Gas) Coal**; elevation, 1040′ B.

				Ft.	In.
1.	Sandstone roof				
2.	**Coal**, splint	1′	1″		
3.	Slate, gray	0	5		
4.	**Coal**, splint	0	8		
5.	Slate, dark	0	6		
6.	**Coal**, splint	0	7		
7.	Slate, gray	1	6		
8.	**Coal**, splint	1	0		
9.	**Coal**, gray splint	0	3		
10.	**Coal**, block, (slate floor)	1	1	7	1

Butts, N. 40° W.; faces, N. 50° E.

The analyses of two samples taken from Nos. 2, 4, 6, 8, 9 and 10, as reported by **Messrs**. Hite and Krak, are given in the table of coal analyses under **No. 96.**

J. T. Hatfield Opening—No. 520 on Map II.

Located on the south side of Cow Creek, 0.7 mile southwest of Pond P. O.; section by Krebs; **Campbell Creek (No. 2 Gas) Coal**; elevation, 1045′ B.

				Ft.	In.
1.	Sandstone, massive				
2.	**Coal**, gas	1′	4″		
3.	Shale, dark	1	4		
4.	**Coal**, gray splint	0	2		
5.	**Coal**, gas	0	6		
6.	Niggerhead	0	2		
7.	**Coal**, gas	0	5		
8.	Shale, dark	0	0½		
9.	**Coal**, splint	0	1		
10.	Shale, gray	1	4		
11.	**Coal**, splint, (slate floor)	2	6	ı	10½

The analysis of a sample collected from Nos. 2, 4, 5, 7, 9 and 11, as reported by **Messrs**. Hite and Krak, is given in the table of coal analyses under **No. 97.**

Wharton Estate Local Mine Opening—No. 521 on Map II.

Located on the south side of Cow Creek, 0.75 mile from its mouth, and 1.5 miles southwest from Pond P. O.; section measured by Krebs; **Campbell Creek (No. 2 Gas) Coal**; elevation, 1020′ B.

			Ft.	In.
Sandstone, massive			40	0
Coal, gas	1′	4″		
Shale, dark	0	4		
Coal, splint	0	1		
Shale, dark	0	0¼		
Coal, splint	0	1		
Coal, gas	0	6		
Niggerhead	0	2		
Coal, splint	0	4		
Shale, dark	6	0		
Coal, impure, (slate floor)	2	6	11	4¼

It will be noted from the above sections that the shale above the lower member of the coal has thickened from 1′ 4″ in Opening No. 520 to 6′ 0″ in Opening No. 521.

Wharton Land Company Opening—No. 643 on Map II.

Located on the west side of Pond Fork, ⅛ mile south of Pond P. O.; section by Krebs and Teets; **Campbell Creek (No. 2 Gas) Coal**; elevation, 985′ B.

			Ft.	In.
Sandstone, massive			21	0
Coal, gas	0′	2″		
Slate	0	0½		
Coal, gas	0	6		
Slate	0	0½		
Coal, splint	0	9		
Shale, dark gray	2	2		
Coal, gas	1	6		
Slate	1	0		
Coal gas, visible	1	6	[illegible]	8

Butts, N. 60° W.; faces, N. 30° E.

The above section is exposed along the road where the coal was formerly mined for local fuel.

Wharton Land Company Opening—No. 644 on Map II.

Located on the south side of Grapevine Branch of Pond Fork, 2.0 miles south of Pond P. O.; section by Krebs; **Campbell Creek (No. 2 Gas) Coal**; elevation, 1144′ B.

	Ft.	In.
Sandstone roof		
Coal, gas, visible	1	6

The Campbell Creek (No. 2 Gas) Coal appears to split into several benches in passing south from Pond P. O. towards Bald Knob, and as far as exposures were found and openings made on Pond Fork up to the extreme head, this bed of coal appears to contain a thick parting of slate between the upper and lower benches. An opening once made near the extreme head of Pond Fork about 0.5 mile southwest from Indian Gap is reported to be about 7 feet thick. This opening had caved in and it was impossible to get a measurement of same. However, an opening has been made on the head of Drews Creek of Marsh Fork, 0.5 mile northeast from Indian Gap, in Marsh Fork District, Raleigh County, where the following section was measured by Krebs:

			Ft.	In.
Gray shale roof				
Coal, gas	3′	0″		
Shale, dark gray, plant fossils	2	6		
Coal, gas	2	2		
Shale	0	4		
Coal, splint	1	4		
Slate	0	1		
Coal, splint	0	1		
Slate	0	1		
Coal, gas	0	2		
Shale, dark	0	8		
Coal, splinty, gas	3	8	14	1
Fire clay floor, (2200′ B.)				

Butts, N. 70° W.; faces, N. 20° E.; opening driven in about 20 feet and coal mined for local fuel.

Pond Fork **Coal** and Land **Company Opening.** **No. 649 on Map II.**

The **Campbell Creek (No. 2 Gas) Coal**, encountered at a depth of 36′ 5″ in **Diamond Core** Test Hole, **E. J. Berwind No. 5 (132)**, located just south of the mouth of Jarrolds Branch of West Fork, shows the following section:

			Ft.	In.
Slate roof				
Coal	0′	9″		
Slate	0	3		
Coal	0	9		
Slate	0	1		
Coal	1	1		
Slate	0	6		

			Ft.	In.
Coal	0′	4½″		
Slate	0	1½		
Coal	2	1		
Slate	0	4		
Coal, (shale floor)	0	4	6	8

E. J. Berwind Diamond Core Test (133)—No. 650 on Map II.

				Ft.	In.
Shale roof					
Coal	1′	0″	Upper Bench	3	0
Shale	0	10			
Coal	1	2			
Shale				20	6½
Slate				0	6½
Coal	0′	1″	Lower Bench	4	4½
Slate	0	1			
Coal, no core	0	8			
Slate	0	1			
Coal	2	2			
Slate	0	0½			
Coal	0	3			
Slate	0	2			
Coal	0	10			

The base of the coal was encountered 65′ 3½″ below the surface.

E. J. Berwind Diamond Core Test (134)—No. 651 on Map II.

Located on the West Fork of Pond Fork, just south of the mouth of Little Ugly Branch; Campbell Creek (No. 2 Gas) Coal.

				Ft.	In.
Shale roof					
Coal and slate	1′	6″	Upper Bench	3	3
Slate	0	8			
Coal, no core	1	1			
Shale				17	5
Slate				0	5
Coal	0′	1″	Lower Bench	2	6
Slate	0	1			
Coal	0	1½			
Slate	0	4			
Coal, no core 11″, (slate floor)	1	10½			

The base of the coal was encountered at a depth of 52′ 6″ from the surface, and the parting between the two benches has decreased from 20′ 0½″ in Opening No. 650 to 17′ 10″ in this core drill hole in a distance of about 1,000 feet to the south.

E. J. Berwind Diamond Core Test (135)—No. 652 on Map II.

Located at the mouth of Petry Fork of Mats Creek, 2.6 miles southeast of Chap P. O.; **Campbell Creek (No. 2 Gas) Coal, 95' 3" deep.**

			Ft.	In.
Coal, no core 8"	2'	7"		
Fire clay	0	4		
Coal, no core	2	4	5	3

E. J. Berwind Diamond Core Test (140)—No. 653 on Map II.

Located in the bed of West Fork of Pond Fork, 3.3 miles south of Chap P. O.; **Campbell Creek (No. 2 Gas) Coal, 117' 11" deep.**

			Ft.	In.
Shale, sandy, roof				
Coal	2'	5 "		
Slate	1	4½		
Coal	0	11		
Slate	0	0½		
Coal, no core, (sandstone floor)	1	5	6	2

The composition and calorific value of the **Campbell Creek (No. 2 Gas) Coal** at the several commercial mines, as well as at prospect openings and country mines, together with an estimate of the probable available coal area of this bed by magisterial districts, will be discussed in detail in a subsequent Chapter of this Report.

THE BROWNSTOWN SANDSTONE.

Underlying the Campbell Creek (No. 2 Gas) Coal from 5 to 10 feet is a medium coarse grained, dark gray sandstone often 25 to 50 feet thick, that has been named the **Brownstown Sandstone** from a village of that name on Kanawha River, (now changed to Marmet), ten miles southeast from Charleston. This sandstone is used for building stone for house foundations, railroad culverts and bridge abutments at several places in Boone County, and makes a very good building material. Its thickness and occurrence in Boone County are given in the general sections on preceding pages of this Report.

THE POWELLTON COAL.

Underlying the Campbell Creek (No. 2 Gas) Coal from 40 to 70 feet is a multiple bedded gas coal, from 2 to 7 feet thick, that has been named the **Powellton Coal** from a mining village of that name on Armstrong Creek in Fayette County, where the coal was first mined and manufactured into coke. The coal has a columnar structure, is soft, low in sulphur, and makes an excellent foundry coke. In Boone County, this coal is thin and of little commercial value in the northern part of the county, so far as development has been made on this bed, or from prospecting and surface exposures. However, in the southern part of the county, this coal reveals sufficient thickness and purity to be a valuable bed.

Washington District.

In testing for the coal beds in Spruce Laurel Fork by Messrs. Chilton, MacCorkle, Chilton and Meany, and others, several of the diamond core test holes penetrate to the Powellton Coal, sections of which will now be given:

Spruce Laurel Diamond Core Test (120)—No. 654 on Map II.

Located on Spruce Laurel Fork, 5.5 miles southeast from Clothier; **Powellton Coal, 279′ 2″ deep.**

	Ft.	In.
Slate roof		
Coal, (fire clay floor)	3	1

Spruce Laurel Diamond Core Test (121)—No. 655 on Map II.

Located on Spruce Laurel Fork at the mouth of Skin Poplar Branch, 5.5 miles S. 42° 30′ E. from Clothier; **Powellton Coal; 283′ 6″ deep.**

	Ft.	In.
Shale roof		
Coal and bone, interlaminated, (shale floor)	2	10

Spruce Laurel Diamond Core Test (124)—No. 656 on Map II.

Located on Spruce Laurel Fork at the mouth of Trough Fork, 5.9 miles S. 37° 30′ E. from Clothier; **Powellton Coal, 302′ 9″ deep.**

			Ft.	In.
Shale roof				
Shale, **coal** streaks	0′	4″		
Coal, (fire clay floor)	1	3	1	7

Crook District.

The Powellton Coal comes out of Pond Fork just north of the mouth of Grapevine Branch and rises to the southeast above the bed of Pond Fork until at the mouth of Skin Fork, 2.0 miles southeast of Echart P. O., the coal is about 150 feet above the bed of the creek.

This coal is mined for local fuel use, about 0.25 mile up Grapevine Branch, where the following section was measured by Krebs:

Wharton Estate Local Mine Opening—No. 657 on Map II.

			Ft.	In.
Sandstone, massive Brownstown				
Coal, gas	2'	6"		
Coal, splint, (slate floor) 1091' L	0	1	2	7

Butts, N. 40° W.; faces, N. 50° E.

Opening driven in about 5 feet; base of coal, 91 feet above the base of the Matewan Coal at this point.

Wharton Land Company Opening—No. 648 on Map II.

Located on the west side of Pond Fork, 0.7 mile south of Bald Knob P. O.; section by Teets; Powellton Coal; elevation, 1220' B.

			Ft.	In.
1. Slate roof				
2. Coal, gray splint	0'	10"		
3. Coal, block	0	6		
4. Coal, gas	1	0		
5. Coal, gray splint, (slate floor)	0	7	2	11

The analysis of a sample collected from Nos. 2, 3, 4 and 5, as reported by Messrs. Hite and Krak, is given in the table of coal analyses under **No. 122.**

J. Q. Dickinson Local Mine Opening—No. 659 on Map II.

Located on the east side of Pond Fork, just north of the mouth of Rocklick Branch; section measured by Teets; Powellton Coal; elevation, 1320' B.

	Ft.	In.
Slate roof		
Coal, gas, (slate floor)	2	6

J. Q. Dickinson Local Mine Opening—No. 660 on Map II.

Located on Rocklick Branch of Pond Fork, 1.2 miles from its mouth; section by Teets; **Powellton Coal**; elevation, 1400' B.

			Ft.	In.
Slate roof				
Coal, gas	1'	4"		
Coal, gray splint	1	2		
Concealed by water	0	6	3	0

Wharton Estate Local Mine Opening—No. 661 on Map II.

Located on the north side of Rocklick Branch of Pond Fork, 0.25 mile from its mouth; section by Krebs; **Powellton Coal**; elevation, 1210' B.

				Ft.	In.
1.	Sandstone				
2.	**Coal**, gas	1'	0"		
3.	Slate	0	4		
4.	**Coal**, gas, visible	1	1	2	5

Butts, N. 40° W.; faces, N. 50° E.

The analysis of a sample collected from Nos. 2 and 4, as reported by **Messrs.** Hite and Krak, is given in the table of coal analyses under **No. 123.**

J. Q. Dickinson Local Mine Opening—No. 662 on Map II.

Located on a branch flowing into Pond Fork from the west, 0.3 mile up said branch, and 2.6 miles south from Bald Knob P. O.; section by Teets; **Powellton Coal**; elevation, 1390' B.

			Ft.	In.
Sandy shale roof				
Coal, gas	0'	1"		
Slaty shale	1	8		
Coal, gas, (slate floor)	2	4	4	1

Butts, N. 40° W.; faces, N. 50° E.; driven in about 15 feet.

Western Pocahontas Coal and Coke Company Prospect Opening—No. 663 on Map II.

Located on the south side of a branch flowing from the west to Lacey Branch of Pond Fork, 1.2 miles north of the Wyoming-Boone County Line; section by Teets; **Powellton Coal**; elevation, 1750' B.

			Ft.	In.
Sandstone, massive, **Brownstown**				
Coal, semi-cannel	1'	11"		
Coal, gas, (slate floor)	1	1	3	0

Butts, N. 40° W.; faces, N. 50° E.

The Powellton Coal does not rise above the bed of West Fork of Pond Fork, but has been encountered in diamond core test holes drilled on the property of E. J. Berwind, sections of which will now be given:

E. J. Berwind Diamond Core Test Hole No. 4 (133). No. 664 on Map II.

Located on West Fork at the mouth of Little Ugly Branch, 1.0 mile south of Chap P. O.; section by driller of well; **Powellton Coal**; 123' 3" deep.

			Ft.	In.
Slate roof				
Coal and slate	1'	0"		
Slate	1	7		
Shale	0	4		
Soft clay	0	4		
Slate	3	5		
Coal	2	0	8	8

E. J. Berwind Diamond Core Test No. 2 (134). No. 665 on Map II.

Located on west side of West Fork, 1.2 miles south of Chap P. O.; **Powellton Coal**, 116' 4" deep.

			Ft.	In.
Slate roof				
Coal and slate	0'	5"		
Coal, no core	0	6		
Slate	2	7		
Fire clay	2	0		
Slate	1	1		
Coal, no core	2	4	8	11

E. J. Berwind Diamond Core Test (135)—No. 666 on Map II.

Located on Mats Creek, 2.0 miles southeast of Chap P. O.; **Powellton Coal**, 125' 7" deep.

	Ft.	In.
Sandstone roof		
Coal, (slate floor)	0	4½

The coal appears to have thinned almost completely away.

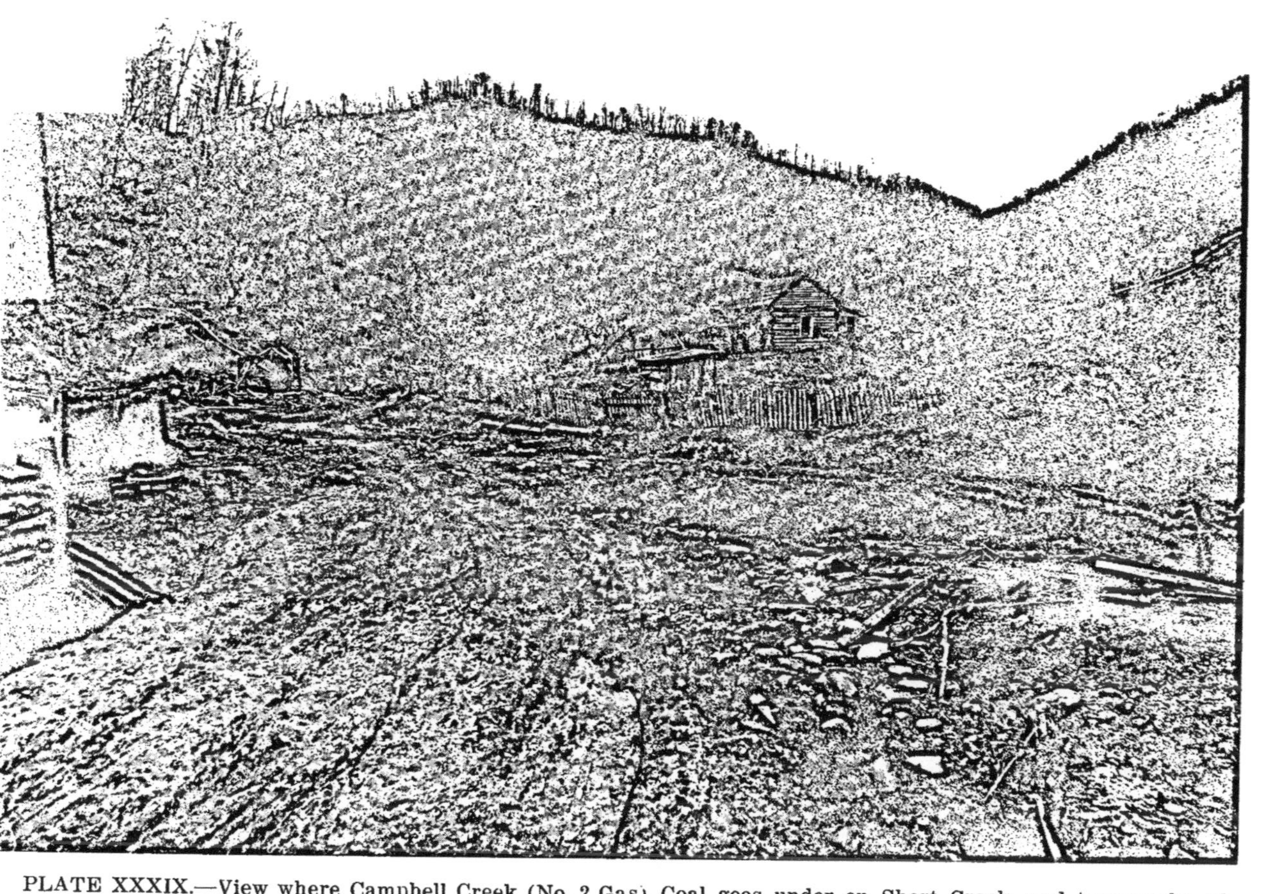

PLATE XXXIX.—View where Campbell Creek (No. 2 Gas) Coal goes under on Short Creek, and topography of Kanawha Series.

E. J. Berwind Diamond Core Test (140)—No. 667 on Map II.

Located on the west side of West Fork of Pond Fork, 3.3 miles south of Chap P. O.; **Powellton Coal, 192' 8" deep.**

	Ft.	In.
Slate roof		
Coal, no core (fire clay floor)	2	0

A further discussion of the Powellton **Coal,** with estimates of its probable tonnage, will be given in a subsequent Chapter.

THE STOCKTON LIMESTONE.

From 75 to 100 feet below the Campbell Creek (No. 2 Gas) Coal, and from 23 to 50 feet above the Eagle Coal, there is often found an impure bed of limestone named the **Stockton Limestone,** from the fact that Mr. Aaron Stockton many years ago made cement from this limestone, near Cannelton, in Kanawha County, on the north side of the Kanawha River, opposite Montgomery. In Boone County, this limestone is impure, and no marine fossils were found in same.

THE BALD KNOB SHALE.

From 50 to 95 feet under the Powellton Coal and separated from same by shale and often massive sandstone, there occurs a dark gray, slaty shale. This shale contains small marine fossils and was first studied by Krebs and Teets at Bald Knob P. O., Boone County.

This shale comes out of Pond Fork at the mouth of Grapevine Branch and rises above the bed of the stream, occurring from 20 to 100 feet above the same, until it goes under about one mile north of Pilot Knob. This shale goes under Skin Fork one mile northwest of Walnut Gap. It usually occurs from nothing to 3 feet above the Matewan Coal and varies in thickness from 1 to 8 feet.

THE MATEWAN COAL.

From 50 to 95 feet below the Powellton Coal and 40 to 60 feet above the Eagle Coal, there often occurs a double bedded seam of coal from 1 to 3 feet in thickness, which has been named the **Matewan Coal** by Hennen and Reger from a village of that name in **Mingo** County. In Boone County, this coal is thin and impure except in the southern part of Crook District, where it reaches its maximum thickness.

This bed rises out of Pond Fork at the mouth of Grapevine Branch and occurs in the hills along this stream from 20 to 100 feet above the bed of the stream to a point about one mile north of Pilot Knob where it goes under water level.

The following section was measured at the mouth of Grapevine Branch:

Wharton Land **Company** Local **Mine Opening.** **No. 643A on Map II.**

Located at the mouth of Grapevine Branch of Pond Fork, 1.5 miles south of Pond P. O.; section by Krebs and Teets; **Matewan Coal**; elevation, 1000' L.

	Ft.	In.
Slate, cannel, marine fossils, roof		
Coal, block	2	1
Butts, N. 40° W.; faces, N. 50° E.		

Wharton Land **Company Exposure—No. 645 on Map II.**

Located on trail leading from Bald Knob P. O. to West Fork, near Bald Knob P. O.; section by Teets; **Matewan Coal**; elevation, 1135' B.

	Ft.	In.
Shale, dark gray		
Shale, dark, limy, marine fossils	1	0
Coal, gas, (fire clay floor)	2	2

Wharton Land **Company Opening—No. 646 on Map II.**

Located on the west side of Pond Fork, 0.9 mile south of Bald Knob P. O.; section by Teets; **Matewan Coal**; elevation, 1210' B.

	Ft.	In.
Slate roof		
Coal, block, (slate floor)	2	9

Wharton Land **Company** Prospect **Opening.** **No. 647 on Map II.**

Located in the bed of Workman Branch of Pond Fork, 1.2 miles almost due north of Bald Knob P. O.; section by Teets; **Matewan Coal**; elevation, 1052' L.

			Ft.	In.
Slate roof				
Coal	1'	0"		
Slate, gray	0	4		
Coal, gas, visible	1	8	3	0

Pond Fork Coal and Land Company Opening. No. 647A on Map II.

Located on east side of Pond Fork, about 0.75 mile south of Bald Knob P. O.; section by Teets; **Matewan Coal**; elevation, 1140' B.

			Ft.	In.
Sandstone, massive				
Slaty shale, marine fossils			7	0
Coal, gas	1'	2"		
Slate, dark	0	2		
Coal, gas, (slate floor)	1	6	2	10

Pond Fork Coal and Land Company Opening. No. 647B on Map II.

Located on east side of Pond Fork, ⅓ mile southwest of mouth of Rocklick Branch; section by Teets; **Matewan Coal**; elevation, 1260' B.

			Ft.	In.
Shale, slaty, gray			5	0
Slate, dark, marine fossils			6	0
Coal, gas	1'	1"		
Coal, impure, slaty	0	2		
Coal, gas, visible	1	5	2	8

Wharton Land Company Local Mine Opening. No. 674 on Map II.

Located on the west side of Pond Fork, 1.1 miles northwest of Echart P. O.; section by Teets; **Matewan Coal**; elevation, 1330' B.

			Ft.	In.
Slate roof				
Coal, splint	1'	5"		
Slate, bone	0	2		
Coal, gray splint, (slate floor)	1	4	2	11

Wharton Land Company Opening—No. 675 on Map II.

Located on the east side of Pond Fork at Echart P. O. and mined by **A. H. Perry** for fuel; section by Teets; **Matewan Coal**; elevation, 1410' B.

			Ft.	In.
1. Slate roof				
2. **Coal**, block	1'	5"		
3. Slate, gray	0	3		
4. **Coal**, block, (slate floor)	2	1	3	9

The analysis of a sample collected from Nos. 2 and 4, as reported by Messrs. Hite and Krak, is given in the table of coal analyses under No. 124.

Wharton Estate Opening—No. 675A on Map II.

Located on the west side of Pond Fork, opposite the mouth of Burnt Camp Branch, about 1.0 mile southeast of Echart P. O.; **Matewan Coal**; elevation, 1500′ B.

			Ft.	In.
Slaty shale, roof, marine fossils				
Coal, gas	1′	1″		
Slate	0	2		
Coal, visible	1	7	2	10

Western Pocahontas Coal and Coke Co. Local Mine Opening. No. 681 on Map II.

Located on the east side of Skin Fork, 1.5 miles from its mouth; section by Teets; **Matewan Coal**; elevation, 1720′ B.

			Ft.	In.
Sandstone, massive			20	0
Slate, cannel, marine fossils			8	0
Coal, gas	1′	0″		
Slaty shale, plant fossils	4	0		
Coal, gas	0	4		
Shale, gray	0	6		
Coal, splint, hard, (slate floor)	1	8	[illegible]	6

Western Pocahontas Coal and Coke Co. Prospect Opening. No. 682 on Map II.

Located on the east side of Skin Fork of Pond Fork, 2.7 miles southeast of Echart P. O.; section by Krebs; **Matewan Coal**; elevation, 1675′ B.

			Ft.	In.
Dark shale			10	0
Slate, cannel, marine fossils			10	0
Coal, gas	1′	2″		
Shale, dark	4	0		
Coal, splint	0	5		
Shale, dark	0	5		
Coal, gas, (slate floor)	0	6	6	6

E. J. Berwind Diamond Core Test Hole (133). No. 682A on Map II.

Located just above the mouth of Little Ugly Branch on Little Ugly; **Matewan Coal; 184′ 8″ deep.**

	Ft.	In.
Shale roof		
Coal, (fire clay floor)	1	0

E. J. Berwind Diamond Core Test Hole No. 1 (140). No. 682B on Map II.

Located on West Fork, 3.3 miles south of Chap P. O.; Matewan Coal, 263' 0" deep.

			Ft.	In.
Slate roof				
Coal, no core 4"	1'	1"		
Slate	3	5		
Coal, no core 8"	1	0		
Coal, bone, (fire clay floor)	0	3	5	9

The composition and calorific value of the coal at the several commercial mines, as well as at prospect openings and country mines, together with an estimate of the probable available coal area of this bed by magisterial districts, will be discussed in a subsequent Chapter of this Report.

THE MATEWAN SANDSTONE.

Underneath the Matewan Coal there appears a medium coarse, buffish-gray sandstone from 20 to 40 feet in thickness. which has been named the **Matewan Sandstone** by Hennen and Krebs. This sandstone occurs in the portion of Boone on Pond Fork and on Little Coal River from a point north of Danville to Greenview along Pond Fork.

This sandstone usually forms abrupt bluffs and massive cliffs.

THE EAGLE "A" COAL.

Underneath the Matewan Sandstone, and overlying the Eagle Coal from 20 to 30 feet, occurs a thin seam of coal from 0 to 2 feet thick, which has been named the **Eagle "A" Coal,**[19] and as far as developments have been made in this bed, it appears to be thin and of little commercial value.

THE EAGLE SANDSTONE.

Underneath the Eagle "A" Coal and the Matewan Sandstone, there occurs a grayish, shaly and massive sandstone that has been named the **Eagle Sandstone.**[20] This bed is from

[19]Hennen and Reger, Logan and Mingo Report; 1914.
[20]Hennen and Reger, Logan and Mingo Report; 1914.

20 to 40 feet in thickness, and when it combines with the **Matewan** Sandstone above, forms cliffs from 70 to 80 feet in thickness.

This bed forms massive cliffs along Coal River from Racine to Peytona, and for nearly a mile on Drawdy Creek in Peytona District. The cliffs range in thickness from 60 to 80 feet at this point.

This sandstone also occurs along Big Coal River, between Seng Creek and Jarrolds Valley. Just north of Whitesville along the railroad grade, the sandstone occurs in massive cliffs from 60 to 70 feet high as shown in the following section:

	Ft.	In.
Gray shale		
Sandstone, massive, forming abrupt cliffs	60	0
Shale, dark	0	4
Coal, gas, **Eagle,** 880′ B		

The **Eagle Sandstone** also forms massive cliffs along Pond Fork, in the southern part of Crook District. The following section was measured just south of Echart P. O.:

	Feet.
Shale and concealed	
Sandstone, massive, forming rugged, abrupt cliffs	75
Coal, Eagle, 1240′ B	

THE EAGLE COAL.

From 140 to 200 feet beneath the Campbell Creek (No. 2 Gas) Coal, there occurs another important bed that has been named the **Eagle Coal** from a village of that name on the south side of Kanawha River, Fayette County, where the coal was first mined and shipped on a commercial scale. The Eagle Coal is a soft, columnar coal, and makes an ideal coke and by-product coal.

The following section, measured at Eagle, gives the type section of this coal:

			Ft.	In.
Sandstone roof				
Coal, gas	0′	3″		
Coal, impure	0	2		
Coal, gas	0	9		
Shale, gray	0	1		
Coal, gas	0	3		
Slate, gray	0	3		
Coal, gas, (slate floor)	2	10	4	7

In the northern part of Boone County, the **Eagle Coal** is thin and of little commercial value, but in the southern part of the county, the coal appears to have attained sufficient thickness and purity to render it valuable.

Peytona District.

Boone and Kanawha Land and Mining Company Opening. No. 668 on Map II.

Located on the north side of Coal River, about 500 feet below the mouth of Short Creek, Racine P. O.; section by Krebs; **Eagle Coal**; elevation, 664' L.

			Ft.	In.
Sandstone roof				
Coal, gas	0'	6½"		
Shale	0	2		
Coal, (fire clay floor)	0	2	0	10½

This coal comes 166 feet under the Campbell Creek (No. 2 Gas) Coal at this point.

In passing down Coal River from this point to Peytona, for a distance of about 2 miles, the **Eagle Coal** rises to the northwest toward the Warfield Anticline. The following opening was measured in a little branch on the north side of the river, just north of Peytona:

Opening—No. 669 on Map II.

			Ft.	In.
Sandstone, massive, **Eagle**				
Coal, gas	0'	2½"		
Shale, dark	0	2		
Coal, gas, (fire clay floor), 700' B.	0	2	0	6½

Peytona Coal Land Company Opening—No. 670 on Map II.

Located on the south side of Coal River and on the north side of Drawdy Creek, about 0.25 mile from its mouth, near Peytona P. O.; section by Krebs; **Eagle Coal**; elevation, 710' B.

	Ft.	In.
Sandstone, massive, **Eagle**	40	0
Coal, gas	0	2
Fire clay	5	0

A sample (No. 427-K) of the **Eagle Fire Clay** from the above opening was taken by Krebs and it yielded the following results on analysis by **Messrs.** Hite and Krak:

	Per cent.
Silica (SiO_2)	66.87
Ferric Iron (Fe_2O_3)	3.52
Alumina (Al_2O_3)	18.31
Lime (CaO)	0.32
Magnesia (MgO)	0.93
Sodium (Na_2O)	0.42
Potassium (K_2O)	2.93
Titanium (TiO_2)	0.66
Phosphoric Acid (P_2O_5)	0.12
Moisture	0.70
Loss on ignition	4.99
Total	99.77

From the above sections it is evident that the **Eagle Coal** is thin and of no commercial value in this portion of Boone County.

Scott District

The **Eagle Coal** is thin and impure in Scott District, as is shown in the following Opening, measured on Rock Creek, 0.8 mile west of Foster P. O.:

Opening—No. 671 on Map II.

	Ft.	In.
Sandstone, massive	50	0
Slate, 6" to	2	0
Coal, gas	0	1
Fire clay, 815' B.	2	0

The Eagle Coal horizon comes out of Little Coal River just north of Danville, and rising rapidly southeastward toward the Warfield Anticline, it is more than 200 feet above the bed of the stream at its crest near Low Gap. The coal is represented only by a trace of dark fire clay and often a thin impure layer of coal.

Sherman District.

The **Eagle Coal** comes out of Coal River about one mile south of the mouth of Seng Creek, and rising rapidly southeastward, the coal is 125 to 150 feet above the bed of the stream at Jarrolds Valley. The following section was measured by Krebs, on the east side of Coal River, 1.0 mile southeast from the mouth of Seng Creek, at an exposure along the Coal River Railroad grading:

Coal River Lumber and Mining Company Exposure.
No. 672 on Map II.

			Ft.	In.
Sandstone, massive, **Eagle**			60	0
Shale, gray			4	0
Coal, impure	0'	4"		
Shale, gray	0	2		
Coal, gray splint	1	4		
Shale, gray	0	6		
Coal, gas, (slate floor), 870' B.	0	6	2	10

Coal River Lumber and Mining Company Exposure.
No. 673 on Map II.

Located on the east side of Coal River, about 0.9 mile southeast from Seng Creek; section by Krebs, at a railroad cutting; **Eagle Coal**; elevation, 880' B.

			Ft.	In.
Sandstone, massive, **Eagle**			60	0
Slate, dark			0	4
Coal, gray splint	0'	4 "		
Slate	0	0½		
Coal, gray splint	0	10		
Slate	0	0½		
Coal, gas	0	8		
Shale, gray	1	0		
Coal, gas, (slate floor)	0	4	3	3

In the northern part of Sherman District, on Trace Fork of Joes Creek, the **Winifrede** Well **No. 1**, drilled by the Carter Oil Company, reports a 5-foot seam of coal at a depth of 390 feet from the surface. However, it is more than possible that the most of this thickness is soft fire clay, as is shown in the section of the Eagle bed at Peytona.

Crook District.

In the northern part of Crook District, the **Eagle Coal** appears thin and is represented by only a thin layer of coal and dark fire clay, while in the southern part of the district, the coal thickens and is a valuable bed from a commercial basis. The horizon of the Eagle Coal dips under Pond Fork about one mile north of Uneeda P. O., and in the Arbogast Well (63), located at the mouth of Robinson Creek, 2.0 miles southeast of Uneeda P. O., the driller reports a 5-foot seam of coal at a depth of 320 feet, which probably correlates with the Eagle Coal.

The Eagle Coal rises out of Pond Fork near Bald Knob P. O., and goes under Pond Fork 0.75 mile southeast of the mouth of Rich Branch, 4.5 miles east of Echart P. O., the rise being very rapid in the last two miles after the coal appears above the bed of the stream.

The following section was measured near Bald Knob P. O., where the coal rises above water level:

Wharton Estate Exposure—No. 657A on Map II.

Located on the east side of Pond Fork, 0.5 mile south of Bald Knob P. O.; section by Krebs; **Eagle Coal**; elevation, 1100' B.

			Ft.	In.
Sandstone, massive				
Shale, full of plant fossils			7	0
Coal, gas	0'	2"		
Shale, gray	2	0		
Coal, gas	1	0		
Shale, dark	0	4		
Coal, gas, (slate floor)	2	0	5	6

Butts, N. 40° W.; faces, N. 50° E.

Wharton Land Company Opening—No. 676 on Map II.

Located on the north side of Burnt Camp Branch of Pond Fork, near its mouth, 0.8 mile southeast from Echart P. O.; section by Teets; **Eagle Coal**; elevation, 1445' B.

				Ft.	In.
1.	Sandstone roof				
2.	**Coal**	0'	4"		
3.	Slate, gray	0	3		
4.	**Coal**	0	4		
5.	Slate, dark	0	1		
6.	**Coal**	0	1		
7.	Slate, gray	0	1		
8.	**Coal**, block	0	10		
9.	Slate, gray	1	0		
10.	**Coal**, gray splint	1	2		
11.	Slate	0	3		
12.	**Coal**, gray splint, (slate floor)	1	9	6	2

Butts, N. 40° W.; faces, N. 50° E.

Analysis of sample collected from Nos. 2, 4, 6, 8, 10 and 12, as reported by Messrs. Hite and Krak, is given in the table of coal analyses under **No. 125.**

Western Pocahontas Coal & Coke Company Prospect Opening—No. 677 on Map II.

Located on the west side of Pond Fork, 0.8 mile southeast of Echart P. O.; section by Teets; **Eagle Coal**; elevation, 1445′ B.

No.				Ft.	In.
1.	Slate roof				
2.	**Coal**	0′	1″		
3.	Slate, dark	0	1		
4.	**Coal**, block	0	8		
5.	Slate, gray	0	2		
6.	**Coal**, gray splint	1	7		
7.	Slate, gray	0	9		
8.	**Coal**	1	1		
9.	Slate, gray	0	2		
10.	**Coal**, splint, (slate floor)	1	11	6	6

Analysis of sample collected from Nos. 2, 4, 6, 8 and 10, as reported by Messrs. Hite and Krak, is given in the table of coal analyses under **No. 126.**

Wharton Land Company Prospect Opening. No. 678 on Map II.

Located on the south side of Pond Fork, about 1.0 mile southeast of Echart P. O.; section by Krebs; **Eagle Coal**; elevation, 1440′ B.

No.				Ft.	In.
1.	Sandstone, massive				
2.	Shale, dark gray			0	4
3.	**Coal**, splint	0′	2″		
4.	Shale, gray	0	0½		
5.	**Coal**, splint	0	2		
6.	Shale, dark	0	2		
7.	**Coal**	0	4		
8.	Shale, dark	0	1		
9.	**Coal**, gas	1	6		
10.	**Coal**, splint	0	5		
11.	Shale, dark	0	11		
12.	**Coal**, gas	1	0		
13.	**Coal**, splint	0	1½		
14.	Shale, dark	0	3		
15.	**Coal**, gas, (slate floor)	1	10	7	0

Analysis of sample collected from Nos. 3, 5, 7, 9, 10, 12, 13 and 15, as reported by Messrs. Hite and Krak, is given in the table of coal analyses under **No. 127.**

Western Pocahontas Coal and Coke Company Opening. No. 679 on Map II.

Located on the east side of Skin Fork of Pond Fork, 1.6 miles southeast of Echart P. O.; section by Teets; **Eagle Coal**; elevation Lower Bench, 1505' B.

				Ft.	In.
Slate roof					
Coal	0'	1"	**Upper Bench**	2	8
Slate, gray	0	5			
Coal	0	4			
Slate	0	3			
Coal, block	1	7			
Slaty shale				2	9
Coal, block	0'	11"	**Lower Bench**	2	9
Slate, gray	0	2			
Coal, block, visible	1	8			

Western Pocahontas Coal & Coke Co. Prospect Opening. No. 680 on Map II.

Located on Skin Fork of Pond Fork, 0.5 mile south of the mouth of Skin Fork; section by Krebs; **Eagle Coal**; elevation, Lower Bench, 1470' B.

				Ft.	In.
Shale, gray				4	0
Shale, dark				0	8
Coal, impure	0'	4"	**Upper Bench**	3	2
Shale, gray	0	5			
Coal, hard	0	5			
Coal, impure	0	6			
Coal, gas	1	6			
Shale, gray				3	0
Coal, gas	1'	0"	**Lower Bench**	2	10
Slate	0	4			
Coal, hard, (slate floor)	1	6			

Western Pocahontas Coal & Coke Co. Opening. No. 683 on Map II.

Located on the south side of Pond Fork, 0.3 mile east of the mouth of Skin Fork; section by Teets; **Eagle Coal**; elevation, 1500' B.

				Ft.	In.
1.	Slaty shale roof				
2.	Slate, dark			0	2
3.	**Coal**	0'	2"		
4.	Slate, gray	0	3		
5.	**Coal**	0	4		
6.	Slate, dark	0	3		
7.	**Coal**, gas	1	5		
8.	Slate, gray	0	3		
9.	**Coal**	0	11		
10.	Slate, gray	0	2		
11.	**Coal**, gas, (slate floor)	0	9	4	6

Analysis of sample collected from Nos. 3, 5, 7, 9 and 11, as reported by Messrs. Hite and Krak, is given in the table of coal analyses under **No. 128.**

Western Pocahontas Coal & Coke Co. Local Mine Opening. No. 684 on Map II.

Located in a small hollow on the south side of Pond Fork, 0.7 mile east of the mouth of Skin Fork; section by Teets; **Mid. Farrell Mine; Eagle Coal;** elevation, 1545' B.

			Ft.	In.
Slaty shale roof				
Coal	0'	4"		
Slate, gray	0	3		
Coal	0	4		
Slate, gray	0	2		
Coal, block	2	1		
Coal, splint, (slate floor)	0	8	3	10

Western Pocahontas Coal & Coke Co. Local Mine Opening. No. 685 on Map II.

Located on the south side of Pond Fork, 1,000 feet northeast from mouth of Lacey Branch; section by Krebs; **Eagle Coal;** elevation, 1570' B.

			Ft.	In.
1. Gray shale			10	0
2. **Coal,** splint	0'	2 "		
3. Shale	0	0½		
4. **Coal,** splint	0	2½		
5. Shale, gray	0	2		
6. **Coal,** gas, (slate floor)	3	6	4	1

Analysis of sample collected from Nos. 2, 4 and 6, as reported by Messrs. Hite and Krak, is given in the table of coal analyses under **No. 129.**

Western Pocahontas Coal & Coke Co. Local Mine Opening. No. 686 on Map II.

Located on Pond Fork, 0.75 mile southwest from the mouth of Farley Branch; section by Krebs; **Eagle Coal;** elevation, 1585' B.

			Ft.	In.
Shale roof				
Coal, splint	0'	11 "		
Slate	0	0½		
Coal	0	1½		
Slate	0	1		
Coal, splint	0	4		
Slate	0	2		
Coal, gas, (slate floor)	3	6	5	2

Western Pocahontas Coal & Coke Co. Local Mine Opening. No. 687 on Map II.

Located on the south side of Pond Fork, 1,000 feet west of Rich Branch; section by Krebs; **Eagle Coal**; elevation, 1750' B.

				Ft.	In.
Shale roof					
Coal	0'	2"			
Slate	0	1			
Coal	0	2			
Slate	0	2			
Coal, gas	0	5			
Coal, impure	0	4			
Coal, gas	1	6			
Slate	0	0¼			
Coal, gas	0	4			
Coal, splint	1	0			
Coal, gas, (slate floor)	1	4	....	5	6¼

Butts, N. 40° W.; faces, N. 50° E.

E. J. Berwind Diamond Core Test Hole No. 5 (132). No. 687A on Map II.

Located on West Fork, at mouth of Jarrolds Branch; **Eagle Coal, 205' 4½" deep.**

				Ft.	In.
Slate roof					
Coal	0'	11½"			
Slate	0	2½			
Coal, (shale floor)	4	3½	....	5	5½

E. J. Berwind Diamond Core Test Hole No. 4 (133). No. 687B on Map II.

Located on West Fork, at the mouth of Little Ugly Branch; **Eagle Coal; 260' 11" deep.**

				Ft.	In.
Slate roof					
Coal	5'	8½"			
Soft clay	1	0			
Coal	7	9			
Coal, no core 2"	1	0	....	15	5½

E. J. Berwind Diamond Core Test Hole No. 2 (134). No. 687C on Map II.

Located near the mouth of Little Ugly Branch; **Eagle Coal**, 246' 9" deep.

			Ft.	In.
Slate roof				
Coal	0'	11"		
Slate	0	2		
Coal, no core 2' 6½", (slate floor)	5	3	6	4

E. J. Berwind Diamond Core Test Hole No. 3 (135). No. 687D on Map II.

Located on Mats Creek near the mouth of Petry Fork, 2.6 miles southeast of Chap P. O.; **Eagle Coal**; 297' 10½" deep.

			Ft.	In.
Slate roof				
Coal	0'	9½"		
Bone	0	2		
Coal	2	11¾		
Slate	0	1		
Coal	0	9¼		
Slate	0	0½		
Coal	0	1		
Slate	0	3½		
Coal and slate	0	3		
Coal, (shale floor)	0	6	5	11½

E. J. Berwind Diamond Core Test Hole No. 1 (140). No. 687E on Map II.

Located on the west side of West Fork of Pond Fork, 3.3 miles south of Chap P. O.; **Eagle Coal**, 320' 2" deep.

			Ft.	In.
Slate roof				
Coal, no core 3"	0'	11 "		
Slate	0	2½		
Coal	4	4		
Coal and slate	1	0	6	8
Coal, (slate floor)	0	2½		

A further discussion of the **Eagle Coal**, with estimate of its probable tonnage, will be given in a subsequent chapter.

THE BENS CREEK SANDSTONE.

Beneath the Eagle Coal, there often occurs a grayish sandstone from 2 to 20 feet in thickness that has been named the **Bens Creek Sandstone**[21], from a stream of that name in Mingo County. In Boone County, where this sandstone is above water level, it is usually displaced by gray sandy shale.

THE BENS CREEK COAL.

From 2 to 30 feet under the Eagle Coal, there often occurs an impure, multiple bedded gas coal, from 6 inches to 3 feet thick, which has been designated the **Bens Creek Coal**[22]. It is more than probable that the bed is a split off the Eagle seam.

THE DECOTA SANDSTONE.

From 10 to 30 feet under the Eagle Coal, there occurs a buffish-gray sandstone, 30 to 60 feet thick, that has been designated the **Decota Sandstone,** from a mining village of that name on Cabin Creek, Kanawha County. In Boone County, this sandstone rises out of Coal River, just north of the mouth of Seng Creek, and forms cliffs 20 to 30 feet high.

The sandstone comes out of Little Coal River near Danville, and rises until at Low Gap, on the crest of the Warfield Anticline, its base is 100 to 150 feet above water level. From this point it dips to the southeast, and goes under Spruce Laurel about one mile north of Greenview.

THE LITTLE EAGLE COAL.

From 60 to 115 feet under the Eagle Coal, there occurs a multiple bedded gas coal from 0 to 3 feet thick that has been named the **Little Eagle Coal.** In Boone County, this bed, where above the surface, is usually thin, and interstratified with fire clay and slate.

The following section was measured on the railroad grade about 1.5 miles southwest from Low Gap Station:

			Ft.	In.
Sandstone, massive, **Decota**				
Shale, dark			6	0
Coal, impure	0′	2″		
Coal, splint	0	3		
Coal, (fire clay floor), 740′ B.	0	2	0	[illegible]

The coal dips rapidly to the southeast under the bed of

[21] Hennen and Reger, Logan and Mingo Report; 1914.
[22] Hennen and Reger, Logan and Mingo Report; 1914.

PLATE XL.—Tobacco barn on head of Big Creek.

Spruce Fork, and it is more than probable that this is the bed of coal noted by the drillers in the **Bennett Well (40)**, 0.25 mile south of Greenview P. O., at a depth of 104 feet, and reported to be 4 feet thick.

THE EAGLE LIMESTONE.

From 50 to 80 feet under the Little Eagle Coal and separated by sandy shale and dark slate, containing marine fossils, occurs a dark, grayish, impure limestone, often full of iron ore nodules, and 1 to 2 feet thick, containing **marine fossils.**

This limestone comes out of Little Coal River, between Danville and Madison, and where it appears on the east side of Spruce Fork, 0.5 mile north of Rucker Branch, the following section was measured by Krebs:

	Feet.
Sandstone, **Decota**	
Shale, gray and sandy	60
Limestone, impure, full of iron ore nodules and marine fossils, Eagle	2
Sandy shale and concealed to railroad grade, 769′ L.	69

The **Eagle Limestone** contains many marine fossils at this point, some of which have been given to Dr. Price, Paleontologist of the Survey, for study.

Near Low Gap Station, the following section was measured by Krebs:

	Feet.
Sandstone, massive, **Decota**	
Dark shale and concealed	65
Dark shale, Eagle, containing marine fossils	2
Dark shale	30
Sandstone and concealed to creek, 890′ B.	160

The **Eagle Limestone** horizon dips under Spruce Fork about 1.0 mile south of Low Gap Station.

The **Eagle Limestone** comes out of Turtle Creek about 1.0 mile south of Turtle Creek P. O., where it is characterized by large concretions, which have assumed the shape of turtle shields, being often 6 feet long, 4 feet wide, and 1 to 2 feet thick, and they probably gave name to the stream.

The following section was measured by Krebs at Turtle Creek:

	Feet.
Shale, gray	20
Limestone, dark gray, containing iron ore nodules and marine fossils, Eagle	1.5
Dark shale	18
Dark shale, containing **marine fossils**, 825′ B.	2

The dark fossiliferous Eagle Shales occur in the bed of Turtle Creek, just north of Olive Branch Church, 0.5 mile south of Turtle Creek P. O., where collections were made by W. Armstrong Price.

On Trace Fork of Big Creek, the following section was measured by Teets, at the mouth of Dog Fork of Trace Fork, 2.5 miles east from Manila P. O.:

	Ft.	In.
Sandy shale	5	0
Limestone, Eagle	2	0
Sandy shale	18	0
Dark shale, marine fossils, 815' L	2	0

The lowest members given in the generalized section of the Kanawha Series were not recognized in Boone County. In several of the wells drilled for oil and gas, coals are reported by the drillers, in the Pottsville Series, which will possibly correlate with some of the New River beds. However, these reports, as to the thickness and quality of the coals, are unreliable, owing to the fact that it is impossible to determine the thickness and quality of the coal with the drills used in sinking oil and gas wells.

DIAMOND DRILL BORINGS.

Several diamond drill borings have been sunk in Boone to test the thickness and character of the coals and other strata. While gathering data in the field for this Report, levels were taken on most of the wells and these have been numbered in connection with the oil and gas test borings for publication in the test and on Map II, the serial number corresponding in each instance. The accompanying table contains the abbreviated logs of 72 borings, as also the elevations on the well mouth, a portion of the records of which could not be obtained.

In the elevation column the letter "B" indicates barometric determinations, checked on near by U. S. G. Survey spirit-level elevations; the letter "L", spirit-levels, both being expressed in feet above tide. The explanations following the table of oil and gas wells apply, if further information is desired concerning any boring:

No.	NAME OF WELL.	Location: County	Location: District	Eleva	[illegible]: Thick-ness	[illegible]: Depth (top)	Ch: Thick-ness	Ch: Depth (top)	Upper Bench: Thick-ness	Upper Bench: Depth (top)	Lower Bench: Thick-ness	Lower Bench: Depth (top)	Th n	D (	Upper Bench: Thick-ness	Upper Bench: Depth (top)	Lower Bench: Thick-ness	Lower Bench: Depth (top)		
100	Forks Cb Cq No 1	Boone	Peytona	680B							1 8	92 8	1 6	144			3 0	227		
101	Forks Cb Co. No 1 (Shot D)	Boone	Peytona	680B			1 8	55 10					5 11½	137 10			3 0	227		
102	Forks Cb Cq No 3	Boone	Peytona	760B							0 5	80 1					5 8	137 1		
103	Forks Cb Cq No 4	Boone	Peytona	860B																
104	Forks Cb Cq No 5	Boone	Peytona	797L																
104"A"	Forks Cb Cq No 2	Boone	Peytona	695B																
105	Horse Cek Coal & Land Co.	Boone	Scott	685																
106	Hos Creek Coal & Land Co.	Boone	Scott																	
107	Horse Creek Coal & Land Co.	Boone	Scott																	
108	Hos Creek Coal & Land Co.	Bo	Scott																	
108"A"	Mandaville Hopkins	Boone	Scott	690B																
109	Boone Cty Coal Corporation No 2	Boone	Washington	797.8L			2 4	21 3	1 9	68 4			3 6	139 10						
110	Bo Cty Cb Corporation No 4	Boone	Washington	840B			1 6	90 4	2 1½	147 10			5 7	215 10						
111	Boone Cty Cb Corporation			835B																
112	Laurel Cb & Land Cq No 9	Boone	Washington	1038L			11 3	77 10					9 1	200 7			6 5½	280 11	5 4	345 9
113	Lac Cb & Land Cq No 10	Boone	Washington	1211L			1 5	262 10			0 1	373					21 7	452 7		
114	Laurel Coal & Land Cq No 3	Boone	Washington	970L					0 2	115 7	0 1	126 1	0 2	144 11			6 5	210 11		
115	Laurel Coal & Ld Cq No 1	Boone	Washington	1006L									6 6	139 9			7 3	236		
116	Laurel Cb & Land Cq No 10	Boone	Washington	1025L			8 4	14 2	26 3	25 10			9 2	132			7 11	226 4		
117	Laurel Cb & Land Cq No 4	Boone	Washington	1065B			2 6	67 4					8 1	175 4			8 2	266 5		
118	Laurel Cb & Lad Cq No 5	Bo	Wa hington	1042L			3 8	24 8					7 6	128 10			7 8	219 8		
119	Laurel Cb & Land Cq No 11	Boone	Washington	1125B									7 4	219 8			8 0	307 3		
120	Laurel Coal & Land Cq No 6	Boone	Washington	1067L									4 8	122	5 11¾	205 7	2 0	222 2	3 1	176 1
121	Laurel Coal & Land Cq No 7	Boone	Wa hington	1091L					0 1	59 6			6 10	123 8	7 0	209 9	2 2½	225 9½	2 10	280 8
122	Laurel Cb & Land Cq No 15	Boone	Washington	1160B			5 10	26 9					6 9	144 11			6 8	222 9		
123	Cassingham Cb & Land Cq No 3	Bo	Washington																	
124	Laurel Coal & Land Cq No 8	Boone	Washington	1123L			0 8	43 0					4 8	157 1			6 6	238 11	1 8	302 9
125	Laurel Coal & Ld Cq No 13	Boone	Washington	1245B			1 0	145 2					7 7	270 6			5 9½	346 4		
126	Laurel Coal & Land Cq No 12	Bo	Wht i	1190B			1 0	20 9					2 7	138 9			5 8	216 6		
127	Laurel Cb & Land Cq No 14	Boone	Washington	1225B			8 4	31 9					4 0	150 10			5 10	236 4		
128	Laurel Coal & Land Cq No 2	Boone	Ws hington	1233L							0 4	96 10	4 6	115 10			6 9	199 2		
129	Boone County Cb Cop No 3	Boone	Washington	1920B																
130	Cassingham Cb & Land Cq No 2	Boone	Washington	1400B			3 5	78	1 10	260 1	2 11½	148 5½	5 2	183 6			1 5	127 10		
131	Cassingham Cb & Land Cq No 1	Boone	Wa hington	1570B	1 0	203 8	0 10½	228 5½	1 11	260 1	0 2	286 0	3 4	348 4			6 3	449 0		
132	Cho & Ashby, No 5	Boone	Crook	1041L							1 0	33 1	2 0	56 1			5 5½	163 6		
133	Crawford & Ashby, No 4	Boone	Crook	1208L									011	8 1	3 0	36 10	4 4½	70 3½	9 8	114 7
134	Chf & Aby No 2	Boone	Crook	1203L											3 3	28 11	2 6	50 0	8 11	116 4
135	Crawford & Ashby, No 3	Boone	Crook	1350B									1 0	24 3			5 3	90 0		
136	Laurel Cb & Land Co., No 4	Boone	Crook	1790B																
137	Laurel Cb & Land Cq Cas. No 2	Boone	Crook	973L									7 0½	171 1			4 9	345 2		
138	Laurel Coal & Lac Cq Ca No 1	Boone	Crook	1035L			3 3	120 6	1 9	148 3	0 5	188 9	6 5	204 8			4 1½	285 8		
139	Laurel Coal & Land Co., Cas. No 3	Boone	Crook	1190B			10 3	175 1			0 8	240 11	9 2	256 5			6 8½	138 8		
140	Crawford & Ashby, No 1	Boone	Crook	1425B									1 2	26 0			7 2	111 9		
141	LaFollette, Robson et al.			780B							1 10½	0 3½	14 5	47 3			6 2	90		
142Lo	Bo County Cb Corporation, No 1	Logan	Logan	304.8L			11 3	126	21 1	142	2 2¼	187 5	4 2	87 11			2 0	354		
143Lo	Coal River Land Co No 1	Logan	Logan	845B			0 0½	128 7½	3 1½	230 2½	1 3	252 8	6 1	263 9						
144Lo	Bo Cty Coal Cpi No 6	Logan	Logan																	
145Lo	Boone County Coal Cpa	Logan	Logan	1290L																
146Lo	Boone County Coal Corporation, No 3	Logan	Logan	880B			2 2	94 0	3 9	177 5	1 3	203 5	1 8	224 8						
147Lo																				
148Lo	B County F Cpa No 6	Logan	Logan	1886B																
149Lo	Boone Cty Cb Cpn No 5	Logan	Logan	1385B			25 3	210 4			6 1	292 6								
150																				
151L in	Standard Fuel Cq No. 1	Lincoln	Washington																	
152K	Winifrede Cb Co.	Kaw v	Cabin Creek	800L																
153K	LaFollette, Robson et al.	Kanawha	Cabin Creek	945B											12 4	133	2 6	155	3 5	197 6
154K	LaFollette, Rb et al.	Kanawha	Cabin Creek	840B											1 6	90	2 8	110 0	2 10	148 8
155	Cassingham Coal & Land Cq No 4	Boone	Washington	1245B																
156	Coal & Land Cq No 5	Boone	Crook	1335B																
157	Boone County Cb Corporation	Bo	Washington	760B					2 6	36 8	1 10	77 7	6 2	107 6						
158Lo	Boone County Cb Corporation	Logan	Logan	940.6L					6 5½	246 3	2 7	307 11	5 9½	359 11						
159Lo	Bo County Cb Corporation	Logan	Logan	835B					3 5	170 11			4 6¼	240 0						
160Lo	Boone County Cb Corporation	Logan	Logan	901.7L					4 2	189 7			1 4	278 11						
161Lo	Bo County Cb Corporation	Logan	Logan	1020B									5 10	239 6	9 3	317 10				
162Lo	Bo County Cb Corporation	Logan	Logan	920B					3 4½	168 4	1 4½	212 6	1 10	227 0						
163Lo	Bo County Coal Corporation	Logan	Logan	1500B					6 3	368 0										
164Lo	Boone County Cb Corporation	Logan	Logan	1090B									3 9½	294 10½	1 7	351 4				
165Lo	Boone County Cb Corporation	Logan	Logan	980B					11 9	214 5	0 10	238 8	0 11	282 11½	1					
166Lo	Bo County Cb Corporation	Logan	Logan	1270B									5 7	364 1						
167Lo	Bo County Cb Cpi	Logan	Logan	1075B									7 8	247 4						
168Lo	Boone County Cb Corporation	Logan	Logan	1900B									6 2	258 10						
169Lo	Boone Cq Cb Corporation	Logan	Logan	1270B					1 0	149 7			7 7½	254 10						

COALS IN THE LOWER POTTSVILLE SERIES.

Mr. C. C. Sharp, a Mining Engineer and Mine Manager, of Athens, Ohio, once had some drilling done at the mouth of Short Creek on Coal River, and in the Foster Well (46) reports the following coals: A two-foot seam at 500 feet and a four-foot seam at 600 feet. The general appearance of the coal taken from the hole was that of a soft gas coal, very much resembling the New River type.

The following table shows the thickness and depth at which coal has been reported by drillers in oil and gas wells in Boone County:

Depth and Thickness of Coals in Lower Pottsville Series.

No. of Well	COAL		COAL	
	Depth, Feet.	Thickness Feet.	Depth, Feet.	Thickness Feet.
28	578	3	...	...
39	648	2	403	3
39A......	250	5	...	...
41	214	2	332	2.
49	648	3	...	...
57	823	3	920	4
58	330	3	...	...
63	903	3	...	...
64	1566	5	...	...

PART III.

Mineral Resources.

CHAPTER VIII.

PETROLEUM AND NATURAL GAS.

All the oil and gas yet discovered in West Virginia, with one or two exceptions, has been produced from sandstone beds, called "Sands" by well drillers. The Greenbrier Limestone or "Big Lime" is producing some dark, heavy oil in northern Calhoun, northeastern Roane, Lincoln and Cabell Counties, and gas in Kanawha, Boone and Lincoln Counties. It is the only calcareous stratum that has produced either oil or gas in West Virginia.

The following table exhibits in descending order the position of the sands, or oil and gas bearing horizons in West Virginia:

The Oil and Gas Horizons of West Virginia.

		Series	Sands
Carboniferous.	Pennsylvanian	Monongahela Series......	Carroll Sand (Uniontown).
		Conemaugh Series.......	Minshall (Connellsville). Murphy (Morgantown). Moundsville (Saltsburg). First Cow Run (Little Dunkard) Sand (Buffalo). Big Dunkard Sand (Mahoning)
		Allegheny Series........	Burning Springs (Upper Freeport) Sand. "Gas Sand" (Lower Freeport).
		Pottsville Series.........	"Gas Sand" of Marion and Monongalia Counties, (Homewood), Second Cow Run of Ohio. "Gas Sand" of Cairo. "Salt Sand" of Cairo. Cairo?
	Mississippian	Mauch Chunk Red Shale.	Maxton, Dawson, Cairo.
		Greenbrier Limestone "Big Lime"	Productive in the southern part of West Virginia.
		Pocono Sandstones......	Keener Sand and Beckett Sand of Milton. Big Injun Sand. Squaw Sand. Wier Sand. Berea Grit.
	Devonian	Catskill Red Beds.......	Gantz Sand. Fifty-foot Sand. Thirty-foot Sand. Stray Sand. Gordon Sand. Fourth Sand. McDonald or Fifth Sand. Bayard or Sixth Sand. Elizabeth or Seventh Sand.
		Chemung and Portage Beds	Warren First or Second Tiona, Speechley Sand. No well defined oil or gas horizons yet discovered in West Virginia.

In Boone County, the highest Sand productive of oil or gas that has been encountered by the drill is the "Salt Sand" in the Pottsville Series; the Big Lime, Big Injun, Squaw and Berea are also producing horizons in the county.

The approximate interval from the Campbell Creek (No. 2 Gas) Coal to the top of the Big Lime varies from 1,000 feet in the northern part of the county to 1,515 feet in the Arbogast well at the mouth of Robinson Creek on Pond Fork, showing a thickening of over 500 feet in less than 15 miles to the south.

DESCRIPTION OF SANDS.

The Salt Sand.

The first productive sand in the area under discussion is the **Salt Sand.** The top of this sand lies from 500 to 700 feet below the horizon of the Campbell Creek (No. 2 Gas) Coal. The thickness of the sand throughout Boone County varies from 300 to 600 feet. A showing of oil and gas was found in the Forks Creek well No. 4 (4) at a depth of 532 feet from the surface.

The Maxton Sand.

The next sand in descending order is the **Maxton** Sand. This sand occurs from 100 to 150 feet below the Salt Sand and in it a showing of gas has been found in several wells in Boone County. In the Coal River **Mining** Company well No. 1 (9) considerable gas was encountered at 960 feet in this sand.

The Big Lime.

The **Big Lime,** known as the **Mountain** or **Greenbrier Limestone,** is persistent under Boone County and varies in thickness from 175 feet to 275 feet. This "Sand" produces gas in nearly every well that has been drilled thus far in Boone County. The "Pay" appears to be about 50 feet from the top of the stratum.

The Big Injun Sand.

The **Big Injun Sand** is easily identified in Boone County, since it comes directly underneath the Big Lime. In some parts of West Virginia, the Big Injun Sand often attains a thickness of 100 to 200 feet. In Boone County, this sand rarely reaches a thickness of 60 feet and thus far has not been very productive in the wells that have been drilled.

The Berea Sand.

The **Berea Sand** occurs from 650 to 750 feet under the top of the Big Lime. It ranges in thickness from 18 to 25 feet and is the great oil producing sand in the Griffithsville, Lincoln County, oil field, from 4 to 6 miles northwest of the Boone-Lincoln County Line. A showing of oil and gas is found in this sand in several of the wells that have been drilled in Boone.

OIL AND GAS DEVELOPMENT.

EARLY HISTORY.

Gas development in Boone County began about 1896, when several wells were drilled near Racine on Short Creek, but as they were only small gas wells, very little development had been done in the county until the years 1913-1914, when test wells were drilled at different points in search of oil and gas. Thus far very little oil has been encountered, but nearly every well drilled contains some gas.

BOONE COUNTY WELL RECORDS.

The main source of information as to thickness of coal and other formations of economic interest where they lie deeply buried below the drainage is the logs or records of the many wells drilled for oil, gas and coal by both individuals and corporations, through whose courtesies the Survey has been able to collect a large number. Levels have been taken on most of these wells while gathering data in the field. The records of the wells in many cases are incomplete in giving the occurrence of the coals encountered, only the most important oil and gas sands being noted. The importance of keeping accurate logs of strata penetrated is plainly set forth in the "Preface" to Volume I(A) of the West Virginia Geological Survey for 1904.

The accompanying table contains the abbreviated records of 65 wells in Boone County and one each in Logan and Kanawha Counties. The wells are numbered consecutively from 1 to 66 and grouped largely by magisterial districts, the serial number in all cases corresponding to the number of the same well on Map II, and generally, where one of these tabulated wells is mentioned in the text, the serial number is added therewith in parentheses:

No.	Name of Well.	Location District.	Owner.	Elev.													Producing Sand	
1	Jhy Dart	Peytona																
2	Joseph Dart	Peytona																
3	Joseph Dart No. 1	Peytona	Eagle Oil & Gas Co.	620B			355	515	192	1028	408	90	1220	27	1713	1780	Big Lime	Gas
4	Forks Ca l Co. No. 4	Peytona	Eagle Oil & Gas Co	665B			340	860	190	1041	376	45	1231			1276	Big Lime	Gas
5	Fk c Ca l Co	Peytona	Clm bs Gas & Fuel Co															
6	Fk c Cal Co	Peytona	Holly & Stephenson	810B														
7	Forks Coal Co	Peytona	Columbus Gas & Fuel Co															
8	Jps Dart No 4	Peytona	Raven Cab Co	715B			430	345	90	1020	305					1063	Big Lime	
9	Cl River Mining Co., No. 1	Peytona	Columbus Gas & Fel Co	819L			528	330	193	1058	239	60	1255	10	1730	1755	Big Lime	Gas
10	Cal River Mining Co., No. 1	Peytona	Walnt Creek Oil Co	775B					160	1144	369			8	1783	1810	Big Lime	2,00
11	Coal Rer Mi 'ng Co., No. 3	Peytona	Columbus Ga & Fuel Co	719L			436	400	154	987	268					1141	Big Lime	Gas
12	Clo River Mining Co., No 2	Peytona	Walnut Creek Oil Co	800B					149	1108	308	70	1257	6	1745	1777	Big Lime	Gas
13	Boone & Kanawha Land & Mining Co., No. 1	Peytona	Chs Ga & Fuel Co	875L					180	1215	340	50	1395	15	1866	1900	Big Lime	Gas
14	Peytona Block Coal Co., No 5	Peytona	Cdu Oil Co	685B														
15	Peytona Block Ca l Co., No. 2	Peytona	Cnd Oil Co	765					280	1120	355	50	1400	4	1874	1897	Big Lime	Gas
16	Forks Coal Co	Peytona	Holly & Stephenson	745B														
17	Joseph Dart, No 6	Peytona	Raven Cal Co	630B														
18	Joseph Dart, No 5	Peytona	Ran Carbon Co	653L														
19	Cal River Mining Co., "A"	Peytona	Columbus Gas & Fuel Co	750					18	1008						1171	Big Lime	Gas
20	Romeo Pickens	Scott	Dr. E. B. Stephenson et al.	754L			585	500	205	1260	506	50	1465	18	1926	1918	Big Lime	Gas
21	Horse Creek Coal & Land Co	Scott	Gpu	680														
22	Horse Creek Coal & Land Co., No. 3	Scott	Bdn & Trees	745B			560	690	26	1360	615	55	1640	24	2115	2139	Berea	Abo
23	Horse Creek Coal & Land Co., No. 4	Scott	Bdn & Trees	860B			595	50	16	1360	500	50	1525	27	2017	2057		Sma
24	Hse Creek Co l & Land Co	Scott	Gap	925B														
25	Pryor & Allen, No 1	Scott	Crude Oil Co	750B			580	350	225	1150	400	60	1375	20	1845	1874	Berea	Sho
26	Jacob Chambers	Scott	Grosscup	768L														
27	Pryor & Allen, No 1	Scott	Columbus Ga & Fuel Co	825B			455	485	213	1080	255	20	1293	15	1796	1822	Big Lime	Sho
28	Pryor & Alln, No 2	Scott	Clm Gas & Fel Co	867L			650	295	205	1145	278	30	1350	12	1818	1861	Big Lime	Gas
29	Rock Creek Land Co	Scott	Columbus Gas & Fuel Co															
30	Mv Hopkins	Scott	Clm Gas & Fuel Co															
31	Mandaville Hopkins, No. 1	Scott	Shields Oil Co	700B			410	290	210	950	250	60	1160	15	1639	1674	Big Lime	Gas
32	Mandaville Hopkins	Scott	Shields Oil Co	765B			250	465	213	1020	255	57	1233	20	1734	1785	Big Lime	Gas
33	Ju lan Hill No. 1	Scott	Shields Oil Co	825B														
34	Lackawanna Coal & Lumber, No. 6	Coo k	Lackawanna Coal & Lumber Co	875B														
35	Lackawanna Coal & Lumber, No. 7	Cko	Lackawanna Coal & Lubr Co	1045														
36	Lackawanna Coal & Lumber	Sherman	Lackawanna Coal & Lbr Co	1090														
37	Lackawanna Coal & Lumber	Sherman	Lacke Coal & Lumber Co	1490														
38	J. M. Hopkins, No. 1	Whi ngton	South Penn Oil Co	730B			708	12	269	916	186	55	1185	36	1669	1892		Dry
39	Little Cal Land Co., No. 1	Washington	Chs Gas & Fuel Co	850					189	1396	546	47	1587	28	2085	2115	Berea	Gas
39"A"	Little Coal Land Co., No. 2	Wgtl	Clm Gas & Fuel Co	833L			500	400	161	1199	366			25	1950	1895		Wel
40	J. M. Bennett, No. 1	Wgtl	Big Creek Develonment Co	750B			770	220	• 300	1337	572	40	1637	31	2105	2705		Dry
41	H. C. Ballard, No. 1	Whi ngton	South Penn Oil Co	740B			260	515	249	985	245	36	1234	24	1726	3028	Berea	Sma
42	T. L. Foster, No 2	Sherman	Chb Natural Gas Co	765B														
43	T. L. Foster, No 1	Sherman	Chb Nha Gas Co	710B														
44	Em Boswell Hrs	Sherman	Fidelity Oil & Ga Co	750B														
45	Bishop Peterkin	Sherm	Fd lity Oil & Ga Co	880B														
46	T. L. Foster, No 3	Sherm	Fd lity Oil & Ga Co	675B														
47	Andrew Porter	Sherman	Crawford & Ashby	672L			380	420	232	1135	463	60	1367	6	18	1902	Big L. and Berea	Gas
48	Peytona Block Co l Co., No. 1	Sherman	Cdu Oil Co	675B			250		250	1092		35	1402				Big Lime	Gas
49	J. E. Tony	Sherman	Crude Oil Co	730														
50	—— Mkf No 1	Sherman	Lackawanna Coal & Lumber Co	685B			580	320	225	1190	505	85	1415	10	18	1907	Big Lime	Gas
51	Mrs. John L. Workman	Sherman	South Penn Oil Co	680B					230	1160	480					1401		Sho
52	Peytona Coal Land Co. No. 3	Sherman	Crude Oil Co	775B			695	580	140	1395	620	56	1535	12	2008	2029		Dry
53	Peytona Coal Land Co. No. 4	Sherman	Crude Oil Co	950B			465	675	230	1470	520	15	1700	3	2180	2702	Big Lime	Gas
54	Lackawanna Cb & Lumber Co. No. 1	Sherman	Lackawanna Coal &Lumb Co	690B	3 6	124	530	758	318	1417	727	55	1735			1820		Sma
55	T. L. Bno No. 1	Sherman	Guyan Oil Co	750B			685	800	181	1603	853	46	1802	20	2259	2290		Littl
56	Wlf rd Coal Co No 1	Sherman	Cnr Oil Co	900B	5	250	472	1055	218	1700	800	22	1948	6	2365	2950		Dry
57	Lackawanna Col & Lumber Co. No. 2	Sherman	Lackawanna Cb & Lumber Co	875B	4	195	556	924	205	1730	855	30	1945	1	2420	2427	Big Lim	Gas
58	Lackawanna Coal & Lumber Co., No. 3	Sherman	Lackawanna Coal & Lumb Co	850B			550	790	228	1880	1030	30	2220			2400	Keener & Big Injun	1.50
59	Lackawanna Coal & Lumber Co	Sherman	Lal c Coal & Lumber Co	938L														
60	LaFollette, Rho et al.	Sherman	Gay Oil Co	720B														
61	Robson-Prichard No 1	Shma	Guyan Oil Co	775B			575	685	281	1724	1039	40	2005	20	2461	2528		Ligh
62	John Tony	Sherman	United Fal Gas Co	760B														
63	D. M. Arbogast, No 1	Ck	Sarver Oil & Gas Co	734L			495	920	183	1625	865	22	1808	17	9	2321	Big Lim & Berea	Gas
64	Sho k & Garrison No 1	Sherman	Columbus Ga & Fuel Co	950B			732	660	240	2053	1103	88	2354	?!	22	2777	Mox . B. I. & Wier	Gas
65	Boone Cty Cal Cmp , No. 1	Lga Dist., Logan Co	Big Creek Development Co	805			765	730	330	1870	1065	40	2200	29	2639	2721	Dry hle	Gas
66	Wlli Coal Co., No 1	Ch i Creek, Kan	Ch ibs Gas & Fuel Co	850B	5 0	115	379	850	239	1815	965	35	2055	38	2482	2520	Berea	75 b

In the elevation column the letter "B" indicates that the elevation of the well mouth was obtained with aneroid, checked with the near by U. S. Geological Survey spirit level elevations; the letter "L," with spirit level, the same being expressed in feet above tide. Depth of the Campbell Creek (No. 2 Gas) Coal and the several Sands also expressed in feet.

The accompanying table furnishes a means of ready reference as to the ownership and elevation of the several wells, along with the depth and thickness of the Campbell Creek (No. 2 Gas) Coal, the depth of the important oil and gas horizons, and the name of the producing sands at most of the wells, but it is essential that the complete logs of a number of these borings be published, not only to preserve them from loss, but for the fund of information they contain as to other sands, as well as the presence or absence of minable coal. The accurate location of any tabulated well is readily determined from its serial or map number, given in the above table, on Map II, and with the heading in parentheses when the complete record is given.

Peytona District.

Peytona District occupies the north end of Boone County. A glance at the structural contours on Map II will show that the southern part of the district is traversed in a northeast-southwest direction by the Warfield Anticline, and the strata are very much warped, giving conditions for the accumulation of oil and gas pools. The wells drilled thus far indicate considerable gas in the **Big Lime**, with a showing of gas in the **Maxton Sand**, and of oil and gas in the **Berea Sand**.

The following is the record of a gas well drilled near the northern part of the district:

Forks **Creek** Coal **Company** No. 4 Well **Record** (4).

Peytona District, 0.5 mile southwest of Dartmont; authority, Columbus Gas and Fuel Company; completed Jan. 31, 1914; elevation, 665′ B.

	Thickness Feet.	Total Feet.
Gravel	40	40
Coal, Campbell Creek (No. 2 Gas)	2	42
Slate	83	125
Sand	45	170
Slate	40	210
Lime	15	225
Slate	105	330
Sand	43	373
Slate	27	400
Sand	15	415
Slate	30	445
Lime	25	470
Sand	45	515
Slate	5	520
Sand	340	860
Slate	8	868
Red rock	12	880
Slate	40	920
Red rock and shells	25	945
Sand, **Maxton**	60	1005
Little Lime	36	1041
Big Lime	190	1231
Sand, **Big Injun**, to bottom	45	1276

Water at 208′; showing of oil at 532′; gas, 1092′, 1184′ and 1225′; 6⅝″ casing, 1051′.

Jos. Dart No. 2 Well Record (3).

Peytona District, on Bull Creek, at Dartmont; authority, Eagle Oil & Gas Company; elevation, 620′ B.

	Thickness Feet.	Total Feet.
Soil	15	15
Sand	82	97
Coal, Eagle	2	99
Sand	81	180
Slate	100	280
Sand	120	400
Slate	8	408
Sand	97	505
Slate	10	515
Sand, **Salt**	355	870
Slate	5	875
Red rock	5	880
Black slate	6	886
Lime	25	911
Red rock	35	946
Slate	7	953
Lime	7	960
Sand, **Maxton**	44	1004
Slate	3	1007
Lime and Pencil Cave	21	1028

		Thickness Feet.	Total Feet.
Big Lime..............80'	Big Lime....	192	1220
Lime 38			
Sand 20			
Lime 54			
Sand, **Big Injun**..............................		90	1310
Slate		60	1370
Sand, **Squaw**..................................		50	1420
Slate and shells................................		293	1713
Sand, **Berea Grit**............................		27	1740
Black slate to bottom..........................		40	1780

8" casing, 1028'; first gas at 1108' in lime formation; second gas at 1146', sandy formation; 1170' tubing; packer set, 1085".

Jos. Dart No. 4 Well Record (8).

Peytona District, 0.5 mile south of Dartmont; authority, F. B Banger; completed, 1912; elevation, 715' B.

		Thickness Feet.	Total Feet.
Gravel		25	25
Sand		75	100
Slate		200	300
Sand		35	335
Slate		10	345
Sand, **Salt**....................................		430	775
Slate		10	785
Black lime..		30	815
Red rock..		25	840
Black lime..		20	860
Sand		10	870
Slate		80	950
Black lime..		10	960
Slate		10	970
Lime ....50'	Big Lime....	90	1060
Sand, gray...............20			
Lime 20			
Unrecorded to bottom..............................		3	1063

Casing, 10", 18'; 8¼", 360'; 6⅝", 980'; hole full of water at 335'; gas well; capacity, 2,500,000 cu. ft. daily.

Coal River Mining Co. "B" No. 1 Well Record (9).

Peytona District, on Falling Rock Fork of Lick Creek, N. 80° E. 2.1 miles from Ashford P. O.; authority, Columbus Gas & Fuel Company; completed, Feb. 25, 1914; elevation, 819' L.

	Thickness Feet.	Total Feet.
Gravel	12	12
Slate	15	27
Sand	133	160
Slate	80	240
Sand	75	315
Slate	15	330
Sand, **First Salt**	255	585
Slate	40	625
Sand, **Second Salt**	233	858
Slate and red rock	99	957
Sand, **Maxton**	60	1017
Little Lime	38	1055
Pencil Cave	3	1058
Big Lime	193	1251
Slate	4	1255
Red sand, **Big Injun**	60	1315
Slate and shells	399	1714
Brown shale	16	1730
Sand, **Berea**	10	1740
Unrecorded to bottom	15	1755

Casing, 8¼", 400'; 6¾", 1062'; hole full of water at 20'; water, 3 bailers, at 450'; hole full of water at 775'; gas in Maxton, 960'; first pay in Big Lime at 1194'; second pay at 1236'; packer, at 1140'; gas well.

This gas well is located about 0.4 mile west from the crest of the Warfield Anticline. Considerable gas was also found in the **M**axton Sand.

Near the head of Roundbottom Creek, another gas well (13) was drilled about 0.6 mile east of the crest of the Warfield Anticline, as follows:

Boone and Kanawha Land and Mining Co. No. 1 Well Record (13).

Peytona District, 1.2 miles N. 10° E. from Peytona P. O., on Roundbottom Creek; authority, Columbus Gas & Fuel Co.; completed, Feb. 5, 1914; elevation, 875' L.

	Thickness Feet.	Total Feet.
Gravel, brown	27	27
Sand, white	53	80
Slate, black	60	140
Sand, white	60	200
Slate, black	40	240
Sand, white	160	400

	Thickness Feet.	Total Feet.
Slate, black	45	445
Sand, white	55	500
Slate, black	25	525
Sand, white	75	600
Slate, black	20	620
Lime, black	40	660
Slate, black	60	720
Sand, **Salt**, white	265	985
Slate, black	30	1015
Sand, white	10	1025
Red rock	45	1070
Slate, white	25	1095
Sand, **Maxton**	45	1140
Little Lime, black	70	1210
Pencil Cave, black	5	1215
Big Lime, white	180	1395
Big Injun, red	50	1445
Slate and shells	414	1859
Shale, brown	7	1866
Berea Grit	15	1881
Slate, gray, to bottom	19	1900

Casing, 10", 115'; 8¼", 500'; 6⅝", 1200'; gas at 1225', 1346', and 1874'; **gas well.**

Peytona Land Company No. 2 Well Record (15).

Peytona District, 0.5 mile S. 10° E. from Peytona P. O.; on Coal Branch; authority, Crude Oil Company; completed, Jan. 7, 1909; elevation, 765' B.

	Thickness Feet.	Total Feet.
Gravel	12	12
Lime	38	50
Sand	40	90
Slate	90	180
Lime	200	380
Sand, **First Salt**	160	540
Slate	10	550
Sand, **Second Salt**	250	800
Slate	50	850
Sand, **Third Salt**	270	1120
Lime	12	1132
Sand	5	1137
Lime	75	1212
Slate, **Pencil Cave**	13	1225
Lime and sand, **Big Lime** (gas)	175	1400
Sand, **Big Injun**	50	1450
Unrecorded	424	1874
Sand, **Berea**	2	1876
Slate to bottom	21	1897

Conductor, 12'; 8¼" casing, 148'; 6½", 1141'; **gas well** in Big Lime.

Coal River Mining Co. "A" No. 3 Well Record (11).

Peytona District, 1.5 miles S. 35° W. from Brushton Station, on Brush Creek, 0.5 mile above mouth of Honey Camp Fork; authority, Columbus Gas and Fuel Co.; completed, Jan. 7, 1914; elevation, 719' B.

	Thickness Feet.	Total Feet.
Clay, yellow	12	12
Lime, gray	88	100
Sand, white	290	390
Slate, black	10	400
Sand, white, **First Salt**	270	670
Slate, black	10	680
Sand, white, **Second Salt**	156	836
Red rock	14	850
Slate, white	65	915
Sand, white	15	930
Slate, white	20	950
Little Lime, black	35	985
Pencil Cave, black	2	987
Big Lime, white, to bottom (gas)	154	1141

Casing, 10", 100'; 8", 390'; 6⅝", 995'; hole full of water at 400'; gas at 1039', 1056', and 1134'; **gas well.**

Coal River Mining Company No. 1 Well Record (10).

Peytona District, S. 30° W. 2.5 miles from Brushton Station, on Brush Creek; authority, Walnut Creek Oil Company; completed, Dec. 4, 1912; elevation, 775' B.

	Thickness Feet.	Total Feet.
Unrecorded	1144	1144
Big Lime	150	1294
Unrecorded	489	1783
Sand, **Berea**	8	1791
Black shale, to bottom	19	1810

Casing, 10", 38'; 8¼", 390'; 6⅝" 1150'; gas at 1176'; capacity, 2,000,000 cu. ft. daily.

Coal River Mining Co. "A" Well No. 4 Record (19).

Peytona District, 2.0 miles S. 55° W. from Brushton Station, on Honey Camp Branch of Brush Creek; authority, Columbus Gas & Fuel Co.; completed, March, 1914; elevation, 750' B.

	Thickness Feet.	Total Feet.
Gravel	10	10
Lime, white	10	20
Slate, black	10	30
Slate and shells, white	60	90
Coal, Eagle	3	93
Slate and shells, white	47	140

	Thickness Feet.	Total Feet.
Lime, white	20	160
Sand	8	168
Slate, black	2	170
Lime, white	20	190
Sand, white	20	210
Lime, white	30	240
Sand	60	300
Slate, black	25	325
Sand, white	65	390
Sand	10	400
Slate, black	16	416
Lime	42	458
Slate and shells	99	557
Lime, white	18	575
Sand, white	50	625
Salt Sand, white	75	700
Slate	10	710
Lime, sandy	120	830
Sand	10	840
Red rock	10	850
Slate, black	10	860
Lime, white	10	870
Red rock	30	900
Sand, Maxton, gray	75	975
Slate, black	5	980
Little Lime	23	1003
Pencil Cave	5	1008
Big Lime	158	1166
Sand, Big Injun, to bottom	5	1171

Water at 160', 300' and 400'; show of gas at 1011', 1071', 1075' and 1120'.

Coal River Mining Co. Well No. 2 Record (12).

Peytona District, S. 60° W. 2.5 miles from Brushton Station, on Honey Camp Fork of Brush Creek, at mouth of Loggy Hollow; authority, Walnut Creek Oil Co.; completed, Dec. 1912; elevation, 800' B

	Thickness Feet.	Total Feet.
Conductor	16	16
Unrecorded	1092	1108
Big Lime	149	1257
Sand, Big Injun	70	1327
Brown shale	405	1732
Unrecorded	13	1745
Sand, Berea	6	1751
Unrecorded to bottom	26	1777

Casing, 10", 37'; 8¼", 390', 6⅝", 1108'; gas at 1210'; smell of oil at 1750'; gas well.

Forks Creek Coal Co. No. 2 Well Record (7).

Peytona District, 6.0 miles S. 5° W. from Brounland on Wilderness Fork of Fork Creek; authority, Columbus Gas & Fuel Co.; completed, July 28, 1914; elevation, 829' L.

	Thickness Feet.	Total Feet.
Gravel	34	34
Slate	60	94
Sand, **First Salt**	536	630
Slate	25	655
Sand, **Second Salt**	295	950
Slate	55	1005
Lime and sand	45	1050
Sand	20	1070
Red rock	20	1090
Lime, shells and slate	85	1175
Sand	20	1195
Broken lime	82	1277
Big Lime	162	1439
Sand, Big Injun	25	1464
Slate and shells	411	1875
Black slate	20	1895
Sand, **Berea Grit**	5	1900
Slate and hard lime shells	69	1969
Slate, soft, to bottom	67	2036

Water to drill with at 780'; hole full of water at 885'; gas at 1324' and 1393'; **gas well.**

Prospective Oil and Gas Territory, Peytona District.

From the foregoing pages it will be readily observed that the search for oil and gas in Peytona District has thus far not been very successful. However, there is yet included within its boundary quite a large acreage that has not been condemned by dry holes. Nearly all the wells thus far have furnished gas in the Big Lime with a showing of gas in the Maxton and a showing of gas and oil in the Berea Sand, and it is more than probable that some small pools of oil or gas will be found in the district, when it has been thoroughly tested.

Scott District.

Scott District lies immediately west of Peytona and from 4 to 8 miles southeast from the Griffithsville oil pool. Several wells have been drilled in the district and nearly all of them

PLATE XLI.—Coalburg Sandstone, northeast of Gordon P. O., on head of Whites Branch.

are producing gas in the Big Lime with a showing of oil and gas in the Berea, but thus far no oil pool has been developed within the bounds of the district.

Romeo Pickens No. 1 Well Record (20).

Scott District, 1.75 miles N. 70° E. from Harless Station on Hewett Creek; authority, Dr. E. B. Stephenson; completed, December, 1912; elevation, 754' L.

		Thickness Feet.	Total Feet.
Conductor		19	19
Sand		16	35
Coal		2	37
Slate		123	160
Sand		20	180
Slate		80	260
Sand		50	310
Slate		90	400
Sand		10	410
Slate		90	500
Sand, **First Salt** 230'	Salt Sand	585	1085
Slate 100			
Sand 30			
Slate 30			
Sand 15			
Slate 35			
Sand, **Second Salt** 145			
Slate		140	1225
Little Lime		35	1260
Big Lime		205	1465
Sand, **Big Injun**		45	1510
Slate and shells		416	1926
Sand, **Berea**		18	1944
Unrecorded to bottom		4	1948

Casing, 10", 19'; 8", 310'; 6⅝", 1285'; showing of oil at 615'; water at 645'; two bailers; big water at 990'; gas at 1334'; showing of oil in Berea; **gas well.**

Pryor and Allen No. 3 Well Record (27).

Scott District, 3.9 miles N. 56° E. from Danville, on Camp Creek; authority, Columbus Gas & Fuel Co.; completed, May 19, 1914; elevation, 825' B.

	Thickness Feet.	Total Feet.
Gravel	16	16
Slate and rock	140	156
Slate	80	236
Lime	45	281
Slate	14	295
Sand	105	400
Slate	30	430

	Thickness Feet.	Total Feet.
Sandy lime	25	455
Slate	30	485
Sand, **Salt**	455	940
Slate	25	965
Lime, white	23	988
Slate and shells	22	1010
Sand, **Maxton**	25	1035
Little Lime	40	1075
Pencil Cave	5	1080
Big Lime	213	1293
Sand, **Big Injun**	20	1313
Slate and shells	472	1785
Shale	10	1795
Sand, **Berea**	15	1810
Slate to bottom	12	1822

Casing, 10", 122'; 8¼", 513; 6⅝", 1120'; water at 560'; gas at 1155', 1226', and 1804-1810'; **gas well.**

Pryor and Allen No. 2 Well Record (28).

Scott District, 2.1 miles N. 55° E. from Danville, on a branch of Rock Creek; authority, Columbus Gas & Fuel Co.; completed, Feb. 3, 1914; elevation, 867' L.

	Thickness Feet.	Total Feet.
Clay, yellow	16	16
Slate, white	34	50
Sand, white	30	80
Slate, white	40	120
Slate, black	120	240
Sand, white	24	264
Lime, white	26	290
Slate, black	5	295
Sand, white	280	575
Coal	3	578
Sand, **Salt,** white	367	945
Slate, white	7	952
Lime, white	6	958
Red rock	7	965
Slate, white	43	1008
Red rock	10	1018
Slate, white	26	1044
Little Lime	97	1141
Pencil Cave, black	4	1145
Big Lime, white (gas)	205	1350
Red rock	30	1380
Sandy shells, gray	80	1460
Slate, white	40	1500
Shells and slate, white	300	1800
Shale, brown	18	1818
Sand, gray, **Berea** (gas)	12	1830
Slate, white, to bottom	31	1860

Casing, 10", 365'; 8", 421'; 6⅝", 1148'; hole full of water at 680'; gas at 1192' and 1824'; **gas well.**

J. M. Hopkins No. 1 Well Record (31).

Scott District, 1.0 mile east from Danville on Hopkins Branch of Little Coal River; authority, Shields Oil Co.; completed, April 5, 1913; elevation, 720' B.

	Thickness Feet.	Total Feet.
Clay	12	12
Sand	100	112
Lime	38	150
Sand	40	190
Slate	5	195
Sand	85	280
Slate	10	290
Sand	50	350
Slate	100	450
Sand	200	650
Slate, black	3	653
Sand	47	700
Slate	15	715
Lime	10	725
Slate, black	1	726
Sand	44	770
Slate	10	780
Red rock	10	790
Slate	30	820
Sand	30	850
Slate	50	900
Lime, black.... 50' } **Big Lime** Lime, white.... 210 }	260	1160
Sand, red, **Big Injun**	60	1220
Slate	419	1639
Slate, brown	15	1654
Sand, **Berea,** to bottom	20	1674

Casing, 10", 140'; 8", 740'; 6½", 973'; hole full of water at 30'; water at 165', 190', 250' and 275'; gas at 800', 1065' and 1664'; **gas well.**

The J. M. Hopkins No. 2 Well (32) is located on Hopkins Branch. This well is also a gas well and is situated 1.2 miles west of the crest of the Warfield Anticline, as determined on the Structural Map with the Campbell Creek (No. 2 Gas) Coal as a base; but the top of the lime indicates that the anticline based on the Big Lime will pass close to the location of this well.

J. M. Hopkins No. 2 Well Record (32).

Scott District, 3,000 feet northeast of J. M. Hopkins No. 1 well, on Hopkins Branch of Little Coal River; authority, Shields Oil Co.; completed, Feb. 6, 1914; elevation, 765' B.

	Thickness Feet.	Total Feet.
Unrecorded	465	465
Sand, **Salt**	250	715
Sandy lime	55	770
Unrecorded	11	781
Sand, **Maxton**	94	875
Unrecorded	96	971
Little Lime	49	1020
Big Lime (gas)	213	1233
Sand, **Big Injun**	57	1290
Unrecorded	110	1400
Black lime	50	1450
Unrecorded	284	1734
Sand, **Berea**	20	1754
Unrecorded to bottom	31	1785

Casing, 10", 88'; 8", 800'; 6⅝", 1057'; water at 40', 170', 570', 590' and 850'; gas at 1071', 1090', 1110', 1145' and 1208'; showing of gas in Berea; **gas well.**

John Hill No. 1 Well Record (30).

Scott District, 0.75 mile northwest of Danville, on north side of Little Coal River; authority, Columbus Gas & Fuel Co.; completed, April, 1914; elevation, 695' L.

	Thickness Feet.	Total Feet.
Gravel	23	23
Lime	27	50
Slate, black	15	65
Sand, white	30	95
Slate, white	25	120
Lime	25	145
Slate, black	15	160
Lime, white	18	178
Slate, white	22	200
Sand	110	310
Slate, black	40	350
Sand, white	50	400
Slate, black	10	410
Sand, white	15	425
Slate, white	45	470
Sand, **Salt**	360	830
Sand, **Maxton**	40	870
Red rock	30	900
Slate, white	60	960
Little Lime, black	45	1005
Pencil Cave, black	5	1010
Big Lime, white	220	1230

	Thickness Feet.	Total Feet.
Sand, **Big Injun**	30	1260
Slate and shells, white	450	1710
Shale, brown	6	1716
Sand, **Berea**	20	1736
Slate, white, to bottom	16	1752

Big Lime gas at 1075′, 1102′, 1169′ and 1200′; Berea gas at 1720′; **gas well.**

Pryor and Allen No. 1 Well Record (25).

Scott District, 0.5 mile southeast of Van Camp Station; authority, Crude Oil Company; elevation, 750′ B.

	Thickness Feet.	Total Feet.
Clay	16	16
Sand, white	84	100
Slate, black	70	170
Sand, white	50	220
Slate, black	130	350
Sand, **First Salt**	250	600
Slate, white	150	750
Sand, **Second Salt**	180	930
Slate, black	70	1000
Red rock	20	1020
Slate, white	20	1040
Red rock	10	1050
Slate, white	20	1070
Sand, **Maxton**, white	40	1110
Slate, black	6	1116
Little Lime	26	1142
Pencil Cave	8	1150
Big Lime	225	1375
Sand, **Big Injun**	60	1435
Slate and shells, white	390	1825
Slate	20	1845
Sand, **Berea**, white	20	1865
Slate to bottom	9	1874

Hole full of water at 70′; show of gas at 400′; 4 bailers water at 540′; show of gas at 600′; hole full of water at 825′; show of gas at 1221′; gas in Berea at 1845′-1851′.

Northwest from Van Camp Station near Julian, **well "21"** on land of Horse Creek Land Company was drilled several years ago, and reported to contain a showing of oil and gas in the Berea Sand and a showing of gas in the Big Lime.

The **Noah Turley No. 1 well (416-L),** located on Laurel Fork of Horse Creek, 1.75 miles north of Woodville P. O., is a gas well in the Berea Sand, with a showing of gas in the Big Lime. The **Horse Creek Land Company No. 1 well (417-L)**, located on Laurel Fork of Horse Creek, is a gas well in the Berea Sand, making about 750,000 cu. ft. of gas daily, with a showing of gas in the Salt Sand.

Horse Creek Coal & Land Co. No. 4 Well Record (23).

Scott District, 1.5 miles southwest from Julian P. O., on Little Horse Creek; authority, Benedum and Trees; elevation, 860' B.

	Thickness Feet.	Total Feet.
Sand	60	60
Unrecorded	40	100
Lime	25	125
Lime shells and slate	125	250
Sand	60	310
Slate	30	340
Sand	70	410
Slate	45	455
Sand	55	510
Gas sand	60	570
Sand, Salt	595	1165
Red rock	15	1180
White slate	45	1225
Lime shells	45	1270
Sand	5	1275
Black lime, slate and Pencil Cave	85	1360
Big Lime	165	1525
Sand, **Big Injun**	50	1575
Sand, **Squaw**	15	1590
Lime shells and slate	160	1750
Slate	250	2000
Brown slate	17	2017
Sand, **Berea**	27	2044
Unrecorded to bottom	13	2057

Hole full of fresh water, 80'; hole full of salt water at 760'; small show of oil and gas at 670'; dry hole; well abandoned.

Horse Creek Coal & Land Co. No. 3 Well Record (22).

Scott District, 0.5 mile west of Mistletoe P. O., on Horse Creek; authority, Benedum and Trees; completed, Feb. 16, 1909; elevation, 745' B.

	Thickness Feet.	Total Feet.
Earth and sand	20	20
Slate	160	180
Sand	45	225
Slate	10	235
Sand	65	300
Slate	20	320

	Thickness Feet.	Total Feet.
Sand	60	380
Slate	40	420
Sand	30	450
Slate	30	480
Sand	60	540
Slate	50	590
Sand	50	640
Slate	50	690
Sand	210	900
Slate	25	925
Sand, **First Salt**	75	1000
Slate	120	1120
Sand, **Second Salt**	130	1250
Lime and slate	110	1360
Big Lime	280	1640
Sand, **Big Injun**	55	1695
Slate and shells	420	2115
Sand, **Berea,** to bottom	24	2139

Casing, 10", 18'; 8", 220'; 6⅝", 1360'; gas at 610'; water at 612'; about 300,000 cu. ft. gas in Berea.

Prospective Oil and Gas Area, Scott District.

This district has not been thoroughly tested for oil and gas and there remains yet quite a large area that has not been condemned by the drill. While the Structure Map based on the upper strata does not precisely give the structure of the oil and gas bearing sands owing to the rapid thickening to the southeast of the Pottsville Series, yet in a general way this map does represent the structure of the lower sands. Nearly all of the wells thus far drilled in Scott District have a strong flow of gas in the Big Lime with a showing of oil and gas in the Berea Sand. It is hardly probable that a large pool of oil will ever be found in this district. However, it is possible that oil will be found in the northern part of the district, east of Julian, if the Berea Sand is not too hard, since well No. 21 at Julian and well No. 416-L near Woodville each had a small showing of oil in the Berea Sand. The northern part of the district on Horse Creek contains quite a large area of land that has not yet been condemned by the drill. As the sands rise rapidly in the southern part of the district toward the crest of the Warfield Anticline, it may be possible that a pool of oil will some time be found in that part of the district.

Washington District.

Washington District lies west and south of Scott, and is traversed in a northeast and southwest direction by the Warfield Anticline in the northern part of the district, and by the Coalburg Syncline in the central part of the same. The strata are considerably warped in this district. Several wells have been drilled in the northern and central portions, but thus far only two have been productive of gas.

H. C. Ballard No. 1 Well Record (41).

Washington District, 1.5 miles southwest from Danville, on Long Branch of Turtle Creek; authority, South Penn Oil Co.; elevation, 740′ B.

	Thickness Feet.	Total Feet.
Conductor	10	10
Lime	30	40
Slate	10	50
Sand	35	85
Slate	5	90
Sand	40	130
Slate	6	136
Sand	34	170
Slate	2	172
Sand	40	212
Coal, Eagle	2	214
Slate	6	220
Sand	40	260
Slate	25	285
Sand	15	300
Slate	15	315
Sand	15	330
Coal	2	332
Slate	18	350
Lime	35	385
Slate	5	390
Lime	125	515
Sand, Salt	260	775
Slate	15	790
Lime	5	795
Slate	8	803
Lime	17	820
Red rock	45	865
Slate	12	877
Lime	8	885
Slate	33	918
Little Lime	52	970
Pencil Cave	15	985
Big Lime	249	1234
Sand, Big Injun	36	1270

	Thickness Feet.	Total Feet.
Slate	80	1350
Lime	60	1410
Slate	221	1631
Lime	26	1657
Slate	13	1670
Lime	20	1690
Slate	35	1725
Sand, **Berea**	24	1749
Slate	5	1754
Lime	61	1815
Slate	10	1825
Lime	15	1840
Slate and shells	115	1955
Lime	55	2010
Slate and shells	385	2395
Lime	95	2490
Slate and shells, to bottom	538	3028

Water at 68'; gas at 1070' and 1190'; a show of gas in Berea at 1729' and 1739'; dry hole; well abandoned.

Little Coal Land Co. No. 2 Well Record (39A).

Washington District, 1.0 mile southwest from Turtle Creek P. O., on Mud Fork of Turtle Creek; authority, Columbus Gas & Fuel Co.; completed, Feb. 2, 1914; elevation, 833' L.

	Thickness Feet.	Total Feet.
Slate	35	35
Lime	5	40
Coal, Eagle?	5	45
Slate	40	85
Sand	65	150
Black slate	50	200
Lime	45	245
Coal	5	250
Slate	20	270
Lime	130	400
Sand, **First Salt**	200	600
Black slate	65	665
Sand, **Second Salt**	85	750
Slate	10	760
Sand, **Third Salt**	140	900
Slate	10	910
Sand, **Maxton**	65	975
Red rock	10	985
White slate	20	1005
Lime	10	1015
Black slate	15	1030
Red rock	10	1040
White slate	40	1080

	Thickness Feet.	Total Feet.
Black lime	20	1100
White slate	10	1110
Black lime	30	1140
White lime	59	1199
Big Lime	161	1360
Unrecorded	490	1850
Sand, Berea	25	1875
Unrecorded to bottom	20	1895

Casing, 8¼", 255'; 6⅝", 512'; showing of gas in Big Lime from 1250'-1257'; gas in top of Berea, about 100,000 cu. ft. daily.

Near the head of Mud River, well No. 39 was drilled at the mouth of Ballard Fork, with a showing of gas in the Big Lime, and a fair gas well in the Berea, by the Little Coal Land Company, as follows:

Little Coal Land Co. No. 1 Well Record (39).

Washington District, 1.4 miles east from Mud P. O., on Mud River, at the mouth of Ballard Fork; authority, C. A. Croft, Madison, W. Va.; completed, Oct. 19, 1913; elevation, 850' L.

	Thickness Feet.	Total Feet.
Conductor	12	12
Sand	50	62
Shale and slate	88	150
Coal	1	151
Sand	49	200
Coal	1	201
Sand	22	223
Slate	7	230
Slate and shale	506	736
Coal	2	738
Sand	122	860
Shale and slate	40	900
Coal, Sewell?	3	903
Shale and slate	147	1050
Sand	70	1120
Coal, Beckley?	6	1126
Sand	89	1215
Slate	5	1220
Lime	15	1235
Sand	55	1290
Red rock	10	1300
Lime	12	1312
Slate	48	1360
Little Lime	20	1380
Slate	16	1396
Big Lime	189	1585
Slate	2	1587

	Thickness Feet.	Total Feet.
Sand, Big Injun	47	1634
Slate	101	1735
Sand	85	1820
Slate and shells	265	2085
Sand, Berea	28	2113
Shale to bottom	2	2115

Gas in Big Lime at 1460'-1495'; gas in Berea after shooting, 1,500,000 cu. ft. daily.

A well was drilled several years ago on the east side of Spruce Fork, about one-fourth mile south of Low Gap Station, just south of the crest of the Warfield Anticline, with the following results:

J. M. Hopkins No. 1 Well Record (38).

Washington District, east side of Spruce Fork of Coal River, 0.25 mile south of Low Gap Station; authority, South Penn Oil Company; elevation, 730' B.

	Thickness Feet.	Total Feet.
Surface	12	12
Sand, Salt	708	720
Slate	20	740
Sand, Maxton	60	800
Slate	15	815
Red rock	75	890
Little Lime	20	910
Pencil Cave	6	916
Big Lime	269	1185
Sand, Big Injun	55	1240
Slate	40	1280
Lime shells	220	1500
Slate	120	1620
Shale	49	1669
Sand, Berea	36	1705
Slate and shells	115	1820
Slate to bottom	72	1892

Show of gas at 1675'-1685' in Berea Sand; dry hole; well abandoned.

J. M. Bennett No. 1 Well Record (40).

Washington District, 0.25 mile south of Greenview P. O., on east side of Spruce Fork of Coal River; authority, Yawkey and Freeman; completed, December, 1913; elevation, 765' B.

	Thickness Feet.	Total Feet.
Alluvium	19	19
Sand	36	55
Lime	45	100
Coal, Eagle	4	104

	Thickness Feet.	Total Feet.
Fire clay	4	108
Slate and lime shells	112	220
Sand	100	320
Slate	10	330
Sand	185	515
Brown shale	95	610
Sand, **Salt**	380	990
Slate	30	1020
Lime, black	45	1065
Red rock and lime shells	48	1113
Red rock	25	1138
Sand	28	1166
Red rock	42	1208
Lime, black	32	1240
Red rock	32	1272
Sand, **Maxton**	65	1337
Lime, **Big Lime**	300	1637
Sand, **Big Injun**	40	1677
Slate and shells	404	2081
Shale, brown	24	2105
Sand, **Berea**	31	2136
Slate and shells to bottom	569	2705
Dry hole; well abandoned.		

The above well is located 2.75 miles almost due south from the J. M. Hopkins No. 1 well (38), and the top of the Big Lime has dipped in that distance from 186 feet below tide in well (38) to 672 feet below tide in well No. 40, a descent of 486 feet, or more than 175 feet per mile.

Boone County Coal Corporation No. 1 Well Record (65-L).

Logan District, Logan County, 0.5 mile south of Clothier, on the east side of Spruce Fork, at the mouth of Bend Branch, and 0.5 mile west of the Boone-Logan County Line; authority, Yawkey and Freeman; completed, January, 1914; elevation, 805' B.

	Thickness Feet.	Total Feet.
Alluvium	24	24
Sand	96	120
Slate	35	155
Sand	5	160
Slate	33	193
Coal, Alma	3	196
Sand	53	249
Coal	3	252
Sand	81	333
Coal, Campbell Creek (No. 2 Gas)	4	337
Slate	8	345
Sand	175	520

	Thickness Feet.	Total Feet.
Slate and shells	210	730
Sand, **First Salt**	270	1000
Slate	60	1060
Sand, **Second Salt**	435	1495
Slate	30	1525
Limestone, black	5	1530
Slate, black	17	1547
Lime, black	10	1557
Red rock	3	1560
Slate and shells	20	1580
Sand	20	1600
Lime, black	25	1625
Slate, white	26	1651
Sand, **Maxton**	94	1745
Red rock	10	1755
Black limy shale	95	1850
Red limy shale	10	1860
Slate, black, Pencil Cave	10	1870
Black limy shale	60	1930
Lime, gray 45′; Lime, black and gray 155; Lime, red 70 } **Big Lime**	270	2200
Sand, **Big Injun**, red	40	2240
Slate and shells	425	2665
Shale, brown	24	2689
Sand, **Berea**	29	2718
Slate to bottom	3	2721

Hole full of water at 1160′ and 1700′; bad caves, 1530′ and 1547′; show of gas in Lime at 2160′; show of gas in top of Berea at 2690′; dry hole; well abandoned.

This well is located in the Coalburg Syncline, about 3.25 miles south from the J. M. Bennett well No. 1 (40). The top of the Big Lime dips 553 feet in that distance, or about 165 feet to the mile.

Prospective Oil and Gas Areas, Washington District.

This district has not been condemned for oil and gas, by the drill, thus far, as very few wells have been drilled within its area, leaving large boundaries yet untested. There is no doubt that gas will be found in the Big Lime, north of the crest of the Warfield Anticline, and possibly near the crest and south of same. The well at Low Gap (38) had a considerable showing of gas in the **Berea Sand.** To the southwest of this district, on Guyandot River, gas wells are found near the

Warfield Anticline, and it is more than possible that this pool will extend into Boone when that portion of the county is developed.

There may also be a small oil pool in the northern part of this district near the head of Mud River, while no doubt there are small pools of gas throughout the northern part of the district.

Sherman District.

Sherman District lies east of Peytona, and also southeast of the Warfield Anticline. Gas has been produced in this district for several years, and was piped to Charleston for local fuel for that city, but now the gas is utilized at Malden in the manufacture of salt from brines. A showing of oil has been found in several of the wells, but thus far no pool has been discovered.

Peytona Land Co. No. 1 Well Record (48).

Sherman District, 1.0 mile south of Racine P. O., on Indian Creek; authority, Crude Oil Company; completed, Nov. 7, 1908; elevation, 675' B.

	Thickness Feet.	Total Feet.
Earth	13	13
Sand	80	93
Slate	90	183
Sand	181	364
Slate	20	384
Lime	156	540
Sand	100	640
Lime	83	723
Sand	70	793
Slate	32	825
Sand	15	840
Lime	20	860
Sand	90	950
Slate	10	960
Sand	132	1092
Big Lime	250	1342
Sand, red, Big Injun	60	1402
Slate to bottom	25	1427

Conductor, 13'; Casing, 8¼", 735'; 6¼", 1097'; gas well in Big Lime.

Peytona Land Co. No. 3 Well Record (52).

Sherman District, 0.5 mile southwest from Sterling P. O., on Indian Creek; authority, Crude Oil Co.; completed, July 17, 1909; elevation, 795' B.

	Thickness Feet.	Total Feet.
Conductor	14	14
Slate	81	95
Sand	305	400
Slate	80	480
Sand	60	540
Slate	40	580
Sand, **Salt**	695	1275
Black lime	50	1325
White lime 45' / Break of slate 25 / White lime 140 } **Big Lime**	210	1535
Sand, **Big Injun**	56	1591
Unrecorded	417	2008
Sand, **Berea**	12	2020
Unrecorded to bottom	9	2029

Water at 20', 80' and 200'; casing, 8¼", 480'; 6⅝", 1180'; a small show of gas in Big Lime from 1415-1445'; dry hole; well abandoned.

Peytona Land Co. No. 4 Well Record (53).

Sherman District, 1.5 miles southwest from Sterling P. O. on Indian Creek; authority, Crude Oil Company; elevation, 950' B.

	Thickness Feet.	Total Feet.
Gravel	24	24
Shale and lime	51	75
Sand	85	160
Lime and shells	90	250
Slate	125	375
Sand, **First Salt**	225	600
Lime	75	675
Sand, **Second Salt**	155	830
Slate	10	840
Sand, Third **Salt**	300	1140
Black lime	20	1160
Sand and lime	50	1210
Sand	120	1330
Slate	20	1350
Sand, **Maxton**	90	1440
Black lime	30	1470
White lime, **Big Lime**	230	1700
Sand, **Big Injun**	15	1715
Slate and shells	465	2180
Sand, **Berea**	3	2183
Slate and shells to bottom	519	2702

Casing, 10", 250'; 6½", 1475'; water at 80' and 945'; **gas well**; capacity, 750,000 cu. ft. daily.

Workman No. 1 Well Record (51).

Sherman District, 0.5 mile southeast of Racine P. O., on north side of Coal River; authority, South Penn Oil Company; elevation, 665' B.

	Thickness Feet.	Total Feet.
Gravel	35	35
Sand	45	80
Slate	115	195
Sand	110	305
Coal	4	309
Sand	61	370
Slate	15	385
Sand	20	405
Slate	10	415
Sand	20	435
Slate	10	445
Sand	140	585
Slate	45	630
Sand	160	790
Lime, shells, etc.	100	890
Sand and lime	80	970
Red rock	20	990
White slate	16	1006
Hard sand, **Maxton**	119	1125
Lime 35' } **Big Lime**	265	1390
Lime and sand 230 }		
Sand, red, **Big Injun** to bottom	11	1401

Show of oil at 1354'; well made from 1 to 2 barrels daily; casing, 8¼", 305'; 6¼", 1125'.

J. E. Tony No. 1 Well Record (49).

Sherman District, south side of Coal River, 1.0 mile southwest from Racine P. O.; authority, Crude Oil Co.; completed, 1913; elevation, 685' B.

	Thickness Feet.	Total Feet.
Gravel, brown	30	30
Sand, white	130	160
Slate, black	30	190
Sand, white	120	310
Slate, black	20	330
Sand, white	90	420
Slate, white	20	440
Sand, white	205	645
Coal	3	648
Lime, black	22	670
Sand, white	180	850
Lime, black	180	1030
Red rock	10	1040
Sand, white, **Maxton**	135	1175
Lime, black	25	1200

	Thickness Feet.	Total Feet.
Slate, cave, black	10	1210
Little Lime, black	37	1247
Big Lime, white	192	1439
Sand, red.....30' } **Big Injun**	61	1500
Sand, white.....31 }		
Slate, black	407	1907
Sand, gray, **Berea**	12	1919
Slate, black, to bottom	16	1935

First water at 120'; water at 300', 315', 380' and 810'; a little gas at 1394'; casing, 10", 400'; 8", 1247'; dry hole; well abandoned.

Porter No. 1 Well Record (47).

Sherman District, 1.5 miles east from Racine P. O., on the north side of Coal River, at mouth of Tony Creek; authority Coal River Oil and Gas Company; elevation, 672' L.

	Thickness Feet.	Total Feet.
Sand	20	20
Gravel	10	30
Lime	7	37
Sand	28	65
Slate	75	140
Sand	20	160
Slate and shells	140	300
Sand	80	380
Slate	40	420
Sand	128	548
Slate	2	550
Sand	250	800
Slate	20	820
Lime	25	845
Sand	100	945
Rock	8	953
Lime	162	1115
Slate	13	1128
Lime	7	1135
Lime, **Big Lime**	232	1367
Sand, **Big Injun**	60	1427
Shells and slate	390	1817
Shale	14	1831
Sand, **Berea**	6	1837
Slate to bottom	65	1902

Casing, 8¼" 380'; 6¼", 1135'; water at 65' and 630'; show of gas in Big Lime at 1193', 1288', and in Berea at 1834'; well good for about 200,000 cu. ft. daily.

Midkiff No. 1 Well Record (50).

Sherman District, 2.0 miles southeast from Racine P. O., on south side of river; authority, Lackawanna Coal & Lumber Co.; completed, 1912; elevation, 685' B.

	Thickness Feet.	Total Feet.
Loose rock and gravel	30	30
Sandstone	30	60
Slate	100	160
Sandstone	58	218
Slate	7	225
Unrecorded	25	250
Sandstone	10	260
Slate	5	265
Sandstone	35	300
Slate	20	320
Sandstone, **First Salt**	285	605
Sandstone and lime	75	680
Sandstone, **Second Salt**	220	900
Sand and lime	90	990
Red rock	30	1020
Sand and lime, **Maxton**	123	1143
Shale	47	1190
Big Lime	225	1415
Sandstone	85	1500
Slate	395	1895
Sand, **Berea**	10	1905
Slate to bottom	2	1907

Water at 605' and 900'; gas in Big Lime and Big Injun; about 1,000,000 cu. ft. daily; **gas well.**

Winifrede Coal Company No. 1 Well Record (56).

Sherman District, 3.75 miles east from Racine P. O., on Trace Fork of Left Fork of Joes Creek; authority, The Carter Oil Company; completed, Jan. 27, 1914; elevation, 900' B.

	Thickness Feet.	Total Feet.
Conductor	9	9
Soil	21	30
Sand	50	80
Coal, Cedar Grove, Upper Bench	2	82
Sand and slate	18	100
Coal, Cedar Grove, Lower Bench	4	104
Slate and lime	146	250
Coal, Campbell Creek (No. 2 Gas)	5	255
Lime	35	290
Sand	30	320
Slate and shells	20	340
Sand	8	348
Slate and shells	37	385
Coal, Eagle	5	390

	Thickness Feet.	Total Feet.
Slate	25	415
Sand	35	450
Lime	45	495
Sand	25	520
Slate and shells	48	568
Lime	27	595
Sand, little gas	35	630
Slate and shells	150	780
Sand, hard	130	910
Slate and shale	105	1015
Sand	10	1025
Slate and shells	30	1055
Sand, **First Salt**	145	1200
Slate	36	1236
Sand and shale	16	1252
Slate	180	1432
Sand	88	1520
Unrecorded	82	1602
Sand, **Maxton**	33	1635
Little Lime	50	1685
Unrecorded	10	1695
Pencil Cave	5	1700
Big Lime	218	1918
Slate	30	1948
Sand, **Big Injun**	22	1970
Unrecorded	15	1985
Sand, **Squaw**	20	2005
Slate and shells	360	2365
Sand, **Berea**	6	2371
Slate and shells. to bottom	579	2950

Show of gas in Lime at 1784'; dry hole; well abandoned.

Lackawanna Coal & Lumber Co. No. 1 Well Record (54).

Sherman District, at Seth, on Coal River, 0.5 mile north of mouth of Laurel Creek; authority, Lackawanna Coal & Lumber Co.; completed, 1912; elevation, 698' B.

	Thickness Feet.	Total Feet.
Loose rock and gravel	50	50
Unrecorded	34	84
Slate	40	124
Coal, Campbell Creek (No. 2 Gas), Upper Bench	3.5	127.5
Slate	32.5	160
Coal, Campbell Creek (No. 2 Gas), Lower Bench	3.5	163.5
Slate	216.5	380
Shale	60	440
Sandstone	230	670
Lime	88	758

		Thickness Feet.	Total Feet.
Sandstone ....302′	Salt Sand....	530	1288
Sand and lime..........210			
Sandstone 18			
Slate		32	1320
Sand and lime..............................		97	1417
Sand and lime, **Big Lime**..................		318	1735
Sand, **Big Injun**..........................		55	1790
Slate		20	1810
Sand, **Squaw**..............................		7	1817
Slate to bottom............................		3	1820

Small showing of gas at 1820′.

T. L. Broun No. 1 Well Record (55).

Sherman District, 0.5 mile northeast from Seth P. O., on a small branch flowing west into Coal River; authority, Guyan Oil Co.; completed, September 1912; elevation, 750′ B.

		Thickness Feet.	Total Feet.
Gravel		5	5
Slate and shells...........................		60	65
Sand		25	90
Black shale................................		10	100
Sand		53	153
Slate		12	165
Lime		10	175
Sand		178	353
Black shale................................		167	520
Sand		170	690
Lime		15	705
Black shale................................		70	775
Lime		25	800
Sand		350	1150
Slate		20	1170
Sand		162	1332
Slate		53	1385
Sand		100	1485
Lime		35	1520
Pencil Cave................................		2	1522
Lime .... 76′	Big Lime....	262	1784
Slate 5			
Lime 181			
Red rock...................................		5	1789
Lime		13	1802
Sand, red, **Big Injun**.....................		46	1848
Slate and shells...........................		393	2241
Black shale................................		18	2259
Sand, **Berea**..............................		20	2279
Slate and shells to bottom.................		11	2290

Water at 65′, 175′, 520′, 670′ and 880′; gas at 160′, 275′ and 520′; showing of oil at 300′; oil at 1752′; well shot and abandoned.

Lackawanna Coal & Lumber Co. No. 2 Well Record (57).

Sherman District, 3.25 miles southwest from Seth P. O., on Sandlick Creek; authority, Lackawanna Coal & Lumber Company; completed, 1912; elevation, 875' B.

	Thickness Feet.	Total Feet.
Loose rock and gravel	50	50
Slate	145	195
Coal, Campbell Creek (No. 2 Gas)	4	199
Slate	31	230
Shale and sandstone	130	360
Sandstone	85	445
Slate and shells	65	510
Sand and lime	42	552
Slate	148	700
Sand, **First Salt**	120	820
Coal?	3	823
Sand, **Second Salt**	97	920
Coal?	4	924
Sandstone, **Third Salt**	316	1240
Coal	5	1245
Sandstone, **Fourth Salt**	235	1480
Slate	60	1540
Slate and shells	5	1545
Sand, **Maxton**	60	1605
Slate	43	1648
Little Lime	33	1681
Sand and lime	45	1726
Pencil Cave	4	1730
Big Lime	205	1935
Slate	10	1945
Sand, **Big Injun**	30	1975
Slate	75	2050
Sandy lime and shells	75	2125
Slate	295	2420
Sand, **Berea**	1	2421
Slate to bottom	6	2427

Gas well in Big Lime.

Lackawanna Coal & Lumber Co. No. 3 Well Record (58).

Sherman District, 4.0 miles south of Seth P. O., on Hopkins Fork of Laurel Creek, at mouth of Lavinia Fork; authority, Lackawanna Coal & Lumber Company; completed, 1913; elevation, 850' B.

	Thickness Feet.	Total Feet.
Gravel	12	12
Coal, Cedar Grove, Lower Bench	3	15
Slate and sand	90	105
Lime shells	65	170
Sand	157	327
Coal. **Eagle**	3	330
Sand and lime	440	770

	Thickness Feet.	Total Feet.
Slate	20	790
Sand	150	940
Lime and sand	130	1070
Sand, **Salt**	280	1350
Black lime	130	1480
Red rock	20	1500
Lime and sand	90	1590
Red rock	10	1600
Sand and lime	55	1655
Sand, **Maxton**	80	1735
Slate and red rock	135	1870
Pencil Cave	10	1880
Big Lime	238	2118
Sand	7	2125
Sandy lime	65	2190
Red rock	35	2225
Sand, **Big Injun**	30	2255
Sand to bottom	145	2400

Show of oil at 950'-980'; show of gas at 1085'; gas at 2120' and 2230'; **gas well.**

Lackawanna Coal & Lumber Co. No. 4 Well Record (59).

Sherman District, 6.2 miles south from Seth P. O., on Hopkins Fork, at mouth of Jarrolds Fork; authority, Lackawanna Coal & Lumber Co.; completed, April 2, 1914; elevation, 935' L.

	Thickness Feet.	Total Feet.
Gravel	10	10
Slate and shells	113	123
Coal, Campbell Creek (No. 2 Gas)	2	125
Sand	60	185
Slate and shells	15	200
Sand	50	250
Slate and shells	400	650
Coal, Lower War Eagle	3	653
Lime	82	735
Sand	130	865
Sand and lime	120	985
Slate and shells	115	1100
Sand	150	1250
Salt Sand	220	1470
Lime and sand	55	1525
Lime	30	1555
Red rock	25	1580
Lime and sand	25	1605
Red rock	120	1725
Sand	15	1740
Red rock	25	1765
Sand, **Maxton**	15	1780
Lime and shells	70	1850
Red rock	30	1880

	Thickness Feet.	Total Feet.
Slate and lime shells	115	1995
Big Lime	280	2275
Sand, **Big Injun**	75	2350
Slate and shells	20	2370
Sand, **Squaw**	30	2400
Sand and shells	160	2560
Slate and shells	98	2658
Sand, **Berea**	32	2690
Slate and shells	64	2754
Sand, **Gantz?**	15	2769
Sand and slate	25	2794
Sand, **50-Foot?**	16	2810
Slate to bottom	40	2850

Gas at 2130′ and 2235′ in Big Lime, and at 2769′.

LaFollette, Robson et al. No. 1 Well Record (60).

Sherman District, 1.0 mile north of Orange P. O., on east side of Coal River at mouth of Jakes Branch; authority, Guyan Oil Company; completed, December, 1912; elevation, 720′ B.

	Thickness Feet.	Total Feet.
Clay	20	20
Gravel	10	30
Lime	20	50
Sand	110	160
Slate	40	200
Sand	90	290
Slate	10	300
Sand	65	365
Slate	45	410
Sand	95	505
Slate	45	550
Sand	40	590
Slate	25	615
Sand	55	670
Slate	15	685
Sand	85	770
Slate	85	855
Sand, **Salt**	340	1195
Coal, Sewell	6	1201
Sand	59	1260
Lime	140	1400
Red rock	20	1420
Lime, gritty	110	1530
Red rock	30	1560
Slate and shells	130	1690
Little Lime	30	1720
Pencil Cave	4	1724
Big Lime	281	2005
Sand, **Big Injun**	40	2045
Slate and shells	143	2188

	Thickness Feet.	Total Feet.
Sand, **Squaw**	117	2305
Slate and shells	141	2446
Brown shale	15	2461
Sand, **Berea**	20	2481
Slate and shells to bottom	47	2528

Water at 955′; showing of gas at 1948-1950′, 2218-2219′, and 2238′; dry hole; well abandoned.

Shonk-Garrison No. 1 Well Record (64).

Sherman District, 3.0 miles southeast from Orange P. O., at mouth of Right Fork of Whiteoak Creek; authority, Columbus Gas & Fuel Company; completed, 1915; elevation, 950′ B.

			Thickness Feet.	Total Feet.
Sand and gravel			30	30
Slate			25	55
Sand			22	77
Slate			21	98
Sand			20	118
Slate			11	129
Lime			5	134
Slate			9	143
Sand			38	181
Slate			14	195
Sand			28	223
Slate			18	241
Sand			102	343
Slate			5	348
Sand			67	415
Slate			17	432
Sand			18	450
Lime			62	512
Slate			8	520
Sand			21	541
Slate and lime shells			21	562
Slate			92	654
Lime			6	660
Sand	266′	**Upper Salt Sand**	290	950
Sand, broken	15			
Sand	9			
Slate			80	1030
Sand	70′	**Lower Salt Sand**	362	1392
Slate	4			
Sand	96			
Slate	10			
Sand	182			
Slate			30	1422
Lime			25	1447
Slate			19	1466
Lime			34	1500
Slate			60	1560

	Thickness Feet.	Total Feet.
Lime	25	1585
Rock, red	25	1610
Sand	47	1657
Rock, red	8	1665
Slate and shells	80	1745
Sand	33	1778
Slate	12	1790
Sand, **Maxton**	110	1900
Rock, red	5	1905
Lime	25	1930
Slate	13	1943
Slate and shells	48	1991
Lime, black 32′ } Little Lime	48	2039
Lime, gray 16 }		
Slate, **Pencil Cave**	14	2053
Big Lime	240	2293
Sand, **Keener**	33	2326
Slate	28	2354
Sand, **Big Injun**	88	2442
Slate	8	2450
Sand 14′ }		
Slate 2 } **Wier**	30	2480
Sand 14 }		
Slate	6	2486
Sand	140	2626
Slate	39	2665
Slate and shells	68	2733
Sand shells, **Berea Grit**	2	2735
Slate and shells to bottom	42	2777

Casing, 12½″, 30′; 10″, 420′; 8″, 1615′; 6⅝″, 1763′; 5 3-16″, 2089′.

Gas in Maxton Sand at 1768′; in Big Injun at 2430′, about 6′ of "pay"; in Wier at 2468′, 2′ of "pay". Filled up with water at 2520′. Water at 71′, 1130′, 1812′ and 1858′. **Gas well.**

A well has recently been drilled in Cabin Creek District. on Longbottom Creek, 2.5 miles northeast of the Boone-Kanawha County Line, which has been producing oil in the Berea Sand. This well is located about 3 miles south of the Coalburg Syncline, as shown by the upper strata. The following is the record of this well:

Williams Coal Company No. 1 Well Record (66).

Cabin Creek District, 2 miles southwest of Miama P. O., on Longbottom Creek; authority, Columbus Gas & Fuel Company; completed, January, 1915; elevation, 850′ B.

	Thickness Feet.	Total Feet.
Sand	25	25
Slate	10	35
Sand	25	60
Slate	30	90
Sand	20	110
Coal, Campbell Creek (No. 2 Gas)	5	115
Slate	35	150
Sand	40	190
Slate	40	230
Sand	40	270
Shells	195	465
Sand	85	550
Slate	5	555
Sand	30	585
Slate	55	640
Sand	125	765
Coal?	4	769
Slate	4	773
Sand	67	840
Slate	10	850
Sand, **Salt**	379	1229
Slate	19	1248
Sand	7	1255
Slate shells	10	1265
Sand, white	13	1278
Slate, cave	2	1280
Sand shells	50	1330
Red rock	60	1390
Sand	50	1440
Red rock	55	1495
Sandy lime	25	1520
Lime, white	55	1575
Sand	5	1580
Red rock	30	1610
Lime shells	5	1615
Slate shells	30	1645
Red rock	50	1695
Sand, **Maxton**	15	1710
Slate	20	1730
Slate, black, cave	5	1735
Lime 45′; Break 3; Lime 25 } **Little Lime**	73	1808
Pencil Cave	7	1815
Lime, black 30′; Lime, white 160; Sand, pink 5; Lime, white 35 } **Big Lime**	230	2045
Sand, **Keener**	10	2055

	Thickness Feet.	Total Feet.
Sand, **Big Injun**	35	2090
Slate and sand	115	2205
Hard shell	10	2215
Slate	185	2400
Lime, hard, black	35	2435
Slate, brown	27	2462
Slate, black	20	2482
Sand, gray, close grained, **Berea**, to bottom	38	2520

Casing, 13", 18'; 10", 277'; 8", 1280'; 6", 1870'; Hole full of water at 1050'; gas at 2499'; first show of oil at 2503'; oil well making 75 barrels per day natural.

Prospective Oil and Gas Areas, Sherman District.

Several wells have been drilled in Sherman District in search for oil and gas, nearly all of which have produced gas, while in a few of them a showing of oil was found, but thus far no oil pools have been developed within the boundary of the district. There yet remains a large acreage that has not been condemned for oil or gas by dry holes. The wells around Racine, in the western part of the district, have been producing gas for a number of years.

The geologic structures seem to indicate that there may be several small pools within the district. However, the strata are fast thickening to the southeast in the Pottsville Series, and it is questionable if the structure of the upper strata will affect very much the lower oil and gas bearing sands. The Coalburg Syncline, running through the district in a southwest and northeast direction, forms a decided trough through the entire length of the district. The wells thus far drilled seem to indicate that this trough is from two to three miles farther north in the upper sand than in the lower sands. The oil encountered in the **Berea Sand** in the Williams well is about three miles south of this trough shown in the upper strata. It is more than probable that the Cabin Creek Pool may extend in a southwest direction to Boone County, thus forming a Berea Oil Pool.

Crook District.

Crook District lies southwest of Sherman, and is crossed in a northeast and southwest direction, in the northern part, by the Warfield Anticline, and north of the center, by the Coalburg Syncline. Its strata are considerably warped. Very little prospect work for oil and gas has been done in the district thus far.

D. M. Arbogast No. 1 Well Record (63).

Crook District, on east side of Pond Fork, at mouth of Robinson Creek, 5.0 miles southeast from Madison; authority, C. E. Sarver Oil and Gas Company; completed, March, 1914; elevation, 734' L.

	Thickness Feet.	Total Feet.
Sand	15	15
Gravel	8	23
Slate	62	85
Sand	15	100
Slate	10	110
Coal, Campbell Creek (No. 2 Gas)	3	113
Slate	12	125
Sand	60	185
Slate	15	200
Lime	10	210
Sand	90	300
Shells	15	315
Coal, Eagle	5	320
Slate	145	465
Sand	40	505
Lime	45	550
Sand	140	690
Slate	20	710
Sand	80	790
Slate	2	792
Sand	108	900
Coal, Sewell?	3	903
Slate and lime	17	920
Sand	275	1195
Slate	20	1215
Lime	15	1230
Sand	185	1415
Red rock	35	1450
Lime shells	123	1573
Slate	7	1580
Lime	10	1590
Shells	35	1625
Big Lime (steel measure)	183	1808
Red rock	22	1830
Slate	20	1850
Shells	125	1975
Black slate	75	2050
Slate and shells	252	2302
Sand, **Berea** (steel measure)	17	2319
Slate to bottom	2	2321

Casing, 10", 67'; 8¼", 915'; 6⅝", 1695'; water at 1005' and 1315'; gas at 1050' and in Big Lime at 1765-1770', 1791-1795'; gas showing in Berea; **gas well.**

The Lackawanna Coal & Lumber Company has recently made two test wells for oil and gas on Whites Branch of West Fork of Coal River. These wells penetrated below the Berea Sand, but encountered no oil and very little gas in any of the sands.

Lackawanna Coal & Lumber Co. No. 6 Well Record (34).

Crook District, on Whites Branch, opposite Gordon P. O.; authority, Lackawanna Coal & Lumber Co.; completed September 1, 1914; elevation, 875' B.

	Thickness Feet.	Total Feet.
Sand and lime, hard, blue	12	12
Lime, hard, dark	6	18
Slate, hard, blue	7	25
Lime, hard, dark	20	45
Sand, hard, white	20	65
Coal, Hernshaw	3	68
Slate, hard, blue	12	80
Lime, hard, dark	40	120
Slate, hard, blue	80	200
Lime, hard, white	10	210
Slate, hard, black	15	225
Lime, hard, white	35	260
Slate, hard, white	40	300
Slate, soft, dark	15	315
Sand, hard, white	40	355
Slate, soft, dark	15	370
Sand, hard, white	90	460
Slate and shells, white	135	595
Sand, white	5	600
Slate, dark	10	610
Coal? Middle War Eagle?	6	616
Slate, dark	9	625
Sand, white	50	675
Slate, dark	75	750
Sand, white	310	1060
Slate and sand, dark.....140' } Salt Sand.... Sand, salt, white........335 }	475	1535
Slate, black	20	1555
Slate and shells, black	45	1600
Slate, soft, black	30	1630
Sand, hard, light gray	60	1690
Slate, hard, black	6	1696
Sand, very hard, gray (base of Pottsville)	123	1819

		Thickness Feet.	Total Feet.
Slate, hard, dark.........18'	Mauch Chunk	235	2054
Slate and shells, hard, white................36			
Sand, hard, red..........20			
Sand, hard, white........46			
Slate, hard, black.........67			
Lime, hard, gray.........17			
Slate and lime, hard, black 31			
Big Lime, hard, gray		216	2270
Lime and shells, hard, dark		33	2303
Sand, **Big Injun**, hard, red		7	2310
Slate and shells, hard, black		375	2685
Slate, soft, black		56	2741
Sand, **Berea**, hard, gray		6	2747
Slate, hard, black		8	2755
Lime, hard, dark		12	2767
Slate, lime and shells, hard, dark to bottom.		32	2799

At 400 feet, 250,000 cu. ft. of gas was encountered, but same blew out. A little gas was encountered at 695 feet, carrying oil. Well abandoned.

Lackawanna Coal & Lumber Co. No. 7 Well Record (35).

Crook District, on headwaters of Whites Branch, near Black White's house, 2.0 miles northeast from Gordon P. O.; authority, Lackawanna Coal & Lumber Company; completed, June 18, 1914; elevation, 1050' B.

		Thickness Feet.	Total Feet.
Slate, broken, black		50	50
Stone, hard, white		25	75
Slate, soft, brown		35	110
Sand, hard, gray		10	120
Slate, soft, gray		30	150
Coal, Hernshaw, soft		5	155
Slate, soft, brown		55	210
Sand, hard, white		5	215
Coal		5	220
Slate, soft, brown		80	300
Sand, hard, gray		60	360
Slate, hard, gray		110	470
Sand, hard, dark		80	550
Sand, hard, gray		60	610
Lime, hard, gray		30	640
Sand, hard, gray		160	800
Slate, soft, black		80	880
Sand, hard, dark		20	900
Slate, hard, dark		60	960
Sand, hard, gray		60	1020
Slate, soft, black		55	1075
Sand, very hard, gray...205'	Salt Sand	625	1700
Sand, hard, gray.........420			
Slate and shells, hard, black		85	1785

		Thickness Feet.	Total Feet.
Sand, hard, gray		58	1843
Slate, soft, black		7	1850
Sand, hard, gray (base of Pottsville)		75	1925
Slate, soft, black........ 2'	Mauch Chunk	237	2162
Sandstone, hard, white **Maxton** 33			
Slate, soft, black........ 8			
Lime, hard, black........ 6			
Slate, soft, black........ 6			
Sand, soft, red........ 10			
Lime and slate, soft, black 29			
Sandstone, soft, red..... 22			
Slate, soft, black........121			
Big Lime, hard, gray		249	2411
Slate and shells, hard, black		22	2433
Sand, **Big Injun**, hard, red		20	2453
Slate and shells, hard, dark		267	2720
Sand, **Wier?** hard, gray		12	2732
Slate and shells, hard, black		146	2878
Shale, soft, brown		14	2892
Sand, **Berea**, hard, gray		8	2900
Lime, hard, black		9	2909
Slate, hard, black		7	2916
Lime, hard, black, to bottom		14	2930

Show of gas, 2546 feet; little more gas at 2700 feet.

Lackawanna Coal & Lumber Co. No. 10 Well Record (36).

Crook District, 3.0 miles west from Whitesville P. O., on Logan Fork of Hopkins Fork of Laurel Creek of Big Coal River; authority. Lackawanna Coal & Lumber Company; completed, 1914; elevation, 1090' L.

	Thickness Feet.	Total Feet.
Sand and gravel	35	35
Sand and slate	45	80
Coal, Alma	3	83
Slate	19	102
Coal, Peerless	4	106
Slate	39	143
Coal, Campbell Creek (No. 2 Gas)	9	154
Slate	21	175
Sand and slate	200	375
Sand, lime and slate	265	640
Red rock	15	655
Slate and shells	65	720
Red rock	25	745
Slate	10	755
Lime and sand	20	775
Slate	10	785
Lime and sand	215	1000
Sandstone	85	1085
Slate and shells	85	1170
Sand and lime	30	1200
Salt sand	90	1290

At 1290 feet, a string of tools was lost, and after fishing for some time and not being able to recover the tools, it was thought best to abandon the hole.

Prospective Oil and Gas Area, Crook District.

This district, thus far, has not been so tested for oil and gas by the drill as to condemn it as a dry territory. It is more than probable that some gas exists in the Big Lime at different points in the district with a possibility of a small pool of oil. The Big Injun Sand may also contain gas at some points within the district.

CHAPTER IX.

THE COAL RESOURCES OF BOONE COUNTY.

The geology, structure and general character of the coals in Boone County have already been described in detail in the preceding pages of this volume, and the purposes of this Chapter are to consider the coal production, the chemical composition and the character of the workable coal beds more in detail, as well as the available coal areas and tonnages still unmined.

STATISTICS OF COAL PRODUCTION.

Coal mining on a commercial scale in Boone County began in 1846, along Coal River at Peytona, where the cannel coal was mined and floated down the river in flat boats. The mining operations were discontinued in 1861, during the Civil War, and at the close of the war several attempts were made to continue the mining operations, but without much success. The real mining operations in Boone County began after the completion of the Coal River Railroad to Clothier, in 1909, and to Peytona in 1910, and the extension of the Cabin Creek Branch Railroad from Kayford to Colcord, in 1909.

Very little coal was mined in 1909, but from that time the annual output of the county has increased rapidly, and if the railroad is extended up Pond and West Forks and on Spruce Laurel Fork, the output will be increased by a very large percentage each year in the future, since Boone County is destined to become one of the great coal producing counties of southern West Virginia, owing to its vast fuel deposits and the purity and quality of same.

The following tables have been compiled from the Annual Report of The Department of Mines of West Virginia, for the year ending June 30, 1913, as given by Hon. Earl A. Henry, Chief of the Department of Mines:

Coal Production of Boone County from 1909 to 1913, Inclusive.

Year	Tons of 2240 Lbs.	Tons of 2000 Lbs.
1909	3,451	3,865
1910	50,566	56,634
1911	69,305	77,622
1912	329,179	368,681
1913	324,594	347,484
Total	777,095	854,286

The same report gives the following table of production of each mine from 1909 to 1913, inclusive:

Coal Production by Mines from 1909 to 1913, Inclusive, Short Tons.

Name of Company.	1909	1910	1911	1912	1913
The Coal River Co.	3,265	33,970	59,366	71,432	73,064
Spruce River Coal Co.	600	22,664	12,755	25,943	37,539
Peytona Block Coal Co.	......	......	2,321	30,775	35,977
Hickory Ash Coal Co.	......	......	......	15,465	45,871
Anchor Coal Co.	......	......	......	59,180	55,355
Webb Coal Mining Co.	......	......	3,180	116,887	67,549
Seng Creek Coal Co.	......	......	......	48,999	32,129
Totals	3,865	56,634	77,622	368,681	347,484

The decrease in production for the year 1913 is due to the fact of a prolonged strike on Cabin and Paint Creeks, in Kanawha County, which also extended into Boone and affected its production.

The same report gives the following table showing the relative rank of Boone County in coal production compared with the other counties of the State:

Order of Counties in the Production of Coal, 1897-1913.

Counties.	1897	1898	1899	1900	1901	1902	1903	1904	1905	1906	1907	1908	1909	1910	1911	1912	1913
Fayette	1	1	1	1	1	1	2	1	2	2	2	2	2	2	2	2	2
McDowell	2	2	2	2	2	2	1	2	1	1	1	1	1	1	1	1	1
Marion	3	3	3	3	3	3	3	4	3	4	4	4	4	4	4	4	3
Kanawha	4	4	4	4	4	4	5	3	4	3	3	3	3	3	3	3	7
Mercer	5	5	5	6	6	6	6	6	6	6	6	6	6	7	8	8	8
Harrison	9	9	8	7	5	5	4	5	5	5	5	5	5	5	5	6	4
Tucker	6	6	7	5	7	7	7	8	8	8	10	10	10	10	10	10	10
Mingo	8	8	9	9	9	8	8	7	7	7	7	7	9	9	9	9	9
Mineral	7	7	6	8	8	9	11	11	12	13	14	13	11	13	15	17	16
Preston	13	11	10	10	10	10	9	10	10	9	9	11	13	11	11	11	11
Taylor	10	10	11	11	11	11	14	16	16	16	17	17	16	15	14	13	12
Marshall	11	12	12	12	13	16	13	14	14	15	15	18	19	16	18	16	15
Randolph	..	..	..	14	14	14	12	13	13	12	13	14	14	14	13	14	14
Barbour	19	19	17	18	12	12	10	9	11	11	11	12	12	12	12	12	12
Raleigh	16	16	16	19	17	13	15	12	9	10	8	8	8	6	6	5	5
Putnam	12	13	13	13	15	15	16	15	15	14	18	16	17	17	17	18	18
Ohio	15	15	14	15	16	17	18	19	21	22	21	21	21	20	20	20	20
Brooke	17	18	18	20	20	20	23	22	17	17	16	15	15	18	16	15	17
Mason	14	14	15	16	18	19	19	20	22	21	22	22	22	22	22	25	25
Grant	..	..	..	..	..	..	22	18	19	19	20	20	20	21	21	23	24
Logan	..	..	..	..	..	..	..	..	20	18	12	9	7	8	7	7	6
Monongalia	20	17	19	17	19	18	17	17	18	20	19	19	18	19	19	19	19
Hancock	18	20	20	21	21	21	20	21	24	24	23	25	26	24	27	..	..
Clay	..	..	..	..	..	..	..	25	23	23	24	27	30	30	25	24	22
Nicholas	..	..	..	..	..	22	24	24	25	25	25	26	27	26	24	29	27
Lincoln	..	..	..	..	..	23	21	23	27	26	30	30	25	28	28	..	30
Braxton	..	..	..	..	..	..	..	..	..	..	26	23	23	23	23	22	23
Wayne	..	..	..	..	..	..	..	..	26	27	29	29	28	29	32	28	28
Upshur	..	..	..	..	..	..	..	..	..	28	27	24	24	25	29	31	29
Greenbrier	..	..	..	..	..	..	..	..	..	..	28	28	29	32	30	27	32
Lewis	..	..	..	..	..	..	..	..	..	..	31	31	32	33	33	30	31
Gilmer	..	..	..	..	..	..	..	..	..	..	32	32	31	31	31	..	..
Wyoming	..	..	..	..	..	..	..	..	..	..	..	33	34	34	..	..	33
Boone	..	..	..	..	..	..	..	..	..	..	..	..	**33**	**27**	**26**	**21**	**21**

COALS IN BOONE COUNTY.

It is difficult to estimate the number of workable coal seams in the area under discussion, since it carries the coals from the Allegheny Series down to and including the Kanawha Series of the Pottsville beds.

The chemical analyses and determinations of heating values, given in the following pages, were made by J. B. Krak, Assistant Chemist of the Survey, under the supervision of Prof. B. H. Hite, Chief Chemist. The coal samples were taken carefully in the field work, and where possible the sample was obtained by digging down a section of the coal entirely across the face of the bed, excluding only such slates and partings as are taken out in mining operations. In the commercial mines, these samples were crushed and quartered down to two or three pounds of coal, placed in tin cans, hermetically sealed, and sent to the Laboratory of the Survey. The same methods of analysis were used as by the Fuel Testing Department of the United States Bureau of Mines and the United States Geological Survey.

The calorific value of the coals analyzed for this Report is expressed in terms of British Thermal Units, the unit of heat measurement more commonly used in the United States. This unit of heat, usually marked B. T. U., represents the amount of heat required to raise one pound of water one degree Fahrenheit in temperature.

The Pittsburgh bed in the Fairmont region gives from 14,000 to 14,400 B. T. U. for each pound of coal, and the New River and Pocahontas Coals run from 15,000 to 15,500 B. T. U.

In many cases both the proximate and ultimate analyses are given, and also the B. T. U. results, as well as the ratio of the total carbon to the oxygen plus ash. It has only recently been insisted upon that oxygen has about the same deteriorating effect as ash in all coal, and the above mentioned ratio, proposed by Dr. David White, of the National Museum and the United States Geological Survey, is the best yet devised for the classification of coals in order to show their relative rank in heating values.

The coals in the different series will now be described in descending order:

COALS IN THE ALLEGHENY SERIES.

The only coal of commercial value in the Allegheny Series of Boone County is the **No. 5 Block Coal.** There may possibly be a small area of the Upper Freeport Coal of sufficient thickness and purity to be of commercial value, in the northern part of Washington District, but the developments in the way of openings are so meager and as that coal bed in this part of West Virginia is usually thin and impure it is impossible to estimate any available coal in Boone County on this seam.

The sections showing the thickness of the different openings in the No. 5 Block Coal have already been given in the preceding pages, and from the measurements of these sections, the following results are obtained:

Quantity of No. 5 Block Coal Available.

It is difficult to make an accurate estimate of the amount of available tonnage of the No. 5 Block Coal bed in Boone, for the reason that this coal occurs high on the hills and very little development has been made to determine the thickness and purity of the seam.

An approximate determination of the area of that portion of Boone County has been made by magisterial districts within the boundary lines shown on the Economic Geologic Map II, and the area of the coal calculated. From the measurements given, the data in the following table for the different magisterial districts have been calculated:

Table Showing Available No. 5 Block Coal in Boone County.

District.	Sq. Mi.	Acres.	Average Thickness.	Cubic Feet.	Short Tons.
Peytona	3.45	2,208	4′ 0″	384,721,920	15,388,876
Scott	14.60	9 344	4′ 6″	1,831,610,880	73,264,435
Washington	7.50	4,800	4′ 6″	940,896,000	37,635,840
Sherman	18.90	12,096	5′ 0″	2.634,508,800	105,380,352
Crook	13.60	8,704	6′ 0″	2,274,877,440	90,995,098
Totals	58.05	37,152	....	8,066,615.040	322,664,601

In order to obtain the above results, it is assumed that one cubic foot of No. 5 Block Coal weighs 80 pounds. These figures are the same as obtain for the weight of the Pittsburgh Coal at the Fuel Testing Plant of the United States Geological Survey at St. Louis, Missouri.

COALS IN THE KANAWHA SERIES.

Boone County contains eleven different seams of coal of sufficient thickness and purity to be of commercial value in the Kanawha Series. To these main beds may be frequently added another, namely, the Lower Bench of the Stockton, and often the Lower Bench of the Winifrede, as well as the Lower Bench of the Campbell Creek (No. 2 Gas), where these seams appear in two benches, making 14 beds in all.

The Stockton-Lewiston Coal is the topmost seam in the Kanawha Series. This bed of coal appears in all the magisterial districts of Boone County.

Quantity of Stockton-Lewiston Coal Available.

In the preceding Chapters of this volume, the sections of the different coal openings in this seam have been given as well as their location and elevation.

An approximate determination of the area of that portion of Boone County has been made by magisterial districts within the boundary lines shown on the Economic Geologic Map II, and the area of the coal calculated. From the preceding measurements given, it is found that the average thickness, tonnage, etc., of the Stockton-Lewiston Coal in Boone County in the several magisterial districts are as given in the table below:

Table Showing Available Stockton-Lewiston Coal in Boone County.

District.	Sq. Mi.	Acres.	Average Thickness.	Cubic Feet.	Short Tons.
Peytona	6 27	4012.8	2′ 0″	349,595,136	13,983,805
Scott	18.95	12128.0	4′ 6″	2,377,330,560	95,093,222
Washington	13.61	8710.4	2′ 0″	758,850,048	30,354,002
Sherman	30.48	19507.2	5′ 6″	4,673,534,976	186.941,399
Crook	20.74	13273.6	5′ 6″	3,180,089,088	127,203,563
Totals	90.05	57,632.0	....	11,339,399,808	453,575,991

In the above calculations, it is assumed that one cubic foot of Stockton-Lewiston Coal weighs 80 pounds.

Quantity of Coalburg Coal Available.

Table Showing Available Coalburg Coal in Boone County.

District.	Sq. Mi.	Acres.	Average Thickness.	Cubic Feet.	Short Tons.
Peytona	7.52	4,812.8	2′ 6″	524,113,920	20,964,557
Scott	......	....	....	....	....
Washington	16.33	10,451.2	2′ 6″	1,138,135,680	45,525,427
Sherman	36.57	23,404.8	5′ 0″	5,097,565,440	203,902,618
Crook	24.89	15,929.6	5′ 0″	3,469,466,880	138,778,675
Totals	85.31	54,598.4	....	10,229,281,920	409,171,277

Quantity of Winifrede Coal Available.

Table Showing Available Winifrede Coal in Boone County.

District.	Sq. Mi.	Acres.	Average Thickness.	Cubic Feet.	Short Tons.
Peytona	19.7	12,608	2′ 6″	1,373,011,200	54,920,448
Scott	40.7	26,048	2′ 8″	3,025,657,536	121,026,301
Washington	42.9	27,456	4′ 0″	4,783,933,440	191,357,337
Sherman	88.3	56,512	5′ 6″	13,539,144,960	541,565,798
Crook	64.6	41,344	5′ 0″	9,004,723,200	360,188,528
Totals	256.2	163,968	....	31,726,467,336	1,269,058,412

Quantity of Chilton Coal Available.

Table Showing Available Chilton Coal in Boone County.

District.	Sq. Mi.	Acres.	Average Thickness.	Cubic Feet.	Short Tons.
Peytona	23.6	15,104	2′ 0″	1,315,860,480	52,634,419
Scott	......	....	....	....	....
Washington	45.1	28,864	4′ 0″	5,029,263,360	201,170,534
Sherman	92.3	59,072	2′ 0″	5,146,352,640	205,854,106
Crook	71.1	45,504	3′ 0″	5,946,462,720	237,858,509
Totals	232.1	148,544	....	17,437,939,200	697,517,568

Quantity of Hernshaw Coal Available.

Table Showing Available Hernshaw Coal in Boone County.

District.	Sq. Mi.	Acres.	Average Thickness.	Cubic Feet.	Short Tons.
Peytona	32.0	20,480	2′ 9″	2,453,299,200	98,131,968
Scott	46.0	29,440	2′ 0″	2,546,812,800	102,592,512
Washington	58.0	37,120	2′ 6″	4,042,368,000	161,694,720
Sherman	98.0	62,720	3′ 6″	9,562,291,200	382,491,648
Crook	91.0	58,240	3′ 6″	8,879,270,400	355,170,816
Totals	325.0	208,000	....	27,502,041,600	1,100,081,664

Quantity of Williamson Coal Available.

From the sections given on preceding pages, the Williamson Coal usually contains considerable impurities, but in several places in Washington District and also in Crook District, this coal bed has attained sufficient thickness and purity to be of commercial value. The writer has estimated that there is possibly 10 square miles of this coal in each of these districts, with an average thickness of 2 feet, and based on this estimate, the following table results:

Table Showing Available Williamson Coal in Boone County.

District.	Sq. Mi.	Acres.	Average Thickness.	Cubic Feet.	Short Tons.
Washington	10.0	6,400	2′ 0″	557,568,000	22,302,720
Crook	10.0	6,400	2′ 0″	557,568,000	22,302,720
Totals	20.0	12,800	....	1,115,136,000	44,605,440

Quantity of Cedar Grove Coal Available.

Table Showing Available Cedar Grove Coal in Boone County.

District.	Sq. Mi.	Acres.	Average Thickness.	Cubic Feet.	Short Tons
Peytona	38.4	24,576	3′ 6″	3,746,856,960	149,874,278
Scott	49.9	31,936	2′ 6″	3.477.830,400	139.113,216
Washington	72.0	46.080	2′ 6″	5.018.112,000	200.724,480
Sherman	105.5	67,520	2′ 0″	5.882.342.400	235 293,696
Crook	107.0	68,480	2′ 0″	5,965,977,600	238,639,104
Totals	372.8	238.592	....	24 091.119.360	963 644.774

Quantity of Alma Coal Available.

Table Showing Available Alma Coal in Boone County.

District.	Sq. Mi.	Acres.	Average Thickness.	Cubic Feet.	Short Tons.
Peytona	10.0	6,400	2′ 0″	557,568,000	22,302,720
Scott	58.9	37,696	3′ 0″	4,926,113,280	197,044,531
Washington	87.0	55,680	3′ 6″	8,488,972,800	339,558,912
Sherman	117.5	75,200	2′ 0″	6,551,424,000	262,056,960
Crook	122.0	78,080	2′ 8″	9,069,538,560	363,781,542
Totals	395.4	253,056	....	29,593,616,640	1,184,744,665

Quantity of Campbell Creek (No. 2 Gas) Coal Available.

It is difficult to estimate the amount of available tonnage of the Campbell Creek (No. 2 Gas) Coal in Boone, for the reason that in a portion of the county this bed is thin, while in other portions of the county the coal attains a thickness of 7 to 8 feet. The average thickness in each magisterial district is based upon the different sections measured in the study of the coal.

Table Showing Available Campbell Creek (No. 2 Gas) Coal in Boone County.

District.	Sq. Mi.	Acres.	Average Thickness.	Cubic Feet.	Short Tons.
Peytona	53.3	34,112	3′ 6″	5,200,715,520	208,028,621
Scott	15.4	9,856	3′ 0″	1,287,982,080	51,519,283
Washington	51.6	33,024	5′ 0″	7,192,627,200	287,705,088
Sherman	130.5	83,520	3′ 6″	12,733,459,200	509,338,368
Crook	108.6	69,504	3′ 6″	10,596,579,840	423,863,193
Totals	359.4	230,016	....	37,011,363,840	1,480,454,552

Quantity of Powellton Coal Available.

From the sections given on the preceding pages of this Report, it is evident that there is no large area of Powellton

Coal of sufficient thickness and purity to be of commercial value in Boone. The writer has estimated the area of this bed in Crook District, and while there may be a small area in Washington District, yet since the latter has not been fully prospected for this seam, no estimate will be made of the Powellton Coal in Washington District:

Table Showing Available Powellton Coal in Boone County.

District.	Sq. Mi.	Acres.	Average Thickness.	Cubic Feet.	Short Tons.
Crook	30.0	19,200	3′ 0″	2,509,056,000	100,362,240

Quantity of Eagle Coal Available.

The Eagle Coal occurs in only one of the magisterial districts of Boone of sufficient thickness and purity to be of commercial value, although it is present as a thin seam in practically every district, but not of commercial thickness except in Crook.

Table Showing Available Eagle Coal in Boone County.

District.	Sq. Mi.	Acres.	Average Thickness.	Cubic Feet.	Short Tons.
Crook	30.0	19,200	3′ 6″	2,927,232,000	117,089,280

Summary of Available Coal in Boone County.

It is possible but not probable that there is more commercial coal in some of the lower beds (New River) throughout a portion of the southern part of the county, but as these coals lie deeply buried it will be necessary to test them with diamond core drills before any estimate of the quantity of coal in these measures can be made.

Table Showing Total Available Coal in Boone County.

Coal Beds	Short Tons.
No. 5 Block	322,664,601
Stockton-Lewiston	453,575,991
Coalburg	409,171,277
Winifrede	1,269,058,412
Chilton	697,517,568
Hernshaw	1,100,081,664
Williamson	44,605,440
Cedar Grove	963,644,774
Alma	1,184,744,665
Campbell Creek (No. 2 Gas)	1,480,454,553
Powellton	100,362,240
Eagle	117,089,280
Total	8,142,970,465

Table Showing Available Coal by Districts.

Peytona District.

Coal Beds.	Short Tons.	
No. 5 Block	15,388,876	
Stockton-Lewiston	13,983,805	
Coalburg	20,964,557	
Winifrede	54,920,448	
Chilton	52,634,419	
Hernshaw	98,131,968	
Cedar Grove	149,874,278	
Alma	22,302,720	
Campbell Creek (No. 2 Gas)	208,028,621—	629,229,692

Scott District.

Coal Beds.	Short Tons.	
No. 5 Block	73,264,435	
Stockton-Lewiston	95,093,222	
Coalburg	....	
Winifrede	121,026,301	
Hernshaw	102,592,512	
Cedar Grove	139,113,216	
Alma	197,044,531	
Campbell Creek (No. 2 Gas)	51,519,283—	779,653,500

Washington District.

Coal Beds.	Short Tons.	
No. 5 Block	37,635,840	
Stockton-Lewiston	30,354,002	
Coalburg	45,525,427	
Winifrede	191,357,337	
Chilton	201,170,534	
Hernshaw	161,694,720	
Williamson	22,302,720	
Cedar Grove	200,724,480	
Alma	339,558,912	
Campbell Creek (No. 2 Gas)	287,705,088—	1,518,029,060

Sherman District.

Coal Beds.	Short Tons.	
No. 5 Block	105,380,352	
Stockton-Lewiston	186,941,399	
Coalburg	203,902,618	
Winifrede	541,565,798	
Chilton	205,854,106	
Hernshaw	382,491,648	
Cedar Grove	235,293,696	
Alma	262,056,960	
Campbell Creek (No. 2 Gas)	509,338,368	—2,632,824,945

Crook District.

Coal Beds.	Short Tons.	
No. 5 Block	90,995,098	
Stockton-Lewiston	127,203,563	
Coalburg	138,778,675	
Winifrede	360,188,528	
Chilton	237,858,509	
Hernshaw	355,170,816	
Williamson	22,302,720	
Cedar Grove	238,639,104	
Alma	363,781,542	
Campbell Creek (No. 2 Gas)	423,863,193	
Powellton	100,362,240	
Eagle	117,089,280	—2,576,233,268
Grand Totals		8,142,970,465

No estimate of the probable tonnage of coal was made of the following seams: Buffalo Creek, Chilton "A," Chilton "Rider," Little Chilton, Dingess, Little Alma and Matewan. These seams occur only at a few points, and the small amount of available coal contained in their beds cannot be accurately estimated.

Of course from the above estimates there will have to be subtracted about 1,000,000 tons, representing the coal that has already been mined in Boone County from the different beds, so that the total coal for the county would be reduced to 8,141,970,465 short tons.

The above estimate of available coal in Boone County does not mean that this amount will eventually be recovered. The percentage of recovery of a coal bed, at present, in West Virginia, varies from 60 to 95 per cent. of the minable coal. However, by the rapid progress that is being made in mining methods in recent years, in the way of saving the largest possible amount of the bed, and the general conditions of the roof and natural mining conditions, the writer is of the opinion that fully 80 per cent. of the above estimate will be finally recovered. Figuring on this basis, the total available coal is reduced to 6,513,576,372 short tons.

The area of Boone County is 506.00 square miles, or 323,840 acres, and from the above available coal tonnage, each acre will contain a little more than 20,000 short tons.

COMPARATIVE CALORIFIC VALUE OF THE COALS IN BOONE COUNTY.

The following table gives the comparative calorific value and fuel ratio of the different coal beds in Boone County, in addition to the analyses already given on preceding pages of this Report:

Analysis No.	MINE	HORIZON.	Condition of Sa[mple]	Moisture.	Volatile Matter.	Fixed Carbon.	Phosphorus.	Ash.	Sulphur.	Carbon.	Hydrogen.	Oxygen.	Nitrogen.		
1	Tuncil Price	No. 5 Block Coal	A. R.	1.14	38.39	50.24	0.018	10.23	0.81	......	......	......	......	......	..
2	D. G. Courtney	No. 5 Block Coal	A. R.	1.54	38.33	55.38	0.004	4.75	0.76	......	......	......	......	......	..
3	Little Coal Land Co.	No. 5 Block Coal	A. R.	2.21	36.42	58.11	0.007	3.26	0.53	......	......	......	......	......	..
4	Little Coal Land Co.	No. 5 Block Coal	A. R.	2.78	35.94	57.04	0.007	4.24	0.77	......	......	......	......	......	..
5	Little Coal Land Co.	No. 5 Block Coal	A. R.	2.20	36.38	56.96	0.005	4.46	0.71	......	......	......	......	......	..
6	Little Coal Land Co.	No. 5 Block Coal	A. R.	2.12	38.87	55.20	0.006	3.81	0.65	......	......	......	......	......	..
7	Asa Williams	No. 5 Block Coal	A. R.	1.58	38.35	55.20	0.005	4.87	0.84	......	......	......	......	......	..
8	Asa Williams	No. 5 Block Coal	A. R.	1.80	36.45	53.43	0.011	8.32	0.80	......	......	......	......	......	..
9	E. J. Berwind	No. 5 Block Coal	A. R.	1.21	33.88	58. 9	0.009	6.12	0.81	79.58	4.97	7.25	1.27	14,080	1
10	Lackawanna Coal & Lumber Co.	No. 5 Block Coal	A. R.	0.72	36.98	60.03	0.031	2.27	0.69	......	......	......	......	......	..
11	W. C. Cook	No. 5 Block Coal	A. R.	1.22	37.71	57.60	0.003	3.47	0.98	......	......	......	......	......	..
12	E. J. Berwind	No. 5 Block Coal	A. R.	1.53	39.14	56.87	0.037	2.46	0.66	......	......	......	......	......	..
13	[illegible]er Lumber Co.	No. 5 Block Coal	A. R.	1.80	38.14	49.67	0.0 2	10.39	0.70	73.79	5.02	8.81	1.29	13,260	1
	Average	No. 5 Block Coal	A. R.	1.68	37.31	55.73	0.017	*5.28	*0.75	76.68	5.00	8.03	1.28	13,670	1
14	Horse Creek Coal & Lumber Co	Stockton-Lewiston Coal	A. R.	2.62	40.36	54.21	0.013	2.81	0.87	79.43	5.48	10.16	1.25	14,190	1
14A	Staunton Grailey	Stockton-Lewiston Coal	A. R.	2.40	38.37	51.99	0.004	7.24	0.92	......	......	......	......	......	..
15	Horse Creek Block Coal Co.	Stockton-Lewiston Coal	A. R.	2.18	39.12	54.23	0.004	4.47	1.20	......	......	......	......	......	..
16	Allen Foster	Stockton-Lewiston Coal	A. R.	1.30	35.39	59.10	0.004	4.21	0.67	78.18	5.37	10.39	1.18	14 190	1
17	Lackawanna Coal & Lumber Co.	Stockton-Lewiston Coal	A. R.	1.28	38.97	53.57	0.011	6.18	0.71	78.85	5.11	8.13	1.02	14,250	1
18	[illegible]y & Freeman	[illegible]n-Lewiston Coal	A. R.	1.52	41.64	53.88	0.004	2.96	0.74	......	......	......	......	......	.
	Average	Stockton-Lewiston Coal	A. R.	1.88	38.98	54.50	0.007	*4.64	*0.85	78.82	5.32	9.56	1.15	14,210	1
19	William Price	Coalburg Coal	A. R.	1.42	40.64	52.10	0.003	5.84	3.02	......	......	......	......	......	..
20	Lackawanna Coal & Lumber Co.	Coalburg Coal	A. R.	0.91	38.67	58.29	0.006	2.13	0.63	84.05	5.38	6.52	1.29	15,120	1
20	Lackawanna Coal & Lumber Co.	Coalburg Coal	A. R.	1.47	35.92	54.49	0.006	8.12	0.74	......	......	......	......	......	..
21	Lackawanna Coal & Lumber Co.	Coalburg Coal	A. R.	0.93	37.52	56.43	0.003	5.12	1.03	......	......	......	......	......	..
22	Lackawanna Coal & Lumber Co.	Coalburg Coal	A. R.	1.14	38.42	56.28	0.015	4.16	1.15	......	......	......	......	......	..
22A	Marion Ferrell	Coalburg Coal	A. R.	1.55	37.70	54.32	0.006	6.43	1.03	......	......	......	......	......	..
23	Lackawanna Coal & Lumber Co.	Coalburg Coal	A. R.	1.44	35.72	60.54	0.003	2.30	0.66	......	......	......	......	......	..
24	Wharton Estate	Coalburg Coal	A. R.	1.33	36.31	55.54	0.014	6.82	1.62	......	......	......	......	......	..
25	[illegible]n Estate	Coalburg Coal	A. R.	1.49	35.86	57.18	0.006	5.47	0.59	79.28	5.12	8.33	1.21	14,120	1
26	E. J. Berwind	Coalburg Coal	A. R.	1.44	36.76	57.40	0.007	4.40	0.78	80.39	5.28	7.81	1.34	14,430	1
26A	E. J. Berwind	Coalburg Coal	A. R.	1.57	33.01	57.05	0.004	8.37	0.65	......	......	......	......	......	..
	Average	Coalburg Coal	A. R.	1.33	36.96	56.33	0.007	*5.38	*1.08	81.24	5.26	7.55	1.28	14,557	1
27	Little Coal Land Co.	Winifrede Coal	A. R.	2.05	38.63	55.83	0.006	3.49	0.50	......	......	......	......	......	..
27A	Little Coal Land Co.	Winifrede Coal	A. R.	1.24	33.75	56.65	0.004	8.16	0.58	......	......	......	......	......	..

Analysis No.	NAME	HORIZON.	Condition of	Moisture.	Volatile Matter.	Fixed Carbon.	Phosphorus.	to Both. Ash.	Sulphur.	Carbon.	Hydrogen.	Oxygen.	Nitrogen.
28	Cassingham	Winifrede Coal	A. R.	1.14	34.16	62.04	0.004	2.66	0.73	.....	.....	.....	.....
29	John Q. Dickinson	Winifrede Coal	A. R.	1.40	40.08	52.48	0.005	6.04	1.48	.....	.....	.....	.....
29A	John Q. Dickinson	[illegible] Coal	A. R.	1.68	35.13	52.88	0.007	10.31	1.15	77.57	5.27	8.26	1.38
30	Anchor Coal Co.	Winifrede Coal	A. R.	1.40	36.52	58.88	0.003	3.20	0.59	82.21	5.41	7.24	1.35
30A	Webb Fuel Co.	Winifrede Coal	A. R.	1.17	35.17	56.71	0.004	6.95	0.67	78.86	5.04	7.21	1.27
31	Seng Creek Coal Co.	Winifrede Coal	A. R.	1.08	40.48	51.50	0.114	6.94	1.29	8.53	5.12	6.87	1.23
32	Squire Gibson	Winifrede Coal	A. R.	1.98	38.42	56.88	0.005	2.72	0.63	.....	.....	.....	.....
33	Lackawanna Coal & Lumber Co.	Winifrede Coal	A. R.	1.46	41.32	52.22	0.002	5.00	1.87	.....	.....	.....	.....
34	Lackawanna Coal & Lumber Co.	Winifrede Coal	A. R.	1.68	37.16	54.23	0.015	6.93	0.84	.....	.....	.....	.....
35	E. J. Berwind	Winifrede Coal	A. R.	0.71	35.02	56.62	0.010	7.65	0.77	79.82	4.99	5.61	1.16
35	E. J. Berwind	Winifrede Coal	A. R.	0.55	30.87	35.39	0.008	33.19	0.66	.....	.....	.....	.....
35A	T. C. Jarrell	Winifrede Coal	A. R.	1.52	39.56	57.08	0.003	1.84	0.81	.....	.....	.....	.....
35B	——— (cannel)	Winifrede Coal	A. R.	1.15	38.16	53.47	0.006	7.22	1.70	76.75	5.01	8.10	1.22
	Average	Winifrede Coal	A. R.	1.35	36.97	54.19	0.013	*7.49	*0.95	78.96	5.14	7.21	1.27
36	Coal River Mining Co.	Chilton Coal	A. R.	1.55	35.24	56.80	0.006	6.41	0.56	78.42	4.80	8.52	1.29
37	Floyd Nelson (cannel)	Chilton Coal	A. R.	0.52	50.92	35.82	0.006	12.74	1.10	73.63	5.91	5.35	1.27
38	Little Coal Land Co.	Chilton Coal	A. R.	1.48	40.12	55.36	0.006	3.04	0.51	.....	.....	.....	.....
39	Yawkey & Freeman	Chilton Coal	A. R.	1.60	38.09	54.94	0.004	5.37	0.62	.....	.....	.....	.....
39A	Ashford Ball	Chilton Coal	A. R.	2.80	39.39	55.05	0.004	2.76	0.69	.....	.....	.....	.....
39	Chilton MacCorkle, Chilton & Meany	Chilton Coal	A. R.	1.53	34.97	56.43	0.004	7.07	0.66	.....	.....	.....	.....
40	Lackawanna Coal & Lumber Co.	Chilton Coal	A. R.	0.92	37.15	57.72	0.006	4.21	1.29	.....	.....	.....	.....
41	Lackawanna Coal & Lumber Co.	[illegible] Coal	A. R.	0.62	38.42	57.31	0.002	3.65	2.12	.....	.....	.....	.....
42	John Q. Dickinson	[illegible] Coal	A. R.	0.50	60.71	30.43	0.004	8.36	0.60	.....	.....	.....	.....
	Average	Chilton Coal	A. R.	1.28	41.67	51.10	0.005	*5.95	*0.91	76.03	5.36	6.93	1.28
43	Wharton Estate	Little Chilton Coal	A. R.	1.17	38.26	57.32	0.003	3.25	0.65	.....	.....	.....	.....
44	Fork Creek Coal Co.	Hernshaw Coal	A. R.	1.66	44.03	47.07	0.007	7.24	3.15	.....	.....	.....	.....
44A	Bedford et al. (cannel)	Hernsl[illegible] Coal	A. R.	0.90	51.80	34.38	0.01	12.92	1.07	.....	.....	.....	.....
45	James Workman	Hernshaw Coal	A. R.	1.82	40.82	54.35	0.011	3.01	0.74	.....	.....	.....	.....
46	Emma Hager	[illegible] Coal	A. R.	1.56	41.03	50.98	0.007	6.43	3.57	.....	.....	.....	.....
47	Chilton MacCorkle, Chilton & Meany	Hernshaw Coal	A. R.	1.37	36.33	54.95	0.004	7.35	1.03	.....	.....	.....	.....
47A	White & Hopkins	Hernshaw Coal	A. R.	1.04	38.26	55.09	0.005	5.61	0.87	.....	.....	.....	.....
48	LaFol[illegible]e, Robson et al	Hernshaw Coal	A. R.	1.13	39.99	52.48	0.004	6.40	1.73	.....	.....	.....	.....
49	LaFollette, Robson et al.	Hernshaw Coal	A. R.	1.23	35.58	55.16	0.004	8.03	0.96	.....	.....	.....	.....
49A	Lackawanna Coal & Lumber Co.	Hernsl[illegible] Coal	A. R.	0.94	39.87	55.05	0.004	4.14	0.76	.....	.....	.....	.....
49B	H. A. Robson	Hernshaw Coal	A. R.	1.07	37.93	57.04	0.003	3.96	0.78	.....	.....	.....	.....

Analysis No.	MINE	HORIZON.	Condition o	Moisture.	Volatile Matter.	Fixed Carbon.	Phosphorus.	Ash.	Sulphur.	Carbon.	Hydrogen.	Oxygen.	Nitrogen.
49C	F. W. Mason	Hernshaw Coal	A. R.	1.02	34.97	59.84	0.022	4.17	0.85				
49D	H. A. Robson	Hernshaw Coal	A. R.	1.11	37.02	54.35	0.004	7.52	0.85				
50	Ira Sutphin	Hernshaw Coal	A. R.	1.09	40.70	54.48	0.004	3.73	0.37				
51	Wharton Estate	Hernshaw Coal	A. R.	1.27	36.24	58.33	0.003	4.16	0.66				
52	Pond Fork Coal & Land Co.	Hernshaw Coal	A. R.	0.91	37.87	55.63	0.003	5.59	1.22				
	Average	Hernshaw Coal	A. R.	1.21	39.49	53.28	0.006	6.02	1.27				
53	Isaac [illegible]	Williamson Coal	A. R.	1.11	41.86	54.41	0.003	2.62	1.77				
54	Ashford Ball	Williamson Coal	A. R.	1.06	36.95	54.23	0.011	7.76	1.36				
55	W. P. Croft (cannel)	Williamson Coal	A. R.	0.66	47.34	40.22	0.015	11.78	1.43				
	Average	Williamson Coal	A. R.	0.95	42.05	49.95	0.010	7.05	1.52				
56	F. G. Kirk	Cedar Grove Coal	A. R.	1.21	40.14	55.99	0.01	2.66	1.00				
57	Minerva Hill	Cedar Grove Coal	A. R.	1.96	37.90	55.71	0.005	4.43	0.78				
58	S. M. Javins	Cedar Grove Coal	A. R.	1.40	39.60	56.59	0.003	2.41	0.78				
58A	Elmer Harless	Cedar Grove Coal	A. R.	1.62	37.12	58.77	0.003	2.49	0.65				
59	Eliza Abshire	Cedar Grove Coal	A. R.	1.24	40.31	54.70	0.004	3.75	1.22	80.17	5.30	8.29	1.27
60	Julia [illegible]	Cedar Grove Coal	A. R.	1.62	42.27	53.78	0.003	2.33	1.35				
61	A. [illegible]. Kerse	Cedar Grove Coal	A. R.	1.40	42.32	51.34	0.006	4.94	1.92				
62	Vernon Vandelinde	Cedar Grove Coal	A. R.	1.20	41.23	54.13	0.004	3.44	1.70				
63	Dora Hopkins	Cedar Grove Coal	A. R.	1.29	41.18	52.59	0.002	4.94	2.84				
64	[illegible]	Cedar Grove Coal	A. R.	1.32	37.08	59.71	0.002	1.89	0.84				
65	Silas Billetts	Cedar Grove Coal	A. R.	1.37	43.80	50.14	0.002	4.69	1.33				
66	Ballard Hager	Cedar Grove Coal	A. R.	1.77	39.10	55.12	0.004	4.01	1.14				
67	D. M. Arbogast	Cedar Grove Coal	A. R.	0.43	56.99	33.90	0.007	8.68	1.85	76.38	6.77	4.91	1.41
67A	R. P. Chew	Cedar Grove Coal	A. R.	0.96	40.30	54.34	0.012	4.40	1.58				
	Average	Cedar Grove Coal	A. R.	1.34	41.38	53.35	0.005	*3.93	*1.36	78.28	6.04	6.60	1.34
68	Julian Hill	Alma Coal	A. R.	0.83	36.32	46.73	0.024	16.12	1.77				
69	Julian Hill	Alma Coal	A. R.	0.61	49.35	36.15	0.011	13.89	1.50				
70	R. B. Allen	Alma Coal	A. R.	2.34	40.11	54.94	0.004	2.61	0.89				
71	Albert Meyers	Alma Coal	A. R.	1.40	40.00	53.47	0.004	5.13	0.82				
71A	Elbert Haner	Alma Coal	A. R.	1.83	38.49	57.03	0.006	2.65	0.67				
72	John Stollings	Alma Coal	A. R.	1.27	41.37	52.21	0.006	5.15	1.57				
73	Kelley Allen	Alma Coal	A. R.	1.06	38.14	53.21	0.004	7.59	1.99	76.76	5.18	7.11	1.37
74	Reuben Connelly	Alma Coal	A. R.	1.91	43.16	49.85	0.005	6.08	1.92				
75	Isaac Bell	Alma Coal	A. R.	1.65	42.78	52.20	0.003	3.37	1.88				
76	J. E. Chapman	Alma Coal	A. R.	1.73	40.44	52.95	0.004	4.88	0.96				
76A	Ed. Fowler	Alma Coal	A. R.	1.48	40.74	50.90	0.004	6.88	1.70				
77	Born Hager	Alma Coal	A. R.	1.32	40.87	53.43	0.004	4.38	1.61				

Analysis No.	MINE	HORIZON.	Condition o	Moisture.	Volatile Matter.	Fixed Carbon.	Phosphorus.	Ash.	Sulphur.	Carbon.	Hydrogen.	Oxygen.	Nitrogen.	
78	Roland Bias	Alma Coal	A. R.	2.10	37. 47	56.33	003	3.83	1.09	......	......	......	......	..
79	James Allen	Alma Coal	A. R.	1.31	42.04	53.14	080	3.51	1.47	......	......	......	......	..
80	Edward Dent	Alma Coal	A. R.	1.50	40.57	53.41	0.003	4.52	1.51	......	......	......	......	..
81	J. D. Dall	Alma Coal	A. R.	1.58	40.96	53.92	0.003	3.54	1.21	......	......	......	......	..
82	Wilburn Estep	Alma Coal	A. R.	1.31	38.74	55.87	0.003	4.08	1.17	......	......	......	......	..
83	Cal ary Bias	Alma Coal	A. R.	1.24	39.96	54.45	0.004	4.35	1.24	80.52	5.55	7.08	1.26	14
84	William Bias	Alma Coal	A. R.	1.23	38.98	56.37	040	3.42	1.13	......	......	......	......	..
85	O. B. Bias	Alma Coal	A. R.	1.37	39.25	55.95	0.003	3.43	1.17	......	......	......	......	..
86	Joe Ellis	Alma Coal	A. R.	1.85	39. 2	56. 9	020	1.94	1.17	......	......	......	......	..
87	Charles Spencer	Alma Coal	A. R.	0.72	38.40	54.70	0.003	6.18	0.76	......	......	......	......	..
88	Quinn Jeffery Heirs	Alma Coal	A. R.	1.23	40.40	55.20	030	3.17	1.18	81.76	5.62	7.08	1.19	1
88A	B. W. Bias	Alma Coal	A. R.	1.20	39.06	55.27	0.003	4.47	1.14	......	......	......	......	..
89	Park Ball	Alma Coal	A. R.	1.20	34.25	48.07	0.007	16.48	3.75	68.01	4.51	6.29	0.96	1
90	Park Ball	Alma Coal	A. R.	0.56	39.86	55.10	0.005	4.48	1.66	......	......	......	......	..
90A	M. Bennett	Alma Coal	A. R.	1.25	39.50	51.78	0.005	·7.47	1.14	77.37	5.30	7.35	1.37	1
91	Spruce River Coal Co	Alma Coal	A. R.	0.93	37.95	53.30	080	7.82	1.80	......	......	......	......	..
91A	Coal River Mining Co	Alma Coal	A. R.	1.44	37.95	55.79	0.008	4.82	0.67	79.43	5.44	8.31	1.33	1
92	Yawkey & Freeman	Alma Coal	A. R.	1.05	40.61	53.48	0.006	4.86	1.06	......	......	......	......	..
92A	Pond Creek Coal & Land Co.	Ima Coal	A. R.	1.09	38.11	58.31	0.025	2.49	1.70	......	......	......	......	..
93	R. L. Curry	Alma Coal	A. R.	1.29	41.98	53.20	006	3.53	1.33	79.70	5.49	8.71	1.24	1
94	Laura Thompson	Alma Coal	A. R.	1.36	42.23	51.27	006	5.14	1.48	79.00	5.52	7.67	1.19	1
95	Grover Green	Alma Coal	A. R.	1.61	35.42	58.80	040	4.17	0.77	......	......	......	......	..
95A	Grant Perry	Alma Coal	A. R.	0.78	34.09	60.92	0.003	4.21	0.73	......	......	......	......	..
95B	John Q. Dickinson	Alma Coal	A. R.	0.72	33.74	62.54	0. 040	3.00	0.51	......	......	......	......	..
	Average	Alma Coal	A. R.	1.29	9.55	53.80	0.005	*5.36	*1.32	77.82	5.32	7.45	1.24	1
96	J. T. Hatfield	Campbell Creek (No. 2 Gas) Coal	A. R.	0.96	35.21	59.40	006	4.43	0.75	82.26	5.04	6.50	1.02	1
96	J. T. Hatfield	Campbell Creek (No. 2 Gas) Coal	A. R.	1.91	37.39	58.17	0.004	2.53	0.73	......	......	......	......	..
97	J. T. Hatfield	Campbell Creek (No. 2 Gas) Coal	A. R.	1.14	34.16	62.04	040	2.66	0.73	......	......	......	......	..
99	ter Kirk	Campbell Creek (No. 2 Gas) Coal	A. R.	0.82	39.39	57.06	0.003	2.73	0.80	......	......	......	......	..
1 0	Boone & Kanawha L. & M. Co.	ll Creek (No. 2 Gas) Coal	A. R.	0.95	39.39	53.60	0.005	5.50	0.77	......	......	......	......	..
101	E. A. Holstein	Campbell Creek (No. 2 Gas) Coal	A. R.	0.64	0.36	49.46	006	9.54	1.72	......	......	......	......	..
101A	Coal River Mining Co	Campbell ek (No. 2 Gas) Coal	A. R.	2.10	36.42	59.55	0.002	1.93	0.66	......	......	......	......	..
102	E. M. Kinder	Campbell Creek (No. 2 Gas) Coal	A. R.	1.19	40.94	55.48	0.004	2.39	0.78	......	......	......	......	..
02A	Coal River Mining Co.	Cam ell Creek (No. 2 Gas) Coal	A. R.	1.87	40.40	. 5	040	4.18	0.65	......	......	......	......	..
103	o. Estep	Campbell Creek (No. 2 Gas) Coal	A. R.	1.14	40.13	56.28	0.002	2.45	1.14	......	......	......	......	..
104	Turley	Campbell Creek (No. 2 Gas) Coal	A. R.	1.89	38.07	53.20	008	6.84	0.67	......	......	......	......	..
04A	Turley	Campbell Creek (No. 2 Gas) Coal	A. R.	1.53	38.88	. 5	0.006	4.01	0.59	......	......	......	......	..
104B	y (Niggerhead)	Campbell Creek (No. 2 Gas) Coal	A. R.	1.06	3 0.27	49. 0	0.019	18.52	046	......	......	......	......	..

Analysis No.	MINE	HORIZON.	Condition of	Moisture.	Volatile Matter.	Fixed Carbon.	Phosphorus.	to Both. Ash.	to Both. Sulphur.	Carbon.	Hydrogen.	Oxygen.	Nitrogen.	
105	[illegible]tis Blake	Campbell Creek (No. 2 Gas) Coal	A. R.	1.87	41.10	52.28	0.005	4.75	1.25	......	......	......	......	..
106	Irvin [illegible]	Campbell Creek (No. 2 Gas) Coal	A. R.	1.90	40.67	53.39	0.011	4.04	0.66	......	......	......	......	..
107	Olsey Perdue	Campbell Creek (No. 2 Gas) Coal	A. R.	0.78	39.38	58.55	0.004	1.29	0.87	......	......	......	......	..
108	Thomas Epling	Campbell Creek (No. 2 Gas) Coal	A. R.	1.32	40.58	55.86	0.003	2.24	0.70	......	......	......	......	..
108A	Peytona Block Coal Co.	Campbell Creek (No. 2 Gas) Coal	A. R.	1.15	38.16	5[illegible]47	0.006	7.22	1.70	76.75	5.01	8.10	1.22	1
109	Mandaville Dolan	Campbell Creek (No. 2 Gas) Coal	A. R.	0.61	38.07	52.37	0.005	8.95	1.21	......	......	......	......	..
110	Rock Creek [illegible]es Co.	Campbell Creek (No. 2 Gas) Coal	A. R.	1.85	40.16	54.37	0.004	3.62	0.89	......	......	......	......	..
112	Samuel Gillispie	Campbell Creek (No. 2 Gas) Coal	A. R.	1.75	37.32	58.46	0.006	2.47	0.73	......	......	......	......	..
113	Hickory Ash [illegible]al Co.	Campbell [illegible]eek (No. 2 Gas) Coal	A. R.	1.43	37.27	57.17	0.009	4.13	1.14	80.13	5.32	8.03	1.25	1
114	Hickory Ash Coal Co.	Campbell Creek (No. 2 Gas) Coal	A. R.	1.15	38.81	52.47	0.007	7.57	1.18	76.70	5.14	7.90	1.51	1
114	Hickory Ash Coal Co.	Campbell Creek (No. 2 Gas) Coal	A. R.	1.08	43.11	52.52	0.005	3.29	2.07	80.83	5.25	7.30	1.26	1
115	G. [illegible] i [illegible]ff	Campbell Creek (No. 2 Gas) Coal	A. R.	1.16	35.90	56.54	0.005	6.40	1.09	......	......	......	......	..
116	Robert Tony	[illegible]bell Creek (No. 2 Gas) Coal	A. R.	1.28	35.96	59.00	0.005	3.76	0.62	80.83	5.16	8.30	1.33	1
117	William Forber	Campbell Creek (No. 2 Gas) Coal	A. R.	1.58	37.22	58.45	0.004	2.75	0.73	......	......	......	......	..
118	Harry Belcher	[illegible]ll Creek (No. 2 Gas) Coal	A. R.	1.65	37.10	58.66	0.002	2.50	0.78	......	......	......	......	..
119	Nancy Kuhn.	Campbell Creek (No. 2 Gas) Coal	A. R.	0.71	35.41	54.93	0.02	8.95	1.35	......	......	......	......	..
120	[illegible]d Creek Coal & Land Co.	Campbell [illegible]ek (No. 2 Gas) Coal	A. R.	1.08	40.39	55.06	0.007	3.47	0.74	82.30	5.48	6.63	1.38	14
121	R. P. Chew	Campbell [illegible]eek (No. 2 Gas) Coal	A. R.	2.18	39.12	54.23	0.040	4.47	1.20	......	......	......	......	..
	Average	Campbell Creek (No. 2 Gas) Coal	A. R.	1.35	38.31	55.51	0.006	*4.83	*0.95	79.97	5.20	7.54	1.28	1
122	Wharton Land Co.	Powellton Coal	A. R.	0.71	36.21	59.10	0.186	3.98	0.66	82.68	5.27	6.16	1.25	1
123	Wharton Estate.	Powellton Coal	A. R.	6.56	32.57	55.93	0.005	4.94	0.56	......	......	......	......	..
	Average	[illegible]ton Coal	A. R.	3.63	34.39	58.52	0.095	*4.46	*0.61	......	......	......	......	..
124	Wharton Land Co.	Eagle Coal	A. R.	1.49	35.86	57.18	0.006	5.47	0.59	79.28	5.12	8.33	1.21	1
125	Wharton Land Co.	Eagle Coal	A. R.	1.42	39.88	54.10	0.080	4.60	0.63	......	......	......	......	..
126	[illegible]rn Pocahontas C. & C. Co.	Eagle Coal	A. R.	1.37	36.42	52.51	0.013	9.70	0.76	......	......	......	......	..
127	Wharton Land Co.	Eagle Coal	A. R.	1.38	31.42	64.35	0.005	2.85	0.62	......	......	......	......	..
128	[illegible]rn Pocahontas C. & C. Co.	Eagle Coal	A. R.	1.66	37.89	53.87	0.003	6.58	0.67	......	......	......	......	..
129	[illegible]rn Pocahontas C. & C. Co.	Eagle Coal	A. R.	1.00	28.43	63.27	0.013	7.30	0.58	......	......	......	......	..
	Average	Eagle Coal	A. R.	1.39	34.98	57.55	0.008	*6.08	*0.64	......	......	......	......	..

*Averages of P[illegible]ate.

Page Reference to Detailed Description and Sections of Coal Openings and Mines Listed in Preceding Table.

Table No.	Map No.	Lab. No.	Name of Owner.	Page of this Report
			No. 5 Block Coal.	
1	6	103-T	Tuncil Price	213
2	7	131-T	D. G. Courtney	214
3	11	301-K	Little Coal Land Co	215
4	12	302-K	Little Coal Land Co	216
5	13	303-K	Little Coal Land Co	216
6	14	304-K	Little Coal Land Co	216
7	17	164-T	Asa Williams	218
8	17	419-K	Asa Williams	218
9	20	185-T	E. J. Berwind	220
10	22	431-K	Lackawanna Coal & Lumber Co	221
11	23	427-K	W. C. Cook	221
12	20	186-T	E. J. Berwind	220
13	5	101-T	Mohler Lumber Co	213
			Stockton-Lewiston Coal.	
14	32	118-T	Horse Creek C. & L. Co	236
14A	25A	100-T	Staunton Grailey	233
15	30	402-K	Horse Creek Block Coal Co	235
16	43	355-K	Allen Foster	240
17	50	158-T	Lackawanna Coal & Lumber Co	243
18	61	119-T	Yawkey & Freeman	247
			Coalburg Coal.	
19	90	422-K	William Price	261
20	94	194-T	Lackawanna Coal & Lumber Co	263
20	94	155-T	Lackawanna Coal & Lumber Co	263
21	96	159-T	Lackawanna Coal & Lumber Co	264
22	99	156-T	Lackawanna Coal & Lumber Co	265
22A	110A	152-T	Marion Ferrell	269
23	108	428-K	Lackawanna Coal & Lumber Co	268
24	111	429-K	Wharton Estate	270
25	115	196-T	Wharton Estate	271
26	119	430-K	E. J. Berwind	273
26A	119	187-T	E. J. Berwind	273
			Winifrede Coal.	
27	142	300-K	Little Land Coal Co	288
27A	143	435-K	Little Land Coal Co	289
28	157	433-K	Cassingham	293
29	166	416-K	John Q. Dickinson	297
29A	167	415-K	John Q. Dickinson	297
30	169	438-K	Anchor Coal Co	298
30A	170	437-K	Webb Fuel Co	299
31	171	190-T	Seng Creek Coal Co	300
32	172	191-T	Squire Gibson	300
33	174	421-K	Lackawanna Coal & Lumber Co	301
34	176	442-K	Lackawanna Coal & Lumber Co	302
35	191	184-T	E. J. Berwind	307
35	191	181-T	E. J. Berwind	307
35A	192A	180-T	T. C. Jarrell	309
35B	192A	164-T	T. C. Jarrell	309

Table No.	Map No.	Lab. No.	Name of Owner.	Page of this Report
			Chilton Coal.	
36	214	400-K	Coal River Mining Co.................	323
37	215	404-K	Floyd Nelson........................	323
38	216	306-K	Little Coal Land Co..................	324
39	218	410-K	Yawkey & Freeman...................	325
39A	221	149-T	Ashford Ball.........................	326
39B	226A	434-K	Chilton, MacCorkle, Chilton and Meany	328
40	237	193-T	Lackawanna Coal & Lumber Co.......	332
41	238	192-T	Lackawanna Coal & Lumber Co.......	332
42	244	423-K	John Q. Dickinson...................	334
			Little Chilton Coal.	
42	251	178-T	Wharton Estate.......................	338
			Hernshaw Coal.	
44	257	301-T	Fork Creek Coal Co..................	341
44A	258	125-T	Bedford et al.........................	341
45	267	138-T	James Workman......................	344
46	273	414-K	Emma Hager.........................	346
47	274	436-K	Chilton, MacCorkle, Chilton and Meany	346
48	286	170-T	LaFollette, Robson et al..............	350
49	287	171-T	LaFollette, Robson et al..............	351
49A	297	157-T	Lackawanna Coal & Lumber Co.......	354
49B	291	163-T	H. A. Robson........................	352
49C	291A	165-T	F. W. Mason.........................	352
49D	291	173-T	H. A. Robson........................	352
50	300	182-T	Ira Sutphin..........................	355
51	303	177-T	Wharton Estate.......................	356
52	309	195-T	Pond Fork Coal & Land Co...........	358
			Williamson Coal.	
53	327	146-T	Isaac Gore...........................	372
54	328	148-T	Ashford Ball.........................	372
55	330	IIA p. 399	W. P. Croft..........................	373
			Cedar Grove Coal.	
56	344	111-T	F. G. Kirk............................	380
57	348	114-T	Minerva Hill.........................	381
58	352	112-T	S. M. Javins.........................	382
58A	362	393-K	Elmer Harless........................	385
59	363	389-K	Eliza Abshire.........................	385
60	369	104-T	Julia Woodrum.......................	387
61	372	121-T	A. J. Kerse..........................	388
62	378	133-T	Vernon Vandelinde..................	390
63	391	132-T	Dora Hopkins........................	394
64	400	130-T	Charles Ball.........................	397
65	401	124-T	Silas Billetts........................	398
66	403	137-T	Ballard Hager.......................	398
67	414	401-K	D. M. Arbogast......................	402

Table No.	Map No.	Lab. No.	Name of Owner.	Page of this Report
			Alma Coal.	
68	421	128-T	Julian Hill	406
69	421	129-T	Julian Hill	406
70	427	397-K	R. B. Allen	408
71	428	126-T	Albert Meyers	408
71A	452	413-K	Elbert Haner	415
72	432	134-T	John Stollings	409
73	434	127-T	Kelley Allen	410
74	436	135-T	Reuben Connelly	410
75	439	409-K	Isaac Bell	411
76	443	408-K	J. E. Chapman	412
76A	448	407-K	Edward Fowler	414
77	450	411-K	Born Hager	415
78	451	412-K	Roland Bias	415
79	458	136-T	James Allen	417
80	463	140-T	Edward Dent	419
81	464	138-T	J. D. Dall	419
82	465	141-T	Wilburn Estep	419
83	466	142-T	Calvary Bias	420
84	467	143-T	William Bias	420
85	468	144-T	O. B. Bias	420
86	473	145-T	Joe Ellis	422
87	474	153-T	Charles Spencer	422
88	475	150-T	Quin Jeffery Heirs	423
88A	477	151-T	B. W. Bias	423
89	479	398-K	Park Ball	424
90	482	308-K	Park Ball	425
90A	483	417-K	M. Bennett	425
91	484	309-K	Spruce River Coal Co	426
91A	485	399-K	Coal River Mining Co	426
92	486	117-T	Yawkey & Freeman	427
92A	511D	172-T	Pond Creek Coal & Land Co	437
93	512	132-T	R. L. Curry	438
94	513	122-T	Laura Thompson	438
95	516	176-T	Grover Green	439
95A	529A	183-T	Grant Perry	443
95B	524A	440-K	John Q. Dickinson	441
			Campbell Creek (No. 2 Gas) Coal.	
96	519	179-T	T. T. Hatfield	488
96	519	188-T	T. T. Hatfield	488
97	520	433-K	T. T. Hatfield	488
99	540	108-T	Walter Kirk	450
100	550	113-T	Boone & Kanawha L. & M. Co	453
101	551	110-T	F. A. Holstein	454
101A	555	385-K	Coal River Mining Co	455
102	556	105-T	F. M. Kinder	455
102A	557	390-K	Coal River Mining Co	456
103	558	106-T	George Estep	456
104	560	386-K	Turley	457
104A	560	388-K	Turley	457
104B	560	387-K	Turley	457

Table No.	Map No.	Lab. No.	Name of Owner.	Page of this Report
105	561	392-K	Curtis Blake	458
106	562	391-K	Irvin Green	458
107	566	107-T	Olsey Perdue	459
108	569	164T	Thomas Epling	460
108A	571A	164-T	Peytona Block Coal Co	462
109	574	109-T	Mandaville Dolan	463
110	579	396-K	Rock Creek Collieries Co	465
112	617	115-T	Samuel Gillespie	478
113	620	169-T	Hickory Ash Coal Co	479
114	621	160-T	Hickory Ash Coal Co	480
114	621	189-T	Hickory Ash Coal Co	480
115	624	174-T	George Midkiff	481
116	625	168-T	Robert Tony	481
117	627	167-T	William Forber	482
118	628	162-T	Harry Belcher	482
119	631	161-T	Nancy Kuhn	484
120	635	175-T	Pond Creek Coal & Land Co	485
121	639	402-K	R. P. Chew	486
			Powellton Coal.	
122	648	198-T	Wharton Land Co	494
123	661	441-K	Wharton Estate	495
			Eagle Coal.	
124	675	196-T	Wharton Land Co	499
125	676	197-T	Wharton Land Co	506
126	677	199-T	Western Pocahontas C. & C. Co	507
127	678	439-K	Wharton Land Co	507
128	683	200-T	Western Pocahontas C. & C. Co	508
129	685	443-K	Western Pocahontas C. & C. Co	509

CHAPTER X.

CLAYS, ROAD MATERIALS, BUILDING STONES, SAND, IRON ORE, FORESTS, AND CARBON BLACK.

THE CLAYS IN BOONE COUNTY.

The clays of Boone County are all of sedimentary origin. They may be divided into two classes; viz, first, the bedded clays, and second, the recent unconsolidated silts or clays of the stream valleys. The former are the most important.

CLAYS IN THE CONEMAUGH SERIES.

There is usually a bed of fire clay underlying the Brush Creek Coal 140 to 160 feet over the No. 5 Block Coal. This bed is from 2 to 3 feet in thickness. The following section in the **core drill hole No. 1 (143-L)**, put down by the Standard Fuel Company, 2.2 miles North 75° West from MacCorkle, exhibits the occurrence of this fire clay bed:

				Ft.	In.
Coal	2′	2″	Brush Creek Coal	3	9
Coal and bone, interlaminated	0	3			
Coal	0	7			
Slate	0	2			
Coal	0	7			
Fire clay	2′	0″	....	13	9
Sandy shale	10	4			
Fire clay	1	5			

This fire clay horizon appears in the northern part of Washington District near the tops of the hills and also in the northwestern portion of Peytona District.

CLAYS IN THE ALLEGHENY SERIES.

Overlying the No. 5 Block Coal from 80 to 110 feet, on top of the East Lynn Sandstone, there often occurs a bed of fire clay from 2 to 5 feet in thickness near the tops of the hills in the northern and northwestern parts of Boone. This bed has never been developed at any point in the county, but has the appearance of a good plastic clay for brick making purposes.

CLAYS IN THE KANAWHA SERIES.

As a rule a bed of impure fire clay occurs under each of the coal beds in the Kanawha Series. This clay varies in thickness from 1 to 5 feet. The clay in this formation has not been mined at any point in Boone County so far as known to the writer. There also occurs a bed of fire clay in Washington District over the Chilton Coal, from 2 to 3 feet thick. This clay occurs just over the coal bed and forms a very bad roof for the coal. The clay is dark gray and hard when the coal is first mined, but as soon as it is exposed to atmospheric agency for a few days, it crumbles and disintegrates, thereby making a dangerous roof.

A sample of this clay was collected in the mine of the Coal Valley Coal Company, in Logan District, Logan County, 0.5 mile south of Clothier P. O., and 0.5 mile south of the Boone-Logan County Line, and the analysis made of same by Messrs. Hite and Krak, under Laboratory No. 420-K, gives the following results:

	Per cent.
Silica (SiO_2)	52.78
Ferric Iron (Fe_2O_3)	6.23
Alumina (Al_2O_3)	23.22
Lime (CaO)	0.54
Magnesia (MgO)	1.94
Sodium (Na_2O)	0.96
Potassium (K_2O)	4.46
Titanium (TiO_2)	0.14
Phosphoric Acid (P_2O_5)	0.12
Moisture	1.24
Loss on ignition	8.74
Total	100.37

There often appears a stratum of fire clay over the Williamson Coal, varying in thickness from 1 to 3 feet. The section measured in the W. P. Croft Opening (330) on page 373 of this Report, gives the thickness of the fire clay as 3 feet.

Underlying the Cedar Grove Coal, there usually occurs a soft bed of gray fire clay from 1 to 3 feet in thickness. This clay is mined in order to increase the height of the haulways where the coal bed is thin. Since it is soft, it is easily mined and removed by the miner. This clay has not been utilized at any point in Boone County for brick making purposes.

RECENT CLAYS.

The recent clays occur in the flood plains of rivers and small streams. These flood plain clays are uncertain in their distribution in the valleys of the larger streams, and there are no means of indicating where they are most likely to occur. They usually range in thickness from 1 to 5 feet. These deposits occur along the valley of Big Coal River and usually contain considerable sand and silt.

ROAD MATERIALS.

Boone County contains very little limestone, so it will be necessary to look to other material than limestone for road building purposes. The roads thus far constructed in the county are made entirely of dirt and no material other than that found where the road was being made has been used.

GRAVEL PITS.

Possibly the best road materials within the county are the gravel deposits along the valley of Coal River and its branches, principally at the mouths of tributary streams. These gravel and sand deposits, if placed on the roads, will make good road material, as has already been proved in several of the counties of West Virginia.

BUILDING STONES.

The sandstones of Boone County should furnish an almost inexhaustible supply of fair building material. The following is a list in descending order of the sandstones that crop to the surface in Boone, and that are available and of sufficient thickness and durability to warrant quarrying for local supply and in other cases for shipping purposes. The thickness, character and distribution of these several ledges are discussed in this Report on pages indicated at the right-hand margin of the list:

	Page.
Allegheny Series:	
Upper Freeport Sandstone	208
East Lynn Sandstone	208
Pottsville Series:	
Homewood Sandstone	226-231
Coalburg Sandstone	250-254
Lower Coalburg	277-278
Upper Winifrede Sandstone	278-284
Lower Winifrede Sandstone	313
Upper Chilton Sandstone	314-315
Lower Chilton Sandstone	336
Hernshaw Sandstone	339
Naugatuck Sandstone	359-360
Williamson Sandstone	360
Upper Cedar Grove Sandstone	375
Middle Cedar Grove Sandstone	403
Lower Cedar Grove Sandstone	403
Logan Sandstone	443
Malden Sandstone	444
Brownstown Sandstone	492
Matewan Sandstone	501
Eagle Sandstone	501
Decota Sandstone	512

SANDS.

Owing to the many sandstone ledges in the Pottsville Series that crop along the bluffs and hillsides of the streams in Boone County, Coal River contains numerous sand bars and deposits of sand along its channel, so that the sand is pumped out of the river and loaded into railroad cars for shipment. Two plants are located near the Forks of Coal River in Kanawha County, about 4 miles north of the Boone-Kanawha County Line. There are several points along Coal River from Brounland to Jarrolds Valley where plants of that nature could be located.

IRON ORE.

There sometimes occurs a deposit of iron ore underneath the Eagle Coal and above the Eagle Limestone. This stratum is usually from 4 to 6 inches thick and has a dark grayish color.

The following section was measured on Turtle Creek, just south of Olive Branch Church, 2 miles south of Turtle Creek P. O., on the land of C. A. Croft:

		Ft.	In.
1.	Coal blossom, Eagle	0	4
2.	Slate, gray	2	0
3.	Iron ore, dark gray	0	4
4.	Sandy shale and concealed	20	0
5.	Limestone, fossiliferous, Eagle, 785' L	1	0

A sample (No. 418-K) was collected from No. 3 of the above section by Krebs and was analyzed by Messrs. Hite and Krak with the following results:

	Per cent.
Silica	3.84
Metallic Iron	49.00
Sulphur	45.84
Moisture and undetermined	1.32
Total	100.00

FORESTS.

There is possibly more hardwood timber at present in Boone County than in any other county in West Virginia. The larger poplars in the forest have been cut and floated down Coal River, but the oak, beech, sugar, maple and hickory yet remain. A glance at the State Survey Map accompanying Volume V, Forestry and Wood Industry, dated February 1, 1911, will show that practically three-fourths of the county still carries forest areas. It is true that there remains very little of virgin forest; however, in the areas that have been cut over, only the soft woods have been taken.

On pages 112-115 of Volume V, Mr. A. B. Brooks gives the following interesting account of the original and present forest conditions and lumber industry of Boone County:

"Original Forest Conditions.

"The forests of Boone County contained little timber except the hardwoods. There was a fringe of hemlocks along many of the streams, and toward the heads of Big and Little Coal Rivers, this species grew in considerable abundance. Pitch pines grew on ridges and southern faces in some parts.

"The following statement in regard to the timber in the virgin forests of the county was written 17 years ago when 'fully one-half of the poplar timber' remained untouched and when 'three-fourths of all other kinds' were standing in their virgin state.

" 'The principal timber trees are poplar, walnut, oak, ash, hickory, maple, beech, birch, lynn, pine, hemlock, cherry, chestnut and locust. Poplar and walnut timber is especially fine in Boone and trees of enormous size are often reported.

" 'A poplar tree was cut in 1889 on the waters of Big Coal River in Boone County, the dimensions of which are given by Maj. Thos. L. Broun, of Charleston, as follows: The tree measured eight feet and nine inches in diameter at the base, and the distance from the ground to the first limb was seventy-one feet. From the tree were cut six logs, each twelve feet in length, the diameters of which at the small end were respectively seven feet, four inches; seven feet; six feet, eight inches; six feet, one inch; five feet, nine inches; and five feet six inches.'

"Wm. C. Reynolds, civil engineer, says: 'I found on West Fork near the head and over one thousand feet above the foot of the mountain, a walnut tree eight feet in diameter and tall enough to furnish half a dozen average length logs.'*

"The Lumber Industry.

"According to Mr. William Thompson, of Madison, 'Up to the year 1830, no timber of any note had been taken out of the county. About this time, however, the people who lived along Little and Big Coal Rivers began to cut the finest of the poplars for flat-boats which they constructed themselves and floated to St. Albans during freshets. From there the boats were towed by steamboats and distributed above at the various salt furnaces along the Great Kanawha. The whole flat-boat was made of poplar except the white oak boat pins. At first the boats were comparatively small and the gunwales were hewed and sawed by hand. The price then was from $2 to $3 a running foot. Later, when the flat-boats were made larger and were loaded with fencing planks and staves, the price was reduced to $1 a foot. There were probably as many as 100 flat-boats built on the Little and Big Coal Rivers each year for 40 years, or more.'

"Beginning about the year 1855, the Coal River Navigation Company put in a system of locks and dams on the Big Coal River from its mouth at St. Albans to the village of Peytona, 30 miles above. The river was thus made navigable for boats and barges, and large quantities of coal, logs, staves, tan bark, and hoop-poles were brought out of the county. During the Civil War, navigation on the river was stopped, the dams were broken, and all improvements greatly damaged. After the war was over, repairs were made and the system remained intact until about the year 1875.

---

*"The Mountain State"—Summers.

"Mr. Julian Hill, of Madison, states that the first steam saw mill in the county was put in operation at Peytona on the Big Coal River about 1840 and was known as the Van Horn mill. Before the completion of the Chesapeake and Ohio branch railroad to Madison in 1907, however, there had been but few steam saw mills in operation in the county and most of the timber cut had been manufactured on mills at St. Albans and other points outside. The Knight Lumber Company, at Sattes, and the Mohler Lumber Company, near St. Albans, are still floating not far from 13,000,000 feet of timber each year down the Coal River from Boone.

"The largest operation located in the county has been that of Peytona Lumber Company, with a single band mill at Peytona on Big Coal River. This plant was put in operation in 1905.

"During the past 5 years, small mills have literally flocked to the forest of the county made accessible by the building of the railroad. There are now no fewer than 44 mills sawing in different sections.

"Since the railroad was extended from Madison to Clothier in 1909, a large band mill belonging to Boone Timber Company has been built just across the line in Logan County. This mill will be supplied in part by logs from Boone County."

"Present Forest Conditions.

"The county is covered largely with cut over forests aggregating about 200,000 acres. Most of this forest land is owned by companies which are holding or operating it for coal or lumber. There are also approximately 10,500 acres of virgin forest land lying principally in four areas. Two virgin tracts lie wholly within the county, one near Peytona and another not far from the northern end of the county on waters of the Little Coal River. In the southern end a portion of a large virgin area extends into the county from Logan, and on the west, part of a 10,000 acre tract on the head of Mud River lies inside the Boone line.

"At least four-fifths of all the woodland still contains some merchantable timber, of which the leading kind is oak.

"The farm land lies in almost continuous belts along the principal streams and in small scattered patches among the hills and mountains. Farms are more numerous in the northern than in the southern end. Much of the rough, hilly land is unfit for agriculture and the timber should be given the right of way.

"The clearing of the narrow, sandy bottoms of the rivers has materially affected the character of their channels. It is stated by old residents of the county that both Big and Little Coal Rivers were one-third narrower 40 years ago than they are at present. Originally the water flowed over solid rock bottoms and at every turn there was a deep, clear pool of water with shallows above and below. Now the deep places are full of sand which has washed from cultivated fields and the stream channels have been widened by the falling in of unprotected banks."

OPERATING SAW MILLS.

At present there are four saw mills operating in Boone County, located as follows: one double Band, eight-foot mill, owned by the Lackawanna Coal & Lumber Company, located at Seth P. O. The capacity of this mill is 80,000 feet B. M.

lumber each day. The company employs about 200 laborers and has timber enough in sight to operate for ten years from 1914. A second double, eight-foot Band mill is located at Clothier. This mill has a capacity of 65,000 feet B. M. daily, and is owned by the Boone Timber Company. The company employs about 150 laborers on the mill and in the woods. The timber used is obtained from Logan County and transported to the mill by rail. The Leatherwood Lumber Company is located at Altman on Little Coal River and operates a single eight-foot Band mill, having a capacity of 40,000 feet B. M. daily. The company employs about 100 laborers in the woods and on the mill. The timber sawn on this mill at present is obtained from Lincoln County. However, the company owns a large area of timber in Washington District, Boone County, where there is enough timber in sight to last at least eight years more.

The Horse Creek Lumber Company owns a single eight-foot Band mill, located at Mistletoe P. O., on Horse Creek. The company employs about 100 men in the mill and in the field and has a daily capacity of 40,000 feet B. M. It has timber in sight for about three years more work.

In addition to the above Band mills in Boone, there are several small circular saw mills located in different parts of the county, cutting lumber, staves and headings. Logs are also being shipped from the county by rail from several large areas of timber.

CARBON BLACK INDUSTRY.

The manufacture of carbon black has become an industry in Boone County, and there is at present one plant engaged in this business at Dartmont. It requires from 700 to 900 cubic feet of natural gas to make one pound of carbon black, which usually sells for from 3 to 8 cents per pound.

RAVEN CARBON COMPANY.

This plant is located on the north side of Coal River at Dartmont, and was established in 1913. The gas used comes from wells Nos. 4, 8, 17 and 18 on Map II. The plant employs 4 men and has a daily capacity of from 1,200 to 1,800 pounds. The main office of the plant is at Kane, Pa., and Mr. Oscar Nelson, of Weston, W. Va., is the Superintendent.

PART IV.

Paleontology.

CHAPTER XI.

NOTES ON THE PALEONTOLOGY OF BOONE COUNTY.

With Descriptions of Fossils from Boone, Logan and Mingo Counties.

By W. Armstrong **Price.**

SCOPE OF THE INVESTIGATION.

The following preliminary report on the invertebrate fossils of Boone County contains a summary and table of the faunal horizons known to outcrop in the area, a list of the fossils described herein, and a register of the localities in Boone County from which fossils have been collected by the writer, a register of localities in Logan and Mingo Counties from which fossils are described in this Report, together with descriptions and illustrations of the fossils which have been studied.

A preliminary report on the paleontology of Logan and Mingo Counties was made in January, 1915, and a list of fossils[1]

1. West Virginia Geol. Survey, Logan-Mingo Report, p. 753; 1914.

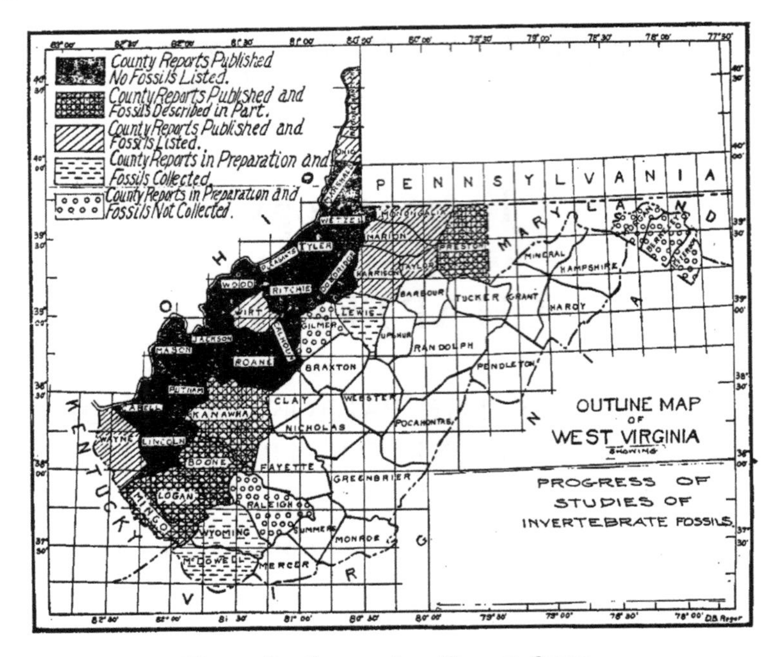

Figure 3. See explanation on figure.

noted during the course of field work was given. It was stated that this list was subject to revision,[2] and it was intended at that time to study all the fossils collected from the three counties named and to publish the results in this Report. Time has not been adequate for this, however, and only a few collections from each county have been studied. The species found in these collections are herein described. It is known that they do not represent at all fully the faunas of the Kanawha Series in the three areas. It has not been possible to give in all cases the relative abundance of the species at the various localities, but this has been done wherever possible.

For descriptions of the various fossiliferous beds the reader is referred to the Chapters of this Report on Stratigraphy, by Mr. Krebs.[3]

The collections from Boone County include small lots brought in by Messrs. Krebs, Teets, and Hennen while engaged in map-

2. Do. p. 750, 1, 19.
3. See above, Part II, Chaps. IV, V, VI, and VII.

ping the strata of the area, and collections made by the writer on a few short trips during the fall and winter of 1913. None of the horizons has been systematically collected from throughout its area of outcrop in the county and fossils from the Dingess Limestone alone have been studied in detail for this Report. The fossils from Logan and Mingo Counties described here are from three horizons, namely, the Buffalo Creek, Dingess, and Eagle Limestones. For descriptions of these horizons in Logan and Mingo Counties and for a complete list of localities from which fossil collections have been made in those areas the reader is referred to the Logan and Mingo Report.[4]

FAUNAL HORIZONS.

The stratified rocks which outcrop in Boone County are included wholly within the upper division of the Carboniferous System; namely, the Pennsylvanian. Of the series of rocks which make up the Pennsylvanian System in West Virginia, the uppermost, or Monongahela Series, and the upper half of the Conemaugh Series are lacking, and if ever present have been removed from the area by erosion. The following series are found exposed in the County:

Conemaugh Series (lower half);
Allegheny Series;
Upper Pottsville, or Kanawha Series.

Series of rocks older than those just named and lying below them are known from deep borings only and will not be discussed here.[5]

The present investigation of the invertebrate fossils of Boone County has been confined to the Kanawha Series. The horizons of the Conemaugh Series which are fossiliferous in other areas are here poorly exposed, of limited extent, and fossils were not observed in them by members of the Survey staff who examined the area of outcrop of Conemaugh rocks. The Alle-

4. Loc. cit., pp. 750-5.
5. See above, Part I, Chap. IV, on "Stratigraphy and General Sections" of Boone County.

gheny Series, so far as known, contains no invertebrate fossils in West Virginia. In the Kanawha Series invertebrate fossils have been found at four different horizons in Boone County. The names of these and their positions in the Series are shown in the following table:

Table of Fossiliferous Members of the Kanawha Series in Boone County.[6]

		Minimum. Feet.	Maximum. Feet.	Total Feet.
1.	Sandstone, Homewood, massive	75	100	100
2.	Shale, sometimes holding coal ("Black Flint" horizon)	0	10	110
3.	Interval		255	365
4.	**Limestone, Buffalo Creek,** gray, hard, lenticular, sometimes containing marine fossils	0	2	367
5.	Interval		363	730
6.	**Limestone, Dingess,** gray, hard, frequently brown and silicious, lenticular, and ferriferous, with marine fossils	0	2	732
7.	Interval		128	860
8.	**Limestone, Seth,** dark gray, shaly, with marine fossils, many *producti*	0	3	863
9.	Interval		224	1087
10.	**Limestone, Campbell Creek,** dark gray, hard, silicious, lenticular, marine fossils found at Bald Knob and High Coal	0	2	1089
11.	Interval		469	1558
12.	**Limestone (and Shale), Eagle—**			
	Shale, black, laminated, with marine fossils	15	20	
	Limestone, dark, brittle, hard, lenticular, with marine fossils	0	2	1605
	Shale, black, iron ore nodules, with marine fossils	10	25	
13.	Interval		228	1833
14.	Coal, Gilbert, double bedded	0	1	1834
15.	Shale, sandy, to bottom of exposed Kanawha Series rocks	5	10	1844

6. Adapted from Generalized Section; pp. 44-7.

List of Fossils from Boone, Logan and Mingo Counties Described in This Report.

Numerals refer to localities from which the fossils were collected. (See *Register of Localities* following). Abbreviations: "a," abundant; "aa," very abundant; "c," common; where no symbol follows the locality number, the species is either rare or its relative abundance is not known.

Fossils.	Marine Horizons. Buffalo Creek.	Dingess.	Eagle.
Crinoidea	....	37c, 38, 57a, 124	...
Lingula umbonata	....	....	121
Orbiculoidea capuliformis	115, 116	117	87c
Derbya crassa	115	38	...
Chonetes variolatus	115c, 116	37	...
Productus semireticulatus	115, 116	55	...
Productus cora	115, 122a	37, 38, 55c, 123	...
Marginifera wabashensis	115a, 122a	37c, 38aa, 57, 123c, 124c	...
Echinoconchus nebraskensis	....	123	...
Echinoconchus symmetricus	115	57	...
Spirifer boonensis ?	115	37c, 38, 57c, 124c	...
Composita sp	....	37c, 38, 124c	...
Astartella gurleyi	....	37	...
Bellerophon crassus var. wewokanus	....	....	87
Euphemus carbonarius	....	....	87
Strophostylus nanus	....	....	87
Conularia crustula ?	....	56-A	...
Griffithides scitulus	....	37	...
Cladodus sp	....	117	...

To attempt to draw conclusions from the above list as to what species are characteristic of the various beds would be unwarranted, because the material examined from the various localities has not been ample; however, from observations of the beds in the field, it was apparent that a great profusion of *Marginifera wabashensis* characterizes certain layers of the Dingess Limestone where it is typically developed. This species is also abundant in one of the collections from the Buffalo Creek Limestone.

Register of Localities.

The following list includes all the localities in Boone County from which fossils have been collected by the writer, together with a few localities from which collections have been made by other members of the Survey and the fossils here described. The name of the fossiliferous bed from which each collection was made follows the description of the locality. An asterisk (*) denotes those localities the collections from which have been studied and are included in this Report:

37.* Washington District, Bearcamp Branch, 0.2 mile south of Mud Fork of Turtle Creek, 4.6 miles southwest of Danville; in bottom of run; Dingess Limestone.

38.* Washington District, Mud Fork of Turtle Creek, 5 miles southwest of Danville; in bed of stream; Dingess Limestone.

39-A Washington District, Mud Fork of Turtle Creek, 4.8 miles southwest of Danville; in road; Dingess Limestone.

39-B. Scott District, Hillside east of Danville; 1090′ B.; not in place; fragment of Dingess Limestone.

57.* Scott District, Roadside near gap between fourth right-hand branch of Rock Creek and Price Branch of Little Coal River; 1165′ L.; 35′ (hand level) below gap, and on Rock Creek side of ridge; Dingess Limestone.

58. 61. Washington District, bed of Low Gap Creek, 1.3 miles southwest of its mouth; not in place; 880′ B.; fragment of Eagle Limestone.

59. Washington District, Turtle Creek at creek level; 760′ B.; 1.1 miles north of Turtletown; Eagle Limestone.

60. Washington District, Turtle Creek, 0.3 mile north of Turtletown; creek level; 825′ B.; Eagle Limestone.

61. (See Locality 58, above).

62. Scott District, 0.5 mile north of U. S. G. S. Bench Mark in town of Madison; roadside east bank of Little Coal River; 769′ L.; Eagle Limestone.

114. Sherman District, C. & O. Ry. cut, 0.8 mile west of High Coal and just north of a fill over a creek flowing into Seng Creek from the southeast; Seth Limestone.

122.* Crook District, west bank of Casey Creek, 0.3 mile above its junction with Pond Fork on point south of mouth of first right-hand branch of the creek; 895′ B.; Dingess Limestone; collection by D. D. Teets.

123.* Crook District, Pond Fork, roadside just north of ford, 1.0 mile north of Van; 775′ B.; Dingess Limestone; collection by C. E. Krebs.

124.* Lincoln County, Harts Creek District, north of road along Big Ugly Creek, 0.7 mile east of mouth of Trace Branch, and 0.2 mile south of Boone County Line; Dingess Limestone; collection by Ray V. Hennen, D. B. Reger, and D. D. Teets.

For convenience of reference, the following descriptive list of localities in Logan and Mingo Counties, fossils from which are described in this Report, is given:

Logan and Mingo Fossil Localities.

Conventions as above.

55.* Mingo County, Hardee District, Right Fork of Laurel Fork of Pigeon Creek, 0.6 mile southeast of Rapp, in bed of stream; Dingess Limestone.

56-A.* Mingo County, Hardee District, Bearpen Branch of Laurel Fork of Pigeon Creek, 1.6 miles west of Canterbury; Dingess Limestone.

87.* Mingo County, Magnolia District, N. & W. Ry. cut, 0.5 mile east of Delorme, at railroad bridge over Tug Fork at Freeburn Colliery; 755' L.; Eagle Shale.

115.* Logan County, Logan District, 1525' B., in left-hand hollow, 0.6 mile below mouth of Conley Branch of Island Creek and 0.5 mile from mouth of hollow (2 miles southeast of Oilville). Buffalo Creek Limestone. Collection by R. M. Gawthrop.

116.* Mingo County, Lee District, 0.6 mile northeast of Rawl on point between Dick Williamson and Murphy Branches of Tug Fork of Big Sandy River; 1395' B.; Buffalo Creek Limestone. Collection by Ray V. Hennen.

117.* Mingo County, Lee District, hillside south of Williamson Creek, coal mine of Superior Thacker Coal Co., 0.25 mile east of mouth of creek; Dingess Limestone. Collection by Ray V. Hennen.

121.* Mingo County, Stafford District, ridge east of Glenalum Junction, on N. & W. Ry., 1055' B., over the railroad tunnel, and 0.2 mile northwest of the Junction; Eagle Shale. Collection by Ray V. Hennen.

DESCRIPTION OF SPECIES.*

*The following abbreviations are used in this Report: "a," abundant; "aa," very abundant; "c," common. Where no symbol follows the description of the locality, the species is either rare or its relative abundance is not known.

INVERTEBRATA.

CRINOIDEA.

Segments of crinoidal columns are common in the Dingess Limestone of Boone County. They are chiefly cylindrical with straight sides and, as represented in our collections, reach a diameter of 12 mm. Three segments forming a portion of a stem are shaped differently from the others; around the middle of each segment on the side wall of the column is a prominent raised ridge from which the surface slopes inward to the upper and lower margins of the segment, the segments being convex on the sides instead of straight; they are very small and composed of colorless, vitreous crystalline, calcareous material in a matrix of fine-grained, calcareous and iron-bearing sandstone. Several small crinoidal plates are also in the collection.

Occurrence.—Kanawha Series, DINGESS LIMESTONE; localities 37, (c); 38; 57, (a); 124.

Collections.—West Virginia Geological Survey and West Virginia University.

BRACHIOPODA.

Genus LINGULA Bruguière.

Lingula umbonata Cox.

Lingula umbonata. Cox, 1857, Kentucky Geol. Survey, vol. iii, p. 575, pl. x, figs. 4, 4a.
Coal Measures: Crittenden, Union and Hancock Counties, Kentucky.

Description.—A single, minute shell 3.5 mm. in length shows all the external features of the shells referred to this species from the Kanawha Black Flint[7] and in addition extremely faint radiating lines extending from the beak to the margin on the central portion of the valve.

Occurrence.—Kanawha Series, EAGLE LIMESTONE. Locality 121.

Collection.—West Virginia Geological Survey.

Genus ORBICULOIDEA d'Orbigny.

Orbiculoidea capuliformis (McChesney).

Discina capuliforma McChesney, 1860, Desc. New Pal. Foss., p. 72.
Coal Measures: Springfield, Illinois; 12 miles northwest of Richmond, Missouri.

Discina capuliformis. McChesney, 1868, Chicago Acad. Sci., Trans., vol. i, p. 23, pl. ii, fig. 20.
Coal Measures: Springfield, Illinois; 12 miles northwest of Richmond, Missouri.

---

7. West Virginia Geol. Survey, 1914, Kanawha County Report, p. 646, pl. i, figs. 3 and 4.

Description.—Several fragmentary specimens of this species are in our collections. The largest has a diameter of 16 mm The features of the shells are well enough exhibited to determine their specific relationship, but as we have no complete specimen a detailed description is not here given. The shells are sub-circular, dome-shaped, the brachial valve having a low convexity, while the pedicle valve is flat and pierced by a lanceolate pedicle slit. Surface of both valves covered with fine, concentric striae.

A specimen from the Eagle Limestone is only 7 mm. in diameter with a convexity of 2.5 mm. and shows the pedicle groove continuing from the beak a little less than one-half the way to the margin at its most distant point where the groove evidently passed into the shell; the walls of the tube by which it was prolonged to the margin have been broken down, leaving an open slit extending to the margin.

Occurrence.—Kanawha Series, BUFFALO CREEK LIMESTONE, Localities 115; 116; DINGESS LIMESTONE, Locality 117; EAGLE LIMESTONE, Locality, 87, (c).

Collection.—West Virginia Geological Survey.

Genus DERBYA Waagen.

Derbya crassa (Meek and Hayden).

Plate xlii, Fig. 1.

Orthis umbraculum? Hall, 1852, Stansb. Expd. Great Salt Lake, p. 412, pl. iii, fig. 6.
Carboniferous: Missouri River above Fort Leavenworth.

Orthis umbraculum? Owen, 1852, Geol. Survey, Wisconsin, Iowa, and Minnesota, pl. v, fig. 11.
Carboniferous: Missouri River, near mouth of Keg Creek, and at Council Bluffs.

Orthis arachnoidea. Roemer (non Phillips), 1852, Kreid. von Texas, p. 89, pl. xi, figs. 9a-b.
Carboniferous: San Saba Valley, Texas.

Orthis arachnoidea. Hall, 1857, Mexican Boundary Survey, pl. xx, fig. 3.

Orthisina crassa. Meek and Hayden, 1858, Acad. Nat. Sci., Philadelphia, Proc., p. 261.
Coal Measures: Leavenworth, Kansas.

Description.—"Shell thick, of medium size, subquadrate, rather compressed; hinge generally a little less than the greatest breadth of the shell, but sometimes equalling it. Front broadly rounded; lateral margins more or less arcuate—in some examples nearly straight. Surface ornamented by numerous straight radiating striae, numbering near the beaks about thirty or forty on each valve, but increasing by the implantation of others between them from one hundred to about one hundred and twenty-four, around the margin; these striae are crossed by numerous fine elevated concentric lines, which are not only quite distinct in the spaces between, but on well preserved specimens are prominent on the striae, to which they impart a sub-crenulate aspect, as seen by the aid of a lens. Adult specimens also generally have several strong concentric imbricating marks of growth.

"Larger or ventral valve nearly flat, cardinal edge sloping a little towards the lateral margins; beak not very prominent or distinct, not incurved, sometimes a little twisted to one side; area rather broad, flat, and inclined obliquely beyond the cardinal edge of the other valve; deltedium thick and prominent.

"Smaller or ventral valve moderately convex in the middle, concave on each side of the umbo, which is generally depressed; mesial tooth strong, and prominent, bifid. Length of a specimen a little above the average size, 1.25 inch; breadth, 1.30 inch; transverse diameter of the two valves, 0.46 inch."—Meek and Hayden, 1858.

The Boone County shells agree with the above description, but are slightly smaller and a little more transverse in outline. The interiors are not exhibited.

Occurrence.—Kanawha Series, BUFFALO CREEK LIMESTONE, Locality, 115; DINGESS LIMESTONE, Locality, 38.

Collection.—West Virginia Geological Survey.

Genus CHONETES Fischer de Waldheim.

Chonetes variolatus d'Orbigny.

Plate xlii, Figs. 7 and 8.

Leptaena variolata. D'Orbigny, 1842, Voyage dans l'Amérique méridionale, vol. iii, p. 49, pl. iv (Pal. Album viii), figs. 10, 11 (under the name of *Productus variolata*).
Carboniferous: Yarbichambi, Bolivia.

Description.—A free translation is as follows: "Dimensions: Apical angle 160 degrees; height, 10 mm.; length, 16 mm.; thickness, 3 mm. Shell much broader than high, much depressed, valves very unequal, the upper valve a little convex, the other concave, beak very short, not extending over the cardinal border. Area almost straight, forming an angle of 160 degrees, and constituting at its ends the greatest width of the shell. From the ends of the area the borders converge toward the front margin, which is very convex, rounded, but indented in the middle by the sinus. Dorsal valve with three depressions; in the middle a sinus becoming very prominent toward the front; the other two on the wings in the higher portion of the shell, disappearing toward the margins. The entire surface of the dorsal valve shows striae scarcely apparent under the lens and very numerous, small pits, arranged in quincunx, which are the scars of small spines; several folds of growth are especially noticeable near the front. There the spines seem to have been very much multiplied." * * * * "I have searched to assure myself that this last characteristic [the great number of small pits] was not an alteration of the shell; but their great depth, and their marked preservation, have convinced me that they should be depended on as a true characteristic of the species."—D'Orbigny, 1842.

Our specimens resemble the above description very nearly; their characters may be summed up as follows: Shell small, depressed, transverse, subquadrate in outline; beak small, not projecting beyond the slightly angulated and mucronate hinge-line, which latter forms the greatest width of the shell; lateral margins sloping slightly toward the front, but nearly at right angles to the anterior and posterior borders; posterior border set with spines along the edge (pedicle valve), four of which have been counted on one side of the beak, in addition to which, judging from the regular spacing of those whose bases are preserved, there must have been three or four more, all traces of which have been lost, due to the breaking of the posterior margin of the shell on which the count was made; anterior margin broadly rounded, only slightly indented by the broad, shallow, poorly-defined median sinus of the pedicle valve; ears of this valve small, triangular, separated from the rest of the shell by slight depressions, borders curving abruptly into the lateral margins; surface covered by numerous fine, regular striae, which increase by bifurcation and number about 80 at the front margin of a shell 7 mm. in height; surface of most pedicle valves exfoliated and showing in the furrows between the ridges very numerous, closely and regularly placed, minute pits; on the inner surface of the valve these seem to be opposed by the rows of very numerous, minute, raised tubercles, which on parts of the shell are elongated and join to form radiating, nodose ridges; these latter are especially pronounced on the anterior and lateral margins of the valves, where they are smaller and more closely arranged. In profile this valve is a broad, low, almost flat-topped dome, the greatest convexity being in the middle laterally, but back of the middle in the anterior-posterior cross-section. Brachial valve slightly concave; like the pedicle valve in its striation and in showing a few concentric growth lines; hinge line straight, slightly raised above the adjacent portions of the shell; the surface of one specimen is only slightly exfoliated and shows the extremely minute, shallow and superficial pits, many of which are slightly elongated toward the front by minute gashes; the pits situated upon the ridges and with no trace of the larger and more numerous pits of the pedicle valve situated between the ridges; interior covered by radiating rows of small tubercles, like those of the pedicle valve; muscle markings only imperfectly seen; median septum, cardinal process, and indistinctly defined lateral diverging ridges of the interior resemble, so far as can be told, de Koninck's[8] figures of the interior of this valve.

These shells differ from the original description of the species

8. De Koninck, Resch. sur les. Animaux Foss., Premiere Partie, 1847, p. 206, pl. XX, fig. 2d.

by d'Orbigny and from that of de Koninck, which is supplemented by a series of illustrations exhibiting all the characters of the species, in the strength of the median sinus, which is said by these writers to be "very prominent toward the front" and "somewhat prominent and very wide." The latter description, of de Koninck, at least suggests that the sinus in shells which had come under his observation was less marked than in those studied by d'Orbigny, and the former's figures show forms with a sinus which would hardly be called "very prominent." Our shells vary from those which are merely flattened in the middle to forms with a broad, shallow, poorly defined sinus. From the fact that other species of the genus which possess a sinus, e. g., *C. granulifer* and *C. verneuilanus*, vary considerably in the prominence with which it is developed, and include some shells in which the sinus is represented merely by a mesial flattening, it has seemed expedient to refer our shells to this species because of the close resemblance in all other features so far as we have been able to ascertain, and especially because the arrangement of the pits and tubercles, and the striae and internal markings of the brachial valve of de Koninck's shells as shown by his illustrations, are similar to those of our shells. In this decision we are following Hall,[9] who found this species near Weston, on the Missouri River, and describes his form as having a "broad, scarcely defined mesial depression" which "gives a straight or slightly sinuous outline in front."

Measurements of eight specimens:

Length. mm.	Height. mm.	Convexity. mm.
18	10	2
16	9	—
15	10	2.5
12	7.5	—
10	9	—
9	6	1
8	5.5	1.25
7	6	—

Ratio of height to length varies from 1:1.1 to 1:1.7.

Occurrence.—Kanawha Series, BUFFALO CREEK LIMESTONE, Localities, 115, (c), 116: DINGESS LIMESTONE, Locality, 37; EAGLE LIMESTONE, Locality, 87.

9. Hall, 1852, Stansbury's Expl. and Surv. Valley Great Salt Lake of Utah. Sen. Exec. Doc., No. 3, Spl. Sess., 1851, App. E., p. 410.

Collections.—West Virginia Geological Survey and West Virginia University.

Genus PRODUCTUS Sowerby.

Productus semireticulatus Martin.

Anomites semireticulatus. Martin, 1809, Petref. Derb., pl. xxxii, figs. 1 and 2; pl. xxxiii, fig. 4.

Productus semireticulatus. De Koninck, 1847, Monog. du Gen. *Productus et Chonetes,* p. 83, pl. xiii, figs. 1a-h; pl. ix, figs. 1a-m; pl. x, figs. 1a-d.

Carboniferous: Bagdad, Harrisville, Cuyahoga, Zanesville, Flint Ridge, Greensburg, and Antrim, Ohio; near Louisville, Kentucky; near St. Louis and St. Charles, Missouri; Leavenworth, on the Missouri River; Long Creek, Crawfordsville, and near New Harmony, Indiana; Sparta, Illinois; Bolivia, South America.

Description.—This widespread and well known species needs no description here. The specimens from Logan and Mingo Counties examined are not perfect, but show the characteristic reticulation in the umbonal region, the small, fairly regular costae, and a few spine bases. They are from 30 to 40 mm. in width at the hinge line.

Occurrence.—Kanawha Series, BUFFALO CREEK LIMESTONE, Localities, 115, 116; DINGESS LIMESTONE, Locality, 55.

Collection.—West Virginia Geological Survey.

Productus cora d'Orbigny.

Plate xlii, Fig. 3.

Productus cora. D'Orbigny, 1842, Voyage dans l'Amérique Méridionale, vol. iii, pt. iv, p. 55.

Carboniferous: Above Patapatoni, on an island in Lake Titicaca; Yarbichambi.

Description.—The shells here referred to *P. cora* have many features in common with the Mississippian species, *P. ovatus* Hall (*P. pileiformis, P. laevicostus, P. coraeformis,* etc., Weller, Illinois State Geol. Surv., Monog. I, 1914, p. 132). The medium size and somewhat narrow shape, the pointed beak and the insignificance of the spine bases, which are found only with difficulty on a few specimens, strongly suggest that species. Our specimens are flattened and the ears in most cases are broken or mashed against the sides of the shell. *P. cora* is quite variable in shape and in the prominence of its spines, and it seems highly improbable that even with perfect specimens at hand specific differences would be found to separate these shells from that species. The variations above noted are characteristic, however, of all the shells of this type from the Kanawha Series which have so far come to hand, which latter can thus be distinguished from specimens of *P. cora* from the higher measures in West Virginia.

Occurrence.—Kanawha Series, BUFFALO CREEK LIMESTONE, Localities, 115; 122, (a); DINGESS LIMESTONE, Localities, 37; 38; 55, (c); 123.

Collections.—West Virginia Geological Survey and West Virginia University.

Genus MARGINIFERA Waagen

Marginifera wabashensis (Norwood and Pratten).

Plate xlii, Figs. 4 to 6.

Productus wabashensis. Norwood and Pratten, 1854, Acad. Nat. Sci., Philadelphia, Jour., 2nd ser. vol. iii, p. 13, pl. i, figs. 6a-d. (Imprint of whole volume, 1855.)
Coal Measures. Near New Harmony, Indiana.

Description.—"Shell small, transverse; dorsal valve gibbous, enrolled on itself; beak small; pointed, and not passing the cardinal border. This valve is divided in its middle by a wide, deep sinus. The surface is covered by about forty ribs, some of which, however, after leaving the visceral part, unite, so as to form by their junction one thicker rib; but this character is not constant, as some specimens possess it while others do not. The ears are thin and enrolled; no tubes have yet been discovered on them. The cardinal border is of about the same breadth as the greatest transverse diameter of the shell. On the surface of the shell itself there are, generally, from six to eight tubes, very thick and long for the size of the shell, one, still attached, being over seven millimeters long. The transverse folds on the visceral disk are wide and shallow, so as to be hardly apparent.

"The ventral valve is very concave, following nearly the movement of the opposite valve; ribs like those of the dorsal, but it has neither tubes nor hollows. It is furnished with a very decided varix, corresponding to the sinus of the dorsal valve.

"Dimensions.—Length, 13 millimeters; breadth, 16 mm.; thickness, 6 mm.

"Ribs.—Forty on the entire surface." Norwood and Pratten, 1854.

From the above description and from shells commonly identified as this species, the shells before us differ in being smaller, in having their spines less coarse, and in lacking a sinus, or if one is present it is shallow and not wide. All have a mesial flattening (or sinus) which widens from the umbonal region toward the front. The above features are apparently not of specific value for the following reasons: the specimens in hand are either internal casts or partially exfoliated shells from a hard sandstone matrix on which the spines are represented only by their very bases and these are also exfoliated, giving them a less coarse appearance than they may have had on the shells when they were entire; the shells are small, are hence less enrolled than those of larger size, and the sinus is but little (if at all) less marked than on the corresponding portion of large shells, if one may judge from the published figures of this species. The submarginal truncation characteristic of the genus is plainly marked on exfoliated dorsal valves.

Measurements of four dorsal valves of representative size:

Width.	Length.	Convexity.
mm.	mm.	mm.
14	9.5	5
12	9	4
14	9.5	5
12.5	9	4

Occurrence.—Kanawha Series, BUFFALO CREEK LIMESTONE, Localities, 115, (aa); 122, (a); DINGESS LIMESTONE, Localities, 37, (c); 38, (aa); 57, 123, (c); 124, (c).

Collections.—West Virginia Geological Survey and West Virginia University.

Genus ECHINOCONCHUS Weller.

It is here proposed to remove from the genus *Productus* the two species *P. nebraskensis* and *P. symmetricus* and to put them in the genus *Echinoconchus* of Weller[10] which contains productoid shells with the external surface "marked by more or less sharply differentiated concentric bands which commonly grow broader in passing from the beak to the outer margins, each band bearing numerous, crowded, fine, appressed, imbricating spines, either subequal or unequal in size, which are produced from elongate, node-like bases."

The two species mentioned above are quite clearly of the type of the genus *Echinoconchus,* though differing slightly from the (Mississippian) species referred to it by Weller. The concentric folds are perhaps not so sharply differentiated in the case of *P. nebraskensis,* and on *P. symmetricus* they become smaller and more closely crowded toward the margin. In the description, the genus, however, is made elastic enough to include these variations.

Echinoconchus nebraskensis (Owen).

Productus nebrascensis. Owen, 1852, Geol. Surv. Wisconsin, Iowa, and Minnesota, p. 584, pl. v, fig. 3.
Carboniferous limestone: Bellevue, Missouri River, Nebraska.

Description.—"In general shape, contour, and sinus, this *Productus* approaches *P. Humboldti;* but it is rather smaller, and differs in the structure of the surface, which, in the dorsal valve, is prominently and interruptedly ribbed, as if by the growth of spines directed downwards, and adhering for some distance to the surface, the intervals being the points where they were thrown off; these intervals are arranged in three or four concentric rows. Ventral valve pitted in numerous concentric rows, with intermediate concentric transverse lamellae." Owen, 1852.

Only a single brachial valve of this species has come to hand. It has a width at the hinge of fifteen millimeters.

Occurrence.—Kanawha Series, DINGESS LIMESTONE, Locality 123.

Collection.—West Virginia Geological Survey.

10. Weller, Illinois State Geol. Surv., Mon. i, 1914, p. 138.

Echinoconchus symmetricus (McChesney).

Productus symmetricus. McChesney, 1860, New Pal. Foss., p. 35. Upper Coal Measures: Lasalle and Springfield, Illinois.
Productus symmetricus. McChesney, 1865, Ill. New Spec. Foss., pl. i, figs. 9a-b.

Description.—"Shell of medium size, depressed, suborbicular; length and breadth about equal. Cardinal line considerably less than greatest width of shell below; extremities rounded; auriculate extensions small, but distinct from the vault of the shell, straight or slightly elevated. Ventral valve quite regularly convex, expanding at the front, and entirely destitute of a mesial sinus; front regularly and broadly rounded; beak large, little incurved, barely passing beyond the hinge line. Dorsal valve regularly concave, of moderate depth.

"Surface marked much as in the preceding species [*P. nebraskensis*], but the spines and broken rib-like striae of the ventral valve are much finer and more thickly set, and the concentric bands are narrower and more closely arranged, becoming crowded on the front of the shell.

"This shell, though closely related to the last described species, differs from it in being a larger shell, less ventricose and always much broader, entirely destitute of mesial sinus, less enrolled, beak not projecting so far beyond the hinge line; in its dorsal valve being less concave, and nearly or quite destitute of the lateral ridges which pass obliquely from the beak to the sides of the valves in that species." McChesney, 1860.

A single brachial valve of this species and a poorly preserved pedicle valve from the area under discussion compose the material so far studied. The figure of this species published in the Preston County Report[11] will serve to illustrate the material in hand.

Occurrence.—Kanawha Series, BUFFALO CREEK LIMESTONE, Locality, 115; DINGESS LIMESTONE, Locality, 57.

Collection.—West Virginia Geological Survey.

---

11. West Virginia Geological Survey, Preston County Report, 1914 p. 499, pl. xliii, fig. 1.

Genus SPIRIFER Sowerby.

Spirifer boonensis Swallow ?

Plate xlii, Fig. 2.

Spirifer boonensis. Swallow, 1860, St. Louis Acad. Sci., Trans., vol. i, p. 646.
Lower Coal Measures: Boone, Randolph, and Monroe Counties, Missouri.

Description.—A dozen specimens of this species, all fragmentary, are of two general sizes. The larger, from 10 to 26 mm. in width at the hinge, show the transverse, slightly convex form previously described from the Black Flint of Kanawha County.[12] Those specimens were flattened by pressure, but these show the convexity to have been at least as great as 3.5 mm. in the largest specimen from Boone County. Several minute forms have the same ribbing as the larger, but their outline is oval or without extended hinge line. I am inclined to regard them as immature forms of the larger shells. The largest specimen has eleven ribs on each side of the varix and shows the faint longitudinal and concentric lines as well as the ribs and growth lines. Several internal casts are included in the material.

As Swallow failed to publish a figure of his species and the author has not seen his type material, our shells are referred to this species only provisionally.

Occurrence.—Kanawha Series, BUFFALO CREEK LIMESTONE, Locality, 115; DINGESS LIMESTONE, Localities, 37, (c); 38; 57, (c); 124, (c).

Collections.—West Virginia Geological Survey and West Virginia University.

---

12. West Virginia Geological Survey, Kanawha County Report, 1914, p. 652, pl. ii, figs. 1-3.

Genus COMPOSITA Bronn.

Composita sp.

Description.—A few small, fragmentary and exfoliated shells of this genus have been found. They are about the size and outline of *C. girtyi* Raymond, but seem on the whole to be less convex than that species, although one or two specimens indicate that the shell may have been almost globose. A sinus is almost entirely absent. The material is too scanty and fragmentary for the specific relations to be definitely determined.

Occurrence.—Kanawha Series, DINGESS LIMESTONE, Localities, 37, (c); 38; 124, (c).

Collection.—West Virginia Geological Survey.

PELECYPODA.

In addition to the pelecypod species about to be described, several other species were found, but are so far represented only by fragmentary specimens or poor internal casts. It has not seemed expedient to describe them until better material is at hand.

Genus ASTARTELLA Hall.

Astartella gurleyi White.

Astartella gurleyi. White, 1878, Acad. Nat. Sci., Philadelphia, Proc., p. 35.

Astartella gurleyi. White, 1880, U. S. Geol. and Geog. Surv. Terr., 12th Ann. Rept., pt. i, p. 166, pl. xlii, figs. 6a, b.
Coal Measures: Danville, Illinois.

Description.—"Shell small, not very gibbous, subtetrahedral in outline; anterior end truncated from the beaks obliquely downward and forward to about midheight of the shell, where the front is sharply rounded to the somewhat broadly rounded basal margin; posterior margin broadly convex or sometimes almost straight and perpendicular, and joining both the dorsal and basal margins by abrupt curves; dorsal margin comparatively short, nearly straight; beaks small; umbones not elevated nor very prominent. An indistinctly defined umbonal ridge extends from each of the umbones to the postero-basal margin, behind which ridge the shell is slightly compressed. Surface marked by concentric furrows, which are separated by sharp linear ridges.

"Length of an average-sized example, 7 millimeters; height from base to beaks, 4½ millimeters." White, 1880.

We have only two specimens of this species, each of which is somewhat fragmentary. The best example lacks a portion of the anterior margin, but is so much like this species in all the features which can be ascertained that it is referred to it with little hesitation.

Length from beak to posterior end of hinge, 3 mm.; height, 3 mm. The latter measurement is inferred from the direction of the concentric ridges on the parts preserved and their probable projections on the portion which has been broken away.

Occurrence.—Kanawha Series, DINGESS LIMESTONE, Locality, 37.

Collection.—West Virginia Geological Survey.

GASTROPODA.

In addition to the species of gastropoda described in the following pages, casts of the interior of two species which are not here described are in our collection, but are in a fragmentary condition and their relationships are uncertain.

Genus BELLEROPHON Montfort.

Bellerophon crassus var. wewokanus Girty.

Bellerophon crassus var. *wewokanus.* Girty, 1911, New York Acad. Sci., Ann., vol. xxi, p. 138.
Wewoka formation: Wewoka and Coalgate quadrangles, Oklahoma.

Description.—"The fossils under consideration are all of small size. They can be described most advantageously by comparing them with *Bellerophon crassus,* to which they are clearly very closely related. For this purpose, it will be best to use the figures and description given by Meek and Worthen, not only because they are the authentic ones, but because although *B. crassus* has appeared in the literature not infrequently, the citations have seldom been based on the observation of good and characteristic specimens. The most essential difference shown by the Wewoka specimens is that the umbilicus instead of being partly open, is so solidly closed that there must have been a continuous imperforate columella. The size is very much smaller; the shape of the aperture more transverse, and the band possibly more elevated." Girty, 1911.

A single specimen of this species is in our collection; it corresponds to the above description.

Occurrence.—Kanawha Series, EAGLE LIMESTONE, Locality 87.

Collection.—West Virginia Geological Survey.

Genus EUPHEMUS McCoy.

Euphemus carbonarius (Cox).

Plate xlii, Fig. 9.

Bellerophon Urii. Norwood and Pratten, 1855, Acad. Nat. Sci., Philadelphia, Jour., 2nd ser., vol. iii, p. 75, pl. ix, figs. 6a-c. (Not *B. urii* Fleming, 1828).
Coal Measures: Galatia and Graysville, Illinois; and 5 miles below New Harmony, Indiana.

Bellerophon carbonarius. Cox, 1857, Kentucky Geol. Surv., vol. iii, p. 562.
Coal Measures: Kentucky.

Description.—A single specimen of this well known species has been collected and it is entirely characteristic of the common form of the species.

Occurrence.—Kanawha Series, EAGLE LIMESTONE, Locality 87.

Collection.—West Virginia Geological Survey.

Genus STROPHOSTYLUS Hall.

Strophostylus nanus (Meek and Worthen).

Platystoma nana. Meek and Worthen, 1860, Acad. Nat. Sci., Philadelphia, Proc., p. 463.
Upper Coal Measures: Springfield, Illinois.
Naticopsis nana. Meek and Worthen, 1861, Acad. Nat. Sci., Philadelphia, Proc., p. 148.
Naticopsis nana. Meek and Worthen, 1866, Geol. Surv. Illinois, Rept., vol. ii, p. 365, pl. 31, figs. 4a, b.
Upper Coal Measures: Springfield, Illinois.

Description.—Shell small, subglobose, wider than high; spire much depressed; volutions about three in number, the last one large, and somewhat ventricose; suture well defined; aperture broadly subovate, somewhat straightened at the inner side, its length nearly equal to seven-eighths of the full axial length of the shell; outer lip thin; inner lip moderately thickened; surface marked by fine lines of growth, which are a little stronger and more uniform on the distal side of the volutions, near the suture, than elsewhere.

"Length, 5 mm.; breadth, 4½ mm." Meek and Worthen, 1866.

One small shell exhibiting the features of the above description, so far as they are preserved on the last volution, seems to be of this species. However, the height of the spire, and the number of volutions are not known, yet the shell closely resembles this widely distributed little species. In size it is very closely the same as the shell described above.

Occurrence.—Kanawha Series, EAGLE LIMESTONE, Locality 87.

Collection.—West Virginia Geological Survey.

Genus CONULARIA Miller.

Conularia crustula White ?

Conularia crustula. White, 1880, U. S. Geol. Geog. Survey, Terr., 12th Ann. Rept., for 1878, pt. i, p. 170' pl. xlii, fig. 4a.
Coal Measures: Kansas City, Missouri; near Taos, New Mexico.

Description.—"Shell rather small, having the usual four-sided pyramidal form; the four sides being equal, and flat or nearly so near the apex, but slightly convex towards the aperture; the four angles distinctly furrowed, and a slender furrow also marks the median line of each side, which furrow is more distinct upon the cast of the interior of the shell than upon the external surface of the test. Surface marked by the numerous transverse raised striae common to this genus, which arch gently forward from each of the four angles; the majority of the striae are continuous across the median line of the sides, and also across the angle-furrows, in crossing which they bend slightly backward.

"Length, 31 millimeters; diameter of aperture, about 16 millimeters." White, 1880.

A fragment of one side of a cone of this species in our collection shows the characteristic striae. As the above description of White's species does not describe the form fully, it is not with certainty that this shell is referred to it. It, however, is the same as the *Conularia* described from Kanawha County by the author.[13]

Occurrence.—Kanawha Series, DINGESS LIMESTONE, Locality 56-A.

Collection.—West Virginia Geological Survey.

Genus GRIFFITHIDES Portlock.

Griffithides scitulus (Meek and Worthen).

Phillipsia (Griffithides) scitula. Meek and Worthen, 1865. Acad. Nat. Sci., Philadelphia, Proc., p. 270.
Upper Coal Measures: Springfield, Illinois.

Phillipsia scitula. Meek, 1872. U. S. Geol. Survey, Nebraska, Final Rept., p. 238, pl. 6, fig. 9.
Upper Coal Measures: Nebraska City, Nebraska; Springfield, Illinois.
Lower Coal Measures: Illinois.

Phillipsia (Griffithides) scitula. Meek and Worthen, 1873. Geol. Surv. Illinois, vol. v, p. 612, pl. 32, fig. 3.
Upper Coal Measures: Springfield, Illinois.

Description.—"Pygidium very convex, smaller than the cephalic shield, forming more than a semicircle; with anterior

13. West Virginia Geological Survey, Kanawha County Report, 1914, p. 653, pl. ii, figs. 4-6.

lateral angles obliquely truncated; posterior outline regularly rounded, with a moderately wide, smooth, depressed, nearly flat or sloping marginal zone, trilobation as in the thorax, strongly defined; mesial lobe prominent, as wide anteriorly as one of the lateral lobes, including its border, distinctly flattened on each side, slightly tapering to an obtuse termination, less than half its own greatest anterior breadth from the posterior edge, segments eleven or twelve, well defined above, but nearly obsolete on the flattened sides. Lateral lobes convex, but distinctly less so than the mesial lobe, horizontally flattened near the latter with an angle along the outer margin of the flattened space, from which the sides slope abruptly to the flattened, smooth border; segments six, simple geniculated or bent down in the middle very distinctly, but terminating abruptly at the rather wide border; each with a minute pustule on the knee. Surface of the glabella and all the segments more or less granular, the granules being larger on the posterior part of the glabella and neck segment than elsewhere.

"Entire length, 0.60 inch; length of pygidium, 0.19 inch; breadth of do., 0.27 inch; length of thorax, 0.18 inch; breadth of do., 0.28 inch; length of cephalic shield, 0.23 inch; breadth of do., 0.32 inch." Meek and Worthen, 1873.

This is a small, elliptic form often found completely enrolled. Meek and Worthen describe the cephalic and thoracic portions in detail, but as we have only a single, badly worn pygidium and know nothing of the upper parts of our form the description of the latter is not quoted here. The pygidium of the one individual found is so badly worn that only its outline, the flattened marginal area, and the presence of three lobes, with transverse segments, and pustules can be determined. It has not been possible to count with certainty the lateral segments, but six at least are present on one side and it is not evident that there were more. The mesial lobe is entirely worn away with only the cast of its inner surface, very poorly preserved, indicating its presence; its relative width cannot be determined, but it may have been greater than that of the lateral segments and the indications are that this was so. Pustules are seen on several segments of each of the lateral lobes. As the specimen resembles *G. scitulus* most closely among described species it is referred to it with little hesitation.

Dimensions: Length of pygidium, 4 mm.; breadth, 6.5 mm.; convexity (at least), 2 mm.

Occurrence.—Kanawha Series, DINGESS LIMESTONE, Locality 37.

Collection.—West Virginia Geological Survey.

VERTEBRATA.

PISCES.

In addition to the fish tooth here described, minute isolated scales have been found in Boone and Mingo Counties in the Dingess Limestone; their description is reserved for a later Report.

Genus CLADODUS Agassiz.

Cladodus sp.

Description.—A single tooth found by Mr. Ray V. Hennen in the Dingess Limestone is of the form generally referred to this genus which is founded on detached teeth. These teeth resemble those of fish of the genus *Cladoselache*, "but there is reason to believe[14] that several primitive genera—perhaps even more than one family or order—were characterized by teeth of this form."

Cones apparently five in number, the central one being slightly more than twice as high as the extreme lateral cones, the two next to the central cone being probably shorter than the others; base transverse to the direction of the cones, constricted in the middle, the constriction widening toward the lower margin, lesser constrictions between the lateral pairs of cones also widening toward the lower margin, and the whole base wider at the top than below, these features producing a series of rounded knobs on the under side of the base, four in number, which are smaller under the extreme pair of cones and larger and more widely spaced under the middle pair; surface of the cones marked

14. Vertebrate Paleontology, by A. S. Woodward, 1898, p. 30.

with fine, sharp, longitudinal striae, which are more numerous near the base, dying out or uniting toward the point—as seen on the central cone on which only a few of the lines reach the point, these becoming almost obsolete on the upper third of the cone.

The specimen from which this description is made is the only one in our possession and is embedded in a sandstone matrix, exposing only one side to view; it is very fragile and in removing the matrix which covered it the lateral cones were broken, and only their outlines can be determined and these not definitely; the cross section of the cones is not known, but it appears to be about that of *Cladodus mortifer* of the Illinois Coal Measures[15] which it apparently resembles in other respects so far as can be told from the fragmentary nature of our specimen and the fact that the species was originally described from a fragmentary tooth. The striae of our form persist further toward the point than in that species as described.

Occurrence.—Kanawha Series, DINGESS LIMESTONE, Locality 117.

Collection.—West Virginia Geological Survey.

---

15. Illinois Geol. Surv., vol. ii, 1866, p. 22, pl. i, fig. 5.

DESCRIPTION OF PLATE.

PLATE XLII.

Page

Fig. 1. *Derbya crassa* (Meek and Hayden)....... x 1.... 599
Buffalo Creek Limestone. Locality 115.

Fig. 2. *Spirifer boonensis* Swallow ?.................... 609
Young individual..x 1.
Dingess Limestone. Locality 124.

Fig. 3. *Productus cora* d'Orbigny......... 604
Pedicle valve, somewhat flattened, with one ear and lateral margin broken away..............................x 1.
Dingess Limestone. Locality 122.

Figs. 4 to 6. *Marginifera wabashensis* (Norwood and Pratten) 605
4. Pedicle valve, from above, surface exfoliated........x 1.
5. Posterior view of the same........................x 1.
Dingess Limestone. Locality 38.
6. Brachial valve, surface exfoliated..................x 1.
Dingess Limestone. Locality 37.

Figs. 7 and 8. *Chonetes variolatus* d'Orbigny............. 601
7. Pedicle valve, surface much exfoliated, showing the minute pits...................................x 1½.
Buffalo Creek Limestone. Locality 116.
8. Pedicle valve of a less convex specimen, exfoliated and almost smooth, showing under a magnifier rows of fine pits radiating from the beak, but not represented on the illustration ....x 1.
Buffalo Creek Limestone. Locality 115.

Fig. 9. *Euphemus carbonarius* (Cox).................. 612
Lip broken; ridges more regular than shown in the illustration...x 1.
Eagle Limestone. Locality 87.

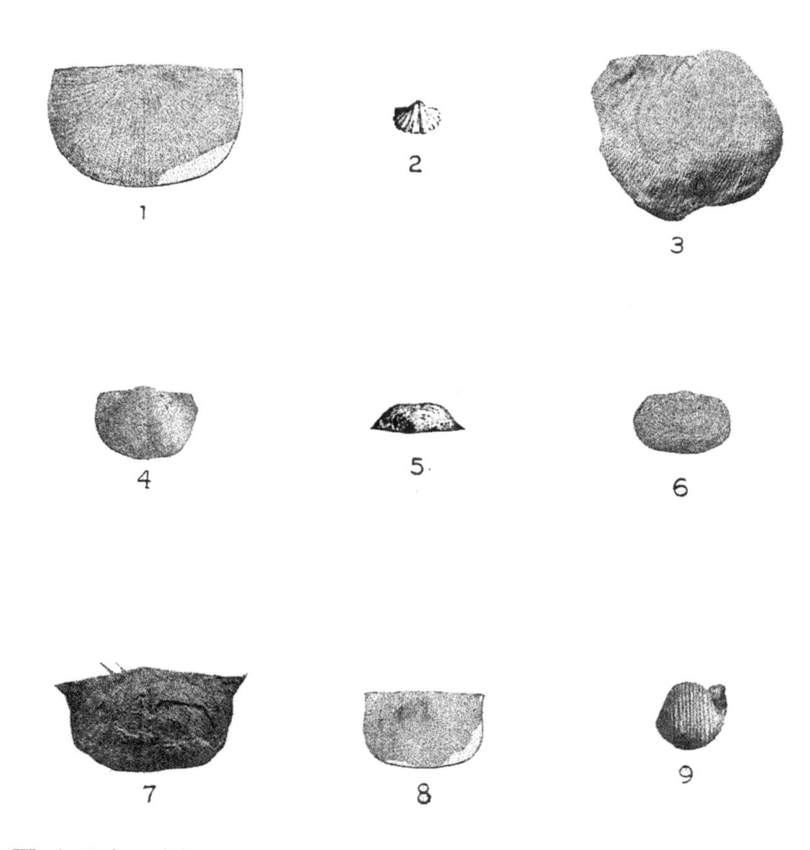

W. A. Price, del.

Fossils from the Kanawha Series.

INDEX TO SPECIES.

(Names in *italic* are synonyms; figures in **black face type** are numbers of pages on which detailed descriptions appear; figures in *italic* denote illustrations.)

Page.

Anomites semireticulatus...........604
Astartella gurleyi............595, **610**

Bellerophon carbonarius............612
Bellerophon crassus...........611, 612
Bellerophon crassus var. wewokanus
..........................595, **611**
Bellerophon Urii.................612

Chonetes 604
Chonetes granulifer................603
Chonetes variolatus......595, **601**, *618*
Chonetes verneuilanus.............603
Cladodus sp.......................595
Cladodus mortifer.................617
Cladoselache 616
Composita girtyi..................610
Composita sp................595, **610**
Conularia 614
Conularia crustula ?.........595, **613**
Crinoidea 595, **597**

Derbya crassa...............595, **599**
Discina capuliforma...............598
Discina capuliformis..............598

Echinoconchus 607
Echinoconchus nebraskensis...595, **607**
Echinoconchus symmetricus...595, **608**
Euphemus carbonarius....595, **612**, *618*

Gastropoda 611
Griffithides scitulus.......595, **614**, 615

Page.

Leptaena variolata.................601
Lingula umbonata............595, **598**

Marginifera wabashensis..595, **605**, *618*

Naticopsis nana...................613

Orbiculoidea capuliformis.....595, **598**
Orthis arachnoidea.................599
Orthis umbraculum ?...............599
Orthisina crassa...................599

Pelecypoda 616
Phillipsia (Griffithides) scitula......614
Phillipsia scitula.................614
Pisces 616
Platystoma nana...................613
Producti 594
Productus....................604, 607
Productus cora......595, **604**, 605, *618*
Productus coraeformis.............604
Productus Humboldti...............607
Productus laevicostus..............604
Productus nebrascensis.............607
Productus nebraskensis........607, 608
Productus ovatus..................604
Productus pileiformis..............604
Productus semireticulatus.....595 **604**
Productus symmetricus........607, 608
Productus variolata................601
Productus wabashensis.............605

Spirifer boonensis ?......595, **609**, *618*
Strophostylus nanus..........595, **613**

APPENDIX

LEVELS ABOVE MEAN TIDE IN BOONE COUNTY.

Coal River Division of the Chesapeake & Ohio Railway.

Distances.	Stations.	County.	Elevation of Low Water	Elevation of Sub-Grade
0.0	St. Albans............	Kanawha ...	....	581.0
1.7	Indian	Kanawha ...	....	578.0
2.5	Calvert	Kanawha ...	....	591.0
4.8	Ferrell	Kanawha ...	....	592.0
6.0	Upper Falls...........	Kanawha ...	....	604.0
7.6	Lincoln	Kanawha ...	....	597.5
10.2	Faqua	Kanawha ...	....	602.2
11.2	Ballard	Kanawha ...	....	604.0
11.9	Alum Creek...........	Kanawha ...	....	609.0
13.0	Forks of Coal.........	Kanawha ...	....	608.0
15.2	Sproul	Kanawha ...	582.0	610.1
17.1	Brounland	Kanawha ...	....	613.7
17.9	Hollyhurst	Kanawha ...	....	622.0
20.2	Emmons	Kanawha ...	....	627.0
21.2	Morris	Kanawha ...	....	629.0
23.2	Dartmont	Boone	....	636.0
24.1	Ashford	Boone	....	641.0
25.6	Brushton	Boone	615.0	643.0
29.2	Peytona	Boone	....	660.0
30.4	Indian Creek..........	Boone	638.0	662.0
31.1	Opposite Racine.......	Boone	644.5	675.0
36.4	Laurel Creek..........	Boone	....	658.6
38.6	Station 2028.00.........	Boone	....	674.5
49.9	Big Coal River.........	Boone	....	790.0

Little Coal River Branch.

Distances.	Stations.	County.	Elevation of Low Water	Elevation of Sub-Grade
15.2	Sproul	Kanawha	....	610.1
16.7	Bluetom	Lincoln	587.0	620.0
17.2	Rolman	Lincoln	588.0	625.0
19.5	Dunlap	Lincoln	597.6	626.0
22.0	MacCorkle	Lincoln	612.0	633.0
24.3	Ivy Branch	Lincoln	....	640.0
26.8	Altman	Lincoln	622.0	650.0
27.5	Julian	Lincoln	....	650.0
28.2	Sayre	Lincoln	....	650.2
30.0	Lory	Boone	....	658.6
32.7	Rock Creek	Boone	639.8	668.0
33.8	Hopkins	Boone	....	671.0
35.2	Danville	Boone	....	674.0
37.3	Madison	Boone	....	687.5

Spruce Fork of Little Coal River Branch.

Distances.	Stations.	County.	Elevation of Low Water	Elevation of Sub-Grade
37.4	Pond Fork Bridge	Boone	681.4	691.0
39.7	Spruce No. 1 Bridge	Boone	682.2	709.7
40.2	Spruce No. 2 Bridge	Boone	686.1	719.7
40.4	Spruce No. 3 Bridge	Boone	694.4	721.5
43.9	Spruce No. 4 Bridge	Boone	720.8	743.2
44.1	Spruce No. 5 Bridge	Boone	722.2	744.7
45.0	Sixmile Creek	Boone	....	750.0
47.1	Hewett Creek	Boone	....	770.5
49.0	Mouth of Laurel	Boone	775.0	790.0
50.2	Rock House Branch	Boone	....	811.3
51.7	Beech Creek	Logan	....	840.2
53.6	Seng Camp Creek	Logan	870.0	880.0

Laurel Fork Branch of Spruce Fork, Little Coal River.

Distances.	Stations.	County.	Elevation of Low Water	Elevation of Sub-Grade
49.0	Mouth of Laurel	Boone	....	790.0
50.7	Little Bend Branch	Boone	....	832.0
51.4	Burnt Cabin Branch	Boone	....	852.0
51.7	Trace Fork	Boone	....	862.0
54.0	End of Location	Boone	....	928.0

Pond Fork Branch of Little Coal River—Projected.

Distances.	Stations.	County.	Elevation of Low Water	Elevation of Sub-Grade
38.1	Workman Branch......	Boone	672.3	....
40.9	Jarrell Branch.........	Boone	707.7	....
46.9	Robinson Creek........	Boone	718.8	....
49.1	West Fork............	Boone	785.4	....
51.6	Big Branch............	Boone	831.1	....
52.1	Opposite Mouth of Casey Creek.........	Boone	841.7	....

Cabin Creek Branch of the Chesapeake & Ohio Railway.

Distances.	Stations.	County	Elevation
0.0	Cabin Creek Junction..	Kanawha	630
2.0	Dry Branch...........	Kanawha	....
3.7	Ronda	Kanawha	674
5.4	Fairfield	Kanawha	710
7.0	Oakley	Kanawha	....
....	Oley	Kanawha	....
10.0	Cane Fork............	Kanawha	....
11.4	Leewood	Kanawha	880
14.0	Acme	Kanawha	1110
20.0	Seng Creek...........	Boone	1480
22.0	Coal River............	Boone	....
24.0	Jarrolds Valley........	Boone	820
30.0	Honaker Lawson.......	Raleigh	1020

ELEVATIONS ABOVE TIDE IN BOONE COUNTY, DETERMINED BY THE UNITED STATES GEOLOGICAL SURVEY.

PEYTONA QUADRANGLE.

From Point Three Miles Up Coal Fork of Cabin Creek West, via Coal River and Laurel Creek to Mouth of Robinson Creek, 1.5 Miles East of Uneeda Post-Office.

Feet.

Ohley, 4.7 miles southwest of, north of road, opposite house, 400 feet northwest of house and hollow, near bend in creek, on rock cliff, chiseled square; painted "983"....... 981.74

Orange post-office, 3.4 miles east of, south of road, 265 feet north of ridge road and along road up mountain, 0.3 mile east of school-house, in large rock; aluminum tablet stamped "1633"......................................1632.017

Orange post-office, 2.5 miles east of, north of road, 80 feet west of Fork's Top of Mountain on outcrop of sand rock, chiseled square; painted "1563" 1562.47

Orange post-office, 1 mile east of, east of road and run, on top of small summit, 0.3 mile west of mountain foot, on outcrop of sand rock, chiseled square; painted "940" 939.60

Orange post-office, 0.5 mile north of, on edge of Whiteoak, south side, 300 feet west of mouth of Little Whiteoak, 0.2 mile east of Whiteoak, at mouth, in rock cliff; aluminum tablet stamped "734" 732.963

Orange post-office, 5 miles west of, 20 feet west of road, at bend of river, where road comes back to river, on large rock boulder, chiseled square; painted "691" 690.30

Orange post-office, 3.2 miles west of, north edge of road, 150 feet west of small hill, 400 feet east of school-house No. 13, opposite Horse Hollow, in rock cliff; aluminum tablet stamped "735" 733.697

Seth post-office, 0.3 mile southeast of, north of road, in bank in front of Laurel Church, opposite mouth of Laurel Creek, 0.1 mile northwest of school-house, in outcrop of rock; aluminum tablet stamped "715" 714.252

Seth, 1.5 miles south of, east of road, west of Laurel Run, opposite hollow, 220 feet east of second house up Laurel Creek, on outcrop of sand rock chiseled square; painted "689" 688.57

Nelson, 200 feet southeast of mouth of Hopkins Fork on Laurel Creek, in face of rock cliff; bronze tablet stamped "734-C" 734.388

Nelson post-office, 300 feet northwest of, west of Laurel Creek 40 feet west of road, 40 feet south of Foot Log, 370 feet south of Road Fork, and opposite hollow, in outcrop of rock in bank; aluminum tablet stamped "732" 731.072

Nelson post-office, 2.0 miles south of, east of log railroad and county road, 0.8 mile south of school-house, 3 feet from ground, in face of rock cliff; aluminum tablet stamped "863" 862.283

Nelson post-office, 3.9 miles southwest of, north of railroad, Laurel Creek, opposite hollow, 0.2 mile west of Skidway, on rock cliff; chiseled square 919.26

Nelson post-office, 5.7 miles southwest of, south of Laurel Creek, 300 feet from foot of mountain, 300 feet from house, 5 feet from barn, in boulder in bank; aluminum tablet stamped "1089" 1087.884

Uneeda post-office, 3.3 miles southeast of, north of Robinson Run, at mouth of hollow, north of house, on outcrop of rock, chiseled square; painted "858" 857.07

Brownstown Southwest Along Lens and Short Creeks to Racine, Thence Northwest Down Coal River, to Cobbs.

	Feet.
Hernshaw, 1 mile south of, south side of road, on south bank of Lens Creek, near A. Hoffman's house, in 3 by 5 foot ledge rock; copper bolt stamped "722-C"	722.185
Peytona, 2 miles northwest of, south side of road down Coal River, 100 feet northwest of Laurel Branch, 1 mile below Whiteoak Branch, in ledge of rock; copper bolt stamped "665-C"	655.401
Cobbs, 200 feet north of mouth of Lick Creek, west side of Coal River road, in rock; copper bolt stamped "648-C"	648.375

Racine Southeast to Comfort, Thence Southwest to Nelson.

	Feet.
Comfort, 0.12 mile north of, 0.5 mile south of mouth of Joes Creek, east of Coal River road, 4 feet above grade in bottom of rock cliff; copper bolt stamped "673-C"	673.488

MADISON QUADRANGLE.

From Robinson Creek West via Pond Fork, Little Coal River, Peter Cave Fork and Sulphur Springs Fork to Griffithsville.

	Feet.
Uneeda post-office, 1.4 miles southeast of, 40 feet north of road, 700 feet east of house, 700 feet east of mouth of Robinson Fork and back of barn, in outcrop of rock in field; aluminum tablet stamped "751"	750.135
Uneeda post-office, 0.5 mile southeast of, 40 feet north of road, north of Pond Run, along narrows, 0.1 mile northeast of hollow, in large rock boulder; aluminum tablet stamped "723"	722.334
Uneeda post-office, 0.8 mile west of, south of road, north of Pond Fork, 400 feet east of school-house, along narrows, on outcrop of rock, chiseled square; painted "718"	717.29
Madison post-office, 2 miles southeast of, west of Pond Fork, 220 feet west of road, 0.2 mile south of bend of run, 0.2 mile north of hollow, in rock cliff; aluminum tablet stamped "697"	696.068
Madison post-office, in west side of foundation of Boone County Jail; aluminum tablet stamped "704"	703.221

Madison post-office, 1.1 miles west of, southeast of railroad, 500 feet southwest of mile post No. 1, on outcrop of sand-rock; chiseled square.................................. 693.12

Danville post-office, 0.5 mile west of, 300 feet west of brick yard, 10 feet west of county road, 10 feet south of railroad, on south abutment of culvert, at mouth of John Hill Branch, chiseled square; painted "686".................. 684.50

Danville post-office, 1.5 miles west of, 90 feet west of Hopkins Station, 500 feet north of mile post No. 3 between house and railroad, in east end of foundation of Mr. Hopkin's cellar; aluminum tablet stamped "679"............. 677.402

Danville post-office, 4.5 miles northwest of, west of railroad, 300 feet west of county road, between houses, in west abutment of culvert; aluminum tablet stamped "666"..... 664.384

Danville post-office, 6.5 miles northwest of, right of railroad, 0.2 mile northwest of mile post No. 8 on rock cliff; chiseled square; painted "668'............................. 668.03

Hill post-office, 600 feet west of, south of railroad, 200 feet west of trestle, 600 feet from road crossing and station, in rock cliff; aluminum tablet stamped "667"............. 665.404

Hill post-office, 2 miles northwest of, west of road, near small summit, 500 feet east of house, on outcrop of sand rock, chiseled square; painted "658"........................... 657.27

Woodville post-office, 300 feet east of, 3 miles west of Hill post-office, 170 feet north of mouth of Laurel Creek, right of road, in rock boulder; copper bolt stamped "673"...... 672.622

Woodville post-office, 2.4 miles west of, 600 feet north of Sulphur Fork, at its mouth, east of road and run, in top of boulder; aluminum tablet stamped "772"................ 770.923

Garretts Bend South, via Sandgap, Sugar Camp Knob and Hill, to Madison.

Feet.

Sand Gap, 500 feet west of forks of road, in huge rock above John A. Midkiff's house; copper bolt stamped "U. S. G. S.-1079" 1078.545

Hill post-office, 3 miles west of, 200 feet above confluence of Laurel Fork and Horse Creek, opposite James McClure's house, near last crossing of Laurel, on east bank, in rock ledge; copper bolt stamped "U. S. G. S.-673"........... 672.622

Hill post-office, 2 miles north of, 1 mile above mouth of Trace Branch on left side of right hand hollow on Trace Branch of Horse Creek; copper bolt stamped "U. S. G. S.-766"... 765.931

Hill post-office, 3 miles southeast of, 0.25 mile north of Camp Creek, on east side of road going up Little Coal River, 300 feet above B. Stolling's, in small ledge of rock; copper bolt stamped "U. S. G. S.-660"............................. 660.170

Single Spur Line up Lick Creek.

Feet.

Danville, 4 miles west of, on right fork of Lick Creek, 25 feet east of creek, between two walnut trees, one of which is blazed, in boulder above coal bank; copper bolt stamped "U. S. G. S.-820" 819.402

LOGAN QUADRANGLE.

Madison, South, up Spruce Fork, to Seng Post-office (Partly Single Spur Line).

Feet.

Seng, 2 miles northwest of, 0.12 mile above mouth of Beech Creek, on north side of Spruce Fork, 300 feet below splash dam, opposite William Coleman's barn, in ledge of rock; copper bolt stamped "U. S. G. S.- 846" 845.309

Hewett, Southwest, to Peck; Thence up Guyandot River to Man.

Feet.

Hewett, 300 feet above mouth of Hewett Creek, opposite school-house, on south side of creek, in ledge of rock; copper bolt stamped "U. S. G. S.-767" 767.029

OCEANA SPECIAL QUADRANGLE.

Mouth of Toney Fork East and Northeast up Big Huff Creek to Echart.

Feet.

Cyclone, 6 miles east of, 600 feet below Lower Gap Branch, east of road up Big Huff Creek, 100 feet above D. H. Cook's store; iron post stamped "1068" 1067.600

BALD KNOB QUADRANGLE.

Mouth of Whiteoak Creek up Coal River to Hazy Creek.

Feet.

Orange, 0.5 mile north of post-office, south side of Whiteoak Creek, 300 feet west of mouth of Little Whiteoak Creek, 0.2 mile east of mouth of Whiteoak Creek, in rock cliff; aluminum tablet stamped "734" 732.963

Mouth of Whiteoak Creek, 1.6 miles south of, 300 feet east of sharp bend in road to east, 2 feet south of road; cross on ledge of rock 748.79

Mouth of Whiteoak Creek, 2.8 miles south of, 3 feet west of river; nail in root of sycamore tree 751.27

Mouth of Whiteoak Creek, 6.3 miles south of, 6 feet east of road, 100 feet south of ford over river, about 0.25 mile south of Seng Creek; cross on rock 788.95

Mouth of Whiteoak Creek, 7.3 miles south of, northwest corner of concrete culvert to Chesapeake & Ohio Railway over Bear Hollow, in top face; bronze tablet stamped "817" 815.983

Jarrolds Valley, 1 mile south of, on northeast corner of concrete abutment of bridge over Little Marsh Fork, in lower face; aluminum tablet stamped "855"........................ 855.404

Jarrolds Valley, 3.2 miles south of, 6 feet west of railroad grade, 150 feet north of culvert, in sharp curve to right; cross on rock... 900.06

Jarrolds Valley, 3.8 miles south of, on east top face of culvert over branch, in concrete; aluminum tablet stamped "918" 916.834

Jarrolds Valley, 5.5 miles south of, about 0.5 mile above Hecla post-office, 6 feet east of road and between same and Marsh Fork; spike in root of beech tree..................... 942.27

Jarrolds Valley, 6.1 miles south of, 20.0 feet west of road, about 10 feet west of Marsh Fork, about 0.25 mile north of Hazy Creek in ledge of rock, near old sycamore snag; bronze tablet stamped "956"................................ 954.950

Echart North Down Pond Fork of Little Coal River, via Bald Knob and Crook, to Mouth of Robinson Creek.

Feet.

Echart, east of Pond Fork of Coal River, opposite mouth of Skin Fork, in ledge of rock; copper bolt stamped "U. S. G. S.-1423" 1422.020

Bald Knob, east side of valley, opposite Eddy Workman's, in boulder; copper bolt stamped "U. S. G. S.-1101"..........1101.060

Bald Knob, 3.5 miles northwest of, north side of Cow Creek, 0.75 mile from mouth of Cow Creek, 800 feet from Jim Gunnoe's, in ledge of rock; copper bolt stamped "U. S. G. S.-1039" 1039.100

Gordon East up West Fork to Browns Branch (Single Spur Line).

Feet.

Gordon, north side of West Fork, 1000 feet east of junction of West Fork and Pond Fork into Little Coal River, in rock; copper bolt stamped "U. S. G. S.-808"...................... 809.014

Gordon, 2.5 miles east of, on West Fork, 800 feet below Browns Branch, west of road, a little above and nearly opposite small water mill, in ledge of rock; copper bolt stamped "U. S. G. S.-884"................................ 884.330

INDEX.

Page

A

Abshire, Eliza, Opening..385, 576, 580
Acreage of Coals in County....565-570
Act of General Assembly of Virginia 2-3
Adkins, Ab., Mine................421
Adkins Branch, Section, 0.5 Mi. up 118
Allegheny Series..............43, 593
Allegheny Series, Chapter VI on... 207-224
Allegheny Series, Clays of........583
Allegheny Series, Coals of.......565-6
Allegheny Series, General Section of.44
Allegheny Series Sandstones of....585
Allen, James, Mine......417, 577, 581
Allen, Kelley, Opening...410, 576, 581
Allen, R. B., Opening....408, 576, 581
Alma "A" Coal...............46, 117
Alma Coal.......................
....404-443, 569, 571, 572, 576-7, 581
Alma Coal.......................
..(See *Coal* by Magisterial Districts)
Alma Coal, Analyses....425, 437, 576-7
Altman, Description.................6
Altman, Section 1 Mi. North.......66
Altman, Section 1 Mi. West.......70
Altman Station Section............69
Ames Limestone..................205
Analyses, Coal:
Alma 425, 437, 576-7
Campbell Creek (No. 2 Gas)......
...452, 457-8, 465, 466, 482, 577-8
Cedar Grove (Thacker).........
...............389, 391, 401, 576
Chilton 324, 325, 575
Coalburg 255, 274, 276, 574
Eagle 578
Hernshaw 342, 575-6
Little Chilton....................575
No. 5 Block.......217, 222, 224, 574
Powellton (Brownstown)........578
Stockton 246, 574
Williamson 576
Winifrede ....
.....289, 298, 299, 304, 308, 574-5
Analyses, Coal, Table of.......574-577
Analyses, Coal, by Nos., Mine Nos., Lab. Nos., etc..............579-582
Analyses. Miscellaneous:
Campbell Creek Limestone.......445
Chilton Fire Clay...............584
Dingess Limestone.........365, 369
Eagle Fire Clay.................504
Iron Ore.......................587
Anchor Coal Co. Mine....298, 575, 579
Anchor Coal Co. Openings.........
.....................220, 242, 268
Anchor Coal Co. Production......562
Anchor P. O. Section............. 91
Anchor P. O. Section 1.5 Mi. S. W. 92
Andrew P. O. Section 2 Mi. West..
.............................65-6
Anticline, Definition.............. 37
Anticline, Warfield...............40-1

Page

Appendix (Levels Above Tide)...620-7
Arbogast, D M., Mines..............
.................355, 402, 576, 580
Arbogast, D. M., Well No. 1 (63)..
..........171-2, 486, 506, 520A, 556
Area 2
Area, Drainage.................14-15
Austin. David, Opening...........464
Available Coal in County......565-572
Average Thickness of Coals....565-570
Axis of Fold, Definition........... 37

B.

Bailey and Dort...............18, 21
Bailey, C. T.............18, 19, 21, 22
Bakerstown (Barton) Coal.....44, 206
Bald Knob Limestone.............
.............46, 184, 186, 189, 497
Bald Knob Quadrangle, Elevations..
.... 626-7
Bald Knob P. O., Section Near...184-6
Bald Knob Section................203
Bald Knob Sections:
1.0 Mi. North 182-4
1.2 Mi. Northwest 182
4.7 Mi. N. 59° W...............179
5.2 Mi. N. 55° W.............177-8
5.7 Mi. N. 47° W...............177
3¼ Mi. Southwest.............147-8
Bald Knob Shale..................497
Ball, Ashford, Openings...........
.............326, 372, 373, 576. 580
Ball, Augustus, Opening...........421
Ball, Blaine, Opening.............326
Ball, Charles, Opening...397, 576, 580
Ball, J. A., Mine..................459
Ball, Park, Mines...424, 425, 577, 581
Ball, Rufus, Opening.............410
Ballard, H. C., Well........520A, 536
Ballard, J. C., Openings..........407
Ballard, John, Opening............417
Ballard, Opening 397
Ballentine, W. I...................299
Banger, F. B......................523
Barker, Jacob, Mine 451
Barker, Rome, Mine...............341
Barker, William, Mine............380
Bear Hollow Sections.........127, 131
Beaver Pond Branch, Description.. 14
Beckley Coal.....................538
Bedford et al., Prospect Openings...
..................337, 341, 575, 580
Belcher, Harry. Opening..482, 578, 582
Bell, Edward. Mine...............399
Bell. Isaac. Mines........411, 576, 581
Belmont (Stockton-Lewiston) Coal..
.................231-250, 574, 579
Bend Branch Section 1 Mi. up:109-110
Benedum & Trees. Wells....520A, 534
Bennett, J. M., Well (40)..........
.....................513, 520A, 539
Bennett, M., Mine.......425, 577, 581
Bens Creek Coal...............47, 512
Bens Creek Sandstone...........511-2
Berea Sand.......................519

Page
Berwind, E. J.. Core Tests.........
443, 490, 491, 492, 496, 497, 500, 501, 510, 511
Berwind, E. J., Mines.............
............220, 273, 307, 574, 579
Berwind, E. J., Prospects..........
222, 248, 249, 274, 275, 276, 277, 306, 308, 309, 310, 333, 334, 359, 375
Bias, Burrell, Mine...............343
Bias, B. W., Mine.......423, 577, 581
Bias, Calvary, Mine......420, 577, 581
Bias, Colway, Mine................372
Bias, David, Mines...........328, 347
Bias, Noah, Mine..................343
Bias, O. B., Mine........420, 577, 581
Bias, Roland, Mine 415, 577, 581
Bias, William, Mine.....420, 577, 581
Big Branch, Description 15
Big Coal River, Description........
......................12, 14, 15-27
Big Coal River, Projected Railroad.. 8
Big Creek Development Co., Wells
..............................520A
Big Creek Station, Section........203
Big Creek, Trace Fork of, Description..................13, 14, 15, 34
Big Injun Sand................518-519
Big Lime 518
Billetts, Silas, Mine......398, 576, 580
Black Band Coal.............284, 339
Blair, J. C.................67, 97, 144
Blake, Curtis, Opening...458, 578, 582
Block Seam........................339
Booker, Noah, Mine................448
Boone & Kanawha Land & Mining Co.:
Mine 340, 453, 577, 581
Prospects ....
348, 377, 378, 379, 380, 381, 400, 401, 449, 450, 451, 452, 476, 477, 478, 503
Wells 520A, 524
Boone County Coal Corporation Core Tests:
No. 1 (142).........98, 100-1, 514A
No. 2 (109)..........97, 427, 514A
No. 3, (129)...........144-5, 514A
No. 3 (146Lo)..........104-5, 514A
No. 4 (110)...........98, 427, 514A
No. 5 (149Lo)............107, 514A
No. 5B (111)........121, 428, 514A
No. 6 (144Lo)..........103-4, 514A
No. 6 (148Lo).......106, 260, 514A
No. 7 (145)...98, 99-100, 259, 514A
No. 9 (158Lo)..........109-10, 514A
No. 11 (169Lo)........119-20, 514A
No. 12 (159Lo)........110-11, 514A
No. 13 (168Lo)...........119, 514A
No. 14 (160Lo)........112-13, 514A
No. 15 (161Lo)........112-13, 514A
No. 16 (166Lo)........117-18, 514A
No. 17 (162Lo)........113-14, 514A
No. 18 (163Lo)........114-15, 514A
No. 18 (165Lo)........116-17, 514A
No. 19 (164Lo)........115-16, 514A
No. 20 (167Lo)...........118, 514A
No. 21 (157)............108, 514A
Boone County Coal Corporation Well
......................98, 101-2, 540
Boone-Logan-Lincoln County Corner Section 1 Mi. N. W............ 88
Boone Timber Co.............589, 590
Boswell, Emma, Heirs, Well.....520A
Bradshaw, Morris, Mine...........348
Brinkley, J. W., Mine............330
Brooks, A. B.....................587

Page
Broun Heirs, Mine................449
Broun Prospect 349
Broun, Thos. L...................587
Broun, T. L., Well.........520A, 548
Brounland Section (Kanawha)..... 60
Brounland Section, 3.5 Mi. S. E... 61
Brown, Garland, Mine.............359
Browns Branch, Description..13, 14, 32
Brownstown Sandstone.............492
Brush Creek Coal..44, 67, 71, 206, 583
Brush Creek, Description....12, 14, 27
Brushton Gaging Station, Records..
..............................16-20
Brushy Fork of Spruce, Section.119-20
Brushy Fork of Spruce, Section 2 Mi. up......................114-15
Buffalo Creek Coal...............278
Buffalo Creek Coal................
..(See *Coal* by Magisterial Districts)
Buffalo Creek Limestone...........
..................45, 278, 595, 597
Buffalo Sandstone................206
Building Fund 4
Building Stones................585-6
Bull Creek of Coal River...12, 14, 27
Bull Creek of Pond Fork...13, 14, 30
Burdette, J. F., Observer......... 21

C

Cabell P. O., Section 2 Mi. South..64
Cabell, Samuel, Openings.....322, 390
Cabin Creek Branch of C. & O. Ry.
...........................8, 622
Cabin Creek Consolidated Coal Co.
..............................295
Cabin Creek District (Kanawha), Well 520A
Callahan, James Morton........... 24
Calorific Value of Coals..........573
Camp Creek, Description.....13, 15, 30
Campbell Creek Coal..............
..(See *Coal* by Magisterial Districts)
Campbell Creek Coal..............
38-9, 446-492, 569, 571, 572, 577-8, 581-2
Campbell Creek Coal, Analyses......
......452, 457-8, 465, 466, 482, 577-8
Campbell Creek Limestone.......444-6
Campbell Creek Limestone..........
(See *Limestone* by Magisterial Districts)
Campbell Creek "Rider" Coal..148, 156
Campbell, M. R...................224
Cantley, T. A., Opening..........405
Carbon Black Industry............590
Carter, Dr. H. Lon Opening......406
Carter Oil Co...........150, 520A, 546
Carpenter, Samuel, Mine..........405
Casey Creek, Description....13, 14, 30
Casey Core Tests:
No. 1 (138)....177-8, 441, 487, 514A
No. 2 (137)......177, 441, 487, 514A
No. 3 (139)......179, 442, 487, 514A
Cassingham Core Tests:
No. 1 (131)....147-8, 434, 475, 514A
No. 2 (130)....145-6, 433, 474, 514A
No. 3 (123)........136-7, 432, 514A
No. 4 (155)........137-8, 475, 514A
No. 5 (156)........180-1, 440, 514A
Cassingham Openings...............
240, 261, 291, 292, 293, 294, 304, 357, 575, 579
Catskill Sandstones............... 43
Cedar Coal........................ 47

Page

Cedar Grove Coal ....
.....376-403, 568, 571, 572, 576, 580
Cedar Grove Coal...................
..(See *Coal* by Magisterial Districts)
Cedar Grove Coal Analyses.........
..................389, 391, 401, 576
Chambers, C. C., Mine............467
Chambers, Fred, Opening..........395
Chambers, Jacob, Well...........520A
Chambers Prospect 467
Chandler, Jacob, Mine.............255
Chap P. O., Sections:
1.3 Mi. Northwest.............192-3
3.0 Mi. Northwest...............192
1.2 Mi. South.................196-8
3.3 Mi. South.................200-1
1.0 Mi. Southeast.............194-6
1.4 Mi. Southeast................194
2.6 Mi. Southeast............198-200
Chapman, J. E., Mines............
.................412, 413, 576, 581
Charleston Natural Gas Co., Wells
..............................520A
Charleston Testing Laboratory.....217
Chemung and Hamilton Shales..... 43
Cherry Pond Mountain............ 36
Chesapeake & Ohio Ry., Cabin Creek Branch 8, 622
Chesapeake & Ohio Ry., Coal River Branch 7-8. 620
Chesapeake & Ohio Ry., Levels.620-622
Chew, R. P., Mine Openings.......
..................406, 486, 578, 582
Chilton "A" Coal...............313-4
Chilton "A" Coal..................
..(See *Coal* by Magisterial Districts)
Chilton Coal......................
.....320-336, 567, 571, 572, 575, 580
Chilton Coal......................
..(See *Coal* by Magisterial Districts)
Chilton Coal, Analyses...324, 325, 575
Chilton Fire Clay, Analysis........584
Chilton, MacCorkle, Chilton & Cassingham......136, 137, 145, 147, 180
Chilton, MacCorkle, Chilton & Meany Core Tests, etc.:
120, 122, 123, 126, 128, 129, 131, 132, 133, 135, 139, 140, 141, 142, 143, 177, 179, 260, 290, 328, 346, 347, 356, 399, 440, 469, 487, 575, 580.
Chilton "Rider" Coal..............
(See *Coal* by Magisterial Districts)
Clarion Coal......................106
Clark and Krebs..................
...........324, 340, 342, 425, 457-8
Clark, James, et al., Shaft.........428
Clark, J. M.......................340
Clays 583-5
Clays in Allegheny Series..........584
Clays in Conemaugh Series.........583
Clays in Kanawha Series.........584-5
Clays, Recent.....................585
Clays, Road Materials, etc., (Chapter X) 583-590
Clear Fork Coal Co...............169
Clear Fork (Raleigh) Section......169
Clothier, Description............. 6
Clothier, Section...........99-102, 203
Clothier, Sections:
1.0 Mi. East 121
0.2 Mi. N. 75° W.............. 98
0.3 Mi. N. 30° W.............. 97
0.7 Mi. N. W...................108
4.0 Mi. South 113-14
4.8 Mi. South 115-16

Page

Clothier, Sections:
8.0 Mi. South 119
1.0 Mi. Southeast 110-11
1.5 Mi. Southeast 109-10
3.0 Mi. S. 11° 30′ E..........104-5
3.0 Mi. Southeast 112-13
4.0 Mi. S. 60° E..............122-3
4.2 Mi. S. 67° E................126
4.6 Mi. Southeast 127
5.0 Mi. S. 47° E................131
5.0 Mi. S. 53° E................128
5.1 Mi. S. 54° E................129
5.2 Mi. S. 82° E..............123-5
5.2 Mi. S. 49° 30′ E..........129-30
5.5 Mi. S. 42° 30′ E..........133-4
5.5 Mi. S. 49° E..............132-3
5.8 Mi. Southeast 137-8
5.8 Mi. S. 32° E..............140-1
5.9 Mi. S. 37° 30′ E..........139-40
6.0 Mi. Southeast 117-18
6.3 Mi. S. 34° E..............141-2
6.3 Mi. S. 40° 30′ E............135
6.8 Mi. S. 70° E................105
6.9 Mi. S. 31° 30′ E...........142-3
7.0 Mi. S. 44° E..............136-7
7.3 Mi. S. 6° E.................106
7.6 Mi. S. 26° E..............144-5
7.7 Mi. S. 39° E................143
8.5 Mi. Southeast 119-20
9.0 Mi. S. 33° E..............145-6
9.5 Mi. Southeast 114-15
10.0 Mi. S. 39° E.............147-8
1.7 Mi. S. 23° W..............102-3
1.9 Mi. Southwest 111-12
4.0 Mi. S. 52° W..............103-4
5.0 Mi. Southwest 116-17
7.0 Mi. Southwest 118

Coal (by Magisterial Districts):

Alma

Crook ....
46, 177, 178, 179, 180, 181, 182, 185, 195, 197, 199, 201, 202, 203, **437-443**, 559, 569, 572, 576-7, 581
Peytona ....
46, 49, 51, 53, 62, 64, 203, **404**, 569, 571, 576-7, 581
Scott ....
46, 74, 78, 79, 81, 84, 203, 210, 252, 392, **404-7,** 569, 571, 576-7 581
Sherman ....
46, 152, 153, 165, 167, 169. 203, **434-7**, 446, 569, 572, 576-7, 581
Washington ....
46, 85, 90, 92, 93, 94, 95, 96, 97, 98, 101, 103, 105, 108, 109, 110, 111, 112, 113, 114, 116, 117, 118, 121, 122, 126, 127. 128, 129, 130, 131, 132, 134, 135, 137, 138, 139, 140, 141, 142, 143, 146, 148, 149, 203, 364, **407-434,** 540, 569, 571, 576-7, 581.

Alma "A"

Washington 46, 117

Bakerstown (Barton)

(General) 44, 206

Beckley

Washington 538

Bens Creek

(General) 47, 512

Brush Creek

Scott 44, 67, 71, 206, 583

Page

Coal (by Magisterial Districts):

Buffalo Creek

Crook 45, 174, 186, 190, 203, 278, 284
Peytona............45, 48, 203, 278
Scott 45, 203, 278, 281
Sherman ...45, 149, 157, 158, 203, 278, 283
Washington 45, 100, 124, 144, 203, 278

Campbell Creek (No. 2 Gas)

Crook 46, 170, 172, 177, 178, 179, 180, 181, 182, 184, 185, 186, 187, 188, 193, 195, 197, 199, 201, 202, 203, 339, 357, 366, 445, 446, **486-492**, 556, 559, 569, 572, 577-8, 581-2
Peytona 46, 49, 50, 51, 52, 53, 54, 55, 56, 57, 58, 62, 63, 64, 65, 66, 203, 209, 227, 251, 314, 315, 404, 445-6, **447-462**, 522, 569, 571, 577-8, 581-2
Scott 46, 74, 78, 80, 81, 84, 203, 252, 316, 445-6, **462-467**, 569, 571, 577-8, 581-2
Sherman 46, 150, 151, 152, 153, 155, 156, 160, 161, 162, 164, 165, 166, 167, 168, 169, 203, 228 282, 318, 319, 367, 445-6, **475-485**, 546, 547, 549, 550, 554, 569, 572, 577-8, 581-2
Washington 46, 85, 86, 90, 91, 92, 93, 94, 95, 101, 103, 108, 109, 113, 122, 125, 126, 127, 128, 129, 130, 131, 132, 134, 135, 137, 138, 139, 140, 141, 142, 143, 146, 148, 149, 203, 317, 445-6, **467-475**, 540, 569, 571, 577-8, 581-2

Campbell Creek (No. 2 Gas) "Rider"

Sherman 156
Washington 148

Cedar

(General) 47

Cedar Grove (Thacker)

Crook 45, 170, 171, 180, 187, 199, 200, 201, 203, 369, **402-3**, 568, 572, 576, 580
Peytona 45, 49, 50, 51, 55, 57, 58, 62, 63, 64, 66 203, **376-86**, 404, 568, 571, 576, 580
Scott 45, 73, 74, 75, 76, 77, 78, 81, 83, 84, 203, **386-96**, 568, 571, 576, 580
Sherman 45, 152, 156, 157, 160, 166, 203, 319, 366, **400-2**, 446, 546, 568, 572, 576, 580
Washington 45, 85, 90, 91, 95, 97, 98, 101, 103, 104, 105, 107, 108, 110, 111, 113, 115, 116, 117, 118, 119, 120, 121, 122, 126, 128, 133, 137, 138, 146, 147, 203, 364, **396-9**, 568, 571, 576, 580

Chilton

Crook 45, 170, 171, 173, 174, 175, 182, 183, 185, 186, 188, 189, 191, 192, 195, 203, 319, **330-6**, 567, 572, 575, 580

Page

Coal (by Magisterial Districts):

Chilton

Peytona 45, 48, 49, 58, 64, 65, 203, 315, **320-2**, 567, 571, 575, 580
Scott 45, 74, 75, 84, 203, 316, 317, **322**, 567, 575, 580
Sherman 45, 153, 156, 157, 159, 160, 164, 167, 203, **329-30**, 567, 572, 575, 580
Washington 45, 85, 88, 89, 90, 91, 93, 96, 100, 104, 108, 109, 111, 114, 115, 145, 203, 317, 318, **323-9**, 567, 571, 575, 580

Chilton "A"

Crook 45, 173, 183, 185, 187, 188, 192, 202, 203, 284, **313**
Peytona 45, 48, 49, 58, 65, 203, **313**, 315
Scott 45, 81, 203, **313**, 315, 316
Sherman 45, 160, 164, 165, 203, 283, **313**, 319
Washington 45, 85, 87, 100, 104, 117, 145, 203, 282, **313**, 314, 317

Chilton "Rider"

Crook 45, 182, 183
Sherman 45, 149
Washington 45, 88, 117

Clarion

Washington 106

Coalburg

Crook 44, 170, 174, 175, 176, 183, 185, 186, 188, 189, 190, 191, 192, 194, 197, 199, 202, 203, 254, **268-277**, 283, 567, 572, 574, 579
Peytona 44, 48, 55, 57, 59, 60, 61, 63, 203, 251, **255-8**, 279, 280, 567, 571, 574, 579
Scott 44, 70, 71, 72, 77, 81, 83, 84, 203, 252, **258-9**, 280, 281, 567, 571, 574, 579
Sherman 44, 149, 150, 153, 158, 160, 161, 162, 164, 165, 166, 167, 168, 203, 253, **261-8**, 282, 283, 567, 572, 574, 579
Washington 44, 85, 90, 96, 99, 106, 124, 144, 203, **259-61**, 282, 567, 571, 574, 579

Coalburg "A"

Crook 190

Coalburg "Rider"

Washington 99

Dingess

Crook.........45, 174, 175, 203, **360**
Peytona.............45, 59, 203, **360**
Scott..................45, 203, **360**
Sherman 45, 203, **360**
Washington 45, 88, 89, 90, 93, 109, 203, 317, **360**

Eagle

Crook 46, 172, 181, 184, 186, 188, 189, 193, 196, 198, 200, 201, 202, 203, **505-511**, 556, 570, 571, 572, 578, 582

Page

Coal (by Magisterial Districts):

Eagle

Peytona 45, 53, 203, **503-4**, 522, 526, 578, 582
Scott 46, 203, **504**, 578, 582
Sherman 46, 151, 167, 169, 203, **504-5**, 546, 549, 578, 582
Washington 46, 95, 133, 203, 536, 537, 578, 582

Eagle "A"

Washington 46, 133, 501

Elk Lick

(General) 44

Fire Creek

Sherman 154

Gilbert

(General) 47

Glenalum Tunnel

Scott 47, 82

Hernshaw

Crook 45, 170, 171, 174, 176, 177, 179, 180, 182, 183, 185, 187, 189, 192, 195, 199, 202, 203, 335, **354-9**, 369, 557, 558, 568, 572, 575-6, 580
Peytona 45, 48, 49, 52, 53, 57, 58, 59, 64, 65, 203, 315, **339-41**, 568, 571, 575-6, 580
Scott 45, 74, 77, 78, 81, 84, 203, 229, 315, 316, **341-2**, 568, 571, 575-6, 580
Sherman 45, 149, 152, 160, 165, 203, 318, **347-54**, 568, 572, 575-6, 580
Washington 45, 88, 93, 95, 96, 99, 102, 107, 111, 112, 115, 117, 119, 125, 127, 203, **342-7**, 568, 571, 575-6, 580

Little Alma

Crook 46, 170, 177
Peytona 46, 62
Scott 46, 74
Washington 46, 96, 103, 105, 114, 125, 127, 137, 138, 148

Little Cedar

Scott 47, 82

Little Chilton ... .

Crook 45, 170, 183, 186, 192, 197, 203, **337-8**, 575, 580
Peytona 45, 48, 203, 315, **336**, 575, 580
Scott 45, 81, 83, 203, 317, **336**, 575, 580
Sherman 45, 154, 156, 157, 168, 203, 319, **337**, 575, 580
Washington 45, 85, 92, 95, 107, 111, 115, 118, 119, 125, 203, 317, **337**, 575, 580

Little Chilton "Rider"

Washington 85

Little Coalburg

Crook 44, 203, 277
Peytona 44, 203, 256, 277
Scott 44, 203, 277
Sherman 44, 203, 277
Washington 44, 99, 106, 124, 144, 203, 277

Page

Coal (by Magisterial Districts):

Little Eagle

Crook 47, 200, 203, 512-13
Peytona and Scott....47, 203, 512-13
Sherman 47, 169, 203, 512-13
Washington 47, 86, 203, 512-13

Lower Cedar Grove

Crook 46, 179, 203, 403
Peytona 46, 203, 403
Scott 47, 78, 203, 391, 403
Sherman 46, 157, 160, 165, 203, 366, 403, 546, 549
Washington 46, 85, 101, 103, 105, 108, 110, 112, 114, 116, 117, 118, 121, 125, 126, 143, 146, 148, 203, 364, 403

Lower Freeport

Scott 67, 71, 208
Washington 87

Lower Kittanning (No. 5 Block)

Crook 44, 170, 171, 175, 182, 184, 190, 192, 194, 196, 198, 203, 211-12, **220-4**, 231, 565, 574, 579
Peytona 44, 48, 54, 60, 61, 63, 68, 203, 211-13, 565, 574, 579
Scott 44 66, 69, 71, 203, 211-12, 213-15, 228, 565, 574, 579
Sherman 44, 150, 156, 157, 158, 159, 161, 164, 165, 166, 168, 203, 211-12, **218-20**, 230, 565, 574, 579
Washington 44, 87, 89, 102, 203, 211-12, **215-18**, 565, 574, 579

Lower War Eagle

Sherman 47, 550

Matewan

Crook 46, 184, 186, 187, 189, 193, 195, 197, 199, 201, 497-501
Peytona, Scott and Sherman...... ....46, 497-8
Washington...........46, 133, 497-8

Middle Kittanning

Crook 44, 198
Scott 44, 68

North Coalburg (See Upper Kittanning)

No. 2 Gas (See Campbell Creek)

No. 5 Block (See Lower Kittanning)

Peerless

Crook 550
Sherman 156, 167

Pittsburgh

(General) 44

Powellton "A"

Crook 46, 187
Peytona, Scott and Sherman..... 46
Washington 46, 140

Powellton (Brownstown) (Hatfield Tunnel)

Crook 46, 182, 186, 187, 193, 195, 197, 199, 201, 203, 494-7, 569, 570, 571, 572, 578, 582
Peytona ..46, 57, 203, 493, 578, 582
Scott 46, 74, 78, 81, 203, 493, 578, 582
Sherman 46, 166, 167, 169, 203, 493, 578, 582

Page

Coal (by Magisterial Districts):

Powellton

Washington ....
46, 86, 94, 95, 123, 133, 134, 140, 203, 493, 578, 582

Powellton "Rider"

Washington 123, 134

Sewell

Crook 172, 556
Scott 78, 82, 83
Sherman 152, 160, 551
Washington 538

Stockton-Lewiston

Crook ....
44, 170, 171, 173, 174, 175, 176, 183, 190, 191, 192, 194, 197, 198, 202, 203, 231-2, **245-9,** 566, 572, 574, 579

Peytona ....
44, 48, 55, 57, 59, 60, 61, 63, 203, **231-3,** 251, 566, 571, 574, 579

Scott ....
44, 66, 68, 70, 71, 72, 73, 84, 203, 228, 231-2, **234-9,** 252, 566, 571, 574, 579

Sherman ....
44, 149, 150, 156, 158, 159, 161, 163, 165, 166, 203, 230, 231-2, **240-5,** 253, 319, 566, 572, 574, 579

Washington ....
44, 85, 87, 90, 102, 106, 123, 124, 127, 144, 203, 231-2, **239-40,** 252, 566, 571, 574, 579

Upper Cedar Grove (See Cedar Grove)

Upper Freeport

Crook and Peytona.........44, 207-8
Scott 44, 67, 71, 207-8
Sherman 44, 158, 207-8
Washington 44, 207-8

Upper Kittanning (North Coalburg)

Crook 44, 196, 211
Peytona 44, 60, 211
Scott and Sherman..........44, 211
Washington 44, 87, 211, 218

Upper Williamson

Washington 145

Williamson

Crook ....
45, 171, 176, 178, 179, 180, 183, 187, 189, 203, 367, 368, 369, 373-5, 568, 572, 576, 580

Peytona ....
45, 49, 55, 59, 65, 203, 361, 370-1, 576, 580

Scott ....
45, 72, 78, 203, 316, 362, 371, 576, 580

Sherman ....
...45, 152, 155, 203, 366, 576, 580

Washington ....
45, 85, 88, 89, 91, 92, 93, 97, 98, 101, 103, 104, 107, 108, 110, 111, 112, 113, 116, 120, 122, 125, 127, 128, 129, 130, 131, 135, 136, 138, 139, 140, 141, 142, 145, 203, 363, 364, 371-3, 568, 571, 575, 580

Winifrede

Crook ....
45, 170, 171, 173, 174, 175, 176, 183 185, 186, 188, 189, 190, 192, 194, 195, 197, 199, 202, 203, 283, 284, **300-13,** 567, 572, 574-5, 579

Peytona ....
45, 48, 49, 51, 52, 55, 56, 57, 58, 59, 61, 63, 64, 65, 203, 279, 280, **284-6,** 567, 571, 574-5, 579

Page

Coal (by Magisterial Districts):

Winifrede

Scott ....
45, 66, 69, 70, 74, 75, 76, 77, 79, 81, 84, 203, 280, 281, **286-8,** 316, 567, 571, 574-5, 579

Sherman ....
45, 149, 150, 151, 153, 154, 157, 158, 159, 160, 162, 163, 165, 166, 167, 168, 169, 203, 282, 283, **294-300,** 318, 319, 567, 572, 574-5, 579

Washington ....
45, 85, 89, 90, 95, 96, 100, 124, 126, 127, 144, 203, 281, 282, **288-94,** 567, 571, 574-5, 579

Coal Analyses, Table of..........574-7
Coal Production 561-3
Coal Production by Counties.......563
Coal Production by Mines.........562
Coal Production by Years.........562
Coal Resources of Boone County...
.... 561-582
Coal River:
At Brushton, 1909............16-17
At Brushton, 1910............ 18
At Brushton, 1911............ 19
At Fuqua, 1911................ 24
At Tornado, 1909.............20-21
At Tornado, 1910.............21-22
At Tornado, 1911.............22-23
Description 7, 12, 14, 15-27
Horsepower Developed by.........26
Coal River Branch of C. & O. Ry..
..........................7-8, 620
Coal River Coal & Mining Co. Opening 484
Coal River Company Production...562
Coal River Land & Mining Co. Opening 485
Coal River Land Co.:
Core Test No. 1 (143Lo).........
..................102-3, 514A, 583
Core Test No. 2 (147Lo)..105, 514A
Coal River Lumber & Mining Co. Exposure 505
Coal River Mining Company:
"A" Well No. 1 (9)......53-4, 520A
"A" Well No. 3 (11).....520A, 526
"A" Well No. 4........520A, 526-7
"B" Well...............520A, 524
Mines ....
323, 426, 455, 456, 575, 577, 580, 581
No. 1 (10) Well..........520A, 526
No. 2 (12) Well..........520A, 527
Prospects 381, 382, 383, 384
Coal River Navigation Co......16, 588
Coal River Siding Section.....168, 203
Coal Valley Coal Co., Clay........584
Coalburg "A" Coal................190
Coalburg Coal ....
.....254-277, 567, 571, 572, 574, 579
Coalburg Coal ....
..(See *Coal* by Magisterial Districts)
Coalburg Coal, Analyses...........
..................255, 274, 276, 574
Coalburg-Kanawha Coal Co., Mine..254
Coalburg "Rider" Coal............ 99
Coalburg Sandstone 250-4
Coalburg Section 203
Coalburg Syncline.......41-2, 149, 170
Coals, Calorific Values of.........573
Coals of the Lower Pottsville......515
Coals in Boone County, Quantity...
.... 564-573
Coals in the Allegheny Series....565-6

Page
Coals in the Kanawha Series...566-573
Cobb P. O., Section 1.5 Mi. East.. 57
Colcord & Edwards...........311, 312
Colcord, F. C.....................311
Cold Fork, Section, Head of.......203
Columbus Gas & Fuel Co., Wells....
49, 50, 53-4, 77-9, 85-6, 520A, 521, 524, 526, 528, 529, 530, 532, 537, 552, 554
Comfort, Section 3.8 Mi. North.149-150
Comparative Calorific Values of Coals 573
Conemaugh Series.............43, 593
Conemaugh Series, Chapter V on..204-6
Conemaugh Series, Clays of........583
Conemaugh Series, General Section. 44
Connelly, Reuben, Opening.........
.....................410, 576, 581
Conners, Jacob, Opening..........463
Contours, Structure..............38-9
Cook, H. F., Mine................346
Cook Mountain....................35-6
Cook, W. C., Mine.......221, 574, 579
Cooper, Blackburn, Mine..........352
Corbett, E. J.....................426
Core Tests by Nos.:
100 ....61-2, 461, 514A
101 ....514A
102 ....63, 461, 514A
103 ....60, 514A
104, 104A, 105, 106, 107, 108...514A
108A 82, 514A
109 ....97, 427, 514A
110 ....98, 427, 514A
111 ....121, 428, 514A
112 ....122-3, 429, 469, 514A
113 ....123-5, 470, 514A
114 ....126, 470, 514A
115 ....127, 429, 470, 514A
116 ....128, 429, 470, 514A
117 ....129, 430, 470, 514A
118 ....129-30, 430, 471, 514A
119 ....131, 430, 471, 514A
120 ......132-3, 431, 472, 493, 514A
121 ......133-4, 431, 472, 493, 514A
122 ....135, 431, 473, 514A
123 ....136-7, 365, 432, 514A
124 ......139-40, 432, 473, 493, 514A
125 ....140-1, 432, 473, 514A
126 ....141-2, 433, 474, 514A
127 ....142-3, 433, 474, 514A
128 ....143, 433, 474, 514A
129 ....144-5, 260, 514A
130 ......145-6, 365, 433, 474, 514A
131 ......147-8, 365, 434, 475, 514A
132 ....192-3, 490, 510, 514A
133 ....
194-6, 443, 491, 496, 500, 510, 514A
134 ......196-8, 491, 496, 511, 514A
135 198-200 329, 492, 496, 511, 514A
136 .....123-5, 155-6, 240, 260, 514A
137 ....177, 441, 487, 514A
138 ....177-8, 441, 487, 514A
139 ....179, 442, 487, 514A
140 ..200-1, 492, 497, 501, 511, 514A
141 ....514A
142Lo 98, 100-1, 514A
143Lo 102-3, 514A 582
144Lo 103-4, 514A
145Lo 98, 99, 100, 259, 514A
146Lo 104-5, 514A
147Lo 105, 514A
148Lo 106, 260, 514A
149Lo 107, 514A
Core Tests by Nos.:
150, 151Lo, 152K, 153K, 154K. 514A
155 ....137-8, 475, 514A
156 ....180-1, 440, 514A
157 ....108, 514A
158Lo 109-10, 514A
159Lo 110-11, 514A
160Lo 111-12, 514A
161Lo 108, 112-13, 514A
162Lo 113-14, 514A
163Lo 114-15 514A
164Lo 115-16, 514A
165Lo 116-17, 514A
166Lo 117-18, 514A
167Lo 109, 118, 514A
168Lo 109, 119, 514A
169Lo 109, 119-20, 514A
Cornell & Murry, Chemists........298
Courtney, D. G., Mines............
.............214, 217, 237, 575, 579
Cow Creek, Description......13, 14, 41
Crawford & Ashby, Authority ......
136, 137, 145, 147, 180, 192, 194, 196, 198, 200
Crawford & Ashby, Core Tests....514A
Crawford & Ashby, Mines.........454
Crawford & Ashby, Wells........520A
Croft & Stolling Core Test (108A)..
..........................82, 514A
Croft, C. A., Authority............538
Croft, C. A. Openings.........214, 215
Croft, C. A., Iron Ore............587
Croft, C. M...................... 81
Croft, S. M...................... 82
Croft, W. P., Clay................585
Croft, W. P., Openings............
.................373, 374, 576, 580
Crook District:
Area 2
Campbell Creek Limestone........
..................46, 184, 444-6
Coalburg Sandstone............253-4
Coals ....
(See *Coal* by Magisterial Districts)
Core Tests....................514A
Dingess Limestone...............
45, 171, 172, 175, 176, 183, 184, 186, 199, 203, 335, 357, 367-370, 375, 595, 596
East Lynn Sandstone............211
Homewood Sandstone...........231
Logan Sandstone 443-4
Malden Sandstone 444
Naugatuck Sandstone........359-360
Population 2
Prospective Oil and Gas Area....560
Sections, General............170-202
Seth Limestone 200, 369
Upper Chilton Sandstone.....319-320
Upper Winifrede Sandstone....283-4
Well Records.........520A, 555-560
Williamson Sandstone...........360
Cross Sections................... 38
Crude Oil Co., Wells..............
..51-2, 520A, 525, 533, 542, 543, 544
Cubic Feet of Coals..........565-570
Cunningham, J. S..................
222, 224, 273, 274, 275, 276, 277, 303, 304, 306, 307, 308, 309, 310, 377, 378, 379, 380, 381, 400, 401, 449, 450 451, 452, 476, 477, 478
Curry P. O. Section, 1.5 Mi. East...
..........................92-3, 203
Curry, R. L., Mine.......438, 577, 581

D.

Page

Dall, J. D., Opening.....419, 577, 581
Daniels, George, Mine.............249
Danville, Description 4
Danville, Section 1 Mi. E........79-80
Danville, Section 1¾ Mi. N. E....77-9
Danville, Section 3¼ Mi. N. E......74
Danville, Section on Price Branch...82
Darlow, Jackson, Mine............405
Dart, Joseph, Openings....286, 447, 448
Dart, Joseph, Well No. 2 (3).......
..............................520A, 522
Dart, Joseph, Wells..............520A
Dart Well No. 1 (1).......54-5, 520A
Dart Well No. 4 (8)...56-7, 520A, 523
Dartmont, Section 4 Mi. N. E.....54-5
Dartmont Station, Section ½ Mi. S...
E. 56-7
Dawes, Buck, Mine................330
Dawson, Terry, Prospects.....371, 459
Decota Sandstone.................512
Demler, Paul, Chemist...............
.........289, 324, 342, 425, 457, 458
Dennison Fork, Section ⅓ Mi. Below
Mouth 145-6
Dennison Fork, Section, Head of...203
Dent, Edward, Opening..419, 577, 581
Department of Mines of W. Va....561
Depth to Lower Pottsville Coals....515
Description, Allegheny Formations...
.... 207-224
Description, Conemaugh Formations.
.... 204-6
Description, Drainage Basins.....15-34
Description of Species (Fossils).....
.... 597-617
Description, General 1-6
Description, Kanawha Series...226-515
Detailed Geologic Structure.......40-2
Development, Historical & Industrial
.... 1-10
Development, Oil and Gas.....519-560
Devonian 43
Diamond Drill Borings (See Core
Tests by Nos.)............514, 514A
Dickens, Leander, Mine...........333
Dickinson, C. C., Mine............447
Dickinson, John O.. Mines..........
297, 334, 441, 575, 576, 577, 579,
580, 581
Dickinson, John Q., Prospects......
.........296, 335, 375, 484, 494, 495
Dilworth, J. B.....................
........224, 246, 273, 274, 276, 298
Dingess Coal 360
Dingess Coal ....
..(See *Coal* by Magisterial Districts)
Dingess Fossils................... 59
Dingess Limestone ....
..............360-370, 595, 596, 597
Dingess Limestone ....
(See *Limestone* by Magisterial Districts)
d'Invilliers, E. V...................
.........224, 273, 276, 298, 436, 482
Discharge Measurements of Coal River:
At Brushton, 1909, 1910 and 1911
.......................16, 18, 19
At Fuqua. 1911................ 24
At Tornado, 1909, 1910 and 1911
........................20, 21, 22
Dolan, Dr. A. J., Opening..........388
Dolan, Geo. W., Mine.............464
Dolan, Mandaville, Mine..463, 578, 582
Dolan, Peter, Mine...............464

Page

Dolan, W. H., Mine...............387
Dort and Bailey.................18, 21
Dotson, Sweet, Mine..............351
Doughton Heirs Mine..............248
Drainage Area 14-15
Drainage Basins, Description.....15-34
Drawdy Creek, Description..12, 14, 27
Dunkard Series, Absence of.......204
Duval District, (Lincoln), Section.70-71

E

Eagle "A" Coal...........46, 133, 501
Eagle Coal ....
.....502-511, 570, 571, 572, 578, 582
Eagle Coal ....
..(See *Coal* by Magisterial Districts)
Eagle Coal, Analyses..............578
Eagle Fire Clay, Analysis..........504
Eagle Limestone 513-14 595, 596
Eagle Limestone ....
(See *Limestone* by Magisterial Districts)
Eagle Oil & Gas Co. Wells..520A, 522
Eagle Sandstone 501-2
Eagle Shale 597
Early History, Oil & Gas Development 519
East Lynn Sandstone..........208-211
Echart, Section 0.9 Mi. North....186-7
Echart, Section 0.8 Mi. S. E.....187-8
Eckels, William, Mine.............390
Edwards and Colcord.........311, 312
Edwards, W. P...................311
Elevations, Railroad...........620-622
Elevations, U. S. Geological Survey
.... 622-627
Elk Lick Coal.................... 44
Elk Run, Description...........12, 14
Elk Run, Section, Head of.........203
Elkins, A. B., Mine...............466
Ellis, Albert, Mine................344
Ellis, J. H., Opening..............421
Ellis, Joe, Opening.......422, 577, 581
Ellishire, Elmer, Mine............422
Emmons P. O., Section 2.0 Mi. S.
W. 63
Emmons P. O., Section 1.2 Mi. West
.............................61-2
Emmons Station Section, 1 Mi. North
.............................58-9
Epling, Frank, Mine..............461
Epling, L. E., Mine...............340
Epling, Thomas, Mine....460, 578, 582
Estep, George, Opening..456, 577, 581
Estep P. O., Section, 1 Mi. N..... 89
Estep P. O., Section 1 Mi. N. W...89
Estep P. O., Section, 3 Mi. W..... 90
Estep, Wilburn, Opening.419, 577, 581
Evan, George, Mines..........400 434
Evans, George C., Mine...........476
Ewart Works, Analysis by.........299

F.

Fall of Streams.................12-13
Farrell, James, Mine..............350
Farrell, Mid, Mine................509
Faunal Horizons..................593-7
Ferrell, Marion, Mines..............
....................269, 303, 574, 579
Fidelity Oil & Gas Co...........520A
Fire Clay, Analyses...........504, 584
Fire Creek Coal...................154
First Geological Survey of Penna..207
Fitzpatrick, G. W., Observer.17, 18, 19

Page
Flora of Allegheny Series.........224
Forber, William, Opening.482, 578. 582
Forest Conditions, Original........588
Forest Conditions, Present.........589
Forests 587-90
Fork Creek, Description.....12, 14, 27
Forks Coal & Land Co. Core Test
........................63, 514A
Forks Coal Co., Core Tests......514A
Forks Coal Co., Wells...........520A
Forks Creek Coal Co., Openings....
212, 213, 232, 233, 255, 256, 257,
258, 284, 285, 286, 321, 322, 341,
575, 580
Forks Creek Coal Co., Well No. 2
(7) 520A, 528
Forks Creek Coal Co., Well No. 4
(4) 520A, 521-2
Formation 2-3
Formations, Table of Geological.... 43
Fossils from Boone, Logan and
Mingo Counties..............591-619
Foster, Allen, Mine......240, 574, 579
Foster, Thomas, Mine............241
Foster P. O., Section 2.0 Mi. E...75-6
Foster P. O., Section 2.4 Mi. N.....75
Foster P. O., Section 1.5 Mi. N. E.
.... 73-4
Foster P. O., Section 2.0 Mi. S.....77
Foster P. O., Section 2.0 Mi. S. E...76
Foster T. L., Wells.............520A
Fowler, Ed., Opening....414, 576, 581
Fuqua Gaging Station, Records..... 24

G.

Gadd, William, Mine..............476
Gage Height, Coal River Stations....
...........17, 18, 19, 20, 21, 22, 24
Gaging Stations.................16-24
Garland Fork, Section 0.7 Mi. Up..119
Garrett, Elsie, Mine..............416
Garrett, M. D., Mine.............416
Garrison & Shonk, Mines......436, 485
Garrison & Shonk, Well.....520A, 552
Gas, Natural, (Chapter VIII)...516-560
Gas, Natural, Development.....519-560
Gas, Natural, Horizons of W. Va...517
Gawthrop, R. M..................597
General Description............... 1-6
General Sections (Chapter IV)..43-203
Generalized Section..............44-7
Geologic Structure, Detailed...... 40-2
Geological Formations, Table of..... 43
Geological Society of America, Bulletin 224
Geology of Boone County (Part II)
.... 37-515
Gibson, Squire, Opening..300, 575, 579
Gilbert Coal..................... 47
Gillispie, Jefferson, Mine.......... 59
Gillispie, Samuel, Mine...478, 578, 582
Glenalum Tunnel Coal..........47, 82
Goddard, H. A...................479
Gordon P. O., Section 2.3 Mi. N. E.
.... 191
Gordon P. O., Section 4.0 Mi. N. E.
.... 190
Gore, Daniel, Mine..............402
Gore, Isaac, Mine........372, 576, 580
Gailey, Staunton, Opening.........
.....................233, 574, 579
Grapevine Branch, Description...13, 14
Gravel Pits.......................585
Green, Cyrus, Mine...............262

Page
Green, Grover, Mine.....439, 577, 581
Green, Irvin, Mine.......458, 578, 582
Greenbrier Limestone..........43, 518
Greenview, Description............ 6
Griffith Bros., Mine..............394
Griffith, Ephraim, Opening.........214
Griffith Heirs, Opening...........463
Griffith, Samuel, Mine............393
Griffith, Section 1 Mi. N. E........
.......................159-161, 203
Grimsley, G. P................... 11
Grosscup Wells..................520A
Guyan Oil Co. Wells....520A, 548, 551
Guyandot Mountain.............. 36

H.

Hager, Ballard, Opening..398, 576, 580
Hager, Barn, Mine.......415, 576, 581
Hager, Emma, Mines..............
.................346, 418, 575, 580
Hager, H., Mine..................345
Hager, Hayes, Opening............398
Hamilton, Patrick, Mine...........402
Handy, M. B......................389
Haner, Elbert, Opening...415, 576, 581
Harless, Daniel, Mine.............453
Harless, E. G., Mine..............452
Harless, Elmer, Mine....385, 576, 580
Harless, Lee, Mine................399
Harless, Wesley, Mine............459
Hatfield, J. H., Mine.............425
Hatfield, J. T., Openings..........
.....................488, 577, 581
Hatfield, U. S., Mine.............416
Henderson, H. A..................
377, 378, 379, 380, 381, 400, 401,
449 450, 451, 452, 476, 477, 478
Hendrickson, J. H., Mine..........531
Hennen and Reger.................
49, 108, 277, 278, 336, 339, 359,
360, 370, 403, 443, 444, 501, 512
Hennen, Ray V....................
........108-9, 250, 278, 501, 596, 597
Hennen Ray V., Section by....98-102
Henry, Earl A....................562
Hernshaw Coal...................
....339-359, 568, 571, 572, 575-6, 580
Hernshaw Coal...................
..(See *Coal* by Magisterial Districts)
Hernshaw Coal, Analyses....342, 575-6
Hernshaw Sandstone 339
Hernshaw, Section 0.6 Mi. S.48-9, 203
Hewett Creek of Spruce Fork.......
.........................13, 14, 33
Hewett P. O., Section 3 Mi. S. W..93-4
Hewitt Creek of Little Coal River...
.........................13, 15, 30
Hickory Ash Coal Co. Mines ....
.................479, 480, 578, 582
Hickory Ash Coal Co., Production..562
Hickory Ash Coal Co., Prospect ...435
High Coal Description 5
High Coal, Section...........166-7, 203
Highways 9-10
Hill, Columbus, Mine.............213
Hill, Geo., Mine..................393
Hill, G. T., Mine.................417
Hill, John, Well No. 1 (30)..520A, 532
Hill, Joseph, Mine................386
Hill, Julian, Authority............589
Hill, Julian, Mine........406, 576, 581
Hill, Julian, Well...............520A
Hill, Kirby, Mine.................345
Hill, Lewis, Mine.................393
Hill, Minerva, Mine......381, 576, 580

Page
Hill, Riley, Mine.................394
Historical & Industrial Development
.... 1-10
History & Physiography...........1-36
History, Early, Oil & Gas Development 519
History of Transportation.........7-10
History of West Virginia.......... 24
Hite and Krak.....................564
Hoge, U. S........................345
Holly & Stephenson Core Test (103)
..........................60, 514A
Holly & Stephenson Wells.......520A
Holmes, John.....................299
Holstein, E. A., Mine....454, 577, 581
Holstein, William, Mine...........405
Homewood Sandstone..........226-231
Hopkins and White, Openings.....
..........................308, 424
Hopkins, Dora, Mine.....394, 576, 580
Hopkins Fork Description..12, 14, 28
Hopkins Fork, Section at Three Forks of.......................162
Hopkins, J. M., Well.............
................520A, 531, 532, 539
Hopkins, Mandaville, Core Test..514A
Hopkins, Mandaville, Mine 466
Hopkins, Mandaville, Wells.........
......................79-80, 520A
Hopkins, Section, 2.0 Mi. S. W.....84
Horizons, Faunal................593-7
Hornickel, George 298
Hornickel, Lute 298
Horse Creek Block Coal Co. Mine..
......................235, 574, 579
Horse Creek Coal & Land Co. Core Tests 514A
Horse Creek Coal & Land Co. Openings 287, 288
Horse Creek Coal & Land Co. Wells
...................520A, 534, 535
Horse Creek Coal Land Co., Openings 214, 234, 235, 236, 574, 579
Horse Creek, Description....13, 15, 29
Horse Creek Lumber Co., Mill.....590
Horsepower, Indicated, Developed by Coal River..................... 26
Horton, A. H...........16, 20, 24, 25
Horton Steel Co., Prospect........310
Hoy, G. C., Observer...........22, 23
Hoy, W. C., Observer............. 24
Hunter, Samuel, Mines.......247, 302

I.

Illustrations of Fossils......facing 618
Index to Species (Part IV).........619
Indian Creek, Description....12, 14, 27
Indian Mounds.................... 7
Indicated Horsepower Developed by Coal River........................26
Industrial Development..........1-10
Industries and Towns............ 4-6
Industry, Carbon Black............590
Investigation of Fossils, Scope of.591-3
Iron Ore..........................587

J.

Jacks Branch, Description....13, 14, 30
Jackson, A. J..................... 16
Jackson, H. J..................... 20
James Branch, Description...12, 14, 31
James Creek, Description....12, 14, 32
Jarrell, Lewis, Mine..............223
Jarrell, T. C., Opening...309, 575, 579
Page
Jarrell, Van, Mines...........296, 355
Jarrold, George, Mine.............332
Jarrold, Manley, Mine.............439
Jarrold, Reed, Core Test near House of 128
Jarrolds Branch, Section at Mouth of 192
Jarrolds Fork, Description.......12, 14
Jarrolds Valley Sections:
4.5 Miles Northwest.............163
5.7 Miles Northwest.............163
3.2 Miles Southwest.........164, 202
Jasper Workman Branch, Description 13
Javin, Frank, Opening.............389
Javins, E. T., Opening............321
Javins, S. M., Mine......382, 576, 580
Jeffery, Description.............. 6
Jeffery P. O., Section 2 Mi. N. E....96
Jeffery, Quinn, Heirs, Mine........
......................423, 577, 581
Jerry Fork Section 122-3
Jigly Branch Sections............135-7
Joes Creek, Description......12, 14, 28
Julian Station, Section Opposite.....69

K.

Kanawha County Sections...........
...............48, 49, 58, 59, 60, 203
Kanawha Series, Clays...........584-5
Kanawha Series, Coals..........56-573
Kanawha Series, Description....225-515
Kanawha Series, Faunal Horizons.593-4
Kanawha Series, General Section.44-47
Kayford, Section 1 Mi. West......366
Keadle, Henry, Mines.........346, 423
Keefer, Daniel, Mine.............461
Kerse, A. J., Mine.......388, 576, 580
Kerse, Patrick, Mine..............387
Kessinger, John, Mines.......408, 409
Key-Rock (Campbell Creek Coal).38-39
Kinder, E. M., Mine.....455, 577, 581
Kinder, John, Mine...............460
Kirk, Charles, Opening............340
Kirk, F. G., Mine........380, 576, 580
Kirk, Walter, Opening....450, 577, 581
Knight Lumber Co.................589
Krak and Hite....................564
Krebs, G. R......................458
Kuhn, Blair, Mine.................351
Kuhn, Nancy, Opening...484, 578, 582

L.

Lackawanna Coal & Lumber Co.:
Mine Openings ....
221, 243, 262, 263, 264, 265, 268, 301, 302, 332, 354, 574, 575, 578, 579, 580
Prospects ....
159, 219, 242, 244, 245, 246, 266, 267, 296, 331, 350, 352, 353, 356, 357, 401, 481, 483
Saw Mill...................589-590
Well No. 1.154-5 483, 520A, 546, 547
Well No. 2. .153-4, 483, 520A, 549
Well No 3...............520A, 549
Well No. 4.....159, 160-1, 520A, 550
Well No. 6...............520A, 557
Well No. 7...............520A, 558
Well No. 10..............520A, 559
LaFollette-Robson et al.:
Core Tests.............155-6 514A
Mines 350, 351, 575, 560

Page
LaFollette-Robson et al.:
Prospects ....
241, 242, 267, 329, 349, 402, 435, 436, 483
Wells 520A, 551
Lantie P. O. Sections:
2 Mi. North....................171-2
4.0 Mi. South....................176
¼ Mi. Southwest................174
2.8 Mi. Southwest................173
3.0 Mi. West.....................174
Laurel Coal & Land Co. Core Tests:
No. 1—(115)...127, 429, 470, 514A
No. 1—(138)..177-8, 441, 487, 514A
No. 2—(137)...177, 441, 487, 514A
No. 3—(114)........126, 470, 514A
No. 3—(139)...179, 442, 487, 514A
No. 4—(117)...129, 430, 470, 514A
No. 4—(136) ....
.......123-5, 155-6, 240, 260, 514A
No. 5—(118) ....
...........129-130, 430, 471, 514A
No. 6—(120) ....
........132-3, 431, 472, 493, 514A
No. 7—(121) ....
........133-4, 431, 472, 493, 514A
No. 9—(112)..122-3, 429, 469, 514A
No. 10—(113)......123-5, 470, 514A
No. 11—(119)....131, 430, 471, 514A
No. 12—(126)..141-2, 433, 474, 514A
No. 13—(125).140-1, 432, 473, 514A
No. 14—(127)..142-3, 433, 474, 514A
No. 15—(122)...135, 431, 473, 514A
No. 15—(128)...143, 433, 474, 514A
No. 16—(116)...128, 429, 470, 514A
No. 18—(124)...... ....
.......139-40, 432, 473, 493 514A
Laurel Creek, Description...12, 14, 28
Laurel Fork Branch, C. & O. Ry., Levels 621
Leatherwood Lumber Co. Mill.....590
Leatherwood Lumber Co. Mine.....214
Leschen Oil & Gas Co. Well (8)..56-7
Levels Above Tide (Appendix)...620-7
Lewis, T. L., Mine...............460
Lewiston (Stockton) Coal......231-250
Lewiston (Stockton) Coal, Analyses
...........................246, 574
Lick Creek of Coal River....12, 14, 27
Lick Creek of Pond Fork....13, 15, 31
Lime, Big.........................518
Limestone, (by Magisterial Districts):
Buffalo Creek
(General) 45 278, 595
Campbell Creek
Crook 46, 184, 444-6
Peytona 46, 444-6
Scott 46, 77, 444-6
Sherman ....
..44, 155, 156, 167, 366, 367, 444-6
Washington 46, 444-6
Dingess
Crook ....
45, 171, 172, 175, 176, 183, 184, 186, 199, 203, 335, 357, 367-70, 375, 595, 596
Peytona 45, 48, 59, 203, 361
Scott.......45, 78, 203, 362, 595, 596
Sherman 45, 154, 167, 203, 365-7
Washington ....
45, 85, 88, 100, 109 127, 136, 145, 203, 253, 282, 362-5, 595, 596
Eagle
Scott..47, 81, 82, 83, 513-14, 595, 596
Sherman............47, 167, 513-14

Page
Limestone, (by Magisterial Districts):
Washington ....
..........47, 94, 95, 513-14, 595, 596
Seth
Crook 200, 369
Sherman **155, 446, 596**
Stockton (Bald Knob)
Crook 46, 184, 186, 189, 497
Limestone, Mountain or Greenbrier.518
Lincoln-Logan-Boone Co. Corner Section, 1 Mi. N. W............... 88
Lincoln County Sections..........70-1
Link-Belt Co......................299
List of Fossils from Boone, Logan and Mingo Counties.............595
Little Alma Coal..................444
Little Alma Coal..................
..(See *Coal* by Magisterial Districts)
Little Cedar Coal...............47, 82
Little Chilton Coal......336-8, 575, 580
Little Chilton Coal..................
..(See *Coal* by Magisterial Districts)
Little Chilton Coal, Analyses.......575
Little Chilton "Rider" Coal.........85
Little Coal Land Co.:
Mines 288, 289, 324, 343, 574, 575, 579
Openings..........215, 216, 217, 335
Well No. 1 (39)..........520A, 538
Well No. 2 (39A)....85-6, 520A, 537
Little Coal River, Description.......
.........................12, 14, 29-34
Little Coal River Branch, C. & O. Ry., Levels.....................621
Little Coalburg Coal...............277
Little Coalburg Coal...............
..(See *Coal* by Magisterial Districts)
Little Eagle Coal..............512-513
Little Eagle Coal..................
..(See *Coal* by Magisterial Districts)
Little Elk Creek, Description.....12, 14
Little Horse Creek, Description......
..........................13, 15, 29
Localities, Register of, Fossil Collections 595-7
Location of Area.................. 1
Logan Fork, Description.........12, 14
Logan-Lincoln-Boone Co. Corner Section, 1 Mi. N. W.................88
Logan County Fossils.....591, 595, 597
Logan County Sections............203
Logan District, Logan Co. Core Tests
.................**107, 109-120, 514A**
Logan District, Logan Co., Wells.520A
Logan Quadrangle, Elevations.....626
Logan Sandstone.................443-4
Loudon District (Kanawha) Section..48
Low Gap Branch, Description.......15
Low Gap P. O., Section............94
Low Gap P. O., Section 2.5 Mi. S. W. 95
Lower Cedar Grove Coal...........403
Lower Cedar Grove Coal............
..(See *Coal* by Magisterial Districts)
Lower Cedar Grove Sandstone....403-4
Lower Chilton Sandstone...........336
Lower Coalburg Sandstone...250, 277-8
Lower Freeport Coal....67, 71, 87, 208
Lower Kittanning Coal (See No. 5 Block) 211-224
Lower Kittanning Coal............
..(See *Coal* by Magisterial Districts)
Lower Pottsville Coals.............515
Lower War Eagle Coal.........47, 550
Lower Winifrede Sandstone........313
Lumber Industry..................588

Page

M.

MacCorkle, Ex-Gov. W. A. ....
120, 122, 126, 128, 129, 131, 132, 133, 135, 139, 140, 141, 142, 143, 177, 179
MacCorkle, Section 2.2 Mi. N. W...67-8
Madison, Description 4
Madison Quadrangle, Elevations.... 624-626
Madison, Section 1 Mile South..... 95
Madison. Section 1.2 Mi. S. E..... 81
Mahoning Sandstone...............206
Malden Sandstone................444
Malden, Type Section, No. 2 Gas Coal 447
Manila Section...................203
Marmet Coal Co...................339
Mason, F. W., Mine......352, 576, 580
Massy, Silas, Mine................268
Matewan Coal................497-501
Matewan Coal.....................
..(See *Coal* by Magisterial Districts)
Matewan Sandstone................501
Mats Creek, Description......12, 14, 32
Mauch Chunk Red Shales..........43
Maxton Sand.....................518
Meade, C. H.....................462
Measurements of Discharge. Coal River 16, 18, 19, 20, 21, 22, 24
Methods of Representing Structure.. 38-39
Meyers, Albert, Mine.....408, 576, 581
Michael, Lorenza, Mine............353
Middle Cedar Grove Sandstone.....403
Middle Kittanning Coal.......... ....44, 68, 198, 211-224
Midkiff, George, Mine....481, 578, 582
Midkiff Well...............520A, 546
Miller, Augustus, Local Mine......236
Miller, Barker, Mine.............412
Miller, Brenny, Mine.............333
Miller, Burrell, Mine.............396
Miller, Columbus Mine............343
Miller, Elisha, Mine..............412
Miller, Joel, Mine................358
Miller, Perry A., Mine............397
Miller, P. K., Mine...............409
Miller, Polly, Mine...............333
Miller, Susan, Mine...............344
Miller, W. C., Mine...............397
Miller, Wm., Mine................308
Mills, Operating Saw...........589-590
Mineral Resources (Part III)..516-590
Mines by Nos.:
1 and 2.......................212
3 and 4.......................213
5 and 6..................213, 579
7 ....214, 579
8 and 9.......................214
10 215
11 ....215, 579
12, 13 and 14...........216, 579
15 217
16 218
17 ....218, 579
18 and 19.....................219
19A 220
20 ....220, 579
21 221
22 and 23...............221, 579
23A 232
24 and 25.....................233
25A 233, 579
26 and 27.....................234
28 and 29.....................235

Page

Mines by Nos.:
30 ....235, 579
31 236
32 ....236, 579
33, 34, and 35.................237
36, 37 and 38..................238
39 and 40 239
41 and 42......................240
43 ....240, 579
44, 45 and 46..................241
47, 48 and 49..................242
50 ....243, 579
51 and 52......................243
53, 54 and 55..................244
56, 57 and 58..................245
59 and 60......................246
61 ....247, 579
62 247
63, 64 and 65..................248
66, 67 and 68..................222
69 and 70......................223
71, 72, 72A and 73.............249
74 255
75, 76, 77 and 78..............256
79, 80, 81 and 82..............257
83, 84, 85 and 86..............258
87 259
88 and 89......................261
90 ....261, 579
91 and 92......................262
93 263
94 ....263, 579
95 263
96 ....264, 579
97 and 98......................264
99 ....265, 579
100 265
101, 102 and 103...............266
104, 105 and 106...............267
107 268
108 ....268, 579
109 and 110....................269
110A 269, 579
111 ....270, 579
112 and 113....................270
114 and 114A...................271
115 ....271, 579
116 and 117....................272
118 273
119 ....273, 579
120 and 121....................274
122, 123, 124 and 125..........275
126 and 127....................276
128 and 129....................277
130, 131, 132 and 133..........285
134, 135 and 136...............286
137, 138 and 139...............287
140 and 141....................288
142 ....288, 579
143 ....289, 579
144 289
145 and 146....................290
147, 148, 149, 150 and 151.....291
152, 153, 154 and 155..........292
156 293
157 ....293, 579
158 293
159 and 159A...................294
160, 161 and 162...............295
163, 164 and 165...............296
166 and 167...............297, 579
168 297
169 ....298, 579
170 ....299, 579
171 and 172...............300, 579
173 301
174 ....301, 579

Page

Mines by Nos.:
175 302
176 ....302, 579
177 302
178 and 179....303
180, 181 and 182....304
183, 184, 185 and 186....305
187 and 188....306
189 and 190....307
191 ....307, 579
191A and 192....308
192A 309, 579
193 and 194....309
195, 196 and 197....310
198, 199 and 200....311
201, 202 and 203....312
204 313
205 314
206 320
207, 208, 209 and 210....321
211, 212 and 213....322
214 and 215....323, 580
216 ....324, 580
217 325
218 ....325, 580
219 and 220....326
221 ....326, 580
222 and 223....327
224, 225 and 226....328
226A 328, 580
227 and 228....329
229, 230, 231 and 232....330
233, 234 and 235....331
236 332
237 and 238....322, 580
239, 240, 241 and 242....333
243 334
244 ....334, 580
245 334
246, 247 and 248....335
249 336
250 337
251 ....338. 580
252 and 253....338
254, 255 and 256....340
257 and 258....341, 580
259 342
260, 261, 262 and 263....343
264, 265 and 266....344
267 ....344, 580
268, 269 and 270....345
271 and 272....346
273 and 274....346, 580
275 347
276, 277, 278 and 279....348
280, 281, 282 and 283....349
284 and 285....350
286 ....350, 580
287 ....351, 580
288 and 289....351
290 352
291 and 291A....352, 580
292 352
293, 294 and 295....353
296 354
297 ....354, 580
298 and 299....355
300 ....355, 580
301 and 302....356
303 ....356, 580
304. 305, 306 and 307....357
308 358
309 ....358, 580
310 358
311, 312 and 313....359
314 360
315 and 316....363

Page

Mines by Nos.:
317 and 318....364
320 367
321 368
322 369
324 and 325....371
326 372
327 and 328....372, 580
329 373
330 ....373, 580
331 374
332 and 333....375
334 and 335....377
336, 337 and 338....378
339, 340 and 341....379
342 and 343....380
344 ....380, 580
345 380
346 and 347....381
348 ....381, 580
349, 350 and 351....382
352 ....382, 580
353, 354, 355 and 356....383
357, 358, 359, 360 and 361....384
362 and 363....385, 580
364 385
365 and 366....386
367 and 368....387
369 ....387, 580
370 387
371 388
372 ....388, 580
373 388
374 and 375....389
376 and 377....390
378 ....390, 580
379, 380 and 381....391
382 and 383....392
384, 385, 386 and 387....393
388, 389 and 390....394
391 ....394, 580
392, 393 and 394....395
395 and 396....396
397, 398 and 399....397
400 ....397, 580
401 ....398, 580
402 398
403 ....398, 580
404, 405 and 406....399
407, 408 and 409....400
410, 411 and 412....401
413 402
414....402, 580
415 402
416, 417, 418 and 419....405
420....406
421 ....406, 581
422 and 423....406
424, 425 and 426....407
427 and 428....408, 581
429 408
430 and 431....409
432 ....409, 581
433 409
434 ....410, 581
435 410
436 ....410, 581
437 and 438....411
439 ....411, 581
440, 441 and 442....412
443 ....412. 581
444, 445 and 446....413
447 414
448 ....414, 581
449 414
450, 451 and 452....415. 581
453, 454 and 455....416

Page

Mines by Nos.:
456 and 457....417
458 ....417, 581
459 417
460, 461 and 462....418
463, 464 and 465....419, 581
466, 467 and 468....420, 581
469, 470 and 471....421
472 422
473 and 474....422, 581
475 ....423, 581
476 423
477 ....423, 581
478 424
479 ....424, 581
480 424
481 425
482 and 483....425, 581
484 and 485....426, 581
486 ....427, 581
487 and 488....427
489 and 490....428
491, 492 and 493....429
494, 495 and 496....430
497, 498 and 499....431
500, 501 and 502....432
503, 504, 505 and 506....433
507, 508 and 509....434
510 and 511....435
511A, 511B and 511C....436
511D 437, 581
512 and 513....438, 581
514 438
515 439
516 ....439, 581
517 439
518 487
518A 440
519 and 520....488, 581
521 489
522 and 523....440
[illegible] 441
524A....441, 581
525 441
526, 527 and 528....442
529 443
529A 443, 581
531 447
532, 533 and 534....448
535, 536, 537 and 538....449
539 450
540 ....450, 581
541 450
542, 543, 544 and 545....451
546 and 547....452
548 and 549....453
550 ....453, 581
551 ....454 581
552 and 553....454
554 455
555 and 556....455, 581
557 and 558....456, 581
559 456
560 ....457, 581
561 and 562....458, 582
563, 564 and 565....459
566 ....459, 582
567 and 568....460
569 ....460, 582
570 and 571....461
571A 462, 582
572 and 573....463
574 ....463, 582
575, 576, 577 and 578....464
579 ....465, 582
580 and 581....466
582 and 583....467

Page

Mines by Nos.:
584, 585, 586 and 587....468
588 and 589....469
590, 591, 592, 593 and 594....470
595 and 596....471
597 and 598....472
599, 601 and 602....473
603, 604, 605 and 606....474
607 and 608....475
609, 610 and 611....476
612, 613 and 614....477
615 and 616....478
617 ....478, 582
618 and 619....479
620 ....479, 582
621 ....480, 582
622 and 623....480
624 and 625....481, 582
626 481
627 and 628....482, 582
629 and 630....483
631 ....484, 582
632, 633 and 634....484
635 ....485, 582
636 and 637....485
638 486
639 ....486, 582
640, 641 and 642....487
643 489
643A 498
644 489
645, 646 and 647....498
647A and 647B....499
648 ....494, 582
649 490
650 and 651....491
652 and 653....492
654, 655 and 656....493
657 494
657A 506
659 494
660 495
661 ....495, 582
662 and 663....495
664, 665 and 666....496
667 497
668, 669 and 670....503
671 504
672 and 673....505
674 499
675 ....499, 582
675A 500
676 ....506, 582
677 and 678....507, 582
679 and 680....508
681, 682 and 682A....500
682B 501
683 ....508, 582
684 509
685 ....509, 582
686 509
687, 687A, 687B....510
687C, 687D and 687E....511

Mingo County Fossils....591, 595, 597
Miscellaneous Items.... 1-6
Mississippian 43
Mistletoe, Description.... 6
Mistletoe P. O., Section 3 Mi. S. W...87
Mistletoe Section, 3 Mi. West.... 71
Mohler Lumber Co. Mill....589
Mohler Lumber Co. Opening....
....213, 574, 579
Monitor Sandstone....443-4
Monongahela Series, Absence of....204
Montgomery, A. R., Authority....
108, 109, 110, 111, 112, 113, 114, 115, 116, 117, 118, 119

Page
Montgomery, A. R., Coal Sections...
..217, 249, 261, 269, 270, 271, 290, 291, 292, 294, 304, 305
Montgomery, Clothier & Tyler.....121
Moore, J. M...................323, 426
Morris, Hamilton, Mine............448
Mounds, Indian 7
Mountain Limestone...............518
Mountain State....................588
Mountains 35-36
Mud Gap, Section 1 Mi. North..... 84
Mud Gap, Section 1 Mi. West......87-8
Mud River, Description......12, 14, 34
Mullens, Jas. P., Mine.............392
Mullens, S. V., Mine...............237

Mc.

McCreath, A. S., Analyses..........
222, 224, 246, 273, 276, 298, 308, 391, 465, 466, 471, 482
McCreath, A. S., & Son, Analyses...
..........................389, 436
McDermott, George M., Opening...450
McKinney, Walker, Mine...........396
McMechen, Albert, Mine...........287

N.

Natural Gas, (Chapter VII)....516-560
Naugatuck Sandstone...........359-360
Nelson, Floyd, Mine......323, 575, 580
Nelson P. O. Sections:
4.0 Mi. Southwest...........157. 158
5.0 Mi. Southwest..............158
2.0 Mi. West..................153-4
Nelson, Oscar, Authority...........590
New River & Pocahontas Cons. Coal Co.......224, 273, 276, 304, 401, 452
Nichol, Thomas....................426
Nicholas, E. V. & I...............246
Niggerhead, Coalburg Coal, Analysis
..............................255
North Carbon Section..............203
North Coalburg Coal (See Upper Kittanning Coal).................211
Nose of Fold, Definition........... 37
Notes on the Paleontology of Boone County (Chapter XI).........591-619
No. 2 Gas Coal....................
..........38-9, 446-492, 569, 571, 572
No. 2 Gas Coal....................
..(See *Coal* by Magisterial Districts)
No. 2 Gas Coal, Analyses............
.......452, 457-8, 465, 466, 482, 577-8
No. 5 Block Coal..................
211-24, 565, 571, 572, 574, 579, 581-2
No. 5 Block Coal..................
..(See *Coal* by Magisterial Districts)
No. 5 Block Coal, Analyses..........
..................217, 222, 224, 574
No. XII Series of Rogers (Chapter VII) 225-515
Numbers on Map II...............212

O.

Oceana Special Quadrangle, Elevations 626
Oil and Gas, (Chapter VIII)...516-560
Oil and Gas Development......519-560
Oil and Gas Horizons of W. Va.....517
Olcott Section (Kanawha).......... 59
Oldhouse Branch, Section 700′ Up...
.............................116-17
Operating Saw Mills............589-590

Page
Orange P. O. Sections:
0.5 Mi. North...................165
4.0 Mi. North...................164
1.8 Mi. Northeast.............150-1
1.5 Mi. Northwest...............156
Order of Counties in Production of Coal 563
Original Forest Conditions..........588
Ottawa, Description................ 6
Ottawa, Section.................... 96

P.

Page Reference to Detailed Description and Sections of Coal Mines..
..........................579-582
Paleontology (Part IV).........591-619
Pardee, Calvin, Openings............291
Pauley, M. P., Opening............411
Payne, C. P., Mine................348
Peerless Coal.............156, 167, 559
Pennsylvanian 43
Perdue, Heirs, Mine...............480
Perdue, Olsey Mine......459, 578, 582
Perry, A. H., Mine................499
Perry, Grant, Opening.....443, 577, 581
Personal Property, Value of........ 4
Peterkin, Bishop, Well...........520A
Petroleum and Natural Gas (Chapter VIII) 516-560
Peytona Block Coal Co. Mine........
......................462, 578, 582
Peytona Block Coal Co., Production.562
Peytona Block Coal Co., Wells....520A
Peytona Coal Land Co. Prospects....
..........................320, 503
Peytona Coal Land Co., Wells....520A
Peytona, Description.............. 5
Peytona District:
Area 2
Campbell Creek Limestone...46, 444-6
Coalburg Sandstone............250-1
Coals ....
(See *Coal* by Magisterial Districts)
Core Tests.....................514A
Dingess Limestone...............
..............45, 48, 59, 203 361
East Lynn Sandstone............209
Homewood Sandstone............227
Logan Sandstone...............443-4
Malden Sandstone...............444
Naugatuck Sandstone.........359-360
Pittsburgh Red Shale............205
Population 2
Prospective Oil & Gas Territory...528
Sections, General 47-66
Upper Chilton Sandstone......314-15
Upper Winifrede Sandstone...279-280
Well Records..........520A, 521-528
Williamson Sandstone............360
Peytona Land Co., Openings........
.....................384, 385, 401
Peytona Land Co., Wells.520A, 542, 543
Peytona Lumber Co., Mill..........589
Peytona P. O., Section 3 Mi. S. W.
..............................64-5
Peytona Quadrangle, Elevations.622-624
Peytona, Section 0.7 Mi. East......51-2
Peytona Well No. 1 (14)....51-2, 520A
Physiography and History.........1-36
Physiography of Area (Chapter II)..
..............................11-36
Pickens, Romeo, Well No. 1 (20)..
.........................520A 529
Pierson, H. C..................383, 384

Page

Pigeonroost Branch, Section 0.5 Mi. Up 115-16
Pigeonroost Branch, Section 1 Mi. Below 113-14
Pittsburgh Coal.... 44
Pittsburgh Red Shale....205
Pocahontas Coal & Coke Co. Opening 249
Pocahontas Cons. Coal Co....310
Pocono Sandstones 43
Polley, Betsey, Mines....249, 269
Polley, Betsey, Section....175
Polly, Green, Mine....343
Pond Creek Coal & Land Co., Opening 485, 577, 581
Pond Fork Branch, C. & O. Ry., Levels 622
Pond Fork Coal & Land Co., Openings..358, 359, 437, 490, 499 576, 580
Pond Fork, Description...12, 14, 30, 31
Pond Fork, Railroad Surveys.... 8
Pond, Section 3 Mi. S. W....180-1
Population 2
Porter, Andrew, Well No. 1 (47)... ....520A, 545
Pottsville Sandstones....586
Pottsville Series....43, 225-515
Powell, Jefferson, Mine....409
Powellton "A" Coal.... ..(See *Coal* by Magisterial Districts)
Powellton Coal.... ....492-7, 569-70, 571, 572, 578, 582
Powellton Coal.... ..(See *Coal* by Magisterial Districts)
Powellton Coal Analyses....578
Powellton "Rider" Coal....123, 134
Present Forest Conditions....589
Price Heirs, Local Mine....234
Price, J. B., Mine....391
Price, J. T., Mine....379
Price, R., Mines....248, 439
Price, Tom, Mine....359
Price, Tuncil, Mine....213, 574, 579
Price, Wm. Armstrong.... ....361, 362, 365, 446, 514, 591
Price, Wm., Openings.... ....176, 261, 306, 574, 579
Products 3
Projected Railroads....8-9
Property Valuation 3-4
Prospective Oil and Gas Areas:
  Crook District....560
  Peytona District....528
  Scott District....535
  Sherman District....555
  Washington District....541-2
Pryor and Allen, Local Mines.... ....238, 239, 259, 371, 387
Pryor and Allen Well (25)..72-3, 520A
Pryor and Allen Well No. 1 (21)... ....520A, 533
Pryor and Allen Well No. 2 (28)... ....77-9, 520A, 530
Pryor and Allen Well No. 3 (27)... ....520A, 529

Q.

Quantity of Coal Available:
  Alma 569, 571, 572
  Campbell Creek (No. 2 Gas).... ....569, 571, 572
  Cedar Grove (Thacker)..568, 571, 572
  Chilton 567, 571, 572
  Coalburg 567, 571, 572

Page

Quantity of Coal Available:
  Eagle 570, 571, 572
  Hernshaw 568, 571, 572
  No. 5 Block....565, 571, 572
  Powellton (Brownstown) 569-70, 571, 572
  Stockton-Lewiston 566, 571, 572
  Williamson 568, 571, 572
  Winifrede 567, 571, 572

R.

Racine, Description.... 5
Racine Sections:
  3.2 Miles North.... 53-4
  1.4 Miles Northeast....49-50
  2.7 Miles Northeast.... 58
  3.0 Miles Northwest....152-3
  0.5 Mile South.... 52-3
Railroad Levels....620-622
Railroads, Projected....8-9
Railroads. Steam.... 7-8
Railroads, Surveys for.... 8-9
Ramage, Description.... 6
Ramage, J. B....426
Ramage Section....364
Rate of Fall of Streams....12-13
Raven Carbon Co. Plant....590
Raven Carbon Co. Wells....520A
Real Estate, Value of.... 4
Recent Clays 584
Reger and Hennen.... 49, 108, 277, 278, 336, 339, 359, 360, 370, 403, 443, 444, 501, 512, 596
Register of Localities, Fossil Collections 595-7
Relief 2
Representing Structure. Methods of.38-9
Republic Section....203
Reynolds Mine Opening....486
Reynolds, Wm. C....588
River Terraces.... 35
Roach Branch, Description.... 14
Road Materials....585
Robertson, John, Mine....342
Robinson Creek, Description....13, 30
Robinson Creek, Section, Mouth of.. ....171-2, 203
Robson, H. A., Mines.351, 352, 576, 580
Robson-Prichard Well....520A
Rock Creek Collieries Co. Mine.... ....465, 578, 582
Rock Creek Colliery Co. Openings... ....389, 391
Rock Creek, Description....13, 15 31
Rock Creek Land Co. Wells....520A
Rock Creek Section, 1.5 Mi. N. W.. 83
Rockhouse Creek, Section 1.8 Mi. from Mouth....111-112
Rockhouse Creek, Section 0.5 Mi. S. of Mouth....110-111
Rocklick Branch, Description....13, 14
Rosecrans, General....16
Ross, O. R....365
Rowland Land Co. Openings.... ....221, 223, 311, 312

S.

Salt Sand....518
Saltsburg Sandstone....205-6
Sand, Berea....519
Sand, Big Injun....518-519
Sand, Maxton....518
Sand, Salt....518
Sandlick Creek, Description...12, 14, 28
Sands 586
Sands, Description of....518-519

Page
Sands. Oil and Gas, of W. Va......517
Sandstone:
Bens Creek..................511-512
Brownstown....................492
Buffalo.........................206
Coalburg....................250-254
Decota.........................512
Eagle........................501-2
East Lynn..................208-211
Hernshaw......................339
Homewood (Roaring Creek)..226-231
Logan (Monitor)..............443-4
Lower Cedar Grove............403-4
Lower Chilton..................336
Lower Coalburg...............277-8
Lower Winifrede................313
Mahoning.......................206
Malden.........................444
Matewan........................501
Middle Cedar Grove.............403
Naugatuck..................359-360
Saltsburg....................205-6
Upper Cedar Grove..............375
Upper Chilton..............314-320
Upper Coalburg.................250
Upper Freeport.................208
Upper Winifrede............278-284
Williamson.....................360
Sandstones. Allegheny & Pottsville..586
Sarver, C. E., Oil & Gas Co. Well...
..........................520A, 556
Saw Mills, Operating..........589-590
School Tax........................ 4
Scope of the Investigation (Fossils)
..............................591-3
Scott District:
Area........................... 2
Campbell Creek Limestone........
....................44, 77, 444-6
Coalburg Sandstone............251-2
Coals..........................
(See *Coal* by Magisterial Districts)
Core Tests....................514A
Dingess Limestone...............
..........45, 78, 203, 362, 595, 596
Eagle Limestone............595, 596
East Lynn Sandstone.........209-210
Homewood Sandstone...........228-9
Logan Sandstone...............443-4
Malden Sandstone................444
Naugatuck Sandstone.........359-360
Pittsburgh Red Shale............205
Population........................ 2
Prospective Oil & Gas Areas......535
Sections, General...............66-84
Upper Chilton Sandstone....315-317
Upper Winifrede Sandstone.....280-1
Well Records..........520A, 528-535
Williamson Sandstone............360
Section, Generalized, Conemaugh,
Allegheny & Kanawha Series...44-47
Sections Cross.................. 38
Sections, General (Chapter IV)...43-203
Sections, Geologic:
Adkins Branch, 0.5 Mi. Up.......118
Altman, 1 Mile North............66
Altman, 1 Mile West.............70
Altman Station..................69
Anchor P. O.....................91
Anchor P. O. 1.5 Mi. S. W.......92
Andrew P. O., 2 Miles West.....65-6
Bald Knob......................203
Bald Knob P. O., Near.........184-6
Bald Knob P. O., 1 Mi. North..182-4
Bald Knob P. O., 1.2 Mi. N. W...182

Page
Sections, Geologic:
Bald Knob P. O., 4.8 Mi. N. 59° W. 179
Bald Knob P. O., 5.2 Mi. N. 55° W. 177-8
Bald Knob P. O., 5.7 Mi. N. 47° W. 177
Bald Knob P. O., 3¼ Mi. S. W..147-8
Bear Hollow...............127, 131
Bend Branch, 1 Mi. Up.....109-110
Big Creek Station...............203
Boone-Logan-Lincoln County Corner, 1 Mi. N. W.................88
Brounland (Kanawha)............60
Brounland, 3.5 Mi. S. E..........61
Brushy Fork of Spruce.......119-120
Brushy Fork of Spruce, 2 Mi. Up
.........................114-115
Cabell P. O., 2 Mi. South........64
Chap P. O., 1.3 Mi. N. W......192-3
Chap P. O., 3.0 Mi. N. W.......192
Chap P. O., 1.2 Mi. South......196-8
Chap P. O., 3.3 Mi. South.....200-1
Chap P. O., 1.0 Mi. S. E.......194-6
Chap P. O., 1.4 Mi. S. E.........194
Chap P. O., 2.6 Mi. S. E.....198-200
Clear Fork (Raleigh).............169
Clothier.................99-102, 203
Clothier, E., N. W., S., S. E. and S. W. (42). (See *Clothier* Sections)
Coal River Siding..........168, 203
Coalburg.........................203
Cobb P. O., 1.5 Mi. East........57
Cold Fork, Head of...............203
Comfort, 3.8 Mi. North......149-150
Curry P. O., 1.5 Mi. East..92-3, 203
Danville, on Price Branch........ 82
Danville, 1 Mile East.........79-80
Danville, 1¾ Mi. N. E.........77-9
Danville, 3¼ Mi. N. E.......... 74
Dartmont, 4.0 Mi. N. E.........54-5
Dartmont Station, ½ Mi. S. E...56-7
Dennison Fork, ⅓ Mi. Below Mouth.........................145-6
Dennison Fork, Head of.........203
Echart, 0.9 Mi. North..........186-7
Echart, 0.8 Mi. S. E...........187-8
Elk Run, Head of...............203
Emmons P. O., 2.0 Mi. S. W.... 63
Emmons P. O., 1.2 Mi. West....61-2
Emmons Station, 1 Mi. North....58-9
Estep P. O., 1 Mi. North........ 89
Estep P. O., 1 Mi. N. W........ 89
Estep P. O., 3 Mi. West......... 90
Foster P. O., 2.0 Mi. East.....75-6
Foster P. O., 2.4 Mi. North..... 75
Foster P. O, 1.5 Mi. N. E......73-4
Foster P O., 2.0 Mi. South...... 77
Foster P. O., 2.0 Mi. S. E....... 76
Garland Fork, 0.7 Mi. Up.......119
Gordon P. O., 2.3 Mi. N. E......191
Gordon P. O. 4.0 Mi. N. E......190
Griffith, 1 Mi. N. E.....159-161, 203
Hernshaw, 0.6 Mi. South...48-9, 203
Hewett P. O., 3 Mi. S. W.......93-4
High Coal................166-7, 203
Hopkins Fork, Three Forks of....162
Hopkins, 2.0 Mi. S. W.......... 84
Jarrolds Branch, Mouth of......192
Jarrolds Valley, 4.5 Mi. N. W....163
Jarrolds Valley, 5.7 Mi. N. W...163
Jarrolds Valley, 3.2 Mi. S. W.....
.........................164, 202
Jeffery P. O., 2 Mi. N. E........ 96
Jerry Fork....................122-3

Page

Sections, Geologic:
Jigly Branch....................135-7
Julian Station, Opposite.......... 69
Kayford, 1 Mi. West.............366
Lantie P. O., 2 Mi. North.....171-2
Lantie P. O., 4.0 Mi. South.......176
Lantie P. O., ¼ Mi. S. W.......174
Lantie P. O., 2.8 Mi. S. W......173
Lantie P. O., 3.0 Mi. West......174
Logan-Lincoln-Boone Co. Corner, 1 Mi. N. W...................88
Low Gap P. O.................... 94
Low Gap P. O., 2.5 Mi. S. W... 95
MacCorkle, 2.2. Mi. N. W.......67-8
Madison, 1 Mile South.......... 95
Madison, 1.2 Mi. S. E........... 81
Manila 203
Mistletoe P. O, 3 Mi. S. W......87
Mistletoe, 3.0 Mi. West......... 71
Mud Gap, 1.0 Mi. North........ 84
Mud Gap, 1.0 Mi. West........87-8
Nelson P. O., 5.0 Mi. S. W,.157, 158
Nelson P. O., 5.0 Mi. S. W.....158
Nelson P. O., 2.0 Mi West....153-4
North Carbon 203
Olcott (Kanawha).............. 59
Oldhouse Branch, 700 ft. Up.116-117
Orange P. O., 0.5 Mi. North....165
Orange P. O., 4.0 Mi. North....164
Orange P. O., 1.8 Mi. N. E....150-1
Orange P. O., 1.5 Mi. N. W.....156
Ottawa 96
Peytona P. O., 3.0 Mi. S. W....64-5
Peytona, 0.7 Mi. East......... 51-2
Pigeonroost Branch, 0.5 Mi. Up.. 115-116
Pigeonroost Branch, 1.0 Mi. Below 113-114
Polley, Betsey..................175
Pond, 3.0 Mi. S. W...........180-1
Racine, 3.2 Mi. North..........53-4
Racine, 1.4 Mi. N. E..........49-50
Racine, 2.7 Mi. N. E......... 58
Racine, 3.0 Mi. N. W.........152-3
Racine, 0.5 Mi. South..........52-3
Ramage 364
Republic 203
Robinson Creek, Mouth of........ ....171-2, 203
Rock Creek, 1.5 Mi. N. W......83
Rockhouse Creek, 1.8 Mi. from Mouth 111-112
Rockhouse Creek, 0.5 Mi. S. of Mouth 110-111
Seng Camp Branch, 1.5 Mi. from Mouth 112-113
Seng, 2.5 Mi. S. E..............105
Seng, 3.0 Mi. S. 35° E..........106
Seng, 4.5 Mi. S. 58° E..........107
Seth 154-5
Seth, 3 Mi. South...............159
Sharples 102-3
Sharples, 1.7 Mi. S. E.........104-5
Sharples, 2.5 Mi. S. 80° W....103-4
Shrewsbury 203
Skin Poplar Branch..135, 137-8, 203
South Carbon....................203
Stow, P. O., 2.9 Mi. North....144-5
Sycamore Fork..................129
Three Forks of Hopkins Fork...162
Tony Creek, Head of...........203
Trough Fork................139-140
Trough Fork, ½ Mi. Up......140-1
Turtle Creek, 1¾ Mi. West.....85-6
Van Camp, 0.5 Mi. S. E.......72-3
Van P. O., 2.0 Mi. South.......175

Page

Sections, Geologic:
Walnut Gap...................188-9
Walnut Gap, 3¼ Mi. North......189
Weveco 203
Whiteoak Branch, 1.0 Mi. Up.117-118
Whitesville, 1.4 Mi. N. W......166
Whitesville, 5.8 Mi. West......161-2
Winifrede 203
Woodville (Lincoln), 2.0 Mi. West 70-1
Workman Branch Near Mouth..203
Workman Knob.................170
Semi-Centennial Commission........24
Seng Camp Branch, Section 1.5 Mi. from Mouth.................112-113
Seng Creek Coal Co. Mine......... ....168, 300, 575, 579
Seng Creek Coal Co. Production...562
Seng Creek, Description.....12, 14, 29
Seng, Section 2.5 Mi. S. E........105
Seng, Section 3.0 Mi. S. 35° E....106
Seng, Section 4.5 Mi. S. 58° E....107
Seth, Description...................6
Seth Limestone.155, 200, 369, 446, 596
Seth Section.....................154-5
Seth, Section 3 Mi. South.........159
Sewell Coal....................... ..(See *Coal* by Magisterial Districts)
Shale, Bald Knob.................497
Sharp, C. C......................515
Sharples Section................102-3
Sharples, Section 1.7 Mi. S. E....104-5
Sharples, Section 2.5 Mi. S. 80° W.. 103-4
Sherman District:
Area 2
Campbell Creek Limestone........ .46, 155, 156, 167, 366, 367, 444-6
Coalburg Sandstone............253
Coals........................... (See *Coal* by Magisterial Districts)
Core Tests....................514A
Dingess Limestone.............. ....45 154, 167, 203. 365-7
East Lynn Sandstone.........210-211
Homewood Sandstone...........230-1
Logan Sandstone.............443-4
Malden Sandstone...............444
Naugatuck Sandstone........359-360
Population 2
Prospective Oil and Gas Area....555
Sections, General............149-169
Seth Limestone........155, 446, 596
Upper Chilton Sandstone....318-319
Upper Winifrede Sandstone....282-3
Well Records.........520A, 542-555
Williamson Sandstone...........360
Shields Oil & Gas Co. Well (31).... ....79-80, 520A
Shields Oil Co. Wells........531, 532
Shonk and Garrison Mines...436, 485
Shonk-Garrison Well.........520A, 552
Short Creek, Description.....12, 14, 27
Shrewsbury Section..............203
Siler, T. E. B...................300
Sixmile Creek, Description...13, 14, 33
Skin Creek, Description............13
Skin Fork, Description.............14
Skin Poplar Branch, Description.13, 14
Skin Poplar Branch Sections........ ....135, 137-8, 203
Smith, "Doc".....................464
Smoot, Daniel, Mine..............394
Smoot, D. J., Mine...............406
Smoot, Mrs. Mary, Mine..........247
Smoot, Milton, Mine..............438

Page

Smoot, W. T., Heirs, Mine........392
Snider, E. B., Authority............
212, 213, 232, 256, 257, 258, 285, 321, 322
Snodgrass, Alice, Mine............348
Snodgrass, Hughey, Mine..........451
Snodgrass, Witcher, Mine.........479
South Carbon Section.............203
South Penn Oil Co. Wells..........
...............520A, 536, 539, 544
Species, Description of (Fossils)....
..........................597-617
Spencer, Charles, Opening..........
......................422, 577, 581
Spruce Fork, Description..12, 13, 32-4
Spruce Fork Branch, C. & O. Ry., Levels 621
Spruce Laurel Core Test Sections of Coal 240
Spruce Laurel Creek, Railroad Surveys 8-9
Spruce Laurel Diamond Core Tests (See Laurel Coal & Land Co. Core Tests)
Spruce Laurel Fork, Description....
.................12, 14, 33-4, 120
Spruce River Coal Co. Mine.........
.................364, 426, 577, 581
Spruce River Coal Co., Production.562
Standard Fuel Co.................67
Standard Fuel Co. Core Test........
...............67-8, 206, 514A, 583
State Auditor.......................3
State Road Engineer.................9
Statistics of Coal Production....561-3
Steam Railroads...................7-8
Stephenson, Dr. E. B. et al. Well..
........................520A, 529
Sterling, Description...............5
Stewart, James....................454
Stockton, Aaron...................497
Stockton-Lewiston Coal.............
.....231-250, 566, 571, 572, 574, 579
Stockton-Lewiston Coal.............
..(See *Coal* by Magisterial Districts)
Stockton-Lewiston Coal, Analyses...
..........................246, 574
Stockton Limestone.................
..............46, 184, 186, 189, 497
Stollings, H. E., Mine............393
Stollings, John, Mine....409, 576, 581
Stones, Building.................585-6
Stow P. O., Section 2.0 Mi. North
...............................144-5
Stratigraphy—Allegheny Series.207-224
Stratigraphy—Conemaugh Series.....
.............................204-206
Stratigraphy—General Sections...43-203
Stratigraphy—Pottsville Series..225-515
Stream Data, Table of...........12-13
Strike, Definition of..............37
Structure, Chapter III...........37-42
Structure Contours...............38-39
Structure, Geologic, Detailed.....40-42
Structure, Methods of Representing
...............................38-39
Sullivan Machinery Co..............81
Surveys for Railroads.............8-9
Sutphin, Ira, Mine.......355, 576, 580
Sutphin Mine......................349
Sutphin, Napoleon, Mine...........424
Sycamore Fork Section.............129
Syncline, Coalburg..............41-42
Syncline, Definition...............37

Page

T

Tables, Showing:
Acreage of Coals............565-570
Average Thickness of Coals..565-570
Coal Analyses...............574-578
Coal Production...............562-3
Cubic Feet of Coals.........565-570
Dail Gage Height of Coal River
..........17, 18, 19, 21, 22, 23, 24
Depths and Thickness of Coals in Lower Pottsville Series.......515
Discharge Measurements of Coal River.....16, 18, 19, 20, 21, 22, 24
Drainage Area.................14-15
Fossiliferous Members of the Kanawha Series................594
Geological Formations..............43
Indicated Horsepower Developed by Coal River.....................26
List of Fossils from Boone, Logan and Mingo Counties...........595
Oil and Gas Horizons of W. Va.517
Quantity of Coal Available..565-572
Stream Data..................12-13
Summarized Record of Diamond Core Tests.....................514A
Summarized Record of Wells...520A
Summary of Available Coal...570-2
Tonnage of Coals............565-572
Tagart, F. W......................426
Tawney John, Mine................395
Tax, School.........................4
Teachers' Fund......................4
Terraces, River....................35
Thickness of Coals............565-570
Thickness of Lower Pottsville Coals
..............................515
Thomas, John, Mine...............413
Thompson, Laura, Mine..438, 577, 581
Thompson, Mary Ann, Opening....237
Thompson, William................588
Three Forks of Hopkins Fork, Section 162
Toney, J. E., Well No. 1 (49).....
...............151-2, 153, 520A, 544
Tonnage of Coals..............565-572
Tony Creek, Description.....12, 14, 28
Tony Creek, Section, Head of....203
Tony, John, Well.151-2, 153, 520A, 544
Tony, Robert, Opening...481, 578,582
Topography of the Land Area....34-6
Tornado Gaging Station, Records.20-23
Towns and Industries..............4-6
Towns, Elevations of..........620-627
Trace Fork of Big Creek, Description
.....................13, 14, 15, 34
Transportation, History of........7-10
Tributaries of Coal River........27-29
Tributaries of Laurel Creek......28-29
Tributaries of Little Coal River..29-34
Tributaries of Pond Fork........30-31
Tributaries of Spruce Fork.......33-34
Tributaries of West Fork..........32
Trough Fork Section.........139-140
Trough Fork, Section ½ Mi. Up...140-1
Turley, Marion, Opening..........383
Turley, Noah, Well...............533
Turley Opening..........457, 577, 581
Turtle Creek, Description....13, 15, 31
Turtle Creek, Section 1¾ Mi. West
...............................85-6

Page

U

United Fuel Gas Co. Well.......520A
U. S. Geological Survey, Elevations 622-627
U. S. Geological Survey, Gaging Stations...........................16-24
Upper Cedar Grove Coal........... ....(See Cedar Grove Coal)
Upper Cedar Grove Sandstone.....375
Upper Chilton Sandstone.......314-320
Upper Coalburg Sandstone.........250
Upper Freeport Coal..............207
Upper Freeport Coal.............. ..(See *Coal* by Magisterial Districts)
Upper Freeport Sandstone.........208
Upper Kittanning (North Coalburg) Coal 211
Upper Kittanning (N. C.) Coal..... (See *Coal* by Magisterial Districts)
Upper Whiteoak Creek, Description 12, 14, 28-29
Upper Williamson Coal...........145
Upper Winifrede Sandstone....278-284

V

Valuation, Property................3-4
Van P. O., Section 2 Mi. South....175
Vancamp, Section 0.5 Mi. S. E..72-73
Vandelinde, B. R., Mine..........407
Vandalinde, Vernon, Mine......... ....290, 576, 580
Venable, M. W., Authority......... ....389, 391, 465, 466
Venango Oil Group................43
Vickers, J., Mine.................413
Voglesang, C. L...................479

W

Wade, J. W., Opening.............340
Walker, Charles, Mine............456
Walker, Mack, Mine..............233
Walnut Creek Oil Co. Wells........ ....520A, 526, 527
Walnut Gap Section.............188-9
Walnut Gap, Section 3¼ Mi. North.189
Warfield Anticline......40-1, 149, 170
Warner, W. H.....................462
Washington District:
Area 2
Campbell Creek Limestone..46, 444-6
Coalburg Sandstone...........252-3
Coals ....(See *Coal* by Magisterial Districts)
Core Tests...................514A
Dingess Limestone............... 45, 85, 88, 100, 109, 127, 136, 145, 203, 253, 282, 362-5, 595, 596
Eagle Limestone............595, 596
East Lynn Sandstone..........210
Homewood Sandstone........229-230
Logan Sandstone..............443-4
Malden Sandstone..............444
Naugatuck Sandstone........359-360
Pittsburgh Red Shale...........205
Population.........................2
Prospective Oil & Gas Areas...541-2
Sections. General............85-149
Upper Chilton Sandstone.....317-318
Upper Winifrede Sandstone....281-2
Well Records.........520A, 536-542
Williamson Sandstone..........360

Page

Washington District (Kanawha) Sections......................58, 59, 60
Water Power Resources.........24-26
Water Ways........................7
Webb Coal Mining Co., Production.562
Webb, Description..................5
Webb Fuel Co., Mine....299, 575, 579
Webb, Lena, Mine................344
Webb, L. M.......................299
Well Records..................519-560
Well Records by Nos.:
1.......................54-5, 520A
2..........................520A
3....................520A, 522-3
4....................520A, 521-2
5 and 6.......................520A
7.....................520A, 528
8..................56-7, 520A, 523
9..................53-4, 520A, 524
10 and 11...............520A, 526
12.......................520A, 527
13...............49, 50, 520A, 524-5
14........................51-2, 520A
15.......................520A, 525
16, 17 and 18..................520A
19.....................520A, 526-7
20.......................520A, 529
21...........................520A
22.....................520A, 534-5
23.......................520A, 534
24...........................520A
25.................72-3, 520A, 523
26...........................520A
27....................520A, 529-530
28..................515, 520A, 530
29...........................520A
30.....................520A, 532-3
31................79-80, 520A, 531
32.......................520A, 532
33...........................520A
34.....................520A, 557-8
35.....................520A, 558-9
36.......................520A, 559
37...........................520A
38.......................520A, 540
39................515. 520A, 538-9
39A..........85-6, 515, 520A, 537-8
40.................513, 520A, 540-1
41.................515, 520A, 536-7
42, 43, 44, 45 and 46..........520A
47.......................520A, 545
48.......................520A, 542
49....151, 152, 153, 515, 520A, 544-5
50.......................520A. 546
51.................153-4, 520A, 544
52 and 53...............520A. 543
54...........154-5, 483, 520A, 547
55.......................520A, 548
56..........150-1, 483, 520A, 546-7
57..............483, 515, 520A, 549
58................515, 520A, 549-50
59..........159, 160-1, 520A, 550-1
60.....................520A, 551-2
61 and 62.....................520A
63..171-2, 486, 506, 515, 520A, 556-7
64.................515, 520A, 552-3
65............98, 101-2, 520A, 540-1
66.....................520A, 554-4
Well Records, Table of..........520A
Weseron, George, Opening........395
West Fork, Description......12, 14, 32
Western Pocahontas Coal & Coke Co. Mines........507, 508, 509, 578, 582

Page
Western Pocahontas Coal & Coke Co. Openings..............495, 500, 510
Weveco Section..................203
Wharton Estate Mines.............. 270, 338, 356, 495, 574, 575, 576, 578, 579, 580, 581, 582
Wharton Estate Openings........... 248, 249, 269, 271, 272, 303 304, 305, 332, 334, 335, 336, 357, 359, 374, 440, 442, 487, 489, 494, 500, 506
Wharton Land Co. Mines.......... ......494, 499, 506, 507, 578, 582
Wharton Land Co. Openings..489, 498
White and Hopkins, Openings.308, 424
White, David....................224
White, I. C...................... 24, 175, 204, 207, 212, 225, 226, 250, 278, 286, 320, 376, 404, 444
White Sidney, Mine..............330
Whiteoak Branch, Section 1 Mi. Up 117-18
Whiteoak Creek, Description....12, 14
Whiteoak Creek, Winifrede Opening 125-6
Whites Branch, Description..13, 14, 32
Whitesville, Description..............5
Whitesville, Section 1.4 Mi. N. W..166
Whitesville, Section 5.8 Mi. West.161-2
Wilkinson, C., Opening..........207-8
Williams, A. H., Mine...........391
Williams, Asa, Mine....218, 574, 579
Williams Coal Co. Well.....520A, 554
Williams, Edgar. Mine............382
Williamson Coal........370-5, 576, 580
Williamson Coal.................... ..(See *Coal* by Magisterial Districts)
Williamson Coal, Analyses.........576
Williamson Sandstone.............360
Wills, Lee, Mine..................330
Winifrede Coal.................... ...284-313, 568, 571, 572, 574-5, 579
Winifrede Coal.................... ..(See *Coal* by Magisterial Districts)
Winifrede Coal, Analyses.......... ......289, 298, 299, 304, 308, 574-5
Winifrede Coal Co., Authority.151, 294
Winifrede Coal Co. Core Test...514A
Winifrede Coal Co. Openings....... ....219, 295, 296
Winifrede Section..................203
Winifrede Well No. 1............. ....150-1, 505, 520A, 546-547
Woodrum, Julia, Mine....387, 576, 580
Woodville, Section 2 Mi. W. (Lincoln)..........................70-71
Workman Branch, Section near Mouth..........................203
Workman, Everett, Mine..........335
Workman, James, Mines........... ....344, 399, 575, 580
Workman Knob Section...........170
Workman, Mrs. John L. Well...... ....520A, 544

Y

Yawkey and Freeman, Mines....... 247-8, 325, 427, 574, 575, 577, 579, 580, 581
Yawkey and Freeman, Openings.... 239, 313, 314, 326, 327, 328, 355, 360, 417, 418, 468, 469
Yawkey and Freeman, Records by.. ....539, 540

Lightning Source UK Ltd.
Milton Keynes UK
UKHW011237310119
336488UK00006B/423/P

9 781528 473194